Scattering by Obstacles and Potentials

Scattering by Obstacles and Potentials

Alexander G Ramm
Kansas State University, USA

World Scientific

NEW JERSEY • LONDON • SINGAPORE • BEIJING • SHANGHAI • HONG KONG • TAIPEI • CHENNAI • TOKYO

Published by

World Scientific Publishing Co. Pte. Ltd.
5 Toh Tuck Link, Singapore 596224
USA office: 27 Warren Street, Suite 401-402, Hackensack, NJ 07601
UK office: 57 Shelton Street, Covent Garden, London WC2H 9HE

Library of Congress Cataloging-in-Publication Data
Names: Ramm, A. G. (Alexander G.) author.
Title: Scattering by obstacles and potentials / by Alexander G. Ramm (Kansas State University, USA).
Description: New Jersey : World Scientific, [2017] | Includes bibliographical references and index.
Identifiers: LCCN 2017024304 | ISBN 9789813220966 (hc : alk. paper)
Subjects: LCSH: Inverse problems (Differential equations) | Scattering (Mathematics)
Classification: LCC QA371 .R336 2017 | DDC 515/.352--dc23
LC record available at https://lccn.loc.gov/2017024304

British Library Cataloguing-in-Publication Data
A catalogue record for this book is available from the British Library.

For any available supplementary material, please visit
http://www.worldscientific.com/worldscibooks/10.1142/10473#t=suppl

Desk Editor: V. Vishnu Mohan

Typeset by Stallion Press
Email: enquiries@stallionpress.com

Printed in Singapore

To Luba

Contents

Preface

This monograph consists of three parts and seven appendices.

Part I, Scattering by Obstacles, is based on the earlier published book by the author [269] and his papers cited in the Bibliography. There are several advances made in Part I compared with [269]. In particular, the inverse scattering problem is studied in the case when the scattering data are non-overdetermined. This is important because such data are the minimal data that allow one to recover the obstacle uniquely. No results were known in this area.

Part II, Scattering by Potentials, deals with direct and inverse scattering problems for compactly supported potentials. The direct scattering problem was studied in many books and papers for much larger class of potentials decaying at infinity. But the inverse scattering problem is studied for compactly supported potentials for the following reason: if the potential decays at infinity sufficiently fast, then the input to the scattering data from the tail of such potential can be considered as noise and cannot be separated from other noises. Practically, one should always consider noisy data when the inverse problem is studied. This simple argument shows that it is natural to assume that the unknown rapidly decaying potential is compactly supported and smooth because the unknown potential can be approximated quite accurately by such potentials and the error of approximation produces an arbitrary small noise in the scattering data if the approximation is sufficiently accurate. The emphasis in Part II is on solving the inverse scattering problem. If the scattering data are fixed-energy data and the potential is real-valued, compactly supported and square integrable, then the author was the first to prove uniqueness theorem (the potential is uniquely defined by the above data). He gave an inversion algorithm for finding the unknown potential when the data are exact and when the data are noisy, and gave error estimates for these algorithms. While uniqueness results were obtained by others after the original publications of the author, no error estimates of the inversion algorithms were given by other authors. The results concerning inverse scattering for fixed-energy data are taken from the author's earlier monograph [356] and from his papers, especially [345]. In Part II, the inverse scattering problem is solved also in the case of non-overdetermined scattering data. The results concerning inverse scattering for non-overdetermined scattering data were obtained by the author [395, 396, 401]. No such results were known.

A numerical method for solving the inverse scattering for non-overdetermined scattering data is proposed.

Part III of this monograph presents the Modified Rayleigh Conjecture (MRC) method for solving scattering problems. The original Rayleigh conjecture was wrong for some obstacles. This was the reason for instability of some engineering methods for numerical solution of direct scattering problems for obstacles. The MRC is a theorem that holds for any bounded obstacle and allows one to develop stable numerical methods for solving scattering problems.

Due to the space limitations, the author did not present in this book his theory of wave scattering by many small bodies and his recipe for creating materials with a desired refraction coefficient. This theory is presented in the monograph [410] and in a series of author's papers cited in this monograph.

The author thanks Springer-Verlag for permission to use large parts of his earlier monographs [269, 356] and the Journals where some of the author's papers, used in this monograph, were published. These papers are referenced in the Bibliography.

Scattering by Obstacles and Potentials

Part I

Scattering by Obstacles

Introduction

In the preface, the topic of this book was described. Consider the problem

$$(\nabla^2 + k^2)u = 0 \quad \text{in } \Omega, \quad k > 0, \quad \Omega \subset \mathbb{R}^3, \tag{0.0.1}$$

$$u = 0 \quad \text{on } \Gamma = \partial\Omega, \tag{0.0.2}$$

$$u = u_0 + v, \tag{0.0.3}$$

where

$$u_0 = \exp\{-ikn \cdot x\}, \tag{0.0.4}$$

$$r\left(\frac{\partial v}{\partial r} - ikv\right) \to 0, \quad r = |x| \to \infty. \tag{0.0.5}$$

The problem (0.0.1)–(0.0.5) is the scattering problem which corresponds to the scattering of the scalar plane wave (0.0.4), propagating in the direction of the unit vector $-n$, by the obstacle $D = \mathbb{R}^3 \setminus \Omega$ with boundary $\Gamma = \partial\Omega$. We consider usually the three-dimensional problem. Some attention will be paid to the two-dimensional problem mostly in connection with the analytic properties of the corresponding scattering matrix (there is a logarithmic branch point at $k = 0$). Unless otherwise stated, the single obstacle D is assumed to be a finite body with a smooth connected boundary $\Gamma = \partial D = \partial\Omega$, and Ω is the exterior domain. Scattering by infinite obstacles and by many finite obstacles will also be studied. The Neumann boundary condition (the unit normal N to the boundary Γ points into Ω)

$$\frac{\partial u}{\partial N} = 0 \quad \text{on } \Gamma, \tag{0.0.6}$$

and the Robin boundary condition,

$$\frac{\partial u}{\partial N} + \eta u = 0 \text{ on } \Gamma, \quad \eta = \eta(s) \in C(\Gamma), \quad \operatorname{Im} \eta(s) \geq 0 \tag{0.0.7}$$

will also be considered instead of (0.0.2).

Problem (0.0.1)–(0.0.5) is of interest in many applications. Most of the results for this problem can be obtained in the same way for electromagnetic wave scattering.

The basic questions about this problem include

(1) existence and uniqueness of the solution to (0.0.1)–(0.0.5),
(2) analytic properties of the solution as a function of k,
(3) properties of the scattering amplitude defined by the formula

$$v = \frac{\exp(ikr)}{r} f(-n, \nu, k) \text{ as } |x| = r \to \infty, \quad xr^{-1} = \nu, \tag{0.0.8}$$

(4) inverse problem of finding the surface of the obstacle from the knowledge of the scattering amplitude (the data will be specified later),
(5) eigenfunction expansions for the Laplacians in exterior domains, construction of the wave operators, properties of the resolvent kernel of the Dirichlet Laplacian in exterior domains,
(6) principles of limiting amplitude and limiting absorption,
(7) numerical solution of exterior boundary value problems,
(8) scattering by small bodies $ka \ll 1$, where a is the diameter of the obstacle,
(9) many-body problems, and
(10) scattering by obstacles with infinite boundaries.

The last problem requires special attention. Even the statement of the scattering problem for obstacles with infinite boundaries is not obvious for a number of reasons. Firstly, one cannot look for a solution of the form (0.0.3) and (0.0.5) because the scattered field may not satisfy the radiation condition (0.0.5).

Example. Take D to be the half-space $x_3 < 0$. In this case, the solution is well known and is of the form $u = \exp\{-ikn \cdot x\} - \exp\{-ik\bar{n} \cdot x\}$, where $n = (n_1, n_2, n_3)$ and $\bar{n} = (n_1, n_2, -n_3)$. The scattered field does not satisfy condition (0.0.5).

Secondly, the uniqueness of the solution to the problem

$$(\nabla^2 + k^2)v = 0 \text{ in } \Omega, \quad v = 0 \text{ on } \Gamma, \quad v \text{ satisfies (0.0.5)} \qquad (0.0.9)$$

does not hold for some domains with infinite boundaries, for example, for the domain $0 \le x_3 \le 1$, or for "bottlenecks", that is, for domains such that $\text{meas}_2\{x : x_3 = R, x \in \Omega\} \to 0$ as $R \to \infty$. Here, meas_2 is the area of the part of the plane $x_3 = R$ which lies in Ω. In such domains, the spectrum of the Dirichlet Laplacian is discrete.

The questions (1)–(10) will be discussed in this book. Not all of these questions can be discussed in detail because one needs several books to do this. There is an extensive literature on the subject. So, the author thinks it is worthwhile to explain what the contents of this book are and to give a very brief review of the literature.

Integral equation methods were used in the study of existence and uniqueness problems (see [37, 110, 111, 124, 159, 474, 478]), but other methods were also of importance. In particular, the limiting absorption method is a powerful tool (see [56, 463, 480]). In [120], a theory based on the conservation of energy for the time-dependent wave equation is given. The author thinks that the easiest way to obtain the basic results of the scattering theory for the Helmholtz equation (0.0.1) is by the way of integral equations, and this way is chosen for our presentation.

The eigenfunction expansions, and the spectral and scattering theories for the Schrödinger equation in the whole space were given in [85] and in domains with infinite boundaries in [181], where a different method was used. This latter method is used in our presentation. The results of Ikebe [85] were generalized by Mishnaevskii [152, 153], Shenk [442], Shenk and Thoe [443], Wilcox [480], and others. Analytic properties of solutions to problem (0.0.1)–(0.0.5) and, more generally, to the problem

$$\begin{aligned}&(l-k^2)v = f \text{ in } \Omega, \quad v = 0 \text{ on } \Gamma,\\ &v \text{ satisfies } (0.0.5), l = -\nabla^2 + q(x),\end{aligned} \tag{0.0.10}$$

were studied in [184] where these properties were linked with the asymptotic behavior for $t \to \infty$ of the solution to the time-dependent problem

$$\begin{aligned}&u_{tt} + lu = 0, x \in \Omega, t > 0; \quad u = 0 \text{ on } \Gamma;\\ &u(x,0) = 0, u_t(x,0) = f(x).\end{aligned} \tag{0.0.11}$$

Unless otherwise stated, it is assumed in Part I that $q(x) = 0$, $l = -\nabla^2$, and the function $f(x) \in L^2(\Omega)$ in (0.0.10) and f in (0.0.11) is compactly supported in Ω. Under these assumptions, one can prove in a simple way, as in [184], that the solution to (0.0.10) (with $q(x) = 0$ or $q(x)$ compactly supported) is a meromorphic function of k with poles k_j in the half-plane $\text{Im}\, k < 0$. These poles are called resonances, or natural frequencies, or complex poles of the resolvent kernel G:

$$[\nabla^2 + k^2 - q(x)]G(x, y, k) = \delta(x - y) \text{ in } \Omega, G = 0 \text{ on } \Gamma, \tag{0.0.12}$$

$$G \in L^2(\Omega) \quad \text{ if } \text{Im}\, k > 0. \tag{0.0.13}$$

The location (especially the values of $\text{Im}\, k_j$) and properties of these poles (their existence, asymptotic behavior as $j \to \infty$, etc.) are of great interest theoretically and practically (see [199, 238, 240, 242, 262, 263]). The location of the poles can be used for identification of the scatterer.

There is a close connection between the resonances (complex poles) and the asymptotic behavior of solutions to problem (0.0.11) as $t \to \infty$. This connection leads to the so-called singularity expansion method (SEM). To describe SEM, let us look for a solution to (0.0.11) of the form $u(x,t) = (2\pi)^{-1} \int_{-\infty}^{\infty} v(x,k) \exp(-ikt) dk$. Then v solves problem (0.0.10) and can be written as

$$v(x,k) = \int_\Omega G(x, y, k) f(y) dy. \tag{0.0.14}$$

This v is a meromorphic function of k. Suppose that its poles k_j satisfy the condition

$$|\mathrm{Im}\, k_1| \le |\mathrm{Im}\, k_2| \le \cdots \le |\mathrm{Im}\, k_j| \to \infty \quad \text{as } j \to \infty. \tag{0.0.15}$$

Then there are only a finite number (N, say) of poles in any strip $-b < \mathrm{Im}\, k \le 0$, $b = \mathrm{const} > 0$. Assume also that

$$|v| < c(b)(1 + |k|)^{-a},\ a > \frac{1}{2},\ |\mathrm{Re}\, k| \to \infty,\ 0 > \mathrm{Im}\, k > -b. \tag{0.0.16}$$

Then one can move the contour in the formula

$$u(x,t) = (2\pi)^{-1} \int_{\infty}^{\infty} \exp(-ikt) v(x,k) dk \tag{0.0.17}$$

down in the complex k plane and obtain the asymptotic SEM expansion

$$u(x,t) = \sum_{j=1}^{N} c_j(x,t) \exp(-ik_j t) + O(\exp\{-|\mathrm{Im}\, k_{N+1}|t\}),\ t \to \infty, \tag{0.0.18}$$

where

$$c_j(x,t) \exp(-ik_j t) = \mathrm{Res}_{k=k_j} \{v(x,k) \exp(-ikt)\},$$
$$c_j(x,t) = O(t^{m_j - 1})$$

as $t \to +\infty$, and m_j is the multiplicity of the poles k_j.

One can prove (0.0.15) and (0.0.16) for some shapes of the obstacle (e.g., for convex smooth obstacles or for star-like obstacles; a star-like obstacle is such an obstacle that there exists a point in D from which every point of Γ can be seen). For trapping obstacles, that is, for obstacles with boundary that can support a standing wave (a ray bouncing between two points of the boundary in geometrical optics approximation), the statment (0.0.15) is not true. For example, for an obstacle consisting of two strictly convex smooth reflecting bodies (i.e., (0.0.2) holds), Ikawa proved that there exist infinitely many poles k_j on the line Im $k_j = -b < 0$, where $b > 0$ is a constant (see [83, 84]). The SEM asymptotic formula (0.0.18) is not valid for such obstacles. It is an interesting open problem whether one can write (under certain assumptions including (0.0.15)) $u = \sum_{j=1}^{\infty} c_j(x,t)\exp(-ik_jt)$, that is, whether one can take $N = \infty$ in (0.0.18). Another interesting problem is to prove existence of infinitely many complex not purely imaginary resonances (a proof of existence of infinitely many purely imaginary resonances is given in this book) and the inverse problem of recovering the scatterer from the knowledge of its complex poles and some information about the corresponding principal parts of $G(x,y,k)$. In [49], one can find a review of the use of complex poles. For some scatterers (for example, star-like), the solution to the problem

$$\begin{aligned} u_{tt} + lu = f(x)\exp(ikt), \quad & x \in \Omega, \quad t > 0, \\ u(x,0) = u_t(x,0) = 0, \quad & u = 0 \quad \text{on } \Gamma \end{aligned} \tag{0.0.19}$$

behaves asymptotically as $t \to +\infty$ like $v(x)\exp(ikt)$, where $v(x)$ is called the limiting amplitude, and solves the stationary problem (0.0.10). There is a close relationship between the limiting amplitude principle and the limiting absorption principle (LAP) which says that the limit $\lim_{\epsilon\downarrow 0}(l - k^2 - i\epsilon)^{-1}f$ exists in a suitable topology and is a solution to (0.0.10). It turns out that the asymptotic of $T^{-1}\int_0^T u(x,t)\exp(-ikt)dt$ as $T \to \infty$ is more naturally related to the LAP than the asymptotic of $u(x,t)\exp(-ikt)$. There is an extensive literature on the limiting absorption and limiting amplitude principles (see [54, 152, 153, 181, 184, 463], and references in these papers). Most of the existing works on the limiting absorption and limiting amplitude principles deal with partial differential operators. But a few results in abstract operator-theoretical settings are also available: Eidus [54] gave a theory for the case when l is an abstract self-adjoint operator on a Hilbert space and the resolution of the identity for l satisfies some smoothness conditions; Ramm [254] gave necessary and sufficient conditions for the validity of the limiting amplitude principle for some linear nonself-adjoint operators. Mishnaevskii [152] gave sufficient conditions for the LAP to hold for an operator Sturm–Liouville equation. His remarkable results generalize the earlier results of Kato [97], Jäger [90], Saito [435], and others. Vainberg [463, 464] justified the limiting absorption and limiting amplitude principles for a wide class of partial differential equations.

A nonclassical approach to the exterior boundary value problem is as follows. Consider the problem

$$(-\nabla^2 - k^2)v = 0 \text{ in } \Omega, \quad v = f \text{ on } \Gamma, \ v \text{ satisfies (0.0.5)}. \quad (0.0.20)$$

Let

$$v = \int_\Gamma g\sigma ds, \quad g = \frac{\exp(ik|x-y|}{4\pi|x-y|}. \quad (0.0.21)$$

Then v solves (0.0.20) if and only if

$$Q\sigma = \int_\Gamma g\sigma ds = f. \quad (0.0.22)$$

The operator Q is compact in $H = L^2(\Gamma)$ but nonself-adjoint. If Q has a complete (in H) set of eigen and root vectors and this set forms a basis for H, then one can solve (0.0.22) using an expansion of σ in root vectors of Q, that is, vectors satisfying the equation $(Q-\lambda_j)^m\psi = 0$, where m is an integer, λ_j is an eigenvalue of Q, and m may be different for various j. If $m = 1$, then ψ is an eigenvector. This method is analogous to solving the equation $A\sigma = f$ with a self-adjoint compact operator A by the formula $\sigma = \sum_{j=1}^\infty \lambda_j^{-1}(f, \psi_j)\psi_j$, where $A\psi_j = \lambda_j\psi_j$ and the set $\{\psi_j\}$ forms an orthonormal basis of H. A nonself-adjoint compact operator may have no eigenvalues (e.g., Volterra's operator, quasinilpotent operators). If the set of root vectors of Q forms a basis of H, then we say that Q has the basisness property. It is an interesting problem to give sufficient conditions for a compact nonself-adjoint operator to have the basisness property. The eigenvalues of Q are functions of k. One can prove (see [212, 226]) that the resonances are precisely the zeros of the eigenvalues of Q. Other characterizations of the resonances and methods for their computation are given in [269].

The inverse diffraction problem consists in finding the obstacles from the scattering data. The scattering data are the values of the scattering amplitude on some subsets of the five-dimensional space $k, \nu, n, 0 < k < \infty, \nu, n \in S^2$, where S^2 is the unit sphere.

A well-known result of M. Schiffer (see [269], M. Schiffer did not publish his result) says that Γ is uniquely defined if the scattering amplitude $f(n_0, \nu, k)$ is known for $a \leq k \leq b, b > a$ and all $\nu \in S^2$. We prove the uniqueness theorem assuming that the scattering data are $f(n_j, \nu, k_0), k_0 > 0, \nu \in \tilde{S}$, where $\tilde{S}$ a solid angle, that is, an open subset of S^2, $\{n_j\}$ is a countable set of unit vectors such that $n_j \neq n_i$ if $j \neq i$. One can prove that if f is known for $\nu \in \tilde{S}$ then f is uniquely defined for all $\nu \in S^2$. The main point in our result is the fact that k is fixed, $k = k_0$, so that the surface is recovered from the fixed energy data. If k_0 is taken sufficiently large and Γ is strictly convex, then an effective algorithm for recovering the surface is given. Our presentation is based on the papers [192, 259, 260]. The method consists of explicit recovery of the support function of the convex scatterer from the high-frequency asymptotics of the scattering amplitude. The support function $a(l)$ of a convex body D is defined to be $a(l) = \sup_{s \in \Gamma}(l, s)$ if the origin is inside D. Here, l is a unit vector. The nonconstructive approach in earlier works was to recover the Gaussian curvature of the convex scatterer from the high-frequency asymptotic of the scattering amplitude, and then Minkowski's theorem (which says that a convex smooth closed surface Γ is uniquely defined by its Gaussian curvature known as a function of the unit normal on Γ) yields the uniqueness theorem for the inverse scattering problem. This approach is not constructive because Minkowski's theorem is a nonconstructive uniqueness theorem which does not provide an algorithm to recover the surface. The inverse scattering problem for nonconvex bodies is of great interest but there are no results concerning constructive solution of this problem presently (except [7]). In the literature, there is a discussion of the inverse scattering problem in the case when the scattering data are known for small k. But for small k ($ka \ll 1$, where a is the diameter of the scatterer), the scattering amplitude is determined basically by the dipole radiation (corresponding to the first term of the expansion of the scattering amplitude in powers of ka). Therefore, for $ka \ll 1$, a body of arbitrary shape scatters waves as an ellipsoid, and practically only the three scalars (semi-axes of the ellipsoid) can be recovered from the scattering amplitude in the case $ka \ll 1$ (see [360]).

Having discussed the basic qualitative questions, such as uniqueness and existence of the solution to the scattering problem, properties of the solution, of the Green function, eigenfunction expansions in solutions to the scattering problem (distorted plane waves, eigenfunctions of continuous spectrum of the Dirichlet Laplacian in the exterior domain), the time-dependent problem and the limiting amplitude and LAP, the singularity and eigenmode expansions, and the inverse scattering problem, we turn to the numerical schemes for solving exterior boundary value problems. The emphasis is on the T-matrix scheme. This scheme can be described as follows. Let $\{\psi_n\}$ be a system of functions satisfying equation (0.0.1) and the radiation condition (0.0.5). Such functions are called outgoing waves. Applying Green's formula to ψ_n and v, the solution of (0.0.20) yields

$$\int_\Gamma \psi_n h ds = f_n, \ \forall n, \quad f_n = \int_\Gamma f \frac{\partial \psi_n}{\partial N} ds, \quad h = \frac{\partial v}{\partial N}, \tag{0.0.23}$$

and N is the outer unit normal on Γ. Let $\{\varphi_j\}$ be a linearly independent and complete system of functions in $L^2(\Gamma) = H$. Substitute $h_m = \sum_{j=1}^m c_j^{(m)} \varphi_j$, where $c_j^{(m)}$ are constants, in place of h in (0.0.23) to obtain

$$\sum_{j=1}^m a_{nj} c_j^{(m)} = f_n, \quad 1 \le n \le m, \quad a_{nj} = \int_\Gamma \psi_n \varphi_j ds. \tag{0.0.24}$$

The system (0.0.24) is to be solved for $c_j^{(m)}$ and it is to be proved that $h_m \to h$ in H as $m \to \infty$, where h solves (0.0.23). If h is known, then the solution of (0.0.20) can be found by the Green formula. Practically, the last step involving integration over the surface can be avoided. This is shown in Chapter 4 where other questions of interest are discussed.

Does $c_j^{(m)} \to c_j$ as $m \to \infty$? Is the convergence uniform in j? Does the equality $h = \sum_{j=1}^\infty c_j \varphi_j$ hold? How does the rate of convergence depend on the choice of the systems $\{\psi_j\}$ and $\{\varphi_j\}$?

There is an extensive literature on the T-matrix approach (see [466]) but for about two decades convergence of this numerical method was not investigated. A proof of its convergence and a theoretical analysis of the method including asnwers to the above question are given in [244, 248] (see also [108]) and presented in Chapter 4. We briefly discuss in Chapter 4 the other major numerical methods: integral equation methods and finite elements methods. In particular, integral equations, which are solvable at the exceptional values of k corresponding to the spectrum of an interior Laplacian, are derived. Several methods of deriving such equations [29, 124, 182, 461] are given in Chapter 1.

Wave scattering by small bodies is a classical topic. The original work was initiated by Rayleigh in 1871. He continued this work until his death. The physics of the problem is well understood: the main term in the scattered field is the dipole radiation. The problem is to compute this dipole radiation for a body of arbitrary shape and to study the dependence of the scattering amplitude on the boundary conditions. This was done by the author and summarized in [239, 360, 410], where references can be found. If the scatterers are small compared with the wavelength $ka \ll 1$, and $d \gg a$, where d is the average distance between the scatterers, then the many-body problem can be greatly simplified. The scattering amplitude in this case is determined by a finite number of functionals of the field, that is, by some numbers rather than functions as in the case of high-frequency scattering. For these numbers, one can derive an algebraic linear system rather than a system of integral equations. This system can be easily solved by iterations as shown in [239]. As in application of the formula for the scattering matrix for wave scattering by a small body of arbitrary shape, one can solve an inverse radiation problem. This problem consists of finding the initial field at the point x_0 from the field, scattered by a small probe placed at the point x_0 in the initial field. This material is presented in Chapter 5 and the presentation is based on [239]. In [37, 62, 134, 475, 476], and references in these papers, the behavior as $k \to 0$ of solutions to exterior boundary value problems was studied. In the two-dimensional case, Green's function has a logarithmic branch point at $k = 0$. In [199], it was shown that

Green's function of the Schrödinger Dirichlet operator with $q(x) \geq 0$ in the exterior domain has a limit as $k \to 0$. If $q = 0$, then Green's function of the exterior Neumann Laplacian does not have a limit as $k \to 0$. These questions are discussed in Section 5.4, as well as the case of the Robin Laplacian.

We do not discuss some ideas of homogenization theory [19, 139, 410, 439]. This theory deals, for example, with the situation when the medium consists of many small bodies and the problem is to describe the medium as a continuous medium (homogenized medium) with some "averaged" effective parameters. We do not discuss the quantum-mechanical scattering by delta-function potentials (see [8, 9] and references therein).

In Chapter 6 we study inverse scattering problems for partial differential equations. These problems are of special interest in geophysics and nondestructive control. A typical example is the following problem. Let $[\nabla^2 + \omega^2 n(x)]u(x, y, k) = -\delta(x - y)$ in $\mathbb{R}^3, n(x) = n_0 + v(x), n_0 = \text{const} > 0, v(x) = 0$ if $|x| > R$, where $R > 0$ is an arbitrary large number, $v(x) \in L^2$. The function $v(x)$ describes an inhomogeneity. The problem is to find $v(x)$ from a knowledge of the field $u(x, y, k)$ on the plane $x_3 = 0$, i.e., for all positions of $x^1 = (x_1, x_2, 0)$ and $y^1 = (y_1, y_2, 0)$ and for $0 < k < k_0$, where k_0 is an arbitrary small number. The data $u(x, y, k)$ can be also given on a sphere or on a cylinder. The problem can be complicated by the assumption that $n_0 = n_1$ if $x_3 > 0$, $n_0 = n_2$ if $x_3 < 0$, where $n_1 \neq n_2$, $n_j = \text{const}, j = 1, 2$. We discuss also the inverse problem for two unknown functions, i.e., the density of the medium and the velocity profile. In this case, the differential equation is

$$\left(\nabla^2 + \omega^2 - \omega^2 a_1(x) - \nabla \cdot a_2(x)\nabla\right) G = -\delta(x - y),$$

and the functions $a_j(x), j = 1, 2$, are to be found from the scattering data. A model problem of the induction logging is also studied. The inverse scattering problem by a compactly supported potential $q(x)$, $x \in \mathbb{R}^3$, in the Born approximation can be reduced to the integral equation for q:

$$\int_{|y| \leq R} \exp\{-ik(s' - s) \cdot y\} q(y) dy = F(k(s' - s)),$$

where $s', s \in S^2, k > 0$ is fixed, and the right side is known. This problem occurs in all the inverse problems studied in Chapter 6. We give an analytic solution to this problem with a prescribed accuracy and the error estimates.

In Chapter 7 the scattering problem in domains with infinite boundaries is discussed. This is a difficult subject. Even the formulation of the scattering problem presents some difficulties as was mentioned in the beginning of this introduction. Some results on the nature of the spectrum of the Laplace and Schrödinger operators in domains with infinite boundaries are given in [53, 69]. In the earlier works [93, 430, 431], the emphasis was on the nature of the spectrum of the Laplacians. An important result of the author is a proof of the absence of positive eigenvalues. This result was obtained for the Dirichlet Laplacian in domains with infinite boundaries which are illuminated everywhere near infinity from a point O placed in D. This assumption can be written as $N \cdot x \leq 0$ on Γ when $|x| > R$ where $R > 0$ is an arbitrary large fixed number. Here, N is the normal to Γ directed into D and x is the radius vector in the coordinate system with the origin at the point O. This assumption is an assumption about the surface Γ near infinity. Rellich's method does not work for the Schrödinger operator. For the first time, the absence of positive eigenvalues for Schrödinger operators with $|q(x)| = o(|x|^{-1})$ as $|x| \to \infty$ in domains with infinite boundaries which are illuminated from some point O only near infinity (that is $N \cdot x \leq 0$ for $|x| > R$, where $R > 0$ is an arbitrarily large number) was proved in [181] (see also [187]). The method used in this book was given in [97] where for the first time similar results were obtained for the Schrödinger operator in the whole space. It is interesting that for the class of infinite boundaries introduced in [181] the problem of absence of positive eigenvalues can be treated with only local assumptions near infinity. For example, this is not possible if the boundary is a half-tube: a perturbation of a finite part of the tube can put a positive eigenvalue on the continuous spectrum of the Dirichlet Laplacian in such domain (see [431]). In [455], the absence of positive eigenvalues was proved for the Dirichlet Laplacian in domains with smooth boundaries such that $\Omega(R) \subset \{x : (|x|^2 - x_1^2) < cx_1^{2l}\}, N(x) \cdot a(x) \leq 0$ for $x \in \Gamma \cap \partial\Omega(R)$, where $R > 0$ is an arbitrarily large fixed constant, $c > 0$, $l < 1$, $\Omega(R) = \Omega \cap \{x : x_1 > R\}, x = (x_1, \dots, x_n), N(x)$ is the normal directed into D at the point $x \in \Gamma$, and $a(x) = (x_1, lx_2, \dots, lx_n)$. Such domains, for example, may be paraboloids. In this case, the assumption $N \cdot x \leq 0$ for $|x| > R$ does not hold (the interior of the paraboloid is Ω). All of these results are very special cases of the theorems proved in [152, 153]. He developed a powerful technique for proving the absence of positive eigenvalues for abstract Sturm–Liouville equations, gave a deep spectral analysis of such equations, proved eigenfunction expansion theorems, a LAP, and showed how these results can be obtained for general second-order elliptic operators in a wide class of domains with infinite boundaries. He considered also various boundary conditions. Before his work, only in [200] for the Schrödinger equation in some domains with infinite boundaries the Robin boundary condition was treated. Proofs of the remarkable results due to Mishnaevskii [152, 153], are lengthy and will not be presented in this book, where we try to keep the technical level of the presentation simple. We give a general investigation of the spectral properties of the Laplacians in domains with infinite boundaries. Necessary and sufficient conditions for the discreteness of the spectrum of the Dirichlet Laplacian L are given. The spectrum of L in semi-infinite tubes is studied. Sufficient conditions for the continuous spectrum of L to fill the ray $(0, \infty)$ and to have no positive eigenvalues are given. The notion of a canonical domain is introduced. A canonical domain is a domain in which Green's function of L exists and admits certain estimates. For example, if Green's function of L can be constructed by the separation of variables method and satisfies certain estimates, then the domain is canonical. A cone is a canonical domain. For such domains, we develop a perturbation theory. This theory allows one to study the spectral properties of L in domains whose boundaries are compact perturbations of the boundary of a canonical domain. We prove the LAP and the eigenfunction expansions theorem for L in such domains. In appendices, some auxiliary material is given. Part of this material is not available in book form.

It is now in order to explain what audience the author had in mind and what is original in this book. This book is intended to serve as an introduction to the subject for an audiencc with various interests: mathematicians, physicists, and engineers interested in the theory. The presentation is essentially self-contained. The author tried to present material in a simple way and avoid technically difficult proofs. Although the book is an introduction to the subject, most of the material is not available in book form and the author's research interests influence the choice of the material considerably. The material in Sections 1.7, 2.2, 3.2–3.5, Chapters 4–9 and Appendices C, D, E, G belongs to the author and was published in his research papers. We discuss scalar wave scattering, but nearly all of our methods and results can be rewritten for electromagnetic wave scattering.

We do not discuss in detail in this book several important topics:

(1) exponential decay of the solutions of the wave equation in the exterior domains;
(2) high-frequency asymptotic of the solutions to scattering problems and rigorous justification of the geometrical optics, except in Section 1.6;
(3) distribution of the complex poles of the scattering amplitudes;
(4) scattering by periodic and rough surfaces, except in Section 9.3;
(5) scattering by moving boundaries.

See references in Bibliographical notes. The bibliography is not complete, but if the reader consults the references, he or she will have reasonably complete information about the subject.

There are practically no intersections between the material and presentation in this book and also in the books by Lax and Phillips [120], Wilcox [480], Reed and Simon [428], etc.

Chapter 1

Scattering by an Obstacle

1.1. Statement of the Problem: Basic Integral Equations

1.1.1. In this section, we introduce some notations, formulate some properties of the potentials of the single and double layers, and discuss the integral equations for solving the exterior boundary value problem (0.0.20) resulting from Green's formula. This discussion is a preliminary step in the study of the solvability of the problem (0.0.20). The classical approach leads to integral equations which are not always solvable for some k. Therefore, one wishes to derive some unconditionally solvable integral equations for solving (0.0.20), equations which are equivalent to the boundary value problem (0.0.20).

The scattering problem (0.0.1)–(0.0.5) is a particular case of the boundary value problem (0.0.20) with $f = -u_0 = -\exp(-ikn \cdot s)$ on Γ. The classical approach to this problem is to look for a solution of the form

$$v(x) = D\mu \equiv \int_\Gamma \mu(t)\frac{\partial g(x,t)}{\partial N_t}dt, \quad g = \frac{\exp(ik|x-y|}{4\pi|x-y|}. \qquad (1.1.1)$$

Then equation (0.0.20) and the radiation condition (0.0.5) are satisfied, and from the boundary condition (0.0.20), one obtains an integral equation for μ (N_t is the exterior normal pointing into Ω):

$$\mu = -A'\mu + 2f, \quad A'\mu = 2\int_\Gamma \mu(t)\frac{\partial g(s,t)}{\partial N_t}dt, \quad s \in \Gamma. \qquad (1.1.2)$$

Here and below, the following well-known formulas from potential theory are used (see [79]):

$$\text{(a) } v_\pm = (D\mu)_\pm = \frac{A'\mu \mp \mu}{2}, \quad \text{(b) } \frac{\partial w}{\partial N_\pm} = \frac{\partial(S\sigma)}{\partial N_\pm} = \frac{A\sigma \pm \sigma}{2}, \tag{1.1.3}$$

$$\frac{\partial(D\mu)}{\partial N_-} = \frac{\partial(D\mu)}{\partial N_+}, \tag{1.1.3'}$$

where

$$w(\sigma) \equiv S\sigma \equiv \int_\Gamma g(x,t)\sigma(t)dt, \ A\sigma = 2\int_\Gamma \sigma(t)\frac{\partial g(s,t)}{\partial N_s}dt, \ s \in \Gamma. \tag{1.1.4}$$

The sign + (−) corresponds to the limiting value on Γ from the interior domain D (exterior domain Ω), and to the upper (lower) sign in the left-hand side of (1.1.3) there corresponds the upper (lower) sign in the right-hand side of (1.1.3). The operator $A' = \bar{A}^*$, where the bar denotes complex conjugation of the kernel of the operator and the operator A^* is adjoint to A in $L^2(\Gamma) \equiv H^0$. Formula (1.1.3) is obtained in [79] under the assumption that μ and σ are continuous and the surface Γ belongs to $C^{1,\lambda}$, that is, in a neighborhood of any point of Γ one can introduce a local cartesian coordinate system such that the x_3-axis is directed along the normal to Γ at this point and the function φ in the equation of the surface $x_3 = \varphi(x_1, x_2)$ has first derivatives satisfying the Hölder condition with exponent λ. A function $\psi(x)$ satisfies the Hölder condition with exponent λ if $|\psi(x) - \psi(y)| \le c|x-y|^\lambda$, $c = \text{const}, 0 < \lambda \le 1$. Formula (1.1.3′) means that

$$\lim_{\epsilon \to +0}\left(\left.\frac{\partial v(x)}{\partial N_s}\right|_{x=s+\epsilon N_s} - \left.\frac{\partial v(x)}{\partial N_s}\right|_{x=s-\epsilon N_s}\right) = 0,$$

where $v(x) = D\mu$, $s \in \Gamma$.

If μ is only continuous, then

$$\lim_{\epsilon \to +0} \left.\frac{\partial v(x)}{\partial N_s}\right|_{x=s\pm\epsilon N}$$

may not exist. These limits exist and are Hölder-continuous if $\mu \in C^{1,\lambda}(\Gamma)$, $\lambda > 0$. In Appendix A, more information on potential theory is given. In particular, the operators A and A^* are compact in the spaces $L^2(\Gamma) = H^0(\Gamma)$ and $C^a(\Gamma)$ if $\Gamma \in C^{1,\lambda}$, where $C^0(\Gamma) = C(\Gamma)$ and $C^a(\Gamma)$ is the space of Hölder-continuous functions with the norm $|f| = \max_{x\in\Gamma} |f(x)| + \sup_{x,y\in\Gamma} |f(x) - f(y)||x - y|^{-a}, 0 < a \leq 1, \mu \in C^{1,\lambda}(\Gamma)$ means that the first derivatives of μ belong to $C^{\lambda}(\Gamma)$.

Therefore, Fredholm's alternative holds for equation (1.1.2). In [31, 61, 104], the potential theory for irregular domains, i.e., domains with nonsmooth boundaries, is given.

1.1.2. If the solution to the problem (0.0.20) exists, then it can be written by Green's formula

$$v(x) = \int_\Gamma \left(v_- \frac{\partial g}{\partial N} - g \frac{\partial v}{\partial N_-} \right) ds, \quad x \in \Omega. \tag{1.1.5}$$

We will show later that the problem (0.0.20) is solvable for any $f \in C(\Gamma)$ or $L^2(\Gamma)$. Therefore, one can always find the solution to (0.0.20) in the form

$$v(x) = Df - Sh, \tag{1.1.6}$$

where

$$v_- = f, \tag{1.1.7a}$$

$$\frac{\partial v}{\partial N_-} = h. \tag{1.1.7b}$$

Let us derive integral equations for f and h. From (1.1.6), (1.1.7a), (1.1.7b), and (1.1.3), it follows that

$$f = \frac{A'f + f}{2} - Qh, \text{ i.e., } Qh = \frac{A'f - f}{2}, \quad Qh = \int_\Gamma g(s,t)h(t)dt, \tag{1.1.8}$$

and

$$h = \frac{\partial(Df)}{\partial N_-} - \frac{Ah - h}{2}, \text{ i.e., } h + Ah = 2\frac{\partial(Df)}{\partial N_-}. \tag{1.1.9}$$

Remark 1.1.1. Condition (1.1.7a) is equivalent to equation (1.1.8). Indeed, we have derived (1.1.8) from (1.1.7a). Conversely, if (1.1.8) holds, then one writes the right-hand side of the first equation (1.1.8) as $(Df)_- - (Sh)_- = v_-$, and obtain (1.1.7a). Condition (1.1.7b) is equivalent to equation (1.1.9). Indeed, we have derived (1.1.9) from (1.1.7b). Conversely, if (1.1.9) holds, then the right-hand side of the first equation (1.1.9) is $\partial v/\partial N_-$, where v is defined by (1.1.6).

Lemma 1.1.1. *If $k^2 \notin \sigma(-\Delta_D), [k^2 \notin \sigma(-\Delta_N)]$, i.e., k^2 is not an eigenvalue of the Dirichlet [Neumann] Laplacian in D, and h is a solution to* (1.1.8) [(1.1.9)] *then the function* (1.1.6) *satisfies equations* (1.1.7a)–(1.1.7b) *and solves problem* (0.0.20).

Proof. (1) Suppose h solves (1.1.8). Then by Remark 1.1.1, equation (1.1.7a) holds for v defined by (1.1.6). Therefore, the function (1.1.6) (with h solving (1.1.8)) solves problem (0.0.20). We wish to show that equation (1.1.7b) holds. Let us write Green's formula (1.1.5) for the solution to problem (0.0.20). We will prove later (Lemma 1.2.2) that the problem (0.0.20) has at most one solution. Therefore, the functions (1.1.5) and (1.1.6) are identical in Ω. Since (1.1.7a) has already been established, we conclude that

$$S\sigma = 0 \text{ in } \Omega, \quad \sigma \equiv \frac{\partial v}{\partial N_-} - h. \tag{1.1.10}$$

If $k^2 \notin \sigma(\Delta_D)$, then $S\sigma = 0$ in D. Therefore, by the jump relation (1.1.3′), $\sigma = 0$. So, equation (1.1.7b) holds.
(2) Suppose h solves (1.1.9). Then by Remark 1.1.1 equation (1.1.7b) holds. Define $v(x)$ by formula (1.1.6) and write Green's formula (1.1.5) for this v. Since the functions (1.1.5) and (1.1.6) are identical in Ω, and (1.1.7b) holds, one concludes that

$$D\mu = 0 \text{ in } \Omega, \quad \mu = f - v_-. \tag{1.1.11}$$

By formula (1.1.3′), $\partial u/\partial N_+ = 0$ where $u(x) = D\mu$. Therefore, $(\nabla^2 + k^2)u = 0$ in D, $\partial u/\partial N_+ = 0$. Since $k^2 \notin \sigma(-\Delta_N)$, one has $u = 0$ in D.

This, (1.1.11) and the jump relation (1.1.3′) imply that $\mu = 0$. So, equation (1.1.7a) holds. □

Remark 1.1.2. If h solves both equations (1.1.8) and (1.1.9), then both equations (1.1.7a) and (1.1.7b) hold with $v(x)$ defined by (1.1.6), and this v solves problem (0.0.20).

Lemma 1.1.2. *There is at most one function which solves both equations* (1.1.8) *and* (1.1.9).

Proof. It is sufficient to prove that $Qh = 0$ and $h + Ah = 0$ imply $h = 0$. These equations mean that $w = Sh$ satisfies the equations $w_+ = 0$, $\partial w/\partial N_+ = 0$. Therefore, $w(x) = 0$ in D. Since $w_+ = w_- = 0$, by Lemma 1.2.2, $w = 0$ in Ω. Thus,

$$h = \frac{\partial w}{\partial N_+} - \frac{\partial w}{\partial N_-} = 0.$$ □

Remark 1.1.3. If $k^2 \in \sigma(-\Delta_D)$ and h solves (1.1.8), then equation (1.1.7b) may not hold. Indeed, if $k^2 \in \sigma(-\Delta_D)$, then dim $N(Q) = r$, where r is the dimension of the eigenspace of $-\Delta_D$ corresponding to the eigenvalue k^2. Therefore, equation (1.1.8) has many solutions. On the other hand, f defines uniquely the solution $v(x)$ to problem (0.0.20) (by Lemma 1.2.2), so that $\partial v/\partial N_-$ is uniquely defined by f. Therefore, if $k^2 \in \sigma(-\Delta_D)$, one can find a (unique) solution to equation (1.1.8) for which (1.1.7b) holds. This solution is the one which solves both equations (1.1.8) and (1.1.9) (see Lemma 1.1.2).

If $k^2 \in \sigma(-\Delta_N)$, then equation (1.1.9) has many solutions. Let $h_j, j = 1, 2$ solve equation (1.1.9), $h_1 \not\equiv h_2, v_j = Df - Sh_j, v = v_1 - v_2, h = h_1 - h_2 \not\equiv 0$. Then $v = -Sh, \partial v/\partial N_- = h, h + Ah = 0$, i.e., $\partial v/\partial N_+ = 0$. Therefore, $v_+ = (Sh)_+ \not\equiv 0, v_+ = v_- \not\equiv 0$ (because if $v_+ = 0$, then $v = 0$ in D and in Ω, and therefore $h = 0$, which is a contradiction) $v_{1-} \not\equiv v_{2-}$ if $h \not\equiv 0$. As above, the only solution to (1.1.9) for which function (1.1.6) satisfies equation (1.1.7a) is the solution which satisfies both equations (1.1.8) and (1.1.9).

We give several proofs of uniqueness and existence of the solution to exterior boundary value problem (in particular, problem (0.0.20)) in Section 1.2. In Section 1.7, we study the possibility of the representations of the double (single) layer potentials as a single (double) layer potential. We can give a simple proof of existence of the solution to the problem (0.0.20), based on equations (1.1.8) and (1.1.9).

Lemma 1.1.3. *Equation* (1.1.3) *is solvable for any* $k > 0$.

Proof. If $k^2 \notin \sigma(-\Delta_D)$, then equation (1.1.8) is uniquely solvable (Lemma 1.7.5) and function (1.1.6) solves problem (0.0.20) (Lemma 1.1.2) and equation (1.1.9) holds. If $k^2 \in \sigma(-\Delta_D)$, then equation (1.1.8) is solvable if and only if

$$\int_\Gamma (A'f - f)\sigma_j dt = 0, \quad 1 \le j \le r, \quad \sigma_j = \frac{\partial u_j}{\partial N_+},$$

where $u_j, 1 \le j \le r$, form a basis in the eigenspace of $-\Delta_D$ corresponding to the eigenvalue k^2 (Lemma 1.7.6). We have $A'f - f = 2v_+(f)$, where $v(f) = Df$ is given by (1.1.3). Thus,

$$\int_\Gamma (A'f - f)\sigma_j dt = 2\int_\Gamma v_+ \frac{\partial u_j}{\partial N_+} dt = 2\int_\Gamma \frac{\partial v}{\partial N_+} u_j dt = 0.$$

Here, we used Green's formula and the boundary condition $u_j = 0$. Therefore, equation (1.1.8) is always solvable, although nonuniquely if $k^2 \in \sigma(-\Delta_D)$. □

Remark 1.1.4. Equation (1.1.9) is solvable if and only if

$$\int_\Gamma \partial(Df)/\partial N_- \mu_j dt = 0, \tag{*}$$

where $u_j, 1 \le j \le r'$, form a basis of $N(I + A')$. If $\mu_j + A'\mu_j = 0$, and $u_j = D\mu_j$, where $D\mu$ is defined in (1.1.1), then $u_{j-} = 0$, and $\mu_j = u_{j-} - u_{j+} = u_{j+}$. Let $v = Df$. Then, by (1.1.3′), $\partial v/\partial N_- = \partial v/\partial N_+$. Thus, $\int_\Gamma \partial v/\partial N_+ u_{j+} dt = \int_\Gamma v \partial u_j/\partial N_+ dt = 0$, since $\partial v/\partial N_+ = \partial v/\partial N_- = 0$. The last equality follows from the fact that $u_j = 0$ in Ω. Therefore, $(*)$ holds for any f (for which $\partial(Df)/\partial N_-$ is defined, i.e., for $f \in C^{1,\lambda}(\Gamma)$), and equation (1.1.9) is always solvable.

Lemma 1.1.4. *The solution to the system of equations* (1.1.8) *and* (1.1.9) *exists and is unique.*

Proof. Uniqueness has been established in Lemma 1.1.2. Let us prove existence. We have proved (Lemma 1.1.3) that problem (0.0.20) has a solution $v(x)$. Let us write formula (1.1.5) for this solution, taking into account that $v_- = f$, and denote $\partial v/\partial N_-$ by h. Then (1.1.6) holds and, as after formula (1.1.7b), one derives equations (1.1.8) and (1.1.9) which are satisfied by $h = \partial v/\partial N_-$. Thus, existence of the solution to the system of equations (1.1.8) and (1.1.9) is established. □

1.1.3. The classical potential theory integral equation for solving (0.0.20) can be obtained by looking for v in the form (1.1.1), applying the boundary condition (0.0.20) and using (1.1.3). This yields

$$\mu = -A'\mu + 2f. \tag{1.1.12}$$

This equation is studied in Section 1.2. The operator A' in (1.1.12) is compact in $L^2(\Gamma)$ and in $C^a(\Gamma), a > 0$ if the surface Γ is smooth. By Fredholm's alternative, (1.1.12) is solvable for any f if and only if $N(I+\bar{A}) = \{0\}$, where $N(A)$ is the null space of a linear operator A. Suppose $\dim N(I+\bar{A}) = r', r' > 0$, and let $\{v_j\}$ form a basis of $N(I+\bar{A}), v_j + \bar{A}v_j = 0, 1 \le j \le r'$, then the necessary and sufficient conditions for solvability of (1.1.12) are

$$(f, v_j) = 0, \quad 1 \le j \le r'. \tag{1.1.13}$$

It is natural to look for an integral equation for solving (0.0.20) that is unconditionally solvable. There are several ways to derive such equations.

1.1.4. The first way is to use a Green's function $g_\epsilon(x, y, k)$ in place of g in (1.1.12). Here, $g_\epsilon(x, y, k)$ is the solution to the following problem:

$$(-\nabla^2 - k^2)g_\epsilon = \delta(x-y) \text{ in } \Omega_\epsilon,\ g_\epsilon = 0 \text{ in } \partial\Omega,\ g_\epsilon \text{ satisfies (0.0.5)}, \tag{1.1.14}$$

$\Omega_\epsilon = \{x : |x| \geq \epsilon\}$, $\partial\Omega_\epsilon$ is the boundary of Ω_ϵ, and $\epsilon > 0$ is so chosen that $N(I + \bar{A}_\epsilon) = \{0\}$, where the operators A_ϵ and A'_ϵ are defined as in (1.1.3), (1.1.4) but with g_ϵ in place of g. The ball $B_\epsilon = \{x : |x| \leq \epsilon\}$ is chosen inside D. If $N(I + \bar{A}_\epsilon) = \{0\}$, then equation (1.1.12) with A'_ϵ in place of A_ϵ is uniquely solvable for any f. In Section 1.2 (see Lemma 1.2.5), it will be shown that it is possible to choose $\epsilon > 0$ such that $N(I + A_\epsilon) = \{0\}$.

1.1.5. The second way to derive an unconditionally solvable integral equation for solving (0.0.20) is to look for a solution of the form

$$v(x) = D\mu + i\eta S\mu, \tag{1.1.15}$$

where $\eta > 0$ is a parameter. Applying the boundary condition (0.0.20) and formula (1.1.3) yields

$$\frac{A'\mu + \mu}{2} + i\eta Q\mu = f. \tag{1.1.16}$$

This is an equation of Fredholm's type. It is unconditionally solvable if the homogeneous equation

$$\frac{\bar{A}\sigma + \sigma}{2} - i\eta\bar{Q}\sigma = 0 \tag{1.1.17}$$

has only the trivial solution. Equation (1.1.17) can be written as

$$\frac{\partial}{\partial N_+}\bar{S}\sigma - i\eta\bar{Q}\sigma = 0. \tag{1.1.18}$$

Thus, the function $u(x) = \bar{S}\sigma$ solves the problem

$$(\nabla^2 + k^2)u = 0 \text{ in } D, \quad \frac{\partial u}{\partial N_+} - i\eta u = 0 \text{ on } \Gamma. \tag{1.1.19}$$

Since $k^2 > 0$ and $\eta > 0$, one obtains from (1.1.19) that $u \equiv 0$ in D, $u = 0$ on Γ and, therefore, $u(x)$ solves the homogeneous exterior problem (0.0.20). Thus, $u(x) \equiv 0$ in Ω, and $\sigma = \partial u/\partial N_+ - \partial u/\partial N_- = 0$. Therefore, equation (1.1.16) is unconditionally solvable (for any $\eta > 0$) and the function $v(x)$ defined by (1.1.15) solves (0.0.20).

1.1.6. The third way to derive an unconditionally solvable integral equation for solving (0.0.20) is to look for the solution of the form

$$v(x) \equiv D\mu + T\rho, \quad T\rho = \int_D g(x,y,k)\rho(y)dy. \tag{1.1.20}$$

Applying boundary condition (0.0.20) yields

$$\frac{A'\mu + \mu}{2} + T\rho = f. \tag{1.1.21}$$

Let us impose a second equation relating μ and ρ:

$$\rho = iT\rho + iD\mu \quad \text{in } D. \tag{1.1.22}$$

The second equation can be chosen in many other ways, but the following two requirements should be met: (1) the system of equations (1.1.21), (1.1.22) is to be of Fredholm's type; (2) the homogeneous system should have only the trivial solution. These requirements are met for equations (1.1.21)–(1.1.22). Indeed, set $\varphi = \binom{\mu}{\rho}$. Then (1.1.21) and (1.1.22) can be written as

$$\varphi + B\varphi = F, \quad F = \begin{pmatrix} 2f \\ 0 \end{pmatrix}, \quad B = \begin{pmatrix} A' & 2T \\ -iD & -iT \end{pmatrix}, \tag{1.1.23}$$

where D is the notation for the operator (1.1.1) and B is a compact linear operator on $H = L^2(\Gamma) \oplus L^2(D)$. Thus, the homogeneous equation (1.1.23) has only the trivial solution. Indeed, the homogeneous equation (1.1.21) shows that $(D\mu)_- + T\rho = 0$. Therefore, the function $v(x)$ defined in (1.1.20) solves the homogeneous problem (0.0.20), and this implies that $v(x) \equiv 0$ in Ω. Equation (1.1.22) can be written as

$$(-\nabla^2 - k^2)v = iv \quad \text{in } D. \tag{1.1.24}$$

Multiply (1.1.24) by $\bar{v}$ and integrate over D to obtain

$$\int_D |\nabla v|^2 dx - \int_\Gamma \frac{\partial v}{\partial N_+}\bar{v}ds - k^2 \int_D |v|^2 dx = i\int_D |v|^2 dx. \tag{1.1.25}$$

Since $v \equiv 0$ in Ω, one has $\partial v/\partial N_- = 0$. This, equation (1.1.3′), and the fact that $T\rho \in C^1(\mathbb{R}^3)$ imply that $\partial v/\partial N_+ = 0$. Therefore, the surface integral in (1.1.25) vanishes and, since the left-hand side of (1.1.25) is real-valued, one concludes that $v \equiv 0$ in D. This implies that $\rho = (-\nabla^2 - k^2)v = 0$ in D and $\mu = (D\mu)_- - (D\mu)_+ = v_- - v_+ = 0$.

1.1.7. The fourth way to derive an unconditionally solvable integral equation for solving (0.0.20) is to consider the function g_M in place of g in the basic equation (1.1.12) of potential theory, where

$$g_M = f + ik \sum_{n=0}^{\infty} \sum_{m=-n}^{n} c_{nm} \psi_{nm}(x) \bar{\psi}_{nm}(y), \tag{1.1.26}$$

where c_{nm} are constants to be chosen soon, and $\psi_{nm}(x)$ are outgoing wave functions

$$\psi_{nm}(x) = h_n^{(1)}(k|x|) Y_{nm}\left(\frac{x}{|x|}\right), \tag{1.1.27}$$

where $h_n^{(1)}(kr)$ are the spherical Hankel functions, $Y_{nm}(\theta, \varphi)$ are the normalized spherical harmonics, and (θ, φ) are the spherical coordinates on the unit sphere S^2. It is known that

$$\begin{aligned} g &= \frac{\exp(ik|x-y|)}{4\pi|x-y|} \\ &= ik \sum_{n=0}^{\infty} \sum_{m=-n}^{n} j_n(k|x|) h_n^{(1)}(k|y|) Y_{nm}\left(\frac{x}{|x|}\right) \bar{Y}_{nm}\left(\frac{y}{|y|}\right), \\ &\quad \text{for } |x| < |y|, \end{aligned} \tag{1.1.28}$$

where $j_n(r)$ are the spherical Bessel functions.

If one looks for a solution to problem (0.0.20) in the form (1.1.1) but with g_M in place of g, then for μ, one obtains equation (1.1.12) with $A' = A'_M$ defined by (1.1.2) with g_M in place of g.

Let us formulate the assumption about c_{nm} which ensure that the homogeneous equation

$$\mu = -A'_M \mu \tag{1.1.29}$$

has only the trivial solution. It follows from (1.1.29), (1.1.3) and the uniqueness of the solution to the homogeneous problem (0.0.20) that

$$v \equiv v_M \equiv \int_\Gamma \mu(t) \frac{\partial g_M(x,t)}{\partial N_t} dt \equiv 0 \quad \text{in } \Omega. \tag{1.1.30}$$

From this and (1.1.3′), it follows that $\partial v/\partial N_+ = 0$. Let B_R be a ball of radius $R > 0$ with center at the origin, $B_R \subset D$, and let $D_R = D \setminus B_R$. One has

$$\begin{aligned} 0 &= \int_{D_R} \left(v\frac{\partial \bar{v}}{\partial N} - \bar{v}\frac{\partial v}{\partial N} \right) ds \\ &= \int_\Gamma \left(v\frac{\partial \bar{v}}{\partial N_+} - \bar{v}\frac{\partial v}{\partial N_+} \right) ds - \int_{\partial B_R} \left(v\frac{\partial \bar{v}}{\partial r} - \bar{v}\frac{\partial v}{\partial r} \right) ds. \end{aligned}$$

Since $\partial v/\partial N_+ = 0$, this implies that

$$0 = \int_{\partial B_R} \left(v\frac{\partial \bar{v}}{\partial r} - \bar{v}\frac{\partial v}{\partial r} \right) ds. \tag{1.1.31}$$

From (1.1.30), it follows that

$$\begin{aligned} v = \Bigg[& ik \sum_{n=0}^{\infty} \sum_{m=-n}^{n} j_n(k|x|) Y_{nm}\left(\frac{x}{|x|}\right) \\ & + ik \sum_{n=0}^{\infty} \sum_{m=-n}^{n} c_{nm} h_n^{(1)}(k|x|) Y_{nm}\left(\frac{x}{|x|}\right) \Bigg] \mu_{nm}, \end{aligned} \tag{1.1.32}$$

where

$$\mu_{nm} \equiv \int_\Gamma \mu(t) \frac{\partial \bar{\psi}_{nm}(t)}{\partial N_+} dt. \tag{1.1.33}$$

From (1.1.31) and (1.1.32), the orthogonality properties of Y_{nm} and the formula for the Wronskian $W[j_n(k|x|) + c_{nm}h_n^{(1)}(k|x|), j_n(k|x|) + \bar{c}_{nm}\bar{h}_n^{(1)}(k|x|)] = (|\frac{1}{2} + c_{nm}|^2 - \frac{1}{4})W[h_n^{(1)}(k|x|), h_n^{(1)}(k|x|)], k > 0$, it follows that

$$0 = \sum_{n=0}^{\infty} \sum_{m=-n}^{n} b_{nm} |\mu_{nm}|^2, \quad b_{nm} = |c_{nm} + 1/2|^2 - 1/4. \tag{1.1.34}$$

If $b_{nm} > 0$, then $\mu_{nm} = 0$. If $\mu_{nm} = 0$ for all $n \geq 0, -n \leq m \leq n$, then $\mu = 0$. Indeed, multiply (1.1.33) by $ikj_n(k|x|)Y_{nm}(x/|x|)$, use (1.1.28) and sum over n, m to obtain

$$u(x) \equiv \int_\Gamma \mu \frac{\partial g(x,t)}{\partial N_t} dt = 0, \quad x \in B_R. \tag{1.1.35}$$

Since $u(x)$ is a solution to Helmholtz's equation in D, (1.1.35) implies that $u(x) \equiv 0$ in D. Therefore, $\partial u/\partial N_+ = 0$ and (1.1.3′) shows that $\partial u/\partial N_- = 0$. Thus, $u(x)$ solves the homogeneous problem

$$(\nabla^2 + k^2)u = 0 \text{ in } \Omega, \quad \frac{\partial u}{\partial N_-} = 0, \quad u \text{ satisfies (0.0.5)}. \tag{1.1.36}$$

The solution to this problem is unique. Therefore, $u(x) = 0$ in Ω, and $\mu = u_- - u_+ = 0$. Thus, if $b_m > 0$ (or $b_{nm} < 0$), then equation (1.1.29) has only the trivial solution.

1.1.8. The fifth way to derive an unconditionally solvable integral equation for solving (0.0.20) is to look for a solution of the form

$$v(x) = D\mu + S\sigma, \tag{1.1.37}$$

where D and S are defined in (1.1.1) and (1.1.4). Applying the boundary condition (0.0.20), one obtains an indeterminate equation, i.e., one equation for two unknown functions:

$$\frac{A'\mu + \mu}{2} + Q\sigma = f, \tag{1.1.38}$$

where Q is defined in (1.1.8). This is an equation in $H = L^2(\Gamma)$ for the two unknown functions μ and σ. One can prove that this equation is solvable. This is done in Section 1.2.

1.1.9. In Section 1.2, we also give a proof of existence of the solution to (0.0.20) using an *a priori* estimate (rather than integral equations), although the proof via integral equation methods is much simpler. Integral equation methods have the great advantage of being applicable for the numerical solution of the problem.

1.1.10. All the methods that we have discussed are applicable without any serious changes to the exterior boundary value problem with the Neumann boundary condition or the Robin boundary condition (impedance boundary condition):

$$\frac{\partial u}{\partial N_-} + \eta u_- = f, \tag{1.1.39}$$

where η is a given constant (possibly complex valued).

1.1.11. Consider the transmission boundary value problem

$$(\nabla^2 + k_0^2)u = 0 \quad \text{in } \Omega, \quad k_0 > 0, \tag{1.1.40}$$

$$(\nabla^2 + k_1^2)u = h(x) \quad \text{in } D, \quad k_1 > 0, \tag{1.1.41}$$

$$u_+ = u_- \quad \text{on } \Omega, \tag{1.1.42}$$

$$\rho\frac{\partial u_+}{\partial N} = \frac{\partial u_-}{\partial N} + f \quad \text{on } \Gamma, \quad \rho = \text{const } > 0, \tag{1.1.43}$$

$$\int_{|s|=R} \left|\frac{\partial u}{\partial r} - ik_0 u\right|^2 ds \to 0 \quad \text{as } R \to \infty. \tag{1.1.44}$$

This is of interest in the problem of wave scattering by a homogeneous penetrable body D. The scattering problem can be formulated as follows: find u which satisfies equations (1.1.40)–(1.1.43) with $f = h = 0$, which is of the form $u = u_0 + v$, where v satisfies the radiation condition (1.1.44) and u_0 is a given incident field which is often of the form $u_0 = \exp\{-ikn\cdot x\}$, the plane wave propagating in the direction of the unit vector $-n$.

1.2. Existence and Uniqueness of the Solution to the Scattering Problems

In this section, we prove that the problems (0.0.20) and (1.1.40)–(1.1.44) are uniquely solvable.

1.2.1. First, let us prove the following well-known lemma.

Lemma 1.2.1. *If $u \in L^2(\Omega_R)$, or $\int_{S_r} |u|^2 ds \to 0$ when $r \to \infty, S_r = \{x : |x| = r\}$, and*

$$(\nabla^2 + k^2)u = 0 \quad \textit{in } \Omega_R = \{x : |x| \geq R\}, \quad k > 0, \tag{1.2.1}$$

then

$$u \equiv 0 \quad \textit{in } \Omega_R \subset \mathbb{R}^3.$$

First proof. Assume that $\lim_{r\to\infty} \int_{S_r} |u|^2 ds = 0$. Solutions to Helmholtz's equation are locally smooth. Multiply equation (1.2.1) by $\bar{Y}_{nm}$ and integrate over the unit sphere S^2 to obtain $(x = (r, \theta, \mu))$:

$$u''_{nm} + \frac{2}{r}u'_{nm} + \left(k^2 - \frac{n(n+1)}{r^2}\right) u_{nm} = 0, \quad u_{nm}(r) \equiv \int_{S^2} u\bar{Y}_{nm} ds, \tag{1.2.2}$$

where $u'_{nm} = du_{nm}/dr$. Thus, $u_{nm} = c_{1n}h_n^{(1)}(kr) + c_{2n}h_n^{(2)}(kr)$, where c_{1n}, c_{2n} are constants and $h_n^{(j)}(kr), j = 1, 2$, are the spherical Hankel functions, $h_n^{(j)}(z) = [\pi/(2z)]^{1/2} H_{n+1/2}^{(j)}(z)$, and $H_p^{(j)}(z)$ are the usual Hankel functions. The Parseval equality

$$\int_{S^2} |u(r, \theta, \varphi)|^2 ds = \sum_{n=0}^{\infty} \sum_{m=-n}^{n} r^2 |u_{nm}(r)|^2 < \infty \tag{1.2.3}$$

and the asymptotic (as $r \to +\infty$)

$$u_{nm} = \frac{(2\pi^{-1})^{1/2} \left\{c_{1n} \exp\left[i\left(kr - \frac{\pi}{2}n - \frac{\pi}{2}\right)\right] + c_{2n} \exp\left[-i\left(kr - \frac{\pi}{2}n - \frac{\pi}{2}\right)\right]\right\}}{kr} + o\left(\frac{1}{kr}\right), \quad r \to \infty, \tag{1.2.4}$$

imply that $c_{1n} = c_{2n} = 0, n \geq 0$. Thus, $u \equiv 0$ in Ω_R.

This proof is valid in $R^r, r \geq 2$. With obvious modification, the proof is valid if $u \in L^2(\Omega_R)$.

Note that the condition $\lim_{r_n \to \infty} \int_{S_{r_n}} |u|^2 ds = 0$ *is not sufficient for the conclusion of Lemma* 1.2.1 *to hold.*

Example. $u = r^{-1} \sin(kr), r_n = n\pi k^{-1} \to \infty, u(r_n, k) = 0$. However, if $\lim_{r \to \infty} \int_{S_r} \{|u|^2 + |\partial u/\partial r|^2\} ds = 0$ and (1.2.1) holds, then $u \equiv 0$.

□

Second proof. From (1.2.1), it follows that if $u \in L^2(\Omega_R)$ then $\nabla u \in L^2(\Omega_R)$. Indeed, define $F(r) = \int_R^r dr r^2 \int_{S^2} |u(r, \omega)|^2 ds, \omega = (\theta, \varphi)$. Then $F(r)$ increases monotonically and $F(\infty) < \infty$. Furthermore,

$$F'(r) = r^2 \int_{S^2} |u|^2 ds > 0; \tag{1.2.5}$$

$$F''(r) = 2r \int_{S^2} |u|^2 ds + 2r^2 \operatorname{Re} \int_{S^2} \frac{\partial u}{\partial r} \bar{u} ds. \tag{1.2.6}$$

Since $F(\infty) < \infty$, and $F' > 0$, clearly either $F''(r) \to 0$ as $r \to \infty$ or there exists a sequence $r_n \to \infty$ such that $F''(r_n) \to 0$ for some sequence $r_n \to \infty$, then (1.2.6) implies that

$$\operatorname{Re} \int_{S_{r_n}} \frac{\partial u}{\partial r} \bar{u} ds \to 0, \quad r_n \to \infty, \quad S_{r_n} = \{x : |x| = r_n\}. \tag{1.2.7}$$

From equation (1.2.1), it follows that

$$\begin{aligned} &\int_{\Omega_{R,r_n}} |\nabla u|^2 dx = k^2 \int_{\Omega_{R,r_n}} |u|^2 dx + \int_{S_{r_n}} \frac{\partial u}{\partial r} \bar{u} ds - \int_{S_R} \frac{\partial u}{\partial r} \bar{u} ds, \\ &\Omega_{R,r} = \{x : R \leq |x| \leq r\}. \end{aligned} \tag{1.2.8}$$

From (1.2.7), (1.2.8), and assumption $u \in L^2(\Omega_R)$, it follows that

$$\int_\Omega (|u|^2 + |\nabla u|^2) dx < \infty. \tag{1.2.9}$$

Therefore, one can apply Green's formula to obtain

$$u(x) = \int_{S_R} \left(g^+ \frac{\partial u}{\partial N} - u \frac{\partial g^+}{\partial N} \right) ds = \int_{S_R} \left(g^- \frac{\partial u}{\partial N} - u \frac{\partial g^-}{\partial N} \right) ds, \tag{1.2.10}$$

where $g^{\pm} = \exp(\pm ik|x-y|)/(4\pi|x-y|), |x| > R$. Let $x = (r, \omega)$. Consider analytic continuation of $u(r, \omega)$ on the complex plane $z = r\ \exp(i\varphi)$. Since S_R is a bounded set, $|x-s| = (r^2 - 2r|s|\cos(x\ s) + |s|^{1/2})$ is analytic in $r, |r| > R, u(z, \omega)$ is analytic near infinity and

$$u(z,\omega) = \frac{\exp(ikz)}{z} f^+(z,\omega) = \frac{\exp(-ikz)}{z} f^-(z,\omega), \tag{1.2.11a}$$

where $f^{\pm}(z, \omega)$ are analytic and bounded near infinity on the complex plane z. One can write explicit expression for $f^{\pm}$ but we do not need this. From (1.2.11a), it follows that $f^+ = f^- = 0$. Indeed, take $z = iy, y \to +\infty$. The left-hand side of (1.2.11a) decays exponentially while the right-hand side grows exponentially unless $f^+ = f^- = 0$. Thus, $u(z, \omega) = 0$ and $u(r, \omega) = 0$. □

Remark 1.2.1. No boundary conditions are imposed on u in Lemma 1.2.1.

Remark 1.2.2. If equation (1.2.1) and the radiation condition $\int_{S_r} |u_r - iku|^2 ds = o(1), r \to \infty$, hold, then $\int_{S_r} |u|^2 ds = O(1), r := |x| \to \infty$.

Proof. From (1.2.1), one gets

$$0 = \int_{R \le |x| \le r} (k^2|u|^2 - |\nabla u|^2) dx + \int_{S_r} \bar{u} u_r ds - \int_{S_R} \bar{u} u_r ds. \tag{1.2.11b}$$

Taking imaginary part, one gets

$$\operatorname{Im} \int_{S_r} \bar{u} u_r ds = \operatorname{Im} \int_{S_R} \bar{u} u_r ds = O(1), \quad r \to \infty. \tag{1.2.11c}$$

From the radiation condition, one gets

$$\int_{S_r} \left(|u_r|^2 + k^2|u|^2\right) ds = 2k \text{ Im} \int_{S_r} \bar{u} u_r ds + o(1), \quad r \to \infty. \tag{1.2.11d}$$

Therefore,

$$\int_{S_r} (|u_r|^2 + k^2|u|^2) ds = O(1), \quad r \to \infty. \tag{1.2.11e}$$

The above argument yields the existence of a finite limit $\int_{R \le |x| \le r} (k^2|u|^2 - |\nabla u|^2) dx$ provided that $\int_{S_r} \bar{u} u_r ds = o(1)$.

If $u|_S = 0$, then the above argument yields the uniqueness theorem:

If $\int_{S_r} |u|^2 ds = o(1)$, *u solves* (1.2.1) *and* $u|_S = 0$, *then* $u = 0$ *in* D'.

Another conclusion:

If $\int_{S_r} |u|^2 ds = o(1)$, *then* $\int_{S_r} |u_r|^2 ds = o(1)$ as $r \to \infty$.

This follows from (1.2.11d). □

Lemma 1.2.2. *Problem* (0.0.20) *has at most one solution.*

Proof. Let us show that the homogeneous problem (0.0.20) has only the trivial solution. Let w solve the homogeneous problem (0.0.20). From the radiation condition (0.0.5) or (1.1.44), one obtains

$$0 = \lim_{R \to \infty} \left\{ \int_{S_R} \left(\left| \frac{\partial w}{\partial r} \right|^2 + k^2 |w|^2 \right) ds \right.$$
$$\left. - ik \int_{S_R} \left(w \frac{\partial \bar{w}}{\partial} - \bar{w} \frac{\partial w}{\partial r} \right) ds \right\}. \tag{1.2.12}$$

From equation (0.0.20) and Green's formula, it follows that

$$\int_{S_R} \left(w \frac{\partial \bar{w}}{\partial r} - \bar{w} \frac{\partial w}{\partial r} \right) ds = \int_{\Gamma} \left(w \frac{\partial \bar{w}}{\partial N} - \bar{w} \frac{\partial w}{\partial N} \right) ds = 0. \tag{1.2.13}$$

Therefore,

$$\lim_{r \to \infty} \int_{S_R} \left(\left| \frac{\partial w}{\partial r} \right|^2 + k^2 |w|^2 \right) ds = 0. \tag{1.2.14}$$

From Lemma 1.2.1 and equation (1.2.14), the conclusion of Lemma 1.2.2 follows. □

1.2.2. Existence of the solution to the boundary value problem (0.0.20) can be established by means of integral equation methods.

Theorem 1.2.3. *Problem* (0.0.20) *is uniquely solvable for any* $f \in L^2(\Gamma)$.

Proof. Uniqueness is proved in Lemma 1.2.2. Existence is proved in Lemma 1.1.3. We give another proof here. To prove existence, assume first that $k^2 \notin \sigma(-\Delta_N)$, that is, k^2 is not an eigenvalue of the Neumann Laplacian in D. Then, looking for a solution to (0.0.20) of the form (1.1.1), one obtains equation (1.1.2). By Fredholm's alternative, this equation is uniquely solvable for μ if and only if the equation $\sigma = -\bar{A}\sigma$ has only the trivial solution $\sigma = 0$. If $\sigma + \bar{A}\sigma = 0$ and $\sigma \not\equiv 0$, then (1.1.3) shows that $w = S\bar{\sigma}$ solves the problem

$$(\nabla^2 + k^2)w = 0 \text{ in } D, \quad \frac{\partial w}{\partial N_+} = 0. \tag{1.2.15}$$

From our assumption $k^2 \notin \sigma(-\Delta_N)$ and (1.2.15), it follows that $w = 0$ in D, $w = 0$ on Ω. Since $w = S\bar{\sigma}$ is continuous across Γ, one sees that w solves the problem

$$(\nabla^2 + k^2)w = 0 \text{ in } \Omega, \ w_- = 0, \ w \text{ satisfies (1.1.44) with } k_0 = k. \tag{1.2.16}$$

Thus, $w = 0$ in Ω by Lemma 1.2.2. Therefore, $\bar{\sigma} = \partial w/\partial N_+ - \partial w/\partial N_- = 0$, and the proof of the existence is complete for the case $k^2 \notin \sigma(-\Delta_N)$.

Suppose $k^2 \in \sigma(-\Delta_N)$. Then our argument is exactly as above but instead of g in (1.1.1) we use g_ϵ defined in (1.1.44). In Lemma 1.2.4 below, we prove that there exists an $\epsilon > 0$ such that the problem (1.2.15) in $D_\epsilon = D \setminus B_\epsilon$ with $w = 0$ on ∂B_ϵ has only the trivial solution. Therefore, the solution to problem (0.0.20) exists, is unique, and can be found in the form (1.1.1) with g_ϵ in place of g. □

Lemma 1.2.4. *For any $k > 0$, there exists an $\epsilon > 0$ such that the problem*

$$(\nabla^2 + k^2)u = 0 \text{ in } D \setminus B_\epsilon,\ u = 0 \text{ on } \partial B_\epsilon,\ \frac{\partial u}{\partial N_+} = 0 \text{ on } \Gamma \quad (1.2.17)$$

has only the trivial solution.

Remark 1.2.3. There are infinitely many such ϵ.

Proof. If problem (1.2.17) has a nontrivial solution for $\epsilon = \epsilon_0$, then for $\epsilon \neq \epsilon_0$ and $|\epsilon - \epsilon_0|$ sufficiently small, problem (1.2.17) has only the trivial solution. This follows from Lemma 1.2.5. □

Lemma 1.2.5. *Let $k_1^2(\epsilon) \le k_2^2(\epsilon) \le \cdots \le k_n^2(\epsilon) \le \cdots$ be the eigenvalues of problem* (1.2.17), *counted according to their multiplicities. If $\epsilon_1 < \epsilon_2$, then $k_n^2(\epsilon_1) < k_n^2(\epsilon_2)$.*

Proof. From the variational principle for eigenvalues, it follows that $k_n^2(a) \le k_n^2(b)$ if $a \le b$. Suppose that $a < b$ and $k_n^2(a) = k_n^2(b)$. Since $k_n^2(a) \le k_n^2(\epsilon) \le k_n^2(b)$ for $a \le \epsilon \le b$, one obtains that $k_n^2 \equiv k_n^2(\epsilon) \le k_n^2(a) \le k_n^2(b), a \le \epsilon \le b$. Pick m such that $k_m^2(a) > k_n^2(a)$ and consider a sequence $a = \epsilon_1 < \epsilon_2 < \epsilon_3 < \cdots < \epsilon_m < b$. Let u_j be the eigenfunction of the problem (1.2.17) in $D_j = D \setminus B_{\epsilon_j}$ corresponding to the eigenvalue k_n^2. Let $u_j = 0$ in the domain $\epsilon_1 \le |x| \le \epsilon_j$, then $u_j \in H^1(D_1)$. The function $u_1, \dots, u_m$ are linearly independent in $L^2(D_1)$. Indeed, if $\sum_{j=1}^m c_j' u_j(x) = 0, c_j' = \text{const}, x \in D_1$, then $u_1(x) = 0$ in $\epsilon_1 < |x|\epsilon_2$, unless $c_1' = 0$. Repeating this argument, one derives that $c_j' = 0, 1 \le j \le m$. Thus, there exists infinitely many linearly independent solutions to (1.2.17) corresponding to $k = k_n^2(a)$. This is a contradiction, since the eigenspaces for (1.2.17) are finite-dimensional. Define $u(x) = \sum_{j=1}^m c_j u_j(x)$ and choose the coefficients c_j so that $(u, u_j)_{L^2(D_1)} = 0, 1 \le j \le m-1, \|u\|_{L^2(D_1)} = 1$. Since u is an admissible function, it follows from the variational principle that

$\int_{D_1} |\nabla u|^2 dx \geq k_m^2(a)$. On the other hand,

$$\int_{D_1} |\nabla u|^2 dx = \sum_{i,j=1}^{m} c_i \bar{c}_j \int_{D_1} \nabla u_i \nabla \bar{u}_j dx$$

$$= \sum_{i,j=1}^{m} c_i \bar{c}_j k_n^2 \int_{D_1} u_i \bar{u}_j dx = k_n^2 \int_{D_1} |u|^2 dx = k_n^2.$$

Thus, $k_m^2 \leq k_n^2(a)$. This is a contradiction. □

1.2.3. Existence of solutions to exterior boundary value problems with Neumann's and impedance boundary conditions can be proved as in Theorem 1.2.3.

1.2.4. Let us give another proof of the existence of the solution to problem (0.0.20) using equation (1.1.38).

Lemma 1.2.6. *Let A and B be linear compact operators on a Hilbert space H, and*

$$N(I + A^*) \cap N(B^*) = \{0\}. \tag{1.2.18}$$

Then the equation $(I + A)\mu + B\sigma = f$ is solvable for any $f \in H$. Condition (1.2.18) *is also necessary for the solvability of the equation $(I + A)\mu + B\sigma = f$.*

Proof. Let $T = I + A$. Since A is linear and compact $R(T)$ is closed, $\dim N(T^*) = \operatorname{codim} R(T) < \infty$. It is clear that equation $T\mu + B\sigma = f$ is densely solvable (i.e., solvable for any f in a dense subset of H) if (1.2.18) holds. Indeed, if $h \in N(T^*) \cap N(B^*)$, $h \neq 0$, then $(f, h) = 0$ is a necessary condition for the solvability of the equation $T\mu + B\sigma = f$. Thus, this equation is not densely solvable. On the other hand, if the set $\{T\mu + B\sigma\}$ is not dense in H, then there exists $h \in H, h \neq 0$, such that $(T\mu + B\sigma, h) = 0$ for all μ and σ. Taking $\sigma = 0$, one sees that $h \in N(T^*)$ and similarly $h \in N(B^*)$. Thus, (1.2.18) does not hold. Since condition (1.2.18) is necessary for dense solvability, it is necessary for everywhere solvability, i.e., solvability for any $f \in H$.

Let us prove its sufficiency. Let P be the orthogonal projection onto $R(T)$. Then

$$T\mu + PB\sigma = Pf, \tag{1.2.19}$$

$$(I-P)B\sigma = (I-P)f. \tag{1.2.20}$$

If (1.2.20) is solvable, then (1.2.19) is solvable and

$$T\mu + B\sigma = f. \tag{1.2.21}$$

Indeed, let σ solve (1.2.20) and let $T\mu = P(f - B\sigma)$. Since the right-hand side in the last equation belongs to $R(T)$, this equation is solvable and

$$T\mu + B\sigma = Pf - PB\sigma + B\sigma = Pf + (I-P)f = f. \tag{1.2.22}$$

To prove that equation (1.2.21) is solvable for any f if (1.2.18) holds, let us use the already established dense solvability of (1.2.21) and, therefore, (1.2.20). But (1.2.20) is a linear equation in a finite-dimensional space, since codim $R(T) < \infty$. If such an equation is densely solvable, then it is everywhere solvable. But then, as was shown above, equations (1.2.19) and (1.2.21) are solvable for any f. □

Let us apply Lemma 1.2.6 to equation (1.1.38). We have to prove that $N(I + \bar{A}) \cap N(Q^*) = \{0\}$. Let $\sigma \in N(I + \bar{A}) \cap N(Q^*)$ and $w = S\bar{\sigma}$. Then $w = 0$ on Γ and $\partial w/\partial N_+ = 0$. By the uniqueness of the solution to Cauchy's problem for elliptic equations, $w = 0$ in D. Since w solves problem (1.2.16), Lemma 1.2.2 says that $w = 0$ in Ω. Thus,

$$\sigma = \frac{\partial w}{\partial N_+} - \frac{\partial w}{\partial N_-} = 0.$$

We have proved the following result.

Proposition 1.2.7. *The solution to problem* (0.0.20) *can be found in the form* (1.1.37).

1.2.5. Another proof of existence of the solution to problem (0.0.20) is given in Section 1.5.

1.2.6. Consider *the transmission boundary value problem* (1.1.40)–(1.1.44).

Lemma 1.2.8. *Problem* (1.1.40)–(1.1.44) *has at most one solution.*

Proof. It is sufficient to prove that the corresponding homogeneous problem has only the trivial solution. From Green's formula, it follows that

$$0 = \lim_{R\to\infty} \int_{|r|=R} \left(u\frac{\partial \bar{u}}{\partial r} - \bar{u}\frac{\partial u}{\partial r} \right) ds - \int_{\Gamma} \left(u\frac{\partial \bar{u}}{\partial N_-} - \bar{u}\frac{\partial u}{\partial N_-} \right) ds. \tag{1.2.23}$$

Furthermore,

$$\int_{\Gamma} \left(u\frac{\partial \bar{u}}{\partial N_-} - \bar{u}\frac{\partial u}{\partial N_-} \right) ds = \rho \int_{\Gamma} \left(u\frac{\partial \bar{u}}{\partial N_+} - \bar{u}\frac{\partial u}{\partial N_+} \right) ds = 0. \tag{1.2.24}$$

Thus,

$$\lim_{R\to\infty} \int_{|r|=R} \left(u\frac{\partial \bar{u}}{\partial r} - \bar{u}\frac{\partial u}{\partial r} \right) ds = 0. \tag{1.2.25}$$

From (1.2.25), (1.1.44), and (1.2.14), one derives (1.2.16) with $k = k_0$. Now, Lemma 1.2.1 says that $u = 0$ in Ω. Therefore, $u = 0$ on $\Gamma, \partial u/\partial N_+ = 0$, and $u = 0$ in D. □

Let

$$\mathcal{H}^q = H^q \oplus \tilde{H}^{q-1/2}, \quad H^q = W^{2,q}(D), \quad \tilde{H}^q = W^{2,q}(\Gamma).$$

The space $\mathcal{H}^q$ consists of the ordered pairs $w = \binom{u}{\sigma}$, where $u \in H^q, \sigma \in \tilde{H}^{q-1/2}, (w_1, w_2)_{\mathcal{H}^q} = (u_1, u_2)_{H^q} + (\sigma_1, \sigma_2)_{\tilde{H}^{q-1/2}}$.

Assume that $h \in H^q, f \in \tilde{H}^{q-1/2}$.

Theorem 1.2.9. *Problem* (1.1.40)–(1.1.44) *is uniquely solvable.*

Proof. Uniqueness follows from Lemma 1.2.8. In order to prove existence, it is sufficient to find a Fredholm-type integral equation for the solution to problem (1.1.40)–(1.1.44) which is equivalent to this problem. Since problem (1.1.40)–(1.1.44) has no more than one solution, the equivalent integral equation has no more than one solution. Since this equation is of Fredholm's type, it is solvable. To derive a suitable integral equation, let us use Green's formula in D:

$$\begin{aligned}\int_{\mathbb{R}^3} &\{g(\nabla^2 + k_0^2)u - u(\nabla^2 + k_1^2)g\}\,dy \\ &= u(x) + \kappa \int_D gu\,dy + \int_D gh\,dy, \quad \kappa := k_0^2 - k_1^2, \end{aligned} \tag{1.2.26}$$

where $g = g(x, y, k_0)$. The left-hand side of this equation can be written as

$$\begin{aligned}\lim_{R\to\infty} &\int_{S_R} \left(g\frac{\partial u}{\partial N} - u\frac{\partial g}{\partial N_-} \right) ds - \int_\Gamma \left(g\frac{\partial u}{\partial N_-} - u\frac{\partial g}{\partial N_-} \right) ds \\ &+ \int_\Gamma \left(g\frac{\partial u}{\partial N_+} - u\frac{\partial g}{\partial N_+} \right) ds \\ &= \int_\Gamma g \left(\frac{\partial u}{\partial N_+} - \frac{\partial u}{\partial N_-} \right) ds \\ &= (1-\rho) \int_\Gamma g\frac{\partial u}{\partial N_+} ds + \int_\Gamma gf\,ds. \end{aligned} \tag{1.2.27}$$

Combine (1.2.26) and (1.2.27) to obtain

$$u(x) = F(x) - \kappa Tu + (1-\rho)S\sigma, \tag{1.2.28}$$

where

$$\sigma = \frac{\partial u}{\partial N_+}, \tag{1.2.29}$$

and

$$F(x) \equiv -Th + Sf. \tag{1.2.30}$$

For any σ, the function u which solves (1.2.28) satisfies (1.1.40)–(1.1.42) and (1.1.44). It satisfies (1.1.43) if and only if

$$(\rho - 1)\frac{\partial F}{\partial N} - \kappa(\rho - 1)\frac{\partial Tu}{\partial N} + (1-\rho)\left[\rho\frac{A\sigma + \sigma}{2} - \frac{A\sigma - \sigma}{2}\right] = f, \tag{1.2.31}$$

or

$$\sigma = \frac{1-\sigma}{1+\sigma}A\sigma - \frac{2\kappa}{1+\rho}\frac{\partial Tu}{\partial N} + \frac{2}{1+\rho}\left(\frac{\partial F}{\partial N} + \frac{f}{1-\rho}\right), \tag{1.2.32}$$

where formula (1.2.3) was used. Equations (1.2.32) and (1.2.29) are equivalent if u satisfies (1.2.28). Let

$$w = \begin{pmatrix} u \\ \sigma \end{pmatrix}. \tag{1.2.33}$$

Then (1.2.28) and (1.2.32) can be written as

$$w = Bw + h_1, \tag{1.2.34}$$

where

$$B = \begin{pmatrix} -\kappa T & (1-\rho)S \\ -\dfrac{2\kappa}{1+\rho}\dfrac{\partial}{\partial N}T & \dfrac{1-\rho}{1+\rho}A \end{pmatrix}, \quad h_1 = \begin{pmatrix} F \\ \dfrac{2}{1+\rho}\left(\dfrac{\partial F}{\partial N} + \dfrac{f}{1-\rho}\right) \end{pmatrix}, \tag{1.2.35}$$

$h_1 \in \mathcal{H}^q$.

It remains to be proved that equation (1.2.34) is a Fredholm-type equation in $\mathcal{H}^q$ and that this equation is equivalent to problem (1.1.40)–(1.1.44). This is done in Lemmas 1.2.10 and 1.2.11. □

Lemma 1.2.10. *Let* $\mathcal{H}^q = H^q \oplus \tilde{H}^{q-1/2}$. *Then* $B : \mathcal{H}^q \to \mathcal{H}^q$ *is compact.*

Proof. It is well known [96] that the imbeddings $H^q \to H^p, \tilde{H}^q \to \tilde{H}^p$ and $H^l \to \tilde{H}^s$ are compact if $q > p$ and $l > s + \frac{1}{2}$. It is also well known that $T : H^q \to H^{q+2}$ is continuous and therefore $\partial/\partial N \circ T : H^q \to H^{q+1}$ is continuous, $A : \tilde{H}^q \to \tilde{H}^{q+2}$ is continuous, $S : \tilde{H}^q \to H^{q+3/2}$ is continuous. From these facts, the conclusion of Lemma 1.2.10 follows. □

Lemma 1.2.11. *Equation* (1.2.34) *is equivalent to problem* (1.1.40)–(1.1.44).

Proof. It was already proved that (1.2.34) follows from (1.1.40)–(1.1.44). Suppose $\binom{u}{\sigma}$ solves (1.2.34), then equations (1.2.28) and (1.2.32) hold. Equation (1.2.28) implies equations (1.1.40)–(1.1.42) and (1.1.44). Equation (1.2.32) implies (1.1.43). □

1.2.7. Consider *the transmission scattering problem*

$$(\nabla^2 + k_0^2)u = 0 \quad \text{in } \Omega, \quad k_0 > 0, \tag{1.2.36}$$

$$(\nabla^2 + k_1^2)u = 0 \quad \text{in } D, \quad k_1 > 0, \tag{1.2.37}$$

$$u_+ = u_- \quad \text{on } \Gamma, \tag{1.2.38}$$

$$\rho\frac{\partial u}{\partial N_+} = \frac{\partial u}{\partial N_-} \quad \text{on } \Gamma, \tag{1.2.39}$$

$$u = u_0 + v, \tag{1.2.40}$$

$$\lim_{R\to\infty}\int_{S_R}\left|\frac{\partial v}{\partial r} - ikv\right|^2 ds = 0, \tag{1.2.41}$$

where $u_0(x)$ is an incident field which satisfies equation (1.2.36) in the whole space, for example,

$$u_0 = \exp(-ik_0 n\cdot x). \tag{1.2.42}$$

The scattered field v satisfies (1.1.40)–(1.1.44) with

$$h = (k_0^2 - k_1^2)u_0, \quad f = (1-\rho)\frac{\partial u_0}{\partial N}. \tag{1.2.43}$$

Thus, Theorem 1.2.9 states that problem (1.2.36)–(1.2.41) has a solution and this solution is unique.

1.2.8. Let us finally sketch yet another approach to the proof of the existence of solutions to exterior boundary value problems.

Let

$$(\nabla^2 + k^2)u = f \text{ in } \Omega\subset\mathbb{R}^3, \quad u|_\Gamma = 0, \quad u \text{ satisfies (1.2.41)}. \tag{1.2.44}$$

Let $f = 0$ if $|x| > R, \Gamma \subset B_R, v = \int_{\mathbb{R}^3} gh\,dy, h = 0$ if $|x| \geq R$, and w be defined as the solution to the problem

$$\nabla^2 w + \lambda^2 w = 0 \text{ in } \Omega \cap B_R,\ w = v \text{ on } \Gamma,\ w = 0 \text{ on } S_R. \quad (1.2.45)$$

We take R such that problem (1.2.45) is uniquely solvable, i.e., λ^2 is not an eigenvalue of the homogeneous problem (1.2.45). Let $\varphi \in C_0^\infty(\Omega \cap B_R), \varphi = 1$ on Γ and $\varphi = 0$ in a neighborhood of S_R. Then the function $u = v - \varphi w$ solves (1.2.44) if and only if

$$\nabla^2 u + k^2 u = h - 2\nabla\varphi\nabla w - w\nabla^2\varphi - (k^2 - \lambda^2)\varphi w = f.$$

This can be written as

$$h - Bh = f, \quad (1.2.46)$$

where

$$Bh = 2\nabla\varphi\nabla w + w\nabla^2\varphi + (k^2 - \lambda^2)\varphi w. \quad (1.2.47)$$

If we show that B is compact in $L^2(\Omega \cap B_R)$ and the homogeneous equation (1.2.46) has only the trivial solution, we conclude that equation (1.2.46) is solvable for any $f \in L^2(\Omega \cap B_R)$, and therefore, problem (1.2.44) is solvable. The mapping $h \to Bh$ is a composition of the mappings $(w \to (k^2-\lambda^2)\varphi w + 2\nabla w\nabla\varphi + w\nabla^2\varphi) \circ (v \to w) \circ (h \to v)$ which acts as follows: $(H^{q+2} \to H^{q+1}) \circ (H^{q+2} \to H^{q+2}) \circ (H^q \to H^{q+2})$, where $H^q = H^q(\Omega \cap B_R)$. Therefore, $h \to Bh$ acts from H^q into H^{q+2}. Since the domain $\Omega \cap B_R$ is compact, the imbedding $i : H^{q+1} \to H^q$ is compact. Therefore, $B : H^q \to H^q$ is compact. If h solves the homogeneous equation (1.2.46), then $v - \varphi w$ solves the homogeneous problem (1.2.44). Thus, $u = 0$ in Ω and $v = \varphi w$ in Ω. Therefore, $v = 0$ in Ω_R, $v = 0$ on S_R, and if one defines $\alpha(x) = v$ in D, $\alpha(x) = w$ in $\Omega \cap B_R$, then

$$(\nabla^2 + k^2)\alpha = 0 \text{ in } D, \quad (\nabla^2 + \lambda^2)\alpha = 0 \text{ in } \Omega \cap B_R. \quad (1.2.48)$$

Multiplying (1.2.48) by $\bar{\alpha}$ and integrating over B_R yields

$$\int_{B_R} (|\nabla\alpha|^2 - p|\alpha|^2)dx = 0, \quad p = \begin{cases} k^2 & \text{in } D, \\ \lambda^2 & \text{in } \Omega \cap B. \end{cases} \quad (1.2.49)$$

If Im $\lambda^2 \neq 0$, then (1.2.49) implies that $\alpha = 0$ in $\Omega \cap B_R$. Therefore, $w = 0$ in $\Omega \cap B_R$ and $v = 0$ in $\Omega \cap B_R$. But then $h = 0$.

1.2.9. Let us prove uniqueness of the solution to *the impedance boundary value problem*

$$(\nabla^2 + k^2)u = 0 \text{ in } \Omega, \quad \operatorname{Im} k \geq 0, \quad \operatorname{Re} k \geq 0, \quad k \neq 0, \tag{1.2.50}$$

$$u_N + \eta(s)u = f \quad \text{on } \Gamma, \quad u_N = \frac{\partial u}{\partial N}, \tag{1.2.51}$$

$$\operatorname{Im}(\bar{k}\eta(s)) \geq 0 \quad \text{on } \Gamma, \tag{1.2.52}$$

$$\int_{S_r} |u_r - iku|^2 ds \to 0, \quad r \to \infty, \quad u_r = \frac{\partial u}{\partial r}. \tag{1.2.53}$$

Lemma 1.2.12. *Problem* (1.2.50)–(1.2.53) *has no more than one solution.*

Proof. Consider the homogeneous problem, i.e., $f = 0$. Condition (1.2.53) yields

$$\int_{S_r} (|u|^2 + |k|^2|u|^2)ds - 2\operatorname{Im}\int_{S_r} \bar{k}\bar{u}u_r ds \equiv \epsilon(r) \to 0, \quad r \to \infty. \tag{1.2.54}$$

Apply Green's formula to u and $\bar{u}$ to obtain

$$0 = \int_{B_r\cap\Omega} \bar{u}(\nabla^2 + k^2)udx = \int_{B_r\cap\Omega} (k^2|u|^2 - |\nabla u|^2)dx + \int_{S_r} \bar{u}u_r ds - \int_\Gamma \bar{u}u_N ds. \tag{1.2.55}$$

From (1.2.55), it follows that

$$-\bar{k}\int_{S_r} \bar{u}u_r ds = \int_{B_r\cap\Omega} (\bar{k}k^2|u|^2 - \bar{k}|\nabla u|^2)dx - \int_\Gamma \bar{k}\bar{u}u_N ds. \tag{1.2.56}$$

If $\text{Im } k \geq 0$, then $\text{Im } \bar{k} \leq 0$. It follows from (1.2.56) that

$$-\text{Im}\left(\bar{k}\int_{S_r} \bar{u}u_r ds\right) = \text{Im}\int_{B_r\cap\Omega}(k|k|^2|u|^2 - \bar{k}|\nabla u|^2)dx$$
$$-\text{Im}\int_{\Gamma} \bar{k}\bar{u}u_N ds. \tag{1.2.57}$$

If

$$-\text{Im}\left(\bar{k}\int_{\Gamma} \bar{u}u_N ds\right) \geq 0, \tag{1.2.58}$$

then the right-hand side of (1.2.57) is nonnegative and

$$-\text{Im}\left(\bar{k}\int_{S_r} \bar{u}u_r ds\right) \geq 0. \tag{1.2.59}$$

It follows from (1.2.54) and (1.2.59) that

$$\int_{S_r}(|u_r|^2 + |k|^2|u|^2)ds \to 0, \quad r\to\infty, \quad k \neq 0. \tag{1.2.60}$$

This implies that $u = 0$ in Ω. It remains to be noted that the homogeneous condition (1.2.51) and inequality (1.2.52) imply that

$$-\text{Im}\left(\bar{k}\int_{\Gamma} \bar{u}u_N ds\right) = \text{Im}\,\bar{k}\int_{\Gamma}\eta(s)|u|^2 ds \geq 0.$$

This completes the proof. □

1.2.10. In this section, we prove that (1.2.50) and (1.2.53) imply that

$$\int_{S_r}|u|^2 ds = O(1), \quad r\to\infty. \tag{1.2.61}$$

This is of interest because often in the literature the radiation condition is stated as two conditions

$$u_r - iku = o(r^{-1}), \quad r\to\infty, \tag{1.2.62}$$
$$u = O(r^{-1}), \quad r\to\infty. \tag{1.2.63}$$

These two conditions imply that Green's formula is applicable in the infinite domain Ω because

$$\int_{S_r} (uv_r - u_r v) ds \to 0, \quad r \to \infty \tag{1.2.64}$$

if u and v satisfy conditions (1.2.62) and (1.2.63). In this case, there is no need to assume that u and v solve equation (1.2.50). One has

$$\int_{S_r} (uv_r - u_r v) ds = \int_{S_r} \{u(v_r - ikv) + v(iku - u_r\} \, ds.$$

Applying Schwarz's unequality and conditions (1.2.62) and (1.2.63) yields (1.2.64). However, if (1.2.62) holds and u solves equation (1.2.50), then (1.2.63) holds automatically. This is the result we want to prove. Note that (1.2.62) implies (1.2.53).

Lemma 1.2.13. *Equations* (1.2.50) *and* (1.2.53) *imply* (1.2.61) *and* (1.2.63) *if* $\text{Im } k \geq 0$.

Proof. It follows from (1.2.54) and (1.2.57) that

$$\begin{aligned} &\frac{1}{2}\epsilon(r) + \text{Im} \int_{\Gamma} \bar{k}\bar{u}u_N ds \\ &\quad = \text{Im} \int_{B_r \cap \Omega} (k|k|^2|u|^2 - \bar{k}|\nabla u|^2) ds \\ &\qquad + \frac{1}{2} \int_{S_r} (|u_r|^2 + |k|^2|u|^2) ds, \quad \epsilon(r) \to 0, \quad r \to \infty. \end{aligned} \tag{1.2.65}$$

If $\text{Im } k \geq 0$, then both terms in the right-hand side of (1.2.65) are non-negative and the left-hand side of (1.2.65) is bounded as $r \to \infty$.

Therefore,

$$\int_{S_r} (|u_r|^2 + |k|^2|u|^2) ds = O(1), \quad r \to \infty. \tag{1.2.66}$$

If $\text{Im } k > 0$, then it follows from (1.2.65) that

$$\int_{\Omega} (|u|^2 + |\nabla u|^2) dx < \infty. \tag{1.2.67}$$

Equation (1.2.66) implies (1.2.61) and the representation formula for u, which follows immediately from Green's formula:

$$u(x) = \int_\Gamma \left\{ u(s)\frac{\partial g(x,s)}{\partial N} - g(x,s)\frac{\partial u(s)}{\partial N} \right\} ds. \tag{1.2.68}$$

From (1.2.68), equation (1.2.63) follows immediately. □

Another proof of Lemma 1.2.13 for the most interesting case $k > 0$ is similar to the proof of Lemma 1.2.1 in Section 1.2. If u satisfies equation (1.2.50), then $u_{nm}(r) \equiv \int_{S^2} u\bar{Y}_{nm} dr$, where Y_{nm} are the spherical harmonics, satisfies Bessel's equation

$$u''_{nm} + \frac{2}{r}u'_{nm} + \left(k^2 - \frac{n(n+1)}{r^2}\right)u_{nm} = 0, \quad u'_{nm} = \frac{du_{nm}}{dr}.$$

Therefore, $u_{nm} = a_{nm}h_n^{(1)}(kr) + b_{nm}h_n^{(2)}(kr)$, where a_{nm} and b_{nm} are constants, and $h_n^{(j)}(z)$, $j = 1, 2$, are the spherical Hankel functions.

Condition (1.2.53) implies that

$$\sum_{n=0}^{\infty}\sum_{m=-n}^{n} |u'_{nm} - iku_{nm}|^2 r^2 \to 0, \quad r \to \infty.$$

Therefore, $r(u'_{nm} - iku_{nm}) \to 0, r \to \infty$ for $n = 1, 2, \dots$.

This implies that $b_{nm} = 0, n = 1, 2, \dots, -n \le m \le n$, and

$$u = \sum_{n=0}^{\infty}\sum_{m=-n}^{n} a_{nm}h_n^{(1)}(kr)Y_{nm}(\theta, \varphi).$$

One has

$$\int_{S_r} |u|^2 ds = r^2 \sum_{n=0}^{\infty}\sum_{m=-n}^{n} |h_n^{(1)}(kr)|^2 |a_{nm}|^2. \tag{1.2.69}$$

If $k > 0$, then (see [75, 8.478])

$$r_1^2|h_n^{(1)}(kr_1)|^2 \le r_2^2|h_n^{(1)}(kr_2)|^2, \quad r_1 \ge r_2, \tag{1.2.70}$$

and one has

$$\int_{S_{r_1}} |u|^2 ds \le \int_{S_{r_2}} |u|^2 ds, \quad r_1 \ge r_2. \tag{1.2.71}$$

This implies (1.2.61). Inequality (1.2.71) is of independent interest: it gives a maximum principle for solutions to equation (1.2.50) satisfying condition (1.2.53). □

Lemma 1.2.14. *If u solves equation* (1.2.50) *in* $\Omega_R, R > 0$, *and satisfies condition* (1.2.53), *then*

$$\max_{r \geq R} \|u\|_{L^2(S_r)} = \|u\|_{L^2(S_R)}, \quad k > 0. \tag{1.2.72}$$

The following simple lemma is often useful.

Lemma 1.2.15. *If*

$$(\nabla^2 + k^2)u = 0 \text{ in } \Omega, \quad k > 0; \quad (1.2.53) \text{ holds} \tag{1.2.73}$$

and

$$\lim_{r \to \infty} \operatorname{Im} \int_{S_r} \bar{u} u_r ds = 0, \tag{1.2.74}$$

then $u = 0$ *in* Ω.

Proof. If $k > 0$, then (1.2.74) and (1.2.54) imply (1.2.60). From this and Lemma 1.2.1, it follows that $u \equiv 0$ in Ω. □

Remark 1.2.4. If

$$\nabla^2 u = 0 \text{ in } \Omega,\ u_N + \eta u = 0 \text{ on } \Gamma,\ \operatorname{Re} \eta \leq 0,\ u(\infty) = 0,$$

then $u \equiv 0$ in Ω.

Proof. One has

$$0 = \int_{\Omega'_R} \bar{u} \nabla^2 u dx = -\int_{\Omega'_R} |\nabla u|^2 dx + \int_\Gamma \eta(r) |u|^2 ds + \int_{S_R} \bar{u} \frac{\partial u}{\partial r} ds, \quad \Omega'_R = \Omega \cap B_R.$$

If $u(\infty) = 0$, then $u = O(|x|^{-1})$ and $|\nabla u| = O(|x|^{-2})$ as $|x| \to \infty$.

Therefore, if $R \to \infty$, one obtains

$$0 = -\int_{\Omega} |\nabla u|^2 dx + \int_{\Gamma} \eta(s)|u|^2 ds.$$

If Re $\eta \leq 0$, this implies that $u = 0$ in Ω. □

1.2.11. In this section, we present yet another approach to the existence of the solution to exterior boundary value problems. This approach is based on *the limiting absorption principle* (LAP). In order to prove the existence of the solution to problem (0.0.20), consider the problem

$$(-\nabla^2 - k^2 - i\epsilon)v_\epsilon = 0 \text{ in } \Omega,\ v_\epsilon \in L^2(\Omega),\ \epsilon > 0,\ v_\epsilon = f \text{ on } \Gamma. \tag{1.2.75}$$

For $\epsilon > 0$, the existence and uniqueness of the solution to problem (1.2.75) is easy to prove. LAP says that if the limit $\lim v_\epsilon = v, \epsilon \to 0$, exists in some norm, then this limit is the soution to problem (0.0.20). The justification of this principle consists of proving the existence of the $\lim v_\epsilon = v$ and in checking that v solves (0.0.20).

Lemma 1.2.16. *Problem* (1.2.75) *is uniquely solvable.*

Proof. Look for a solution of the form (1.1.1) with $g(x, y, \sqrt{k^2 + i\epsilon})$ in place of $g(x, y, k)$. The argument in Section 1.2 shows that the solution of this form does exist and is unique because the equation $v + \bar{A}_\epsilon v = 0$ has only the trivial solution $v = 0$. Here, A_ϵ is defined by formula (1.1.4) with $g(x, y, \sqrt{k^2 + i\epsilon})$ in place of $g(x, y, k)$. To prove that $v = 0$ consider $w = S_\epsilon \bar{v}$, where $S_\epsilon v$ is defined by formula (1.1.4) with $g(x, y, \sqrt{k^2 + i\epsilon})$ in place of $g(x, y, k)$.

Then

$$(\nabla^2 + k^2 + i\epsilon)w = 0 \text{ in } D, \quad \frac{\partial w}{\partial N_+} = 0 \text{ on } \Gamma. \tag{1.2.76}$$

Since the spectrum of the Neumann Laplacian in D is real, equation (1.2.76) implies that $w = 0$ in D. Thus, $w_+ = 0, w_- = 0$ and $(\nabla^2 + k^2 + i\epsilon)w = 0$ in Ω. Therefore, $w = 0$ in Ω and $\bar{v} = \partial w/\partial N_+ - \partial w/\partial N_- = 0$. □

Remark 1.2.5. One can prove that

$$(\nabla^2 + k^2 + i\epsilon)w = 0 \text{ in } \Omega, \quad w_- = 0, \quad \epsilon > 0 \tag{1.2.77}$$

implies that $w = 0$ in Ω as follows. From the representation of $w = S_\epsilon \bar{v}$, it is seen that $w \in L^2(\Omega)$ if $\epsilon > 0$, and $|w| \le c\exp(-a|x|)$, $c = \text{const} > 0$, $a = a(\epsilon, k) > 0$. Multiply (1.2.77) by $\bar{w}$ and integrate over $B_R \setminus D$ to obtain

$$\int_{B_R\setminus D}(k^2|w|^2 + i\epsilon|w|^2 - |\nabla w|^2)dx + \int_{S_R}\frac{\partial w}{\partial N}\bar{w}ds = 0. \tag{1.2.78}$$

Take $R \to \infty$. The surface integral vanishes in the limit and (1.2.78) yields

$$\int_\Omega(k^2|w|^2 + i\epsilon|w|^2 - |\nabla w|^2)dx = 0. \tag{1.2.79}$$

Take the imaginary part to obtain $\int_\Omega |w|^2dx = 0$. Thus, $w = 0$ in Ω.

Theorem 1.2.17 (LAP, Limiting Absorption Principle). *There exists the limit in* $L_{2,1+\delta}(\Omega) := L_{2,1+\delta}$ *where* $\|v\|^2_{L_{2,1+\delta}} = \int_\Omega |v|^2(1+|x|)^{-(1+\delta)}dx \equiv \|v\|^2$, *(*$\delta > 0$ *is arbitrary),*

$$\lim_{\epsilon\to 0} v_\epsilon = 0. \tag{1.2.80}$$

The function v *solves problem* (0.0.20). *In any compact strictly inner subdomain* $\tilde{\Omega} \subset \Omega$ *convergence in* (1.2.80) *holds in* $C^m(\tilde{\Omega}), m = 1, 2, \dots$.

Proof. (i) Suppose first that $\sup\|v_\epsilon\| \le c, 0 < \epsilon \le 1$. Then $\|\Delta v_\epsilon\| \le c_1$. If $f \in H^{l+2}(\Gamma)$, then

$$\|v_\epsilon\|_{H^{l+5/2}(B_R\setminus D)} \le c(R,k)(c + \|f\|_{H^{l+2}(\Gamma)}). \tag{1.2.81}$$

This inequality is a particular case of the *a priori* estimates for elliptic operators (see [20]). In particular, if one takes $l = -\frac{1}{2}$ in (1.2.81) then it follows from (1.2.81) that $\|v_\epsilon\|_{H^2(B_R\setminus D)} \le c_1$, where c_1 does not depend on ϵ. From the embedding theorem (see [20]), it follows that there is a subsequence, which is denoted again v_ϵ such that

$$v_\epsilon \to v \text{ in } H^l(B_R \setminus D),\ l < 2,\ \epsilon \to 0,\ v_\epsilon \to v \text{ in } H^l(\Gamma),\ l < 3/2.$$

Therefore, one can pass to the limit $\epsilon \to 0$ in Green's formula for v_ϵ and obtain

$$v(x) = \int_\Gamma \left(f\frac{\partial g}{\partial N} - g\frac{\partial v}{\partial N} \right) ds. \tag{1.2.82}$$

This formula shows that v satisfies Helmholtz's equation in Ω and the radiation condition. Passing to the limit in the boundary condition shows that $v = f$ on Γ. Thus, the limit function does exist and solves problem (0.0.20). Since the solution to problem (0.0.20) is unique, every subsequence v_ϵ converges to v and (1.2.80) holds in $C^m(\tilde{\Omega}), m = 1, 2, 3, \dots,$ and in $H^{l'+5.2}(B_R \setminus D)$ for any $l' < l$, where $f \in H^l(\Gamma)$. From the Green's formula for v_ϵ, it follows that

$$|v_\epsilon| \leq \frac{c}{1 + |x|}, \tag{1.2.83}$$

where $c > 0$ does not depend on ϵ. Estimate (1.2.83) and the local convergence of v_ϵ to v (convergence on compact domains) show that (1.2.80) holds in $L_{2,1+\delta}$ with arbitrary $\delta > 0$.
(ii) Suppose now that

$$\|v_{\epsilon_n}\| \to \infty, \quad \epsilon_n \to 0. \tag{1.2.84}$$

Consider $w_n = v_n\|v_n\|^{-1}, v_n \equiv v_{\epsilon_n}$. Clearly $\|w_n\| = 1, w_n = f\|v_n\|^{-1}$ on Γ, so that $\|w_n\|_{H^l(\Gamma)} \to 0$ as $n \to \infty$. The argument in part (i) of the proof shows that $w_n \xrightarrow[L_{2,1+\delta}]{} w, n \to \infty$, where w solves the homogeneous problem (0.0.20) since $w = 0$ in Ω. On the other hand, passing to the limit in $L_{2,1+\delta}$ in the equation $\|w_n\| = 1$ yields $\|w\| = 1$. This contradiction proves that assumption (1.2.84) cannot hold. LAP is thus proved. □

Remark 1.2.6. Theorem 1.2.17 says that the problem

$$\begin{aligned}(\nabla^2 + k^2)v = 0 \text{ in } \Omega, \quad k > 0; \quad v = f \text{ on } \Gamma;\\ v(x, k + i\epsilon) \in L^2(\Omega), \quad \epsilon > 0\end{aligned} \tag{1.2.85}$$

has a solution $v = \lim v(x, k+i\epsilon)$, $\epsilon \to +0$, and this solution is unique.

1.3. Eigenfunction Expansion Theorem

1.3.1. Consider the solutions to the scattering problem

$$\left(\nabla^2 + k^2\right) u = 0 \text{ in } \Omega,\ k > 0,\ u = 0 \text{ on } \Gamma,\ u = \exp\{-ikn \cdot x\} + v, \tag{1.3.1}$$

$$\lim_{R\to\infty} \int_{S_R} \left| \frac{\partial v}{\partial r} - ikv \right|^2 ds = 0. \tag{1.3.2}$$

The solution to problem (1.3.1)–(1.3.2) does exist and is unique. This was proved in Section 1.2. The aim of this section is to study analytic properties of this solution, to derive this solution from the resolvent kernel of the Dirichlet Laplacian in Ω, which is a self-adjoint nonnegative operator in $L^2(\Omega)$, and to prove the eigenfunction expansion theorem. The resolvent kernel is defined as the solution to the problem.

$$\begin{aligned}(\nabla^2 + k^2)G(x, y, k) = -\delta(x - y) \text{ in } \Omega,\\ G = 0 \text{ on } \Gamma,\ G \text{ satisfies (1.3.2)}.\end{aligned} \tag{1.3.3}$$

From the results in Section 1.2, it follows that problem (1.3.3) is uniquely solvable and G can be represented as

$$\begin{aligned}G(x, y, k) = g - \int_\Gamma g(x, s, k)\mu(s, y, k)ds,\\ \mu = \frac{\partial G(s, y, k)}{\partial N_s}, \quad s \in \Gamma,\end{aligned} \tag{1.3.4}$$

where $g = \exp(ik|x - y|)/(4\pi|x - y|)$. Taking the normal derivative on Γ and using formula (1.1.3) one obtains from (1.3.4) the equation for μ:

$$\mu = -A\mu + 2\frac{\partial g}{\partial N_s}, \quad y \in \Omega, \tag{1.3.5}$$

where A is defined in (1.1.4). It was shown in Section 1.1 that the solution to equation (1.3.5), which satisfies the equation

$$g(s, y, k) = \int_\Gamma g(s, s', k)\mu(s', y, k)ds', \tag{1.3.6}$$

is unique. Equation (1.3.6) is equivalent to the boundary condition (1.3.3) on Γ.

1.3.2. Note that

$$g(x, y, k) = \frac{\exp(ik|y|)}{4\pi|y|}\exp(-ikn \cdot x)\left(1 + O\left(\frac{1}{|y|}\right)\right) \\ \text{as } |y| \to \infty, \ \frac{y}{|y|} = n, \tag{1.3.7}$$

and the factor $\exp(-ikn \cdot x)$, the plane wave, is the solution to the scattering problem in the free space (in this case, there is no scattering, no outgoing wave v). Actually, $o(1)$ in (1.3.7) is $O(|y|^{-1})$. For the basic theorem below, it is sufficient to have $o(1)$ in (1.3.7). We prove that the solution to the scattering problem (1.3.1) can be constructed in the same way.

Lemma 1.3.1. *The following formula holds:*

$$G(x, y, k) = \frac{\exp(ik|y|)}{4\pi|y|}u(x, n, k)\left(1 + O\left(\frac{1}{|y|}\right)\right), \ |y| \to \infty, \ \frac{y}{|y|} = n, \tag{1.3.8}$$

where $u(x, n, k)$ *is the solution to problem* (1.3.1), *and* D *is a bounded domain.*

Proof. It follows from (1.3.4) and (1.3.7) that

$$G(x, y, k) = \frac{e^{ik|x|}}{4\pi|x|}\left\{\exp(-ikn \cdot y) \right. \\ \left. - \int_\Gamma \exp(-ikn \cdot s)\mu(s, y, k)ds\right\}\left(1 + O\left(\frac{1}{|x|}\right)\right) \tag{1.3.9}$$

as $|x| \to \infty$, $x/|x| = n$. Let us denote by $\tilde{u}(y, n, k)$ the expression in the braces and show that $\tilde{u}$ solves problem (1.3.1). If $u(x, n, k)$ solves (1.3.1), then Green's formula for v yields

$$v(x,n,k) = -\int_\Gamma \left(G\frac{\partial v}{\partial N} - v\frac{\partial G}{\partial N} \right) ds + \lim_{R\to\infty} \int_{S_R} \left(G\frac{\partial v}{\partial r} - v\frac{\partial G}{\partial r} \right) ds$$

$$= \int_\Gamma v\frac{\partial G}{\partial N} ds = -\int_\Gamma u_0(s,n,k)\mu(s,x,k)ds, \tag{1.3.10}$$

where

$$u_0(x,n,k) = \exp(-ikn \cdot x). \tag{1.3.11}$$

Thus,

$$u(x,n,k) = u_0(x,n,k) - \int_\Gamma u_0(s,n,k)\mu(s,x,k)ds = \tilde{u}(x,n,k), \tag{1.3.12}$$

where μ is defined in (1.3.4). □

Let us use the symmetry $G(x, y, k) = G(y, x, k)$. It follows from (1.3.4) that

$$G(x,y,k) = G(y,x,k) = g(x,y,k) - \int_\Gamma g(x,s,k)\mu(s,y,k)ds. \tag{1.3.4$'$}$$

Therefore, as $|y| \to \infty$, $y|y|^{-1} = n$, one obtains

$$G(x,y,k) = \frac{\exp(ik|y|)}{4\pi|y|} \Bigg\{ \exp(-ikn \cdot x) - \int_\Gamma g(x,s,k)\frac{\partial u(s,n,k)}{\partial N_s} ds \Bigg\} (1 + o(1)). \tag{1.3.8$'$}$$

Comparing (1.3.9), (1.3.8′), and (1.3.12) yields

$$u(x,n,k) = u_0(x,n,k) - \int_\Gamma u_0(s,n,k)\mu(s,x,k)ds$$

$$= u_0(x,n,k) - \int_\Gamma g(x,s,k)\frac{\partial u(s,n,k)}{\partial N_s} ds, \tag{1.3.12$'$}$$

where u_0 is given by (1.3.11).

In particular,

$$\int_{\Gamma} \exp(ikn\cdot s)\frac{\partial G(s,x,k)}{\partial N_s}ds = \int_{\Gamma} g(x,s,k)\frac{\partial u(s,n,k)}{\partial N_s}ds. \tag{1.3.12''}$$

In the following theorem, an analogue of the Fourier transform is constructed which gives an eigenfunction expansion in solutions to the scattering problem for the Dirichlet Laplacian in Ω.

Theorem 1.3.2. *Let $f(x) \in L^2(\Omega)$. Define*

$$Ff = \tilde{f}(\kappa) = \frac{1}{(2\pi)^{3/2}}\int_{\Omega} f(x)\overline{u(x,\kappa)}dx, \tag{1.3.13}$$

where $u(x,\kappa) = u(x,n,k)$ is the solution to the scattering problem given in (1.3.12), *the bar stands for complex conjugation.*
Then

$$F^{-1}\tilde{f} = f(x) = \frac{1}{(2\pi)^{3/2}}\int_{\mathbb{R}^3} \tilde{f}(\kappa)u(x,\kappa)d\kappa = F^*\tilde{f}. \tag{1.3.14}$$

Here, $\kappa = kn$, $n \in S^2$, $k \geq 0$, the integrals in (1.3.13) *and* (1.3.14) *converge in L^2, and the mapping $F : f \to \tilde{f}$ is an isometric isomorphism of $L^2(\Omega)$ onto $L^2(\mathbb{R}^3)$,*

$$f = F^{-1}\tilde{f}, \quad F^{-1} = F^*, \quad \text{and} \quad F(-\nabla^2)F^{-1} = k^2, \tag{1.3.15}$$

that is, the Dirichlet Laplacian in Ω is unitarily equivalent to the operator of multiplication by k^2.

Proof. Step 1. The spectral theorem for self-adjoint operator says that

$$\frac{1}{\pi}\int_0^{\infty} \operatorname{Im} G(x,y,\sqrt{\lambda+i0})d\lambda$$
$$= \frac{2}{\pi}\int_0^{\infty} \operatorname{Im} G(x,y,k)kdk = \delta(x-y), \tag{1.3.16}$$

where $G(x,y,k)$ is the resolvent kernel of the Dirichlet Laplacian in Ω, $\lambda = k^2$. We compute $\operatorname{Im} G(x,y,k)$ in terms of the solutions to the scattering problem and then equation (1.3.16) will lead to the eigenfunction expansion theorem. Applying Green's formula to $G(x,y,k)$ and $\overline{G(x,y,k)}$ yields

$$\frac{1}{\pi}\operatorname{Im} G(x,y,k) = \frac{1}{2\pi i}\int_{|r|=R} ds \left\{ G(x,s,k)\,\overline{\frac{\partial G(s,y,k)}{\partial |s|}} - \overline{G(s,y,k)}\,\frac{\partial G(x,s,k)}{\partial |s|} \right\}. \tag{1.3.17}$$

If $R \to \infty$, it follows from (1.3.17) and (1.3.18) that

$$\begin{aligned}\frac{1}{\pi}\operatorname{Im} G(x,y,k) &= \frac{1}{2\pi i}\int_{S^2} d\omega \frac{2ik}{16\pi^2} u(x,\omega,k)\overline{u(y,\omega,k)} \\ &= \frac{k}{16\pi^3}\int_{S^2} d\omega u(x,\omega,k)\overline{u(y,\omega,k)}, \quad \omega = \frac{s}{|s|}.\end{aligned} \tag{1.3.18}$$

From (1.3.16) and (1.3.18), one obtains the completeness relation for the eigenfunctions $u(x,k) = u(x,n,k)$:

$$\frac{1}{(2\pi)^3}\int_{\mathbb{R}^3} u(x,\kappa)\,\overline{u(u,\kappa)}d\kappa = \delta(x-y), \tag{1.3.19}$$

where $d\kappa = k^2 dk d\omega$, and the integral is understood as the kernel of the identity operator in $L^2(\mathbb{R}^3)$, that is in the same sense as (1.3.16).

Step 2. Define the operator

$$Ff \equiv \tilde{f}(\kappa) = \frac{1}{(2\pi)^{3/2}}\int_{\Omega} f(x)\overline{u(x,\kappa)}dx \tag{1.3.20}$$

on the set $C_0^\infty(\Omega)$. Let us show that (1.3.14) holds if $f \in C_0^\infty$, and that F is an isometry. We have

$$\begin{aligned}&\lim_{N\to\infty}\frac{1}{(2\pi)^{3/2}}\int_{|\kappa|\le N} dk\tilde{f}(\kappa)u(x,\kappa) \\ &= \lim_{N\to\infty}\frac{1}{(2\pi)^3}\int_{\Omega} dy f(y) \\ &\quad\times\int_{|\kappa|\le N} u(x,\kappa)\overline{u(y,\kappa)}d\kappa = f(x),\end{aligned} \tag{1.3.21}$$

where (1.3.19) was used. Thus, the eigenfunction expansion theorem is proved for $f \in C_0^\infty(\Omega)$. The operator F is isometric on $C_0^\infty(\Omega)$:

$$\begin{aligned}\|Ff\|^2 &= \frac{1}{(2\pi)^3}\int_{\mathbb{R}^3} |\tilde{f}(\kappa)|^2 d\kappa \\ &= \lim_{N\to\infty}\int_{|\kappa|\le N} d\kappa \int_\Omega\int_\Omega dxdy f(x)\overline{f(y)}\cdot\overline{u(x,\kappa)}u(y,\kappa) \\ &= \int_\Omega\int_\Omega dxdy f(x)\overline{f(y)}\delta(x-y) = \int_\Omega |f|^2 dx. \qquad (1.3.22)\end{aligned}$$

Let us extend $F : L^2(\Omega) \to L^2(\mathbb{R}^3)$ by continuity. Since $C_0^\infty(\Omega)$ is dense in $L^2(\Omega)$ the extended operator, which is denoted by F also, is closed, defined on all of $L^2(\Omega)$, and isometric. Therefore, F is invertible, its range $R(F)$ is closed, and to complete the proof one should demonstrate that F is surjective, that is, $R(F) = L^2(\mathbb{R}^3)$.

Step 3. It is clear that the adjoint operator $F^* : L^2(\mathbb{R}^3) \to L^2(\Omega)$ is of the form

$$F^*h = \int_{\mathbb{R}^3} d\kappa h(\kappa)u(x,\kappa), \quad h \in C_0^\infty(\mathbb{R}^3). \qquad (1.3.23)$$

Since F is an isometry, $R(F)$ is closed. Therefore, $R(F) = L^2(\mathbb{R}^3)$ if $N(F^*) = \{0\}$, where $N(A)$ is the null space of the operator A. In order to prove that $F^*h = 0$, $h \in L^2(\mathbb{R}^3)$, which implies that $h = 0$, we need some lemmas. □

Lemma 1.3.3. *If $h \in L^2(\mathbb{R}^3), 0 < a < b < \infty$, and*

$$f(x) = \int_a^b dkk \int_{S^2} d\omega h(\kappa)u(x,\kappa), \quad \kappa = k\omega, \qquad (1.3.24)$$

then

$$(E_b - E_a)f = f, \qquad (1.3.25)$$

where E_λ is the spectral family of the Dirichlet Laplacian in $L^2(\Omega)$.

Proof. Let $\xi^2 = \kappa^2 + i\epsilon, \epsilon > 0$. Green's formula yields

$$u(x,\kappa) = (\kappa^2 - \xi^2) \int_\Omega G(x,y,\xi) u(y,\kappa) dy. \tag{1.3.26}$$

From (1.3.24) and (1.3.25), it follows that

$$\int_\Omega G(x,y,\xi) f(y) dy = \int_a^b dkk \int_{S^2} d\omega h(\kappa) u(x,\kappa)(k^2 - \xi^2)^{-1}. \tag{1.3.27}$$

Since

$$\begin{aligned}(E_b - E_a)f &= \frac{1}{\pi}\int_a^b \operatorname{Im} G(\sqrt{\lambda + i0}) f d\lambda \\ &= \frac{1}{\pi}\int_a^b \frac{G(\sqrt{\lambda + i0})f - G(\sqrt{\lambda - i0})f}{2i} d\lambda, \end{aligned} \tag{1.3.28}$$

where $G(z)$ is the operator in $L^2(\Omega)$ with the kernel $G(x, y, \sqrt{z})$ and $\operatorname{Im} \sqrt{z} > 0$,

$$\lim_{\epsilon \downarrow 0} \frac{1}{\pi} \operatorname{Im} \frac{1}{k^2 - \kappa^2 - i\epsilon} = \delta(k^2 - \kappa^2) \tag{1.3.29}$$

and $k > 0$, equation (1.3.25) follows from (1.3.24), (1.3.27), and (1.3.29). □

Lemma 1.3.4. $N(F^*) = \{0\}$.

Proof. If

$$f(x) \equiv F^* h = \int_{\mathbb{R}^3} d\kappa h(\kappa) u(x,\kappa) = 0, \tag{1.3.30}$$

then for any $0 < a < b < \infty$ one has

$$0 = (E_b - E_a)f = \int_a^b dkk \int_{S^2} d\omega h(\kappa)u(x,\omega,k). \qquad (1.3.31)$$

Therefore,

$$0 = \gamma(x,k) \equiv \int_{S^2} d\omega h(k,\omega)u(x,\omega,k), \quad x \in \Omega, \qquad (1.3.32)$$

for almost all $k > 0$. Consider the identity

$$\begin{aligned} 0 &= \int_{S_r} \left(\gamma(y,k)\frac{\partial g(x,y,k)}{\partial|y|} - g(x,y,k)\frac{\partial \gamma}{\partial|y|}\right) dS_r \\ &= \int_{S^2} d\omega h(k,\omega) \int_{S_r} \left(u(y,\omega,k)\frac{\partial g}{\partial|y|}\right. \\ &\quad \left. - g(x,y,k)\frac{\partial u(u,\omega,k)}{\partial|y|}\right) dS_r \\ &= \int_{S^2} d\omega h(k,\omega)\left\{\int_{S_r}\left(u_0\frac{\partial g}{\partial|y|} - g\frac{\partial u_0}{\partial|y|}\right) dS_r\right. \\ &\quad \left. + \int_{S_r}\left(v\frac{\partial g}{\partial|y|} - g\frac{\partial v}{\partial|y|}\right) dS_r\right\}. \end{aligned} \qquad (1.3.33)$$

Taking $r \to \infty$ one obtains from (1.3.33) the identity

$$0 = \int_{S^2} d\omega h(k,\omega)u_0(x,\omega,k), \quad x \in \Omega \qquad (1.3.34)$$

for almost all $k > 0$, where $u_0 = \exp(-ik\omega \cdot x)$ and the radiation condition for v was used: the first integral in the braces on the right-hand side of (1.3.34) is equal to u_0 while the second goes to zero as $r \to \infty$. Multiplying (1.3.34) by $k^2 dk$ and integrating in k over $(0,\infty)$ one obtains that the Fourier transform of $h(k,\omega) = h(\kappa)$ is zero. Therefore, $h = 0$. □

Let us finish the proof of Theorem 1.3.2. It remains to be proved that (1.3.15) holds.

Step 4. Let $\tilde{f} \in C_0^\infty(\mathbb{R}^3)$. Then $-\nabla^2 F^{-1}\tilde{f} = F^{-1}(k^2\tilde{f})$. Applying F to this equation and using the density of $C_0^\infty(\mathbb{R}^3)$ in $L^2(\mathbb{R}^3)$ one obtains (1.3.15). □

1.3.3. In the same way, the eigenfunction expansion theorem can be proved for the Neumann Laplacian in Ω and for the Laplacian corresponding to the Robin boundary condition $\partial u/\partial N + \eta u = 0$ on Γ, $\operatorname{Im}\eta = 0$. The condition $\operatorname{Im}\eta = 0$ rather than $\operatorname{Im}\eta \geq 0$ is imposed for the corresponding operator to be self-adjoint.

1.4. Properties of the Scattering Amplitude

1.4.1. The scattering amplitude is defined by the formula

$$f(-n,\nu,k) = \lim \gamma^{-1}(r)v(x,n,k), \quad r = |x| \to \infty, \quad xr^{-1} = \nu, \tag{1.4.1}$$

where v is defined in (1.3.1) and $\gamma(r) = r^{-1}\exp(ikr)$.
It follows from (1.3.9) and (1.3.10) that

$$f(-n,\nu,k) = -(4\pi)^{-1}\int_\Gamma \exp(-ikn\cdot s)\frac{\partial u(s,\nu,k)}{\partial N}ds, \tag{1.4.2}$$

while (1.4.1), (1.3.1), (1.3.12′), and (1.3.12″) imply

$$f(-n,\nu,k) = -(4\pi)^{-1}\int_\Gamma \exp(-ik\nu\cdot s)\frac{\partial u(s,n,k)}{\partial N}ds. \tag{1.4.2′}$$

It follows from (1.4.2) and (1.4.2′) that $f(-n,\nu,k) = f(-\nu,n,k)$. An alternative proof of this reciprocity relation is given below as a part of the proof of the following theorem in which the bar denotes complex conjugate.

Theorem 1.4.1. *The following formulas hold:*

$$f(n,\nu,-k) = \overline{f(n,\nu,k)}, \quad k > 0, \tag{1.4.3}$$

$$f(n,\nu,k) = f(-\nu,-n,k), \tag{1.4.4}$$

$$\frac{f(n,\nu,k) - \overline{f(\nu,n,k)}}{2i} = \frac{k}{4\pi}\int_{S^2} f(n,\omega,k)\overline{f(\nu,\omega,k)}d\omega. \tag{1.4.5}$$

In particular,

$$\operatorname{Im} f(n,n,k) = \frac{k}{4\pi}\int_{S^2} |f(n,\omega,k)|^2 d\omega. \tag{1.4.6}$$

Equation (1.4.6) *is called the optical theorem.*

Note that $f(n,\nu,k)$ *is meromorphic in* k *on all of the complex plane (see Theorem* 1.4.3 *below). This and the condition* (1.4.3) *imply*

$$f(n,\nu,-k_1+ik_2) = \overline{f(n,\nu,k_1+ik_2)}, \tag{1.4.3$'$}$$

where k_1 *and* k_2 *are real numbers. Equation* (1.4.3$'$) *implies that the zeros and poles of the scattering amplitude occur in pairs symmetric with respect to the imaginary axis on the complex* k*-plane.*

Proof. (1) Equation (1.4.3) follows from the real-valuedness of the Helmholtz equation and the boundary condition (1.3.1). Indeed, if u solves (1.3.1), then

$$u(x,n,k) = u_0 + v, \quad u_0 = \exp(-ikn\cdot x), \tag{1.4.7}$$
$$v = f(-n,n',k)\gamma(r), \quad r = |x| \to \infty, \quad n' = xr^{-1}. \tag{1.4.8}$$

Furthermore, $u(x,n,-k)$ solves the same problem as $\overline{u(x,n,k)}$, namely $u(x,n,-k)$ solves (1.3.1) and

$$v(x,n,-k) = f(-n,n',-k)\bar{\gamma}(r), \quad r\to\infty, \quad xr^{-1} = n'. \tag{1.4.9}$$

Since $\overline{u_0(x,n,k)} = u_0(x,n,-k)$ and since the solution to this exterior boundary value problem is unique, one concludes that

$$u(x,n,-k) = \overline{u(x,n,k)} \text{ in } \Omega. \tag{1.4.10}$$

Equation (1.4.3) follows from (1.4.10) immediately.

(2) Let us prove (1.4.4). Let $u_j = u(x, n_j, k), j = 1, 2$, solve problem (1.3.1) with $n = n_j$,

$$u_j = u_{0j} + v_j = u_{0j} + f_j\gamma + O(r^{-2}), \quad r \to \infty, \quad \omega = xr^{-1}, \tag{1.4.11}$$

where $u_{0j} = \exp(-ikn_j \cdot x)$ and $f_j = f(-n_j, \omega, k)$.
Green's formula yields

$$\begin{aligned}
0 &= \lim_{r\to\infty} \int_{S_r} \left(u_1 \frac{\partial u_2}{\partial r} - u_2 \frac{\partial u_1}{\partial r}\right) ds \\
&= \lim_{r\to\infty} \left\{ \int_{S_r} \left(u_{01} \frac{\partial u_{02}}{\partial r} - u_{02} \frac{\partial u_{01}}{\partial r}\right) ds \right. \\
&\quad + \int_{S_r} \left(v_1 \frac{\partial v_2}{\partial r} - v_2 \frac{\partial v_1}{\partial r}\right) ds \\
&\quad \left. + \int_{S_r} \left(u_{01} \frac{\partial v_2}{\partial r} - v_2 \frac{\partial u_{01}}{\partial r}\right) ds + \int_{S_r} \left(v_1 \frac{\partial u_{02}}{\partial r} - u_{02} \frac{\partial v_1}{\partial r}\right) ds \right\}.
\end{aligned} \tag{1.4.12}$$

The first integral in the right-hand side of (1.4.12) is equal to zero since u_{0j} solve the Helmholtz equation everywhere in the ball $B_r = \{x : |x| \le r\}$. The limit of the second integral in the right-hand side of (1.4.12) is zero because v_j satisfies the radiation condition. Therefore,

$$\begin{aligned}
0 &= \lim_{r\to\infty} \int_{S_r} (u_{01} ik\gamma f_2 + f_2\gamma ikn_1 \cdot \omega u_{01} + f_1\gamma u_{02}(-ikn_2 \cdot \omega) \\
&\quad - u_{02} ik\gamma f_1) ds \\
&= \lim_{r\to\infty} ikr^2\gamma \int_{S^2} [f_2 u_{01}(1 + n_1 \cdot \omega) - f_1 u_{02}(1 + n_2 \cdot \omega)] d\omega.
\end{aligned} \tag{1.4.13}$$

Let us use the following formula which is proved later as Lemma 1.4.2:

$$\begin{aligned}
&\int_{S^2} \exp(ikn \cdot \omega r) f(\omega) d\omega \\
&\quad = \frac{2\pi i}{k} [\bar{\gamma} f(-n) - \gamma f(n)] + o\left(\frac{1}{r}\right), \quad r \to \infty,
\end{aligned} \tag{1.4.14}$$

where $f \in C^1(S^2)$. We will prove later that $f(n,\omega,k)$ is analytic in ω in the sense specified at the end of this section. From (1.4.7), (1.4.13), and (1.4.14), one obtains

$$\begin{aligned}0 = \lim_{r\to\infty} ikr^2\gamma\frac{2\pi i}{k}[&\bar{\gamma}f(-n_2,n_1,k)(1+n_1\cdot n_1)\\&-\gamma f(-n_2,-n_1,k)(1-n_1\cdot n_1) - \bar{\gamma}f(-n_1,n_2,k)(1+n_2\cdot n_2)\\&+\gamma f(-n_1,n_2,k)(1-n_2\cdot n_2) + o(r^{-1})]. \qquad (1.4.15)\end{aligned}$$

Since $n_1\cdot n_1 = n_2\cdot n_2 = 1$, one concludes from (1.4.15) that $f(-n_2,n_1,k) = f(-n_1,n_2,k)$, which is equation (1.4.4). □

Let us prove (1.4.14).

Lemma 1.4.2. *Formula* (1.4.14) *holds if* $f \in C^1(S^2)$.

Proof. Let $\omega = (\theta,\phi)$, $n\cdot\omega = \xi = \cos\theta$. Then

$$\begin{aligned}&\int_{S^2}\exp(ikrn\cdot\omega)f(\omega)d\omega\\&\quad= \int_0^{2\pi}d\varphi\int_{-1}^1 d\xi\,\exp(ikr\xi)f(\arccos\xi,\phi)\\&\quad= \int_0^{2\pi}d\varphi\left[\frac{\exp(ikr)f(0,\phi)-\exp(-ikr)f(\pi,\phi)}{ikr}+o\left(\frac{1}{r}\right)\right]\\&\quad= \frac{2\pi}{ik}[\gamma f(n)-\bar{\gamma}f(-n)]+o\left(\frac{1}{r}\right), \quad r\to\infty.\end{aligned}$$

One can write (1.4.14) for $r\to\infty$ as

$$\exp(ikrn\cdot\omega) \sim 2\pi ik^{-1}[\bar{\gamma}\delta(n+\omega)-\gamma\delta(n-\omega)], \qquad (1.4.16)$$

where $\delta(n-\omega)$ is the delta function. In $\mathbb{R}^2$ one has, for $r\to\infty$,

$$\exp(ikr\theta\cdot\omega) \sim \sqrt{\frac{2\pi}{k}}\left[e^{-i\frac{\pi}{4}}\delta(\theta-\omega)\frac{e^{ikr}}{\sqrt{r}} + e^{i\frac{\pi}{4}}\delta(\theta+\pi-\omega)\frac{e^{-ikr}}{\sqrt{r}}\right].$$

□

(3) Let us prove formulas (1.4.5) and (1.4.6). Formula (1.4.6) is a particular case of (1.4.5). In order to prove (1.4.5), start with the identity:

$$
\begin{aligned}
0 &= \int_{S_r} \left(u_1 \frac{\partial \bar{u}_2}{\partial r} - \bar{u}_2 \frac{\partial u_1}{\partial r} \right) ds \\
&= \int_{S_r} \left[\left(u_{01} \frac{\partial \bar{u}_{02}}{\partial r} - \bar{u}_{02} \frac{\partial u_{01}}{\partial r} \right) \right. \\
&\quad + \left(v_1 \frac{\partial \bar{u}_{02}}{\partial r} - \bar{u}_{02} \frac{\partial v_1}{\partial r} \right) + \left(u_{01} \frac{\partial \bar{v}_2}{\partial r} - \bar{v}_2 \frac{\partial u_{01}}{\partial r} \right) \\
&\quad \left. + \left(v_1 \frac{\partial \bar{v}_2}{\partial r} - \bar{v}_2 \frac{\partial v_1}{\partial r} \right) \right] ds \equiv T_1 + T_2 + T_3 + T_4. \qquad (1.4.17)
\end{aligned}
$$

Green's formula shows that $T_1 = 0$. The remaining integrals can be written as in formula (1.4.13)

$$
\begin{aligned}
0 = \lim_{r \to \infty} \int_{S_R} & (\gamma f_1 ikn_2 \cdot \omega \bar{u}_{02} - ik\gamma f_1 \bar{u}_{02} \\
& -ik\bar{\gamma} \bar{f}_2 u_{01} + \bar{\gamma} \bar{f}_2 ikn_1 \cdot \omega u_{01} - 2ik|\gamma|^2 f_1 \bar{f}_2) ds. \qquad (1.4.18)
\end{aligned}
$$

Applying formula (1.4.14) to this equation, one obtains

$$
\begin{aligned}
0 &= \lim_{r \to \infty} \int_{S_r} [ik\gamma(n_2 \cdot \omega - 1) f_1 \bar{u}_{02} \\
&\quad + ik\bar{\gamma}(n_1 \cdot \omega - 1) \bar{f}_2 u_{01} - 2ik|\gamma|^2 f_1 \bar{f}_2] ds \\
&= \lim_{r \to \infty} \left[4\pi|\gamma|^2 r^2 f(-n_1, -n_2, k) - 4\pi|\gamma|^2 r^2 \overline{f(-n_2, -n_1, k)} \right. \\
&\quad \left. - 2ik|\gamma|^2 r^2 \int_{S^2} f_1 \bar{f}_2 d\omega \right].
\end{aligned}
$$

Thus,

$$
\begin{aligned}
& 4\pi f(-n_1, -n_2, k) - 4\pi \overline{f(-n_2, -n_1, k)} \\
& = 2ik \int_{S^2} f(-n_1, \omega, k) \overline{f(-n_2, \omega, k)} d\omega. \qquad (1.4.19)
\end{aligned}
$$

This formula is equivalent to (1.4.5).

We use the most widely accepted definition of the scattering amplitude $f(n, \nu, k)$. This scattering amplitude does not satisfy the symmetry relation $f(n, \nu, k) = f(\nu, n, k)$ in general. In [120], another definition of the scattering amplitude is used. Their scattering amplitude satisfies the symmetry relation but does not satisfy the reciprocity relation (1.4.4), it is analytic in the lower half-plane $\operatorname{Im} k \leq 0$ of the complex k-plane, while our scattering amplitude is analytic in the upper half-plane $\operatorname{Im} k \geq 0$. Sometimes the scattering amplitude is denoted $A(\beta, \alpha, k)$, $A(\beta, \alpha, k) = f(-n, \nu, k)$, $\nu = \beta, -n = \alpha$. □

1.4.2. The scattering amplitude is often called the radiation pattern by engineers. It is a function of five variables $n, \nu, k \in S^2 \times S^2 \times R_+$, where $R_+ = (0, \infty)$. Let us study its properties as a function of k. We assume in this chapter that the scatterer is a bounded body with a smooth boundary.

Theorem 1.4.3. *The scattering amplitude is a restriction on R_+ of a function $f(n, \nu, k)$ which is analytic in the half-plane* $\operatorname{Im} k > 0$, *and meromorphic on the whole complex k-plane.*

Proof. Because of (1.4.2), it is sufficient to prove that $\partial u(s, \nu, k)/\partial N$ has these properties as a function of k, or that $G(x, y, k)$ has these properties as a function of k (see formula (1.3.8), for example). From formula (1.3.4), it follows that G has these properties if and only if $\mu = \partial G/\partial N_s$ has. From formula (1.3.5) and Lemma 1.4.4 below, it follows that μ does have these properties. □

Lemma 1.4.4. *Let $B(k)$ be a bounded linear mapping of a Banach space X into a Banach space Y which is analytic in k in a connected domain Δ of the complex k-plane.*

Assume that $B(k) \in \operatorname{Fred}(X, Y)$, *that is* $R(B(k)) = \overline{R(B(k))}$, $r = \dim N(B(k)) = \dim N(B^*(k)) < \infty$. *Then either $B^{-1}(k)$ does not exist at any point $k \in \Delta$, or $B^{-1}(k)$ does exist at all points of Δ except for a discrete set k_j. In a neighborhood of any point $k \neq k_j$, $k \in \Delta$, the operator $B^{-1}(k)$ is analytic. The points k_j are poles of $B^{-1}(k)$ and the Laurent coefficients B_p, $-m_j \leq p \leq$*

-1 of $B^{-1}(k) = \sum_{p=-m_j}^{\infty} B_p(k-k_j)^p$ around k_j are finite rank operators. This conclusion remains valid if $B(k)$ is meromorphic in k provided that the coefficient b_0 in the expansion $B(k) = (k-k_0)^{-m_0}\sum_{p=0}^{\infty} b_p(k-k_0)^p$ around any pole k_0 is a Fredholm operator, $b_0 \in \text{Fred}(X, Y)$.

Remark 1.4.1. The operator $B(k) = I + A(k)$ in equation (1.3.5) is analytic in k on the whole complex plane (which is Δ in this case). Since $A(k)$ is compact in $C(\Gamma)$, the operator $B(k) \in \text{Fred}(C(\Gamma))$, where $\text{Fred}(X) \equiv \text{Fred}(X, X)$. Finally, $B^{-1}(k)$ does exist for $\text{Im}\, k > 0$. Thus, Lemma 1.4.4 is applicable to equation (1.3.5).

To prove Lemma 1.4.4, we need Lemma 1.4.5.

Lemma 1.4.5. *Let $B \in \text{Fred}(X, Y)$. Then there exists a finite rank operator K such that $B + K$ is an isomorphism between X and Y.*

Proof. Let $\{\phi_j\}$, $1 \le j \le r$ be a basis of $N(B)$, $\{\tilde{\phi}_j\} \in X^*$, $1 \le j \le r$, be a system of functionals which is biorthogonal to $\{\phi_j\}$,

$$(\phi_i, \tilde{\phi}_j) = \delta_{ij} = \begin{cases} 1, & i = j, \\ 0, & i \ne j, \end{cases}$$

$\{\psi_j\}$, $1 \le j \le r$, be a basis of $N(B^*)$, and $\{\tilde{\psi}_j\}$, $1 \le j \le r$, be a system of elements of Y which is biorthogonal to $\{\psi_j\}$, $B : X \to Y$, $B^* : Y^* \to X^*$, X^* is the adjoint space for X, that is the space of linear bounded functionals on X, and (ϕ, ψ) is the value of the functional ψ at the element ϕ. Define

$$Kf = \sum_{j=1}^{r} (f, \tilde{\phi}_j)\tilde{\psi}_j. \tag{1.4.20}$$

Let us prove that the operator $B+K$ is an isomorphism of X onto Y. The operator $B + K$ is injective. Indeed, if $(B+K)f = 0$ then $Bf = -Kf = -\sum_{j=1}^{r}(f, \tilde{\phi}_j)\tilde{\psi}_j$. Necessary conditions for the solvability of this equation are

$$(Kf, \psi_j) = 0, \quad 1 \le j \le r. \tag{1.4.21}$$

Since $\{\phi_j\}$ and $\{\tilde{\phi}_j\}$ are biorthogonal, system (1.4.21) implies that

$$(f, \tilde{\phi}_j) = 0, \quad 1 \le j \le r \text{ and } Bf = 0. \tag{1.4.22}$$

If $Bf = 0$, then $f = \sum_{j=1}^{r} c_j \varphi_j$ and $(f, \tilde{\phi}_j) = c_j = 0$, where the biorthogonality of $\{\phi_j\}$ and $\{\tilde{\phi}_j\}$ is used. Thus, $f = 0$ and $B + K$ is injective. Since $R(B)$ is closed and K is a finite rank operator, one concludes that $R(B+K)$ is closed. Indeed, assume $Bu_n + Ku_n \to f$. Then, since $(Bu_n + Ku_n, \psi_j) = (u_n, \tilde{\phi}_j) \to (f, \psi_j)$ as $n \to \infty$, one sees that Ku_n converges. Therefore, Bu_n converges. Since $R(B)$ is closed, $\lim Bu_n = Bu$. One can choose u to be the limit of u_n because $\dim N(B) < \infty$. To do this, decompose X into a direct sum $X = N(B) \oplus X_r$ (here we use the basic assumption $\dim N(B) < \infty$), and take $u_n \in X_r$. The operator B_r, which is the restriction of B onto X_r, is injective, $B_r u_n = Bu_n$. Consider $B_r : X_r \to R(B) = R(B_r)$. Since $R(B)$ is closed, the Banach theorem about inverse operators says that $B^{-1} : R(B) \to X_r$ is continuous. Therefore, $Bu_n \to Bu$ implies $u_n \to u$. Furthermore, $Ku_n \to Ku$ and $Bu = f - Ku$. Thus, $R(B+K)$ is closed. To prove that $B+K$ is surjective, take any $y \in Y$ and consider the equation

$$Bu = y - Ku = y - \sum_{j=1}^{r} (u, \tilde{\phi}_j)\tilde{\psi}_j. \tag{1.4.23}$$

Since $R(B)$ is closed, this equation is solvable if and only if $(y - Ku, \psi_m) = 0$, $1 \le m \le r$. Since $(\tilde{\psi}_j, \psi_m) = \delta_{jm}$, one obtains

$$0 = (y, \psi_m) - (u, \tilde{\phi}_m), \quad 1 \le m \le r. \tag{1.4.24}$$

Let

$$u = u_r + \sum_{j=1}^{r} c_j \varphi_j, \tag{1.4.25}$$

where $u_r \in X_r$, c_j are constants and $\{\phi_j\}$, $1 \leq j \leq r$ is the basis of $N(B)$. Note that $Bu = Bu_r$. Conditions (1.4.24) can be written as

$$(y, \psi_m) - (u_r, \tilde{\phi}_m) = \sum_{j=1}^{r} c_j(\phi_j, \tilde{\phi}_m) = c_m, \quad 1 \leq m \leq r. \tag{1.4.26}$$

Therefore, equation (1.4.23) can be written as a solvable equation for u_r:

$$\begin{aligned} Bu_r &= y - \sum_{j=1}^{r}(u_r, \tilde{\phi}_j)\tilde{\psi}_j - \sum_{j=1}^{r}\left(\sum_{m=1}^{r}[(y, \psi_m) - (u_r, \tilde{\phi}_m)]\phi_m, \tilde{\phi}_j\right)\tilde{\psi}_j \\ &= y - \sum_{j=1}^{r}(u_r, \tilde{\phi}_j)\tilde{\psi}_j - \sum_{j=1}^{r}[(y, \phi_j) - (u_r, \tilde{\phi}_j)]\tilde{\psi}_j \\ &= y - \sum_{j=1}^{r}(y, \psi_j)\tilde{\psi}_j. \end{aligned} \tag{1.4.27}$$

If u_r is the solution to (1.4.27) (there is only one solution to (1.4.27) in X_r), then (1.4.25) solves (1.4.23) provided that $c_j = (y, \psi_j) - (u_r, \tilde{\phi}_j)$. Note that our proof of surjectivity does not use the closedness of $R(B + K)$ and gives an independent proof of this closedness. □

Proof of Lemma 1.4.4. Suppose that $B^{-1}(k_0)$ exists, $k_0 \in \Delta$. Then $B^{-1}(k)$ exists and is analytic in k in some neighborhood $|k - k_0| < \delta$. Indeed $B(k) = B(k_0)[I + B^{-1}(k_0)(B(k) - B(k_0))]$. Thus,

$$B^{-1}(k) = \sum_{j=0}^{\infty}\left\{B^{-1}(k_0)[B(k) - B(k_0)]\right\}^j B^{-1}(k_0) \qquad \text{if } \|B^{-1}(k_0)\|\|B(k) - B(k_0)\| < 1. \tag{1.4.28}$$

If $B(k)$ is analytic, then formula (1.4.27) shows that $B^{-1}(k)$ is analytic in a neighborhood $|k - k_0| < \delta$, where $\delta > 0$ is sufficiently small so that the inequality in (1.4.28) holds. Suppose now that $B^{-1}(k_0)$ does not exist. Let us show that in this case either k_0 is a pole of $B^{-1}(k)$ and the Laurent coefficients of $B^{-1}(k)$ around k_0 are finite rank operators, or $B(k)$ is not invertible in a neighborhood of k_0. Since $B(k_0) \in \text{Fred}(X, Y)$ the operator $B(k_0) + K$ is invertible by Lemma 1.4.4, where K is a finite rank operator which does not depend on k (it depends on k_0). Thus,

$$B^{-1}(k) = (B(k_0) + K + B(k) - B(k_0) - K)^{-1} = (I - \Gamma K)^{-1}\Gamma, \tag{1.4.29}$$

where

$$\Gamma = \Gamma(k) = (B(k_0) + K + B(k) - B(k_0))^{-1}. \tag{1.4.30}$$

Since $B(k_0) + K$ is an isomorphism and $\|B(k) - B(k_0)\| \to 0$ as $|k - k_0| \to 0$, the operator $\Gamma(k)$ is analytic in a neighborhood $|k - k_0| < \delta$ if $\delta > 0$ is sufficiently small. We will prove that the operator $(I - \Gamma K)^{-1}$ is either analytic in $0 < |k - k_0| < \delta$ and has a pole at $k = k_0$, or does not exist in the neighborhood $|k - k_0| < \delta$. Consider the equation

$$(I - \Gamma K)u = f, \quad u = f + \sum_{j=1}^{r} (u, \tilde{\phi}_j)\Gamma\tilde{\psi}_j, \tag{1.4.31}$$

where $r = \dim N(B(k_0))$ and we use the notations introduced in the proof of Lemma 1.4.5, $B = B(k_0)$. From (1.4.31), it follows that

$$u = f + \sum_{j=1}^{r} c_j\eta_j, \quad \eta_j = \Gamma(k)\tilde{\psi}_j, \tag{1.4.32}$$

$$\begin{aligned} c_m &= f_m + \sum_{j=1}^{r} c_j a_{mj}, \quad a_{mj} \equiv (\eta_j(k), \tilde{\psi}_m), \\ f_m &= (f, \tilde{\psi}_m), \quad 1 \le m \le r. \end{aligned} \tag{1.4.33}$$

Equations (1.4.33) and (1.4.31) are equivalent. The solution to equation (1.4.31) is given by formula (1.4.32) if and only if c_m, $1 \le m \le r$, solve (1.4.33). The solution to (1.4.33) can be computed by the formula

$$c_m = \frac{\Delta_m}{\Delta_0}, \quad 1 \le m \le r, \quad \Delta_0 = \det(\delta_{mj} - a_{mk}(k)), \tag{1.4.34}$$

where Δ_m is the determinant of the matrix $\delta_{mj} - a_{mj}(k)$ in which the m-th column is substituted by the column vector $\{f_m\}$, $1 \le m \le r$.

Since $\Gamma(k)$ is analytic in the neighborhood $|k-k_0|<\delta$, the functions Δ_m, $0\le m\le r$, are analytic in this neighborhood. Therefore, either $\Delta_0(k)\equiv 0$ in the disk $C_\delta : |k-k_0|<\delta$, or it has only isolated zeros of finite order.

In the first case, the system (1.4.33) is not solvable at any $k\in C_\delta$ and some $\{f_m\}$. In the second case, the system (1.4.33) is solvable for all k except $k=k_j$, where the set $\{k_j\}$ is the discrete set of zeros of $\Delta_0(k)$. Taking $\delta>0$ sufficiently small, one can assume that in the second case all the points k_j lie outside of the region $0<|k-k_0|\le\delta$. Since equation $(I-\Gamma(k_0)K)u=0$ has nontrivial solutions, the function $\Delta_0(k)$ has a zero at $k=k_0$. Let m_0 be the order of this zero. Then it follows from (1.4.34) that the coefficients $c_m=c_m(k)$ have a pole at $k=k_0$. The order of this pole is not more than m_0 and is equal to m_0 for some choices of f_m. From formulas (1.4.33) and (1.4.34), it is seen that the Laurent coefficients of the operator $(I-\Gamma(k)K)^{-1}$ are finite rank operators. The last statement of Lemma 1.4.4 follows from the formula $B^{-1}(k)=(k-k_0)^{m_0}\{\sum_{p=0}^{\infty}b_p(k-k_0)^p\}^{-1}$ and the already proven part of Lemma 1.4.4. □

Remark 1.4.2. In [429, Problem 34, pp. 108 and 369], it is stated that the assumption, that the coefficients A_p, $-m_0\le p\le -1$, of the operator $A(k)=\sum_{p=-m_0}^{\infty}A_p(k-k_0)^p$ are finite-rank, is crucial for the validity of Theorem XIII.13 in their book. Lemma 1.4.4 shows that the Fredholm property of A_{-m_0}, rather than the assumption that A_{-m_0} is finite-rank, is important.

Remark 1.4.3. The set of poles of the scattering amplitude is the same as the set of poles of the resolvent kernel G. These poles are called resonances in the physics literature and natural frequencies (sometimes scattering frequencies) in the engineering literature. The location and properties of these poles (resonances) are of interest theoretically and practically.

The fact that the resonances lie in the half-plane $\operatorname{Im} k<0$ does not depend on the geometry of the surface Γ. On the other hand, most of what is known about resonances is proved for convex or star-like scatterers. The scatterer D is called star-like (star shaped) if there exists a point $0 \in D$ such that any point of Γ can be seen from 0. The set of convex scatterers is a proper subset of the set of star-like scatterers.

For convex and star-like obstacles (and, more generally, for nontrapping obstacles), it is proved that

There is a strip

$$-a < \operatorname{Im} k \leq 0 \tag{1.4.35}$$

which is free from complex poles and

$$|\operatorname{Im} k_j| \to \infty \text{ as } j \to \infty. \tag{1.4.36}$$

Because of (1.4.3), the poles are symmetrical with respect to the imaginary axis: if $k_j = a_j - ib_j$, $b_j > 0$ is a pole of multiplicity m_j, then $-a_j - ib_j$ is a pole of the same multiplicity. A nontrapping obstacle is such an obstacle that singularities of the solution to the nonstationary problem

$$E_{tt} = \nabla^2 E \text{ in } \Omega, \quad t > 0, \tag{1.4.37}$$

$$E(t, x, y) = 0 \text{ on } \Gamma, \quad t > 0, \quad y \in \Omega, \tag{1.4.38}$$

$$E(0, x, y) = 0, \quad E_t(0, x, y) = \delta(x - y) \tag{1.4.39}$$

propagate to infinity as $t \to +\infty$. This condition is difficult to check. There are some alternative conditions which are formulated in terms of the stationary problem, but these are also difficult to check (see [144, 158]).

Condition (1.4.36) implies that in any half-plane $\operatorname{Im} k > -b$, there is a finite number of complex poles. Some properties of these poles are given in Appendix G.

1.4.3. Let us study the properties of the scattering amplitude as a function of ν. The basic result is the following.

Theorem 1.4.6. *If $f(n, \nu, k)$ is known for a fixed $n = n_0$, fixed $k = k_0$ and all ν in a solid angle $\tilde{S}^2$, then $f(n_0, \nu, k_0)$ is uniquely defined on all of the unit sphere S^2.*

Proof. From (1.4.2′) and the compactness of Γ, it follows that $f(n, \nu, k_0)$ is an entire function $F(\nu_1, \nu_2, \nu_3)$ of the three complex variables ν_1, ν_2, ν_3 where $\nu_j, 1 \leq j \leq 3$, are the Cartesian components of vector ν. This function is known on the manifold which can be described in an appropriate coordinate system as $-a \leq \nu_1 \leq a, -a \leq \nu_2 \leq a, \nu_3 = \sqrt{1 - \nu_1^2 - \nu_2^2}$, where $a > 0$ is a small number. The function $F(\nu_1, \nu_2, \sqrt{1 - \nu_1^2 - \nu_2^2} \equiv \phi(\nu_1, \nu_2)$ is analytic in the domain $|\nu_1|^2 + |\nu_2|^2 < 1$ and is known in a neighborhood of the origin. By the uniqueness of analytic continuation, this function is uniquely defined in the region $|\nu_1|^2 + |\nu_2|^2 \leq 1$. Therefore, $F(\nu_1, \nu_2, \nu_3)$ is uniquely defined on $S^2 = \{\nu : \nu_1^2 + \nu_2^2 + \nu_3^2 = 1\}$. □

1.5. The S-matrix and Wave Operators

1.5.1. Let F_0 be the usual Fourier transform, $F_0 : L^2(\mathbb{R}^3_x) \to L^2(\mathbb{R}^3_k)$, $F_0 f = (2\pi)^{-3/2} \int_{\mathbb{R}^3} f(x) \exp(ikn \cdot x) dx$, F_+ be the Fourier transform corresponding to formulas (1.3.13), (1.3.14), $F_+ : L^2(\Omega) \to L^2(\mathbb{R}^3_k)$, and F_- be the Fourier transform corresponding to the solutions of the problem

$$(\nabla^2 + k^2) u_-(x, n, k) = 0 \quad \text{in } \Omega, \quad k > 0, \tag{1.5.1}$$

$$u_- = 0 \quad \text{on } \Gamma, \tag{1.5.2}$$

$$u_- = \exp(-ikn \cdot x) + v_-, \tag{1.5.3}$$

$$\int_{S_r} \left| \frac{\partial v}{\partial r} + ikv_- \right|^2 ds \to 0, \quad r \to \infty. \tag{1.5.4}$$

The function u_- are "incoming" solutions to the scattering problem, while $u_+ = u$, solutions to (1.3.1)–(1.3.2), are "outgoing" solutions. One can prove formulas (1.3.13)–(1.3.14) with $u_-(x, k)$ in place of $u(x, k)$ in the same way as in the proof of Theorem 1.3.2 in Section 1.3.

Define the wave operators by the formula

$$\Omega_\pm = F_0^* F_\pm, \tag{1.5.5}$$

and the S operator by the formula

$$S = \Omega_+^* \Omega_- = F_+^* F_-. \tag{1.5.6}$$

Since F_0, F_+ and F_- are unitary operators, the operators $\Omega_\pm$ and S are unitary, $\Omega_\pm : L^2(\Omega) \to L^2(\mathbb{R}_x^3)$, $S : L^2(\Omega) \to L^2(\Omega)$.

It follows from (1.3.15) that

$$F_\pm H F_\pm^* = k^2 = F_0 H_0 F_0^*, \tag{1.5.7}$$

where H is the Dirichlet Laplacian in $L^2(\Omega)$, and H_0 is the Laplacian in $L^2(\mathbb{R}^3)$. Therefore,

$$F_0^* F_\pm H = H_0 F_0^* F_\pm, \tag{1.5.8}$$

or

$$\Omega_\pm H = H_0 \Omega_\pm. \tag{1.5.9}$$

This is the intertwining relation for the wave operators. In the same way, one can prove that

$$\Omega_\pm f(H) = f(H_0)\Omega_\pm, \tag{1.5.10}$$

where $f(H)$ is a function of the self-adjoint operator H.

The physical meaning of the S operator is clear from the formula

$$u_+ = S u_-. \tag{1.5.11}$$

Since $u_\pm$ does not belong to $L^2(\Omega)$, one should explain the mathematical meaning of (1.5.11), but from the physical point of view the S operator transforms incoming solutions to the scattering problem into outgoing solutions. The wave operators relate solutions in the free space, $u_0 = \exp(-ikn \cdot x)$, and the solution $u_\pm$ to the scattering problem

$$\Omega_+ u_+ = u_0, \quad \Omega_- u_- = u_0. \tag{1.5.12}$$

Formula (1.5.11) follows from (1.5.12) and (1.5.6): apply Ω_+^* to the first formula (1.5.12) and use the unitarity of Ω_+, $\Omega_+^* \Omega_+ = I$. Therefore, it is sufficient to explain the mathematical meaning of

(1.5.12). Let $h(\kappa) \in C_0^\infty(\mathbb{R}^3_\kappa)$, F_+ be defined by formula (1.3.13). The first equality (1.5.12) means that

$$\Omega_+ F_+^* h = F_0^* h, \quad \forall h \in C_0^\infty(\mathbb{R}^3_k). \tag{1.5.13}$$

This formula is equivalent to (1.5.12) and has a clear mathematical meaning since $F_+^* h \in L^2(\Omega)$, $F_0^* h \in L^2(\mathbb{R}^3_x)$, $\Omega_+ : L^2(\Omega) \to L^2(\mathbb{R}^3)$. A similar argument holds for the second equation (1.5.12).

1.5.2. We now turn to the definition of the S-matrix. Physicists often define the S-matrix $S(k)$ as follows. For a fixed $k > 0$, the functions $u_\pm = u_\pm(x, -n, k)$ have asymptotics

$$\begin{gathered} u_\pm = \exp(ikn \cdot x) + f_\pm(n, \nu, k)\gamma^\pm, \quad \gamma^\pm = \frac{\exp(\pm ikr)}{r}, \\ r = |x| \to \infty, \quad xr^{-1} = \nu. \end{gathered} \tag{1.5.14}$$

Formula (1.4.16) shows that (1.5.14) can be written for $r \to \infty$ as

$$\begin{aligned} u_+ &= \frac{2\pi i}{k}\delta(n+\nu)\gamma^- - \left[\frac{2\pi i}{k}\delta(n-\nu) - f_+(n,\nu,k)\right]\gamma^+ \\ &:= \psi_-\gamma^- - \psi_+\gamma^+. \end{aligned} \tag{1.5.15}$$

Define the kernel of the scattering matrix by the formula

$$\psi_+(n) = \int_{S^2} S(n, -n', k)\psi_-(n')dn'. \tag{1.5.16}$$

This is equivalent to

$$\begin{aligned} u_+ &= Su_- \quad \text{as } r \to +\infty, \\ u_+(r\theta', \theta, k) &= \int_{S^2} S(\theta'', \theta, k) u(r\theta', -\theta'', -k)d\theta'', \\ u_- &= u(x, -\theta, -k). \end{aligned}$$

It follows from (1.5.15) and (1.5.16) that

$$\frac{2\pi i}{k}\delta(n-\nu) - f_+(n,\nu,k) = \frac{2\pi i}{k} S(n,\nu,k). \tag{1.5.17}$$

Thus,

$$S(n, \nu, k) = \delta(n - \nu) - \frac{k}{2\pi i} f_+(n, \nu, k). \tag{1.5.18}$$

The S operator (1.5.6) is an operator in $L^2(\Omega)$. The S-matrix $S(k)$ is an operator in $L^2(S^2)$ with the kernel (1.5.18), i.e.,

$$S(k) = I - \frac{k}{2\pi i} F, \tag{1.5.19}$$

where I is the identity, and F is an operator in $L^2(S^2)$ with the kernel $f_+(n, \nu, k)$. Since $f_+(n, \nu, k)$ is a smooth function of n and ν, the operator F belongs to the trace-class operators. Note that $f_+(n, \nu, k) = f(n, \nu, k)$ where $f(n, \nu, k)$ is defined by (1.4.1).

1.6. Inequalities for Solutions to Helmholtz's Equation for Large Frequencies

These inequalities were proved in [4] for $D \subset \mathbb{R}^2$. The proof holds for $D \subset \mathbb{R}^3$ and we follow his arguments closely.

Let $D \subset \mathbb{R}^3$ be a bounded domain with a smooth boundary Γ. Assume that

$$x \cdot N \geq \beta > 0 \text{ on } \Gamma, \tag{1.6.1}$$

where N is the unit normal to Γ pointing into Ω, $\Omega = \mathbb{R}^3 \setminus D, \Omega_R = \{x : |x| \geq R\}$, β is a constant, and the origin is inside D. Let

$$\|u\| = \|u\|_{L^2(\Omega)}, \; \|u\|_R = \|u\|_{L^2(\Omega_R)}, \; \|u\|_{\partial\Omega} = \|u\|_{L^2(\partial\Omega)}.$$

Theorem 1.6.1. *Let u solve the problem*

$$(\nabla^2 + k^2)u = f \text{ in } \Omega, \quad k > 0, \tag{1.6.2}$$

$$u = \mu \text{ on } \Gamma, \tag{1.6.3}$$

$$r\left(\frac{\partial u}{\partial r} - iku\right) \to 0, \quad r \to \infty, \tag{1.6.4}$$

and

$$\|rf\| < \infty. \tag{1.6.5}$$

Then

$$\|\nabla u\|_R^2 + k^2\|u\|_R^2 \leq CR^{1+\epsilon}\{\|rf\|^2 + \|\partial_t\mu\|_{\partial\Omega}^2 + (1+k^2)\|\mu\|_{\partial\Omega}^2\}, \tag{1.6.6}$$

and

$$\|r^{-1/2-\epsilon}\nabla u\|^2 + k^2\|r^{-1/2-\epsilon}u\|^2 \leq C(1+\epsilon^{-1})\{\|rf\|^2 + \|\partial_t\mu\|_{\partial\Omega}^2 + (1+k^2)\|\mu\|_{\partial\Omega}^2\}, \tag{1.6.7}$$

where $\partial_t\mu$ is the tangential derivative on $\partial\Omega$, c denotes various constants which do not depend on k, R and ϵ.

In order to prove inequalities (1.6.6)–(1.6.7), one needs two lemmas.

Lemma 1.6.2. *If* (1.6.2) *holds, then*

$$\int_{\partial\Omega_R}[2\operatorname{Re}(ru_r\bar{u}_N) - (|\nabla u|^2 - k^2|u|^2)N\cdot x + 2u\bar{u}_N]ds = \int_{\Omega_R}[|\nabla u|^2 + k^2|u|^2 + 2u\bar{f} + 2\operatorname{Re}(u_r r\bar{f})]dx. \tag{1.6.7'}$$

Here, u_r and u_N are the radial and normal derivatives of u, and the bars denote complex conjugation.

Proof. Note that

$$\nabla(u_r) = \nabla(\nabla u \cdot xr^{-1}) = (\nabla u)_r + r^{-1}\nabla u - xr^{-2}u_r. \tag{1.6.8}$$

The Green formula and (1.6.8) imply that

$$\int_{\partial\Omega_R} r(u_r\bar{u}_N + \bar{u}_r u_N)ds = \int_{\Omega_R} r(u_r\Delta\bar{u} + \bar{u}_r\Delta u)dx + \int_{\Omega_R}\{r[(\nabla u)_r\cdot\nabla\bar{u} + (\nabla\bar{u})_r\cdot\nabla u] + 2|\nabla u|^2\}dx. \tag{1.6.9}$$

This formula, the identity

$$\int_{\partial\Omega_R} N\cdot x(|\nabla u|^2-k^2|u|^2)ds$$
$$=\int_{\Omega_R}\{r[(|\nabla u|^2)_r-k^2(|u|^2)_r]+3(|\nabla u|^2-k^2|u|^2)\}dx, \tag{1.6.10}$$

and equation (1.6.2) imply

$$\int_{\partial\Omega_R}[r(u_r\bar{u}_N+\bar{u}_r u_N)-N\cdot x(|\nabla u|^2-k^2|u|^2)]ds$$
$$=\int_{\Omega_R}[2\,\mathrm{Re}(u,r\bar{f})+3k^2|u|^2-|\nabla u|^2]dx$$
$$=\int_{\Omega_R}[2\,\mathrm{Re}(ru_r\bar{f})+|\nabla u|^2+k^2|u|^2]dx$$
$$+2\int_{\Omega_R}[k^2|u|^2-|\nabla u|^2]dx$$
$$=\int_{\Omega_R}[|\nabla u|^2+k^2|u|^2$$
$$+2u\bar{f}+2\,\mathrm{Re}(r\bar{f}u_r)]dx-2\int_{\partial\Omega_R}u\bar{u}_N ds. \tag{1.6.11}$$

This formula is equivalent to (1.6.7′). □

Lemma 1.6.3 ([157]). *If* (1.6.1)–(1.6.5) *hold, then*

$$\frac{\beta}{4}\int_\Gamma |u_N|^2 ds+\int_\Omega[|\nabla u|^2-|u_r|^2+\frac{1}{2}|u_r-iku+r^{-1}u|^2]dx$$
$$\le c\|rf\|^2+c\beta^{-1}\int_\Gamma\{(s\cdot t)^2+|s\cdot N||u_t|^2+k^2|u|^2\}ds, \tag{1.6.12}$$

where $k\ge\delta>0$, $c=const>0$, β is the constant from (1.6.1), *the origin is inside D and a ball $|x|\le a$ belongs to D for some positive a, the constant c does not depend on k and β but depends on δ and a.*

Proof of Theorem 1.6.1. Let

$$y(R) = \int_{\Omega_R} (|u_r| + k^2|u|^2)dx, \quad y' = \int_{S_R \cap \Omega} (|u_r|^2 + k^2|u|^2)ds. \tag{1.6.13}$$

Let us write formula (1.6.7) as

$$\begin{aligned} F &= \int_{S_R \cap \Omega} [r(|u_r|^2 + k^2|u|^2) + 2\operatorname{Re} u\bar{u}_r]ds \\ &= \int_{\Omega_R} (|u_r|^2 + k^2|u|^2)dx \\ &\quad + \int_{\Omega_R} [|\nabla u|^2 - |u_r|^2 + 2\operatorname{Re}(u\bar{f}) + 2\operatorname{Re}(ru_r\bar{f})]dx \\ &\quad + \int_{S_r \cap \Omega} r(|\nabla u|^2 - |u_r|^2)ds - \int_{\Gamma \cap \partial\Omega_R} [2\operatorname{Re}(ru_r\bar{u}_N) \\ &\quad - (|\nabla u|^2 - k^2|u|^2)N \cdot x + 2\operatorname{Re} u\bar{u}_N]ds. \end{aligned} \tag{1.6.14}$$

From (1.6.14) and the elementary inequality $ab \le \epsilon a^2 + (4\epsilon)^{-1}b^2$, $\epsilon > 0$, one obtains

$$\begin{aligned} F \le y(R) + \int_{\Omega_R} &[|\nabla u|^2 - |u_r|^2 + \epsilon(|u|^2 + |u_r|^2) \\ &+ (4\epsilon)^{-1}(|f|^2 + |rf|^2)]dx + F_1, \end{aligned} \tag{1.6.15}$$

where F_1 is the sum of the surface integrals in (1.6.14). Therefore,

$$F \le (1+\epsilon)y(R) + J_1(R) + J_2(r), \tag{1.6.16}$$

where

$$J_2 = \int_{\Omega_R} [|\nabla u|^2 - |u_r|^2 + (4\epsilon)^{-1}(1+r^2)|f|^2]dx. \tag{1.6.17}$$

Let

$$\psi(r) = \max\left(\frac{r}{2}, r-1\right),$$

$$\eta(r) = \begin{cases} 2r^{-3-2\epsilon}, & 0 < r \le 2, \\ (r-1)^{-2-\epsilon}2^{-2-2\epsilon}, & 2 \le r < \infty. \end{cases} \tag{1.6.18}$$

Then

$$(\psi\eta)' = -(1+\epsilon)\eta. \tag{1.6.19}$$

Note that

$$\begin{aligned} \psi(R)y'(R) &\le (R-1)y' = Ry' - y' \\ &\le R\int_{S_R\cap\Omega}(|u_r|^2 + k^2|u|^2)ds \\ &\quad - \int_{S_R\cap\Omega}(|u_r|^2 + k^2|u|^2)ds \\ &\le \int_{S_R\cap\Omega}[r(|u_r|^2 + k^2|u|^2) + 2\,\mathrm{Re}(u\bar{u}_r)]ds \end{aligned} \tag{1.6.20}$$

provided that

$$R \ge 2, \quad k \ge 1. \tag{1.6.21}$$

If

$$R \le 2, \quad k \ge 1, \tag{1.6.22}$$

then $\psi(R) = R/2$,

$$\begin{aligned} \psi(R)y'(R) = Ry' - \frac{R}{2}y' \le \int_{S_R\cap\Omega}&[r(|u_r|^2 + k^2|u|^2) \\ &+ 2\,\mathrm{Re}(u\bar{u}_r)]ds. \end{aligned} \tag{1.6.23}$$

From (1.6.20), (1.6.23), and (1.6.16), it follows that

$$\psi(R)y'(R) \le F \le (1+\epsilon)y(R) + J_1(R) + J_2(R). \tag{1.6.24}$$

Multiply (1.6.24) by η and use (1.6.19) to obtain

$$(\eta\psi y)' \leq \eta(J_1 + J_2). \tag{1.6.25}$$

Therefore,

$$y(R) \leq \eta^{-1}(R)\psi^{-1}(R)\int_a^R \eta(r)(J_1(r) + J_2(r))ds. \tag{1.6.26}$$

Here, $a > 0$ is chosen so that $y(a) = 0$. This choice is possible because $S \cap \Omega = \emptyset$ for sufficiently small $a > 0$. One has

$$\begin{aligned}\int_a^R \eta J_2 dr &\leq \sup_{a\leq r\leq R}|J_2|\int_a^R \eta dr \leq c\sup_{r\geq a}|J_2(r)| \\ &\leq c\epsilon^{-1}\int_\Omega (1+r^2)|f|^2dx + c\int_\Gamma |u_t|^2 ds + ck^2\int_\Gamma |u|^2ds,\end{aligned} \tag{1.6.27}$$

where inequality (1.6.12) was used.

Furthermore, using inequality (1.6.12) again, one obtains

$$\begin{aligned}\left|\int_a^R \eta J_1 dr\right| &\leq \int_0^R dr\eta(r)\int_{S_r\cap\Omega} r(|\nabla u|^2 - |u_r|^2)ds \\ &\quad + \int_a^R \eta dr \sup_{a\leq r\leq R}\int_{\Gamma\cap S_r}[2r|u_r u_N| + |\nabla u|^2 + k^2|u|^2 \\ &\quad + 2|uu_N|]ds \leq c\|rf\|^2 + c\int_\Gamma |u_r|^2 ds + ck^2\int_\Gamma |u|^2 ds.\end{aligned} \tag{1.6.28}$$

The constants c in (1.6.27) and (1.6.28) depend on β but not on $k \geq \delta > 0$ and R. Notice that

$$\eta^{-1}(R)\psi^{-1}(R) \leq cR^{1+\epsilon}, \quad R \geq 2. \tag{1.6.29}$$

Inequalities (1.6.26)–(1.6.29) imply (1.6.6). Inequality (1.6.7) follows from (1.6.26) and (1.6.29). □

Remark 1.6.1 (Maximum principle for solutions to Helmholtz's equation). If $D = B_a$ is a ball of radius a, then the solution to problem (1.6.2)–(1.6.4) with $f = 0$ satisfies the following estimate (a maximum principle for solutions to Helmholtz's equation):

$$\max_{r\geq a}\int_{S_r}|u(r,\omega)|^2 ds_r=\int_{S_a}|u(a,\omega)|^2 ds_a, \tag{1.6.30}$$

where $ds_r = r^2 d\omega$, and $d\omega$ is the element of the area of the unit sphere S^2.

Proof. If $f=0$, then

$$\begin{aligned} u&=\sum_{n=0}^{\infty}\sum_{m=-n}^{n} c_{nm}h_n^{(1)}(kr)Y_{nm}(\omega),\\ \int_{S^2}|u(r,\omega)|^2 d\omega&=\sum_{n=0}^{\infty}\sum_{m=-n}^{n}|c_{nm}|^2|h_n^{(1)}(kr)|^2, \end{aligned} \tag{1.6.31}$$

where Y_{nm} are the spherical harmonics normalized in $L^2(S^2)$, and $h_n^{(1)}(kr)$ are the spherical Hankel functions. The following properties of the spherical Hankel functions (see [75, 8.478])

$$r^2|h_n^{(1)}(kr)|^2\leq a^2|h_n^{(1)}(ka)|^2 \quad \text{if } r\geq a \quad \text{and} \quad n\geq 0, \tag{1.6.32}$$

and (1.6.31) yield (1.6.30). □

1.7. Representation of Solutions to Helmholtz's Equation

There is an extensive literature on the equivalence of the integral equations of potential theory and the corresponding boundary value problems. The results of this section show, for example, that the solution to the interior Dirichlet problem can be obtained in the form of a single-layer potential as easily as in the standard form of a double-layer potential. These results belong to the author.

1.7.1. The aim of this section is to give conditions necessary and sufficient for a potential of double layer v to be equal to a potential of single layer w, where v and w are defined by the formulas (1.1.1) and (1.1.4), respectively. The basic results can be formulated as the following theorems.

Theorem 1.7.1. *For any $w(\sigma)[v(\mu)]$, there exists a $v(\mu)[w(\sigma)]$ such that $w(\sigma) = v(\mu)$ in the interior domain D. The $v(\mu)[w(\sigma)]$ is uniquely determined.*

Theorem 1.7.2. *Suppose that $w(\sigma)$ is given. Then a necessary and sufficient condition for this $w(\sigma)$ to be identically equal to a $v(\mu)$ in Ω is*

$$(w, \bar{h}_j) = \int_\Gamma w h_j dt = 0, \quad 1 \leq j \leq r', \quad (I + A)h_j = 0, \tag{1.7.1}$$

where the system $\{h_j\}$ forms a basis of $N(I + A)$, and A is defined by the formula (1.1.4).

Suppose that $v(\mu)$ is given. Then a necessary and sufficient condition for this $v(\mu)$ to be identically equal to a $w(\sigma)$ in Ω is

$$(\mu, \bar{\sigma}_j) = 0, \quad 1 \leq j \leq r, \quad A\sigma_j - \sigma_j = 0, \tag{1.7.2}$$

where the system $\{\sigma_j\}$ forms a basis of $N(A - I)$, and A is defined by formula (1.1.4).

Condition (1.7.1) *can be written as a condition on σ:*

$$\int_\Gamma \sigma Q h_j dt = 0, \quad 1 \leq j \leq r', \quad (I + A)h_j = 0, \tag{1.7.1$'$}$$

where Q is defined in (1.1.8), *or*

$$\int_\Gamma \sigma \mu_j dt = 0, \quad 1 \leq j \leq r', \tag{1.7.1$''$}$$

where μ_j are defined in Lemma 1.7.4.

Corollary 1.7.3. *If the problem*

$$(\nabla^2 + k^2)u = 0 \text{ in } D, \quad k > 0, \quad u_+ = f, \tag{1.7.3}$$

is solvable then the solution can be represented as $u = w(\sigma)$ *and also as* $u = v(\mu)$. *The same is true for the problem*

$$(\nabla^2 + k^2)u = 0 \text{ in } D, \quad k > 0, \quad \frac{\partial u}{\partial N_+} = f. \tag{1.7.4}$$

1.7.2. In this section, we give some lemmas which are used in the proof of the Theorems 1.7.1 and 1.7.2.

Lemma 1.7.4. *Let problem* (4_0), *i.e. problem* (1.7.4) *with* $f = 0$, *have* r' *linearly independent solutions* u_j, $1 \leq j \leq r'$. *Then the equation*

$$A'\mu + \mu = 0 \tag{1.7.5}$$

has precisely r' *linearly independent solutions* $\mu_j = u_j^+$, *and* $u_j = v(\mu_j)$.

Let problem (3_0), *i.e., problem* (1.7.3) *with* $f = 0$, *have* r *linearly independent solutions* U_j, $1 \leq j \leq r$. *Then the equation*

$$A\sigma - \sigma = 0 \tag{1.7.6}$$

has precisely r *linearly independent solutions* $\sigma_j = \partial U_j/\partial N_+$, *and* $U_j = w(\sigma_j)$.

Lemma 1.7.5. *Assume that* $\Gamma \in C^\infty$. *If* (3_0) *has only the trivial solution, then the operator* Q *defined by* (1.1.8), $Q : \tilde{H}^q \to \tilde{H}^{q+1}$ *is an isomorphism for any real* q.

Lemma 1.7.6. *If* (3_0) *has* r *linearly independent solutions, then the equation*

$$Q\sigma = f \tag{1.7.7}$$

is solvable if and only if

$$\int_\Gamma f\sigma_j dt = 0, \quad 1 \leq j \leq r$$

where σ_j *are defined in Lemma* 1.7.4.

Proof of Lemma 1.7.4. Both parts of the Lemma can be proved similarly. Let us prove the first part. We prove that to each solution to problem (4_0) there corresponds a solution to (1.7.5) and vice versa, and this correspondence preserves linear independence.
Let u solve (4_0). Green's formula yields

$$u(x) = -\int_\Gamma u_+ \frac{\partial g(x,t)}{\partial N_t} dt = -v(u_+). \tag{1.7.8}$$

Let $\mu = u_+$. Then $v(\mu)$ solves problem (4_0). Thus, by (1.1.3′), one obtains

$$0 = \frac{\partial v}{\partial N_+} = \frac{\partial v}{\partial N_-}. \tag{1.7.9}$$

Since the homogeneous exterior Neumann problem has only the trivial solution satisfying the radiation condition, one concludes that

$$v(\mu) = 0 \quad \text{in } \Omega. \tag{1.7.10}$$

Therefore, by (1.1.3), μ solves equation (1.7.5). Thus, every nontrivial solution to (4_0) generates a nontrivial solution to (1.7.5) by the formula $\mu = u_+$.

Let $\mu \neq 0$ solve (1.7.5). Then the potential $v(\mu)$ solves the problem

$$(\nabla^2 + k^2)v = 0 \text{ in } \Omega, \quad \frac{\partial v}{\partial N_-} = 0, \quad k > 0, \tag{1.7.11}$$

and v satisfies the radiation condition. Therefore, $v = 0$ in Ω, $0 = \partial v/\partial N_- = \partial v/\partial N_+$, and the function $u(x) = v(\mu) \neq 0$ solves problem (4_0) in D. Thus, every nontrivial solution to (1.7.5) generates a nontrivial solution to (4_0). The number of linearly independent solutions to (4_0) and (1.7.5) is the same. This follows from the fact that the correspondence between solutions to (4_0) and (1.7.5) preserves linear independence. Indeed, suppose that the system $\{u_j\}$, $1 \leq j \leq r'$, of solutions to (4_0) is linearly independent, that is,

$$\left\{\sum_{j=1}^{r'} c_j u_j(x) = 0 \text{ in } D,\ c_j = \text{const}\right\} \Rightarrow \forall c_j = 0. \tag{1.7.12}$$

Then the system $\{u_j\}$, $1 \leq j \leq r'$, $\mu_j = u_j^+$, of solutions to (1.7.5) is linearly independent. To prove this, assume that

$$\sum_{j=1}^{r'} c_j \mu_j = 0. \tag{1.7.13}$$

Then

$$0 = v\left(\sum_{j=1}^{r'} c_j \mu_j\right) = \sum_{j=1}^{r'} c_j v(\mu_j) = \sum_{j=1}^{r'} c_j u_j. \tag{1.7.14}$$

From (1.7.12) and (1.7.14), it follows that $\forall c_j = 0$. The first part of Lemma 1.7.4 is proved.

To prove the second part, let U solve (3_0). Green's formula yields

$$U(x) = \int_\Gamma g \frac{\partial U}{\partial N_+} dt. \tag{1.7.15}$$

Let $\sigma = \partial U/\partial N_+$. Then $w(\sigma)$ solves problem (3_0). Since $0 = w_+ = w_-$, and the exterior homogeneous Dirichlet problem

$$(\nabla^2 + k^2)U = 0 \text{ in } \Omega, \quad k > 0, \quad U_- = 0 \tag{1.7.16}$$

has only the trivial solution satisfying the radiation condition, one concludes that $U(x) = 0$ in Ω. Therefore,

$$0 = \frac{\partial U}{\partial N_-} = A\sigma - \sigma$$

so that σ solves (1.7.6). Thus, every nontrivial solution to (3_0) generates a nontrivial solution to (1.7.6). If $\sigma \neq 0$ solves (1.7.6), then $\partial w(\sigma)/\partial N_- = 0$. This implies that $w = 0$ in Ω. Therefore, $0 = w_- = w_+$. So, $U(x) = w(\sigma)$ solves problem (3_0) and $U \neq 0$ in D. Thus, every nontrivial solution to (1.7.6) generates a nontrivial solution to (3_0). This correspondence between the solutions to (3_0) and (1.7.6) preserves linear independence. A proof is left to the reader. Lemma 1.7.4 is proved. $\square$

The differentiation of the weakly singular integral $Q\sigma = \int_S g(s,t)\sigma(t)dt$ with respect to s yields the singular integral

$$\nabla(Q\sigma) = \nabla\left(\int_S \frac{\sigma(t)dt}{4\pi|s-t|} + B\sigma\right), \quad B\sigma := \int_S \frac{e^{ik|x-t|}-1}{4\pi|x-t|}\sigma(t)dt. \tag{1.7.17}$$

The operator ∇B has a sufficiently smooth kernel to improve smoothness. The singular operator is

$$\partial_m \int_S \frac{\sigma(t)dt}{4\pi|s-t|} = \int_S \sigma(t)\partial_m \frac{1}{|s-t|}dt, \quad \partial_m = \frac{\partial}{\partial s_m}. \tag{1.7.18}$$

Here, we have used the known formula for differentiation of weakly singular integrals and took into account that S is a closed surface (see [147, p. 242]). The singular operator is

$$H\sigma := \int_S \sigma(t)\frac{t_m - s_m}{|s-t|^3}dt. \tag{1.7.19}$$

This operator is bounded in $L^2(S)$ (see [147, p. 278]). Therefore,

$$\|Q\sigma\|_{H^1(S)} \le \|\nabla Q\sigma\|_{L^2(S)} \le c\|\sigma\|_{L^2(S)}. \tag{1.7.20}$$

This implies that $Q : \tilde{H}^q \to \tilde{H}^{q+1}$ is a bounded operator.

Proof of Lemma 1.7.5. If the problem (3_0) has only the trivial solution, then Q is injective. Indeed, if $Q\sigma = 0$ and $\sigma \neq 0$, then $w(\sigma)$ solves (3_0) and $w(\sigma) \neq 0$ in D. Indeed, if $w = 0$ in D, then $\sigma = \partial w/\partial N_+ - \partial w/\partial N_- = -\partial w/\partial N_-$. But $w(\sigma) = 0$ in Ω since it solves (1.7.16) and satisfies the radiation condition. Thus, $\partial w/\partial N_- = 0$, and $\sigma = 0$. It remains to be proved that $Q : \tilde{H}^q \to \tilde{H}^{q+1}$, and that Q is surjective. We give a new proof of that.

Let us prove the following claim.

Claim: *The operator $Q\sigma := \int_S g(s,t)\sigma(t)dt$ is an isomorphism of $\tilde{H}^q$ onto $\tilde{H}^{q+1}$ if k^2 is not a Dirichlet eigenvalue of the Laplacian in D: $k^2 \notin \sigma(-\Delta_D)$.*

If q here is an arbitrary real number, then we assume $S \in C^\infty$. To prove the above claim, we use the known elliptic estimate and the trace theorem:

$$c\|\sigma\|_{\tilde{H}^{q-3/2}} \le c\|\nabla u\|_{H^{q-1}} \le \|u\|_{H^q} \le c\|u\|_{\tilde{H}^{q-1/2}} = c\|T\sigma\|_{\tilde{H}^{q-1/2}}. \tag{1.7.21}$$

Here, we have used the elliptic estimate $\|u\|_{H^q} \le c\|u\|_{\tilde{H}^{q-1/2}}$ which holds if $k^2 \notin \sigma(-\Delta_D)$, and the known formula $\sigma = u_N^+ - u_N^-$ which, together with the trace theorem, yields the extreme left inequality. Inequality (1.7.21) implies that $T : \tilde{H}^q \to \tilde{H}^{q+1}$, and that the range $R(T)$ is closed in $\tilde{H}^{q+1}$. Therefore, if $k^2 \notin \sigma(-\Delta_D)$ then $R(T) = \tilde{H}^{q+1}$ and T is an isomorphism of $\tilde{H}^q$ onto $\tilde{H}^{q+1}$, and T is of Fredholm type from $\tilde{H}^q$ into $\tilde{H}^{q+1}$ in the general case, that is, regardless of the assumption $k^2 \notin \sigma(-\Delta_D)$.

The claim is proved.

Lemma 1.7.5 is proved. □

The above proof of Lemma 1.7.5 is much shorter than the original proof in [269] because it uses elliptic estimates (1.7.21) and potential theory estimates (1.7.20). The original proof was based on the theory of pseudodifferential operators. From our proof, it follows that the operator $Q : \tilde{H}^q \to \tilde{H}^{q+1}$ is an isomorphism if $k^2 \notin \sigma(-\Delta_D)$. If $k^2 \in \sigma(-\Delta_D)$, then $Q : \tilde{H}^q \to \tilde{H}^{q+1}$ is of Fredholm type.

Proof of Lemma 1.7.6. If $Q\sigma = f$, $Q\sigma_j = 0$, then $(f, \bar{\sigma}_j) = (Q\sigma, \bar{\sigma}_j) = (\sigma, \overline{Q\sigma_j}) = 0$, where we used the symmetry of the kernel of Q. This proves the necessity. Let us prove the sufficiency. The operator $Q : \tilde{H}^q \to \tilde{H}^{q+1}$ is a Fredholm operator with index $Q = \text{ind}Q = \dim N(Q) - \text{codim}\, R(Q) = \dim N(Q) - \dim N(Q^*) = 0$. Therefore, there are $r = \dim N(Q)$ orthogonality conditions

$$(f, q_j)_{\tilde{H}^{q+1}} = 0, \quad 1 \le j \le r, \tag{1.7.22}$$

(where q_j are some functions), which are necessary and sufficient for $f \in R(Q)$. Consider Q as an operator from $\tilde{H}^{-1}$ onto $\tilde{H}^0$. Then conditions (1.7.22) can be written as

$$(f, \tilde{q}_j)_{\tilde{H}^0} = 0, \quad 1 \leq j \leq r, \tag{1.7.23}$$

where $\{\tilde{q}_j\}$, $1 \leq j \leq r$ is a linearly independent system of functions in $\tilde{H}^0$. Since the system $\{\bar{\sigma}_j\}$, $1 \leq j \leq r$, is linearly independent in $\tilde{H}^0$ and conditions

$$(f, \bar{\sigma}_j)_{\tilde{H}^0} = 0, \quad 1 \leq j \leq r, \tag{1.7.24}$$

are necessary for $f \in R(Q)$, one concludes that

$$\operatorname{span}\{\tilde{q}_1, \ldots, \tilde{q}_r\} = \operatorname{span}\{\bar{\sigma}_1, \ldots, \bar{\sigma}_r\}. \tag{1.7.25}$$

Therefore, conditions (1.7.24) are also sufficient for $f \in R(Q)$. □

1.7.3. In this section, we prove Theorems 1.7.1 and 1.7.2.

Proof of Theorem 1.7.1. Assume first that

$$r = 0, \tag{1.7.26}$$

i.e., problem (3_0) has only the trivial solution. If

$$w(\sigma) = v(\mu) \quad \text{in } D, \tag{1.7.27}$$

then

$$A'\mu - \mu = 2w(\sigma). \tag{1.7.28}$$

If $v(\mu)$ is given, then (1.7.28) is an equation for σ. By Lemma 1.7.5, this equation is uniquely solvable for σ if (1.7.26) holds.

If σ is its solution, then

$$w_+ = w_-. \tag{1.7.29}$$

From (1.7.29), it follows that (1.7.27) holds, and σ is determined uniquely. If $w(\sigma)$ is given, then (1.7.28) is an equation for μ. This equation is uniquely solvable if (1.7.26) holds. Indeed, by Lemma 1.7.4 if $r = 0$ then equation (1.7.6) has only the trivial solution. Therefore, the homogeneous equation (1.7.28) has only the trivial solution. Thus, equation (1.7.28) is uniquely solvable by Fredholm's alternative, μ is uniquely determined, and, as before, (1.7.27) holds.

Assume now that $r > 0$. If $w(\sigma)$ is given, then (1.7.28) is an equation for μ. This equation is solvable in $\tilde{H}^0$ if and only if

$$(w(\sigma), \bar{\sigma}_j)_{\tilde{H}^0} = 0, \text{ where } \bar{A}\bar{\sigma}_j - \bar{\sigma}_j = 0, \quad 1 \leq j \leq r. \tag{1.7.30}$$

Here, we used Fredholm's alternative, the equation $A'^* = \bar{A}$, where $\bar{A}$ is the operator whose kernel is the complex conjugate of the kernel of A, and the fact that A and $\bar{A}$ have null spaces of the same dimension r. In fact, if $\sigma \in N(\bar{A})$ then $\bar{\sigma} \in N(A)$. By Lemma 1.7.4, $\bar{\sigma}_j = \partial U_j / \partial N_+$. Therefore,

$$(w(\sigma), \sigma_j)_{\tilde{H}^0} = \int_\Gamma w(\sigma) \frac{\partial U_j}{\partial N_+} ds = \int_\Gamma \frac{\partial w}{\partial N_+} U_j ds = 0 \tag{1.7.31}$$

since $U_j = 0$ on Γ. Thus, equation (1.7.28) is solvable for μ. The potential $v(\mu)$ satisfies the condition

$$v_+(\mu) = w(\sigma) \quad \text{on } \Gamma. \tag{1.7.32}$$

The requirement (1.7.27) determines μ uniquely.

Indeed, if

$$v(\mu_1) = v(\mu_2) = w(\sigma) \quad \text{in } D, \tag{1.7.33}$$

then

$$v(\mu) = 0 \text{ in } D, \quad \mu = \mu_1 - \mu_2. \tag{1.7.34}$$

Thus, $\partial v / \partial N_+ = 0$. Since $\partial v / \partial N_- = \partial v / \partial N_+$, one has $\partial v / \partial N_- = 0$.

Therefore, $v(\mu) = 0$ in Ω since it solves problem (1.7.11). Thus, $\mu = v_- - v_+ = 0$, and $\mu_1 = \mu_2$.

If $v(\mu)$ is given, then (1.7.28) is an equation for σ. By Lemma 1.7.6, this equation is solvable if and only if

$$\int_\Gamma (A'\mu - \mu)\sigma_j ds = 0, \quad 1 \leq j \leq r, \tag{1.7.35}$$

where σ_j solves (1.7.6). One has

$$\int_\Gamma (A'\mu - \mu)\sigma_j ds = \int_\Gamma (A\sigma_j - \sigma_j)\mu ds = 0, \tag{1.7.36}$$

so that (1.7.35) holds. Therefore, equation (1.7.28) is solvable for σ. The requirement (1.7.27) determines σ uniquely. Indeed, if

$$w(\sigma_1) = w(\sigma_2) = v(\mu) \text{ in } D, \tag{1.7.37}$$

then

$$w(\sigma) = 0 \text{ in } D, \quad \sigma = \sigma_1 - \sigma_2. \tag{1.7.38}$$

Therefore, $w_-(\sigma) = w_+(\sigma) = 0$, $w(\sigma) = 0$ in Ω, $\sigma = \partial w/\partial N_+ - \partial w/\partial N_- = 0$, and $\sigma_1 = \sigma_2$. Theorem 1.7.1 is proved. □

Proof of Theorem 1.7.2. (1) Suppose that $v(\mu)$ is given and

$$w(\sigma) = w(\mu) \quad \text{in } \Omega. \tag{1.7.39}$$

Then

$$Q\sigma = w_-(\sigma) = v_-(\mu) = \frac{A'\mu + \mu}{2} \tag{1.7.40}$$

is an equation for σ. By Lemma 1.7.6, this equation is solvable for σ if and only if

$$\int_\Gamma (A'\mu + \mu)\sigma_j ds = 0, \quad 1 \le j \le r, \quad \sigma_j = A\sigma_j. \tag{1.7.41}$$

Condition (1.7.41) can be written as

$$\int_\Gamma \mu\sigma_j ds = 0, \quad 1 \le j \le r, \tag{1.7.42}$$

which is condition (1.7.2). If (1.7.40) holds, then (1.7.39) holds since the exterior Dirichlet problem has at most one solution satisfying the radiation condition. The second statement in Theorem (1.7.2) is proved.

(2) Suppose that $w(\sigma)$ is given and (1.7.39) holds. Then (1.7.40) is an equation for μ. By Fredholm's alternative, this equation is solvable if and only if condition (1.7.1) holds. If (1.7.40) has a solution, then

(1.7.39) holds. This proves the first statement in Theorem 1.7.2. To prove (1.7.1″), one writes

$$0 = \int_\Gamma w(\sigma)h dt = \int_\Gamma \sigma w(h_j) dt = \int_\Gamma \sigma v_+(\mu_j) ds$$
$$= \int_\Gamma \sigma \frac{A'\mu_j - \mu_j}{2} ds = \int_\Gamma \sigma \mu_j ds,$$

where we used Lemma 1.7.4. □

Remark 1.7.1. Our argument shows that if $v(\mu)$ is given, $r > 0$, and (1.7.42) holds, then equation (1.7.40) is solvable for σ and the general solution to (1.7.40) is of the form

$$\sigma = \sigma_0 + \sum_{j=1}^{r} c_j \sigma_j, \tag{1.7.43}$$

where σ_0 is a particular solution to (1.7.40), c_j are arbitrary constants, and σ_j, $1 \le j \le r$, is a basis of $N(Q)$. The set of r linearly independent solutions to equation (1.7.6) forms a basis of $N(Q)$.

Indeed, if $\sigma \in N(Q)$ then $w(\sigma) = 0$ in Ω. Thus, $0 = \partial w/\partial N_- = A\sigma - \sigma/2$, so that σ solves (1.7.6). Conversely, if σ solves (1.7.6), then $\partial w(\sigma)/\partial N_- = 0$. Thus, $w(\sigma) = 0$ in Ω so that $Q\sigma = w_-(\sigma) = 0$, and $\sigma \in N(Q)$.

Therefore, equation (1.7.40) has many solutions. Nevertheless, $w(\sigma)$ for any solution to (1.7.40) is equal to $v(\mu)$ in Ω as we proved earlier. This can also be seen from the fact that $w_-(\sigma_j) = 0$ implies $w(\sigma_j) = 0$ in Ω.

If $w(\sigma)$ is given, $r' > 0$, and (1.7.1) holds, then equation (1.7.40) is solvable for μ, and its general solution is

$$\mu = \mu_0 + \sum_{j=1}^{r'} c_j \mu_j, \tag{1.7.44}$$

where c_j are constants. For any μ, the function $v(\mu)$ is the same in Ω, $v(\mu) = w(\sigma)$ in Ω. This was proved earlier and can also be seen from the fact that

$$v(\mu_j) = 0 \quad \text{in } \Omega. \tag{1.7.45}$$

To prove (1.7.45), notice that μ_j solves (1.7.5) and therefore $v_-(\mu_j) = 0$. Thus, $v(\mu_j)$ solves (1.7.16), and so (1.7.45) holds.

Proof of Corollary 1.7.3. If (1.7.3) is solvable, then its solution can be written (by Green's formula) as

$$u(x) = \int_\Gamma \left(g\frac{\partial u}{\partial N_+} - u\frac{\partial g}{\partial N_+} \right) dt = w\left(\frac{\partial u}{\partial N_+}\right) - v(u_+). \tag{1.7.46}$$

By Theorem 1.7.1, the right-hand side of (1.7.46) can be represented as $v(\mu)$ and as $w(\sigma)$. The same argument holds for problem (1.7.4). □

1.7.4. In this section, we assume that $D \subset \mathbb{R}^2$, $\Gamma = \partial D$, $g_0 = (1/2\pi)\ln(1/|x-y|)$; $S\sigma = \int_\Gamma g_0(x,t)\sigma(t)dt$; $Q\sigma = \int_\Gamma g_0(s,t)\sigma(t)dt$, $s \in \Gamma$; v, A_0 and A_0' are defined as in Section 1.1 with $g = g_0$. In $\mathbb{R}^3$, the equation

$$\int_\Gamma \frac{\sigma(t)dt}{4\pi|s-t|} = 0, \quad s \in \Gamma$$

implies that $\sigma = 0$. In $\mathbb{R}^2$, the analogous equation $Q\sigma = (1/2\pi)\int_\Gamma \ln(1/|s-t|)\sigma(t)dt = 0$, $s \in \Gamma$, may have nontrivial solutions for some Γ.

Example. $\sigma = 1$, $\Gamma = \{t : |t| = R\}$. For a suitable $R > 0$, one has

$$\int_{|t|=R} \ln\frac{1}{|s-t|}dt = 0. \tag{1.7.47}$$

One can prove that $\dim N(Q) = 1$. This holds even in the case when Γ consists of several components (see Section 5.4). We wish to prove the following proposition, which will be useful in Section 5.4.

Proposition 1.7.7. *The operator* $Q : \tilde{H}^q \to \tilde{H}^{q+1}$ *is an isomorphism if and only if* $N(Q) = \{0\}$.

Proof. The proof consists of three parts. First, one shows that Q acts from $\tilde{H}^q$ into $\tilde{H}^{q+1}$. This can be done as in the proof of Lemma 1.7.5. Secondly, one notices that if Q_0 is an isomorphism then $N(Q) = \{0\}$. Finally, one proves that if $N(Q) = \{0\}$, then Q is surjective. We now turn to the proof of this last claim.

Let us prove that if $N(Q) = \{0\}$ then Q is an isomorphism between $\tilde{H}^q$ and $\tilde{H}^{q+1}$. Since Q acts from $\tilde{H}^q$ into $\tilde{H}^{q+1}$, we need to prove that $R(Q) = \tilde{H}^{q+1}$. Take any $f \in \tilde{H}^{q+1}$ and look for the solution of the problem

$$\Delta u = 0 \text{ in } \Omega, \quad u_- = f \text{ on } \Gamma \tag{1.7.48}$$

of the form

$$u = c \ln \frac{1}{|x - x_0|} + v, \tag{1.7.49}$$

where c is a constant, $x_0 \in D$ and $v = D\mu$ is the double-layer potential.

Then

$$f = c \ln \frac{1}{|s - x_0|} + \frac{A'\mu + \mu}{2}, \quad s \in \Gamma. \tag{1.7.50}$$

The equation $A\sigma + \sigma = 0$ has precisely one linearly independent solution $\sigma = \text{const} \neq 0$. Therefore, (1.7.50) is solvable if and only if

$$\int_\Gamma f ds = c \int_\Gamma \ln \frac{1}{|s - x_0|} ds.$$

This equation determines c uniquely if and only if $\int_\Gamma \ln(1/|s - x_0|)ds \neq 0$. The last inequality holds for some $x_0 \in D$ if and only if $N(Q) = \{0\}$. Indeed, suppose that $\int_\Gamma \ln \frac{1}{|s-x|} ds = 0$ for all $x \in D$. Then $Q\,1 = 0$. So, $N(Q) \neq \{0\}$ contrary to the assumption. We have proved that if $N(Q) = \{0\}$ then for any given $f \in \tilde{H}^{q+1}$ the problem (1.7.48) is solvable and its solution can be found in the form (1.7.49) where x_0 is a point in D, and $v = D\mu$, so that $\Delta v = 0$ in Ω, $v(\infty) = 0$.

The idea of the proof that $R(Q) = \tilde{H}^{q+1}$ is as follows. Compute $h = \partial u/\partial N_-$ where u is given by (1.7.49). Thus, u solves the problem

$$\Delta u = 0 \text{ in } \Omega, \quad \frac{\partial u}{\partial N_-} = 0, \quad u(\infty) = 0. \tag{1.7.51}$$

Look for the solution of (1.7.51) of the form

$$u(x) = Q\sigma. \tag{1.7.52}$$

Then

$$\frac{A\sigma - \sigma}{2} = h. \tag{1.7.53}$$

This equation is uniquely solvable since $N(A-I) = \{0\}$. The problem (1.7.51) has at most one solution, so the function (1.7.52) (with σ which solves (1.7.53)) is the same as the function (1.7.49). Thus,

$$Q\sigma = f. \tag{1.7.54}$$

Since $\sigma = \partial u/\partial N_+ - \partial u/\partial N_-$ and $\partial u/\partial N_\pm \in \tilde{H}^q$ provided that $f \in \tilde{H}^{q+1}$, one concludes that $\sigma \in \tilde{H}^q$. The proof is complete. □

Chapter 2

The Inverse Scattering Problem

2.1. Statement of the Problem and Uniqueness Theorems

2.1.1. The problem is to find the shape of the obstacle and the boundary condition on its surface Γ from the scattering data. The surface Γ may consist of several connected closed components. The data are the values of the scattering amplitude on a subset in the space $S^2 \times S^2 \times \mathbb{R}_+$ of the variables n, ν, k; $n, \nu \in S^2$, $k > 0$. This subset can be chosen in a number of ways:

$$n = n_0 \text{ is fixed}, \ \nu \in \tilde{S}^2, \quad 0 \le a \le k \le b, \quad b > a, \tag{2.1.1}$$

where $\tilde{S}^2$ is a solid angle, i.e., an open set in S^2,

$$n \text{ and } \nu \text{ run through a subset of } S^2, k_0 > 0 \text{ is fixed}, \tag{2.1.2}$$

$$n_0 \text{ is fixed}, \ k_0 > 0 \text{ is fixed}, \ \nu \in \tilde{S}^2. \tag{2.1.3}$$

The boundary condition on the surface Γ of the obstacle is assumed to be one of the three types

$$u = 0 \quad \text{on } \Gamma, \tag{2.1.4}$$

$$\frac{\partial u}{\partial N} = 0 \quad \text{on } \Gamma, \tag{2.1.5}$$

$$\frac{\partial u}{\partial N} + \eta(s)u = 0 \quad \text{on } \Gamma, \tag{2.1.6}$$

where

$$\operatorname{Im}\eta \geq 0, \quad \eta \in C(\Gamma). \tag{2.1.7}$$

Condition (2.1.7) is sufficient for the exterior impedance problem or Robin problem

$$(\nabla^2 + k^2)u = 0 \quad \text{in } \Omega, \quad \operatorname{Im} k \geq 0, \quad k \neq 0, \tag{2.1.8}$$

$$u \text{ satisfies (2.1.6), (2.1.7) and the radiation condition,} \tag{2.1.9}$$

to have only the trivial solution $u = 0$ in Ω. If $k = 0$, Im $\eta = 0$, $\eta \leq 0$ and $u = O(r^{-1})$ as $r \to \infty$, then the problem

$$\nabla^2 u = 0 \text{ in } \Omega, \quad \frac{\partial u}{\partial N} + \eta u = 0 \text{ on } \Gamma, \quad \eta \geq 0, \quad u = O(r^{-1}) \tag{2.1.10}$$

has only the trivial solution as well. If $\eta > 0$, this is no longer true.

Example. $\Omega = \{x : |x| \geq 1\}$, $u = r^{-1}, \eta = 1$.

Note that from Theorem 1.4.6, it follows that if $f(n_0, \nu, k_0)$ is known for all $\nu \in \tilde{S}^2$, then it is uniquely determined for all $\nu \in S^2$.

2.1.2. Let us prove Lemma 2.1.1.

Lemma 2.1.1. *The scattering data* (2.1.1) *determine* Γ *and the boundary condition* (2.1.4), (2.1.5), *or* (2.1.6) *uniquely.*

Proof. Suppose there are two obstacles D_1 and D_2 with the same scattering amplitude $f(n_0, \nu, k)$, $\nu \in S^2$, $0 \leq a \leq k \leq b$, $b > a$, and the boundary condition is any one of (2.1.4)–(2.1.6), the same for D_1 and D_2.

Assume first that $D_1 \cap D_2 = \emptyset$. Let $u_j(x, n_0, k)$ solve problem (0.0.1)–(0.0.5) with $D = D_j$, $j = 1, 2$. Consider the function $u = u_1 - u_2$ in Ω_R, where $R > 0$ is sufficiently large so that $(D_1 \cup D_2) \subset B_R$. Since $u_j = u_{0j} + f\gamma + o(r^{-1})$ as $r \to \infty$, $\gamma = r^{-1}\exp(ikr)$, one sees that $u = o(r^{-1})$ as $r \to \infty$. Since u solves equation (1.2.1) and $ru \to 0$ as $r \to \infty$, one obtains from Lemma 1.2.1 that $u = 0$ in Ω_R.

By the unique continuation property, $u = 0$ in $\Omega_{12} = \mathbb{R}^3 \setminus (D_1 \cup D_2)$. In particular, $u = 0$ on $\Gamma_1 \cup \Gamma_2$. Therefore,

$$u_1 = u_2 \quad \text{in } \Omega_{12}, \quad u_1 = 0 \quad \text{on } \Gamma_2.$$

Thus, u_1 solves problem

$$(\nabla^2 + k^2)u_1 = 0 \text{ in } D_2, \quad u_1 = 0 \text{ on } \Gamma_2, \quad a \leq k \leq b, \quad b > a \tag{2.1.11}$$

and $u_1 \neq 0$ in D_2 because otherwise $u_1 = 0$ in $\Omega_1 = \mathbb{R}^3 \setminus D_1$, which is impossible since

$$|u_1| = |u_0 + v_1| \geq |u_0| - |v_1| = 1 - O\left(\frac{1}{r}\right) > 0 \quad \text{as } r \to \infty. \tag{2.1.12}$$

Equation (2.1.11) shows that the Dirichlet Laplacian in D_2 has a continuous spectrum, since any number k^2, $a \leq k \leq b$ is an eigenvalue of this Dirichlet Laplacian. This is a contradiction since the spectrum of the Dirichlet Laplacian in a compact domain is known to be discrete. This contradiction proves that the assumption $D_1 \neq D_2$, $D_1 \cap D_2 = \emptyset$, does not hold.

If $D_1 \neq D_2$ but $D_1 \cap D_2 \neq \emptyset$, then consider a connected component Δ of the set $D_1 \setminus (D_1 \cap D_2)$. The boundary of Δ consists of $\tilde{\Gamma}_2 \cup \tilde{\Gamma}_1$, where $\tilde{\Gamma}_j$ is a subset of Γ_j, $j = 1, 2$. One derives as above that the spectrum of the Dirichlet Laplacian in Δ is continuous. This contradiction proves that $D_1 = D_2$. In the above argument, we assume that the Dirichlet condition on Γ_j holds. For other boundary condition, a contradiction can be derived from the existence of a continuum of orthonormal eigenfunction in D_1 since $L^2(D_1)$ is separable. Note that there are bounded domains in which the Neumann Laplacian has continuous spectrum. Finally, if the scattering amplitude $f(n_0, \nu, k)$ defines the obstacle uniquely, as was proved above, and defines $u(x, n_0, k)$ uniquely in Ω, then the function η in the boundary condition (2.1.6) is also uniquely defined:

$$\eta = -\frac{1}{u}\frac{\partial u}{\partial N} \text{ on } \Gamma. \tag{2.1.13}$$

Notice that if u satisfies boundary condition (2.1.6) and η is a bounded piecewise continuous function then u cannot vanish on an open set $\tilde{\Gamma}$ because if $u = 0$ on $\tilde{\Gamma}$, then it follows from (2.1.6) that $\partial u/\partial N = 0$ on $\tilde{\Gamma}$. This and the uniqueness of the solution to the Cauchy problem for elliptic equations imply that $u = 0$ in Ω which is impossible. The solution u cannot vanish on a closed subset Γ_1 of Γ which is dense in an open subset of Γ. Indeed, if Γ_1 is dense in some open subset $\tilde{\Gamma}$ of Γ, then, by continuity of u, one concludes that $u = 0$ on $\tilde{\Gamma}$, and this leads to a contradiction as above. If Γ_1 is nowhere dense in Γ, then for every point $s \in \Gamma_1$ one finds a sequence of points $s_n \to s$, $u(s_n) \neq 0$. At the points s_n, the function η is uniquely defined by (2.1.13). At the point s, the function η is uniquely defined by continuity: $\eta(s) = \lim \eta(s_n)$. Therefore, both the scatterer and the function $\eta(s)$, $s \in \Gamma$ are uniquely determined by the data (2.1.1). □

Remark 2.1.1. The uniqueness theorem that we used at the end of the proof, and which says that a solution of the Helmholtz equation with the zero Cauchy data on an open set of a smooth surface Γ is identically zero in Ω, is a particular case of the Holmgren uniqueness theorem (see [91, p. 83]).

Let us specify the data (2.1.2). Let $\tilde{S}_0^2$ denote a sequence of distinct unit vectors n_j (i.e., $n_j \neq n_i$ if $j \neq i$) with a limit point n_0. Assume that the scattering data $f(n, \nu, k)$ are given for

$$n \in \tilde{S}_0^2, \quad \nu \in \tilde{S}^2, \quad \text{and } k = k_0 > 0 \text{ fixed.} \tag{2.1.2$'$}$$

Theorem 2.1.2. *The scattering data* (2.1.2′) *determine* Γ *and the boundary condition* (2.1.4), (2.1.5) *or* (2.1.6) *uniquely.*

Proof. The proof goes as in Lemma 2.1.1 and we conclude that $u_j = u_j(x, n_j, k)$ solves the problem

$$(\nabla^2 + k_0^2)u_j = 0 \quad \text{in } \Delta, \quad j = 1, 2, \ldots \tag{2.1.14}$$

$$u_j \text{ satisfies a homogeneous boundary condition on } \partial\Delta. \tag{2.1.15}$$

The system of functions $\{u_j\}$ is linearly independent in $L^2(\Delta)$. This is proved below. Therefore, k_0^2 is an eigenvalue of infinite multiplicity of the Dirichlet Laplacian defined by the boundary conditions (2.1.15) in a bounded domain Δ. Since this Laplacian has eigenvalues of finite multiplicity only, we have a contradiction which proves that $D_1 = D_2$.

Let us prove that the system $\{u_j\}$ is linearly independent in $L^2(\Delta)$. Assume that

$$\sum_{j=1}^{J} c_j u_j = 0 \quad \text{in } L^2(\Delta), \tag{2.1.16}$$

where $c_j = \text{const}$ and J is an arbitrary integer.

The unique continuation property of the solutions to Helmholtz's equation and (2.1.16) imply that

$$\sum_{j=1}^{J} c_j u_j = 0 \quad \text{in } \Omega_R. \tag{2.1.17}$$

Thus,

$$\sum_{j=1}^{J} c_j \{\exp(-ikn_j \cdot s) + O(|x|^{-1})\} = 0 \quad \text{in } \Omega_R, \tag{2.1.18}$$

and

$$c_m\{1+O(|x|^{-1})\}+\sum_{j\neq m} c_j\{\exp[ik(n_m-n_j)\cdot x]+O(|x|^{-1})\} = 0 \text{ in } \Omega_R. \tag{2.1.19}$$

Since $n_j \neq n_i$ if $i \neq j$, one has

$$\min |n_m - n_j| > 0, \quad j \neq m. \tag{2.1.20}$$

Let

$$n_m - n_j = a_{mj} \cdot n_{mj}, \quad a_{mj} = |n_m - n_j|, \quad n_{mj} = (n_m - n_j)a_{mj}^{-1}. \tag{2.1.21}$$

Note that

$$\begin{aligned}
&\int_{|x-x_0|\le a} \exp(ika_{mj}n_{mj}x)dx \\
&\quad = \exp(ika_{mj}n_{mj}\cdot x_0)\int_{|y|\le a}\exp(ika_{mj}|y|\cos\theta)dy \\
&\quad = \exp(ika_{mj}n_{mj}\cdot x_0) \\
&\qquad \times \int_0^a drr^2\left(\frac{\pi}{2rka_{mj}}\right)^{1/2}4\pi\left(\frac{2}{\pi ka_{mj}r}\right)^{1/2}\sin(ka_{mj}r) \\
&\quad = 4\pi\exp(ika_{mj}n_{mj}\cdot x_0)\frac{1}{ka_{mj}}\int_0^a r\sin(ka_{mj}r)dr = O(a).
\end{aligned} \tag{2.1.22}$$

Integrate (2.1.19) over the ball $|x - x_0| \le a$, choose x_0 large, so that $O(|x|^{-1}) < 1$, and take a large. Then the first term in (2.1.19) after integration is $O(|c_m|a^3)$ while the sum after integration is $O(a)$ according to (2.1.22). Therefore, $c_m = 0$. Since m is arbitrary, we proved that $c_j = 0$, $1 \le j \le n$. □

Remark 2.1.2. The argument in Lemma 2.1.1 shows that there are no two (or more) disjoint obstacles with the same scattering data $f(n_0, \nu, k_0)$, $\nu \in S^2$. Notice that k is fixed here, $k = k_0$.

2.1.3. *Inverse obstacle scattering with non-overdetermined data*

In this section, it is proved that the scattering amplitude $A(\beta, \alpha_0, k_0)$, known for all $\beta \in S^2$, where S^2 is the unit sphere in $\mathbb{R}^3$, and fixed $\alpha_0 \in S^2$ and $k_0 > 0$, determines uniquely the surface S of the obstacle D and the boundary condition on S. The boundary condition on S is assumed to be the Dirichlet, or Neumann, or the impedance one. The uniqueness theorem for the solution of multidimensional inverse scattering problems with non-over-determined data was not known for many decades. A detailed proof of such a theorem is given in this section for inverse scattering by obstacles for the first time. It follows from our results that the scattering solution vanishing on the boundary S of the obstacle cannot have closed surfaces of zeros in the exterior of the obstacle different from S.

The uniqueness theorems for the solution of multidimensional inverse scattering problems with non-overdetermined scattering data were not known since the origin of the inverse scattering theory, which goes, roughly speaking, to the middle of the last century. A detailed proof of such a theorem is given in this section for inverse scattering by obstacles for the first time. To have a uniqueness theorem for inverse scattering problems with non-overdetermined data is of principal interest because these are the minimal scattering data that allow one to uniquely recover the scatterer. In [395, 396, 401], such theorems are proved for the first time for inverse scattering by potentials. The result, presented in this subsection was announced in [420], where the ideas of its proof were outlined. In this section, the arguments are given in detail, parts of the paper [420] and the ideas of its proofs are used, two new theorems (Theorems 2.1.4 and 2.1.5) are formulated and proved, and it is pointed out that from these results it follows that the scattering solution vanishing on the boundary S of the obstacle cannot have closed surfaces of zeros different from S in the exterior of the obstacle.

The data is called non-overdetermined if it is a function of the same number of variables as the function to be determined from these data. In the case of the inverse scattering by an obstacle, the unknown function describes the surface of this obstacle, in $\mathbb{R}^3$, so it is a function of two variables. The non-overdetermined scattering data is the scattering amplitude depending on a two-dimensional vector. The exact formulation of this inverse problem is given below.

Let us formulate the problem discussed in this section. Let $D \subset \mathbb{R}^3$ be a bounded domain with a connected C^2-smooth boundary S, $D' := \mathbb{R}^3 \setminus D$ be the unbounded exterior domain and S^2 be the unit sphere in $\mathbb{R}^3$. The smoothness assumption on S can be weakened.

Consider the scattering problem:

$$(\nabla^2 + k^2)u = 0 \quad \text{in } D', \quad \Gamma_j u|_S = 0, \quad u = e^{ik\alpha\cdot x} + v, \qquad (2.1.23)$$

where the scattered field v satisfies the radiation condition:

$$v_r - ikv = o\left(\frac{1}{r}\right), \quad r := |x| \to \infty. \qquad (2.1.24)$$

Here, $k > 0$ is a constant called the wave number and $\alpha \in S^2$ is a unit vector in the direction of the propagation of the incident plane wave $e^{ik\alpha\cdot x}$. The boundary conditions are assumed to be either the Dirichlet (Γ_1), or Neumann (Γ_2), or impedance (Γ_3) type:

$$\Gamma_1 u := u, \quad \Gamma_2 u := u_N, \quad \Gamma_3 u := u_N + hu, \qquad (2.1.25)$$

where N is the unit normal to S pointing out of D, u_N is the normal derivative of u on S, h = const, $\operatorname{Im} h \geq 0$, h is the boundary impedance, and the condition $\operatorname{Im} h \geq 0$ guarantees the uniqueness of the solution to the scattering problem (2.1.23)–(2.1.24).

The notations of this subsection are these: $\beta = \nu$ is the direction of the scattered field and $\alpha = -n$ is the direction of the incident plane wave.

The scattering amplitude $A(\beta, \alpha, k)$ is defined by the following formula:

$$v = A(\beta, \alpha, k)\frac{e^{ikr}}{r} + o\left(\frac{1}{r}\right), \quad r := |x| \to \infty, \quad \frac{x}{r} = \beta, \qquad (2.1.26)$$

where $\alpha, \beta \in S^2$, β is the direction of the scattered wave, and α is the direction of the incident wave.

For a bounded domain D, one has $o(\frac{1}{r}) = O(\frac{1}{r^2})$ in formula (2.1.26). The function $A(\beta, \alpha, k)$, the scattering amplitude, can be measured experimentally. Let us call it *the scattering data.* It is known (see [269, p. 25]) that the solution to the scattering problem (2.1.23)–(2.1.24) does exist and is unique.

The inverse scattering problem (IP) *consists of finding* S *and the boundary condition on* S *from the scattering data.*

Schiffer was the first to prove in the sixties of the last century that if the boundary condition is the Dirichlet one then the surface

S is uniquely determined by the scattering data $A(\beta, \alpha_0, k)$ known for a fixed $\alpha = \alpha_0$, all $\beta \in S^2$, and all $k \in (a, b)$, $0 \leq a < b$.

Schiffer did not publish his proof. This proof can be found, for example, in [269, p. 85], and the acknowledgement of Schiffer's contribution is [269, p. 399].

Ramm was the first to prove that the scattering data $A(\beta, \alpha, k_0)$, known for all β in a solid angle, all α in a solid angle and a fixed $k = k_0 > 0$ determine uniquely the boundary S *and the boundary condition on* S. This condition was assumed of one of the three types Γ_j, $j = 1, 2$ or 3, (see [269, Chapter 2], for the proof of these results). By subindex zero, fixed values of the parameters are denoted, for example, k_0, α_0. By a solid angle in this section, an open subset of S^2 is understood.

In [269, p. 62], it is proved that for smooth bounded obstacles, the scattering amplitude $A(\beta, \alpha, k)$ is an analytic function of β and α on the noncompact analytic variety $M := \{z | z \in \mathbb{C}^3, z \cdot z = 1\}$, where $z \cdot z := \sum_{m=1}^{3} z_m^2$. The unit sphere S^2 is a subset of M. If $A(\beta, \alpha, k)$ as a function of β is known on an open subset of S^2, it is uniquely extended to all of S^2 (and to all of M) by analyticity. The same is true if $A(\beta, \alpha, k)$ as a function of α is known on an open subset of S^2. By this reason, one may assume that the scattering amplitude is known on all of S^2 if it is known in a solid angle, that is, on open subsets of S^2 as a function of α and β.

In papers [321, 322], a new approach to a proof of the uniqueness theorems for inverse obstacle scattering problem (IP) was given. This approach is used here.

In paper [252], the uniqueness theorem for IP with non-overdetermined data was proved for strictly convex smooth obstacles. The proof in [252] was based on the location of resonances for a pair of such obstacles. These results are technically difficult to obtain and they hold for two strictly convex smooth obstacles with a positive distance between them.

The purpose of this section is to prove the uniqueness theorem for IP with non-overdetermined scattering data for arbitrary S. By the boundary condition, any of the three conditions Γ_j are understood below, but the argument is given for the Dirichlet condition for definiteness.

Theorem 2.1.3. *The surface S and the boundary condition on S are uniquely determined by the data $A(\beta)$ known in a solid angle.*

Theorem 2.1.4. *If $A_1(\beta) = A_2(\beta)$ for all β in a solid angle, then it is not possible that $D_1 \neq D_2$ and $D_1 \cap D_2 = \emptyset$.*

Theorem 2.1.5. *If $A_1(\beta) = A_2(\beta)$ for all β in a solid angle, then it is not possible that $D_1 \neq D_2$ and $D_1 \subset D_2$.*

Corollary. *It follows from Theorems* 2.1.4 *and* 2.1.5 *that the solution to problem* (2.1.23)–(2.1.24) (*the scattering solution*) *cannot have a closed surface of zeros except the surface S, the boundary of the obstacle.*

In this section, some auxiliary material is formulated and Theorems 2.1.3–2.1.5 are proved. The corollary is an immediate consequence of these theorems. Theorems 2.1.3–2.1.5 and the corollary are our main results.

Let us explain the logic of the proof of Theorem 2.1.3. Its proof is based on the assumption that there are two different obstacles, D_1 with the surface S_1 and D_2 with the surface S_2, that generate the same non-overdetermined scattering data. This assumption leads to a contradiction which proves that $S_1 = S_2$. If it is proved that $S_1 = S_2$, then the type of the boundary condition (of one of the three types (2.1.26)) can be uniquely determined by calculating u or $\frac{u_N}{u}$ on S.

There are three cases to consider. The first case, when S_1 intersects S_2, is considered in Theorem 2.1.3. The second case, when S_1 does not intersect S_2 and does not lie inside S_2, is considered in Theorem 2.1.4. The third case, when S_1 does not intersect S_2 and it lies inside S_2, is considered in Theorem 2.1.5.

Our results show that these cases cannot occur if the non-overdetermined scattering data corresponding to S_1 and S_2 are the same. They also show that a scattering solution cannot have a closed surface of zeros except S.

The presentation of Section 2.1.3 is self-contained and independent so some of the proofs, given earlier in this book are repeated.

Proof of Theorems 2.1.3–2.1.5. Let us formulate some lemmas which are proved by the author, except for Lemma 2.1.8, which is proved in Section 1.2.1 as Lemma 1.2.1. Another proof of Lemma 2.1.8, based on a new idea, is given in [236].

Denote by $G(x, y, k)$, the Green's function corresponding to the scattering problem (2.1.23)–(2.1.24). The parameter $k > 0$ is assumed fixed in what follows. For definiteness, we assume below the Dirichlet boundary condition, but our proof is valid for the Neumann and impedance boundary conditions as well. If there are two surfaces S_m, $m = 1, 2$, we denote by G_m the corresponding Green's functions of the Dirichlet Helmholtz operator in D'_m. □

Lemma 2.1.6 (see Lemma 1.3.1). *One has:*

$$G(x, y, k) = g(|y|)u(x, \alpha, k) + O\left(\frac{1}{|y|^2}\right), \quad |y| \to \infty, \quad \frac{y}{|y|} = -\alpha. \tag{2.1.27}$$

Here, $g(|y|) := \frac{e^{ik|y|}}{4\pi|y|}$, $u(x, \alpha, k)$ is the scattering solution, that is, the solution to problem (2.1.23)–(2.1.24), $O\left(\frac{1}{|y|^2}\right)$ is uniform with respect to $\alpha \in S^2$, and the notation $\gamma(r) := 4\pi g(r) = \frac{e^{ik|r|}}{|r|}$ is used below.

The solutions to equation (2.1.23) have the unique continuation property:

If u solves equation (2.1.23) *and vanishes on a set $\tilde{D} \subset D'$ of positive Lebesgue measure, then u vanishes everywhere in D'.*

Formula (2.1.27) holds if y is replaced by the vector $-\tau\alpha + \eta$, where $\tau > 0$ is a scalar and η is an arbitrary fixed vector orthogonal to $\alpha \in S^2$, $\eta \cdot \alpha = 0$. If $\eta \cdot \alpha = 0$ and $y = -\tau\alpha + \eta$, then $\frac{|y|}{\tau} = 1 + O\left(\frac{1}{|\tau|^2}\right)$ as $\tau \to \infty$. The relation $|y| \to \infty$ is equivalent to the relation $\tau \to \infty$, and $g(|y|) = g(\tau)(1 + O(\frac{1}{|\tau|}))$.

Denote by $D_{12} := D_1 \cup D_2$, $D'_{12} := \mathbb{R}^3 \setminus D_{12}$, $S_{12} := \partial D_{12}$, $\tilde{S}_1 := S_{12} \setminus S_2$, that is, $\tilde{S}_1$ does not belong to D_2, $B'_R := \mathbb{R}^3 \setminus B_R$, $B_R := \{x : |x| \leq R\}$, $\mathcal{D}^{12} := (D_1 \cap D_2)' = \mathbb{R}^3 \backslash D^{12}$. The number R is sufficiently large, so that $D_{12} \subset B_R$. Let S^{12} denote the intersection of S_1 and S_2. This set may have positive two-dimensional Lebesgue measure or it may have two-dimensional Lebesgue measure zero. In the first case, let us denote by $L \subset S^{12}$ the line such that in an arbitrary small neighborhood of every point $s \in L$ there are points of S_1 and of S_2. The line L has two-dimensional Lebesgue measure equal to zero. Denote by S_ϵ the subset of points on S_{12} the distance from which to L is less than ϵ. The two-dimensional Lebesgue measure m_ϵ of S_ϵ tends to zero as $\epsilon \to 0$.

A part of our proof is based on a global perturbation lemma, Lemma 2.1.7 below, which is proved in [322], see there formula (4). A similar lemma is proved for potential scattering in [356], see there formula (5.1.30). For convenience of the readers, a short proof of Lemma 2.1.7 is given below.

Lemma 2.1.7. *One has:*

$$\begin{aligned}&4\pi[A_1(\beta,\alpha,k) - A_2(\beta,\alpha,k)]\\&\quad = \int_{S_{12}} [u_1(s,\alpha,k)u_{2N}(s,-\beta,k) - u_{1N}(s,\alpha,k)u_2(s,-\beta,k)]ds,\end{aligned} \tag{2.1.28}$$

where the scattering amplitude $A_m(\beta,\alpha,k)$ corresponds to obstacle S_m, $m = 1,2$.

Proof. Denote by $G_m(x,y,k)$, the Green's function of the Dirichlet Helmholtz operator in D'_m, $m = 1,2$. Using Green's formula, one obtains

$$\begin{aligned}&G_1(x,y,k) - G_2(x,y,k)\\&\quad = \int_{S_{12}} [G_1(s,x,k)G_{2N}(s,y,k) - G_{1N}(s,x,k)G_2(s,y,k)]ds.\end{aligned} \tag{2.1.29}$$

Pass in (2.1.29) to the limit $y \to \infty$, $\frac{y}{|y|} = \beta$, and use Lemma 2.1.6 to get:

$$\begin{aligned}&u_1(x,-\beta,k) - u_2(x,-\beta,k)\\&\quad = \int_{S_{12}} [G_1(s,x,k)u_{2N}(s,-\beta,k) - G_{1N}(s,x,k)u_2(s,-\beta,k)]ds.\end{aligned} \tag{2.1.30}$$

Use the formula

$$u_m(x, -\beta, k) = e^{-ik\beta\cdot x} + A_m(-\alpha, -\beta, k)\frac{e^{ik|x|}}{|x|} + O\left(\frac{1}{|x|^2}\right),$$
$$|x| \to \infty, \ \frac{x}{|x|} = -\alpha, \tag{2.1.31}$$

pass in equation (2.1.30) to the limit $x \to \infty$, $\frac{x}{|x|} = -\alpha$, use Lemma 2.1.6 and get

$$4\pi[A_1(-\alpha, -\beta, k) - A_2(-\alpha, -\beta, k)]$$
$$= \int_{S_{12}} [u_1(s, \alpha, k)u_{2N}(s, -\beta, k) - u_{1N}(s, \alpha, k)u_2(s, -\beta, k)]ds. \tag{2.1.32}$$

The desired relation (2.1.28) follows from (2.1.32) if one recalls the known reciprocity relation

$$A(-\alpha, -\beta, k) = A(\beta, \alpha, k),$$

which is proved, for example, in [269, pp. 53–54].

Lemma 2.1.7 is proved. □

Remark 2.1.3. In (2.1.29), the Green's formula is used. The surface S_{12} may not be smooth because it contains the intersection S^{12} of two smooth surfaces S_1 and S_2, and this intersection may not be smooth. However, the integrand in (2.1.29) is smooth up to the boundary S_{12} and is uniformly bounded because x and y belong to the exterior of D_{12}. The integral (2.1.29) is defined as the limit of the integral over $S_{12} \setminus S_\epsilon$ as $\epsilon \to 0$ (where S_ϵ was defined above Lemma 2.1.7). This limit does exist since m_ϵ, the two-dimensional Lebesgue measure of S_ϵ, tends to zero as $\epsilon \to 0$ while the integrand is smooth and uniformly bounded on S_{12}. Consequently, the integral (2.1.29) is well defined. This argument also shows that formula (2.1.32) is valid for the domain D_{12} if the surfaces S_1 and S_2 are smooth and the functions u_1 and u_2 are smooth and uniformly bounded up to S_1 and S_2, respectively. This argument justifies the applicability of formula (2.1.29).

The following lemma was proved earlier as Lemma 1.2.1.

Lemma 2.1.8. *If* $\lim_{r\to\infty}\int_{|x|=r}|v|^2 ds = 0$ *and* v *satisfies the Helmholtz equation* (2.1.23) *in* B'_R, *then* $v = 0$ *in* B'_R.

The following lemma is used in our proof.

Lemma 2.1.9 (Lifting lemma). *If* $A_1(\beta,\alpha,k) = A_2(\beta,\alpha,k)$ *for all* $\beta,\alpha \in S^2$, *then* $G_1(x,y,k) = G_2(x,y,k)$ *for all* $x, y \in D'_{12}$. *If* $A_1(\beta,\alpha_0,k) = A_2(\beta,\alpha_0,k)$ *for all* $\beta \in S^2$ *and a fixed* $\alpha = \alpha_0$, *then* $G_1(x,y_0,k) = G_2(x,y_0,k)$ *for all* $x \in D'_{12}$ *and* $y_0 = -\alpha_0\tau + \eta$, *where* $\tau > 0$ *is a number and* η *is an arbitrary fixed vector orthogonal to* α_0, $\alpha_0 \cdot \eta = 0$.

Proof. The function

$$w := w(x,y) := G_1(x,y,k) - G_2(x,y,k) \tag{2.1.33}$$

satisfies equation (2.1.23) in D'_{12} as a function of y and also as a function of x, and w satisfies the radiation condition as a function of y and also as a function of x. By Lemma 2.1.6, one has:

$$w = g(|y|)[u_1(x,\alpha,k) - u_2(x,\alpha,k)] + O\left(\frac{1}{|y|^2}\right), \quad |y| \to \infty,\ \alpha = -\frac{y}{|y|}. \tag{2.1.34}$$

Using formulas (2.1.23) and (2.1.26), one gets:

$$u_1(x,\alpha,k) - u_2(x,\alpha,k) = \gamma(|x|)[A_1(\beta,\alpha,k) - A_2(\beta,\alpha,k)] + O\left(\frac{1}{|x|^2}\right), \quad |x| \to \infty,\ \ \beta = \frac{x}{|x|}, \tag{2.1.35}$$

because, for $m = 1, 2$ and $\gamma(|x|) := \frac{e^{ik|x|}}{|x|}$, one has:

$$u_m(x,\alpha,k) = e^{ik\alpha\cdot x} + A_m(\beta,\alpha,k)\gamma(|x|) + O\left(\frac{1}{|x|^2}\right), \quad |x| \to \infty,\ \ \beta = \frac{x}{|x|}. \tag{2.1.36}$$

If $A_1(\beta,\alpha,k) = A_2(\beta,\alpha,k)$, then equation (2.1.35) implies

$$u_1(x,\alpha,k) - u_2(x,\alpha,k) = O\left(\frac{1}{|x|^2}\right). \tag{2.1.37}$$

Since $u_1(x,\alpha,k) - u_2(x,\alpha,k)$ solves equation (2.1.23) in D'_{12} and relation (2.1.37) holds, it follows from Lemma 2.1.8 that $u_1(x,\alpha,k) = u_2(x,\alpha,k)$ in B'_R. By the unique continuation property for the solutions to the Helmholtz equation (2.1.23), one concludes that $u_1 = u_2$ everywhere in D'_{12} and even in $\mathcal{D}^{12} := \mathbb{R}^3 \setminus D^{12}, D^{12} = D_1 \cap D_2$. Consequently, formula (2.1.34) yields

$$w(x,y) = O\left(\frac{1}{|y|^2}\right), \quad |y| > |x| \geq R. \tag{2.1.38}$$

Since the function w solves the homogeneous Helmholtz equation (2.1.23) in the region $|y| > |x| \geq R$, it follows by Lemma 2.1.8 that $w = w(x,y) = 0$ in this region and, by the unique continuation property, $w = 0$ everywhere in D'_{12}. Thus, the first part of Lemma 2.1.9 is proved.

Its second part deals with the case when $\alpha = \alpha_0$, where α_0 is fixed. Let us prove that if

$$A_1(\beta) := A_1(\beta,\alpha_0,k) = A_2(\beta,\alpha_0,k) := A_2(\beta) \qquad \forall \beta \in S^2, \tag{2.1.39}$$

then

$$w(x,y_0) = 0, \tag{2.1.40}$$

where $x \in D'_{12}$ is arbitrary, $y_0 = -\tau\alpha_0 + \eta$, $\alpha_0 \in S^2$ is fixed, $\tau > 0$ is a number and η is an arbitrary fixed vector orthogonal to α_0, $\eta \cdot \alpha_0 = 0$.

From (2.1.39), it follows that $u_1(x,\alpha_0) = u_2(x,\alpha_0)$ for all $x \in D'_{12}$. Let us derive a contradiction from the assumption that (2.1.40) is not valid, or, which is equivalent, that $S_1 \neq S_2$.

The Green's formula yields

$$G_1(x,y_0) = g(x,y_0) - \int_{S_1} g(x,s) G_{1N}(s,y_0) ds,$$

where $g(x, y_0) = \frac{e^{ik|x-y_0|}}{4\pi|x-y_0|}$, and a similar formula holds for G_2 with the integration over S_2. Consequently,

$$G_1(x, y_0) - G_2(x, y_0) = \int_{S_2} g(s, x) G_{2N}(s, y_0) ds - \int_{S_1} g(s, x) G_{1N}(s, y_0) ds, \tag{2.1.41}$$

where ds is the surface area elements.

Let $y_0 \to \infty$, $y_0/|y_0| = -\alpha_0$ and take into account that if $u_1(x, \alpha_0) = u_2(x, \alpha_0)$ for all $x \in D'_{12}$ then $u_1(x, \alpha_0) = u_2(x, \alpha_0) := u(x, \alpha_0)$ for all $x \in \mathcal{D}^{12} := (D_1 \cap D_2)'$ by the unique continuation principle for the solutions to homogeneous Helmholtz equation. Therefore, equation (2.1.41) and Lemma 2.1.6 yield

$$\int_{S_2} g(s, x) u_N(s, \alpha_0) ds = \int_{S_1} g(s, x) u_N(s, \alpha_0) ds, \quad \forall x \in \mathcal{D}^{12}. \tag{2.1.42}$$

The right-hand side of (2.1.42) is an infinitely smooth function when x passes the part of S_2 which lies outside of D_1 while the normal derivative of the left side has a jump $u_N(s, \alpha_0)$ in such a process. This is a contradiction unless $u_N(s, \alpha_0) = 0$ on S_2. However, $u = 0$ on S_2 and if $u_N(s, \alpha_0) = 0$ on S_2 then, by the uniqueness of the solution to the Cauchy problem for the Helmholtz equation, one concludes that $u = 0$ in D'_2. This is impossible since $\lim_{x\to\infty} |u(x, \alpha_0)| = 1$. This contradiction proves that $D_1 = D_2 := D$, $S_1 = S_2 := S$, and $G_1(x, y_0) = G_2(x, y_0) := G(x, y_0)$, where G is the Green's function of the Dirichlet Helmholtz operator for the domain D', and G satisfies the radiation condition at infinity.

Thus, the proof of the relation $G_1(x, y_0) = G_2(x, y_0)$ is completed and the second part of Lemma 2.1.9 is proved.

Lemma 2.1.9 is proved. □

Lemma 2.1.10. *One has*

$$\lim_{x\to t} G_{2N}(x, s, k) = \delta(s - t), \quad t \in S_2, \tag{2.1.43}$$

where $\delta(s - t)$ denotes the delta function on S_2 and $x \to t$ denotes a limit along any straight line nontangential to S_2.

Proof. Let $f \in C(S_2)$ be arbitrary. Consider the following problem: W solves equation (2.1.23) in D'_2, W satisfies the boundary condition $W = f$ on S_2, and W satisfies the radiation condition. The unique solution to this problem is given by the Green's formula:

$$W(x) = \int_{S_2} G_{2N}(x,s)f(s)ds. \tag{2.1.44}$$

Since $\lim_{x \to t \in S_2} w(x) = f(t)$ and $f \in C(S_2)$ is arbitrary, the conclusion of Lemma 2.1.10 follows.

Lemma 2.1.10 is proved. □

Let us point out the following implications:

$$G(x,y,k) \to u(x,\alpha,k) \to A(\beta,\alpha,k), \tag{2.1.45}$$

which hold by Lemma 2.1.6 and formula (2.1.36). The first arrow means that the knowledge of $G(x,y,k)$ determines uniquely the scattering solution $u(x,\alpha,k)$ for all $\alpha \in S^2$, and the second arrow means that the scattering solution $u(x,\alpha,k)$ determines uniquely the scattering amplitude $A(\beta,\alpha,k)$.

The reversed implications also hold:

$$A(\beta,\alpha,k) \to u(x,\alpha,k) \to G(x,y,k). \tag{2.1.46}$$

These implications follow from Lemmas 2.1.6, 2.1.8, 2.1.9 and formula (2.1.36).

Let us explain why the knowledge of $u(x,\alpha,k)$ determines uniquely $G(x,y,k)$. If there are two G_m, $m = 1,2$, to which the same $u(x,\alpha,k)$ corresponds, then $w := G_1 - G_2$ solves equation (2.1.23) in D'_{12} and, by Lemma 2.1.6, $w = O(\frac{1}{|x|^2})$. Thus, by Lemma 2.1.8, $w = 0$, so $G_1 = G_2$ in D'_{12}. This implies, as in the proof of Theorem 2.1.3 below, that $D_1 = D_2 := D$.

Similar implications for $\alpha = \alpha_0$ fixed are formulated after the proof of Theorem 2.1.3.

Proof of Theorem 2.1.3. If $A_1(\beta) = A_2(\beta)$ for all β in a solid angle, then the same is true for all $\beta \in S^2$, so one may assume that $A_1(\beta) = A_2(\beta)$ for all $\beta \in S^2$.

Let us assume that $A_1(\beta) = A_2(\beta)$ for all β but $S_1 \neq S_2$. We want to derive from this assumption a contradiction. This contradiction will prove that the assumption $S_1 \neq S_2$ is false, so $S_1 = S_2$.

If $A_1(\beta) = A_2(\beta)$ for all $\beta \in S^2$, then Lemma 2.1.9 yields the following conclusion:

$$G_1(x, y_0) = G_2(x, y_0), \quad \forall x \in D'_{12}, \tag{2.1.47}$$

where $k > 0$ and $\alpha_0 \in S^2$ are fixed and $y_0 = -\alpha_0\tau + \eta$, $\tau > 0$, $\eta \cdot \alpha_0 = 0$, η is an arbitrary fixed vector orthogonal to α_0. This is the key point in the proof of Theorem 2.1.3. We assume in the proof of Theorem 2.1.3 that S_1 intersects S_2. Other cases are treated in Theorems 2.1.4 and 2.1.5.

If $S_1 \neq S_2$, then one gets a contradiction: let y_0 approach a point $t \in S_2$ which does not belong to S_1 along the ray $-\tau\alpha_0 + \eta$. Then, on one hand, $G_1(x, t) = G_2(x, t) = 0$ for all $x \in D'_{12}$, and, on the other hand, $G_1(x, t) = O(\frac{1}{|x-t|})$, so that $|G_1(x, t)| \to \infty$ as $x \to t$. This contradiction proves that $S_1 = S_2$.

If $S_1 = S_2 := S$, then $D_1 = D_2 := D$ and $u_1(x, \alpha_0, k) = u_2(x, \alpha_0, k) := u(x, \alpha_0, k)$ for $x \in D'$, and, consequently, the boundary condition on S is uniquely determined: if $u|_S = 0$, then one has the Dirichlet boundary condition Γ_1, otherwise calculate $\frac{u_N}{u}$ on S. If this ratio vanishes, then one has the Neumann boundary condition Γ_2, otherwise one has the impedance boundary condition Γ_3, and the boundary impedance $h = -\frac{u_N}{u}$ on S, so the boundary condition is uniquely determined by the non-overdetermined scattering data.

Theorem 2.1.3 is proved. □

One may give different proofs of Theorem 2.1.3. For example, if $S_1 \neq S_2$ and S_1 intersects S_2 then, by analytic continuation, the scattering solutions $u_m(x, \alpha_0)$, $m = 1, 2$, admit analytic continuation to the exterior of the domain $D^{12} = D_1 \cap D_2$. The boundary of this domain has edges. If a point t belongs to an edge, then the gradient of the solution to the homogeneous Helmholtz equation is singular when $x \to t$. On the other hand, this t belongs to a smooth boundary S_1 or S_2, so that the above gradient has to be smooth. This contradiction proves that $S_1 = S_2$ in the case when S_1 intersects S_2. In Remark 2.1.8 another proof of Theorems 2.1.3 and 2.1.5 is given.

Let us formulate the implication similar to the one given before the proof of Theorem 2.1.3. If $y = y_0 = -\tau\alpha_0 + \eta$, $\tau > 0$ is an arbitrary number, α_0 is a fixed unit vector, and η is an arbitrary fixed vector orthogonal to α_0, then

$$G(x, y_0, k) \to u(x, \alpha_0, k) \to A(\beta, \alpha_0, k), \tag{2.1.48}$$

where α_0 is a free unit vector, that is, a vector whose initial point is arbitrary.

The reversed implications also hold:

$$A(\beta, \alpha_0, k) \to u(x, \alpha_0, k) \to G(x, y_0, k). \tag{2.1.49}$$

The first of these implications follows from Lemma 2.1.8 and the asymptotic of the scattering solution, while the second follows from Lemmas 2.1.6 and 2.1.9.

We have assumed implicitly that D_1 and D_2 have a common part but none of them is a subset of the other, that is, S_1 intersects S_2. Let us discuss the two remaining possibilities.

The first possibility is that $D_1 \neq D_2$ and $D_1 \cap D_2 = \emptyset$.

Proof of Theorem 2.1.4. If $A_1(\beta) = A_2(\beta)$ in a solid angle, then $A_1(\beta) = A_2(\beta)$ in S^2. This implies that $u_1(x, \alpha_0) = u_2(x, \alpha_0)$ in D'_{12}. Since $u_1(x, \alpha_0)$ is defined in D_2 and satisfies there the Helmholtz equation (2.1.23), the unique continuation property implies that $u_2(x, \alpha_0, k)$ is defined in D_2 and satisfies there the Helmholtz equation. Consequently, $u_2(x, \alpha_0, k)$ is defined in $\mathbb{R}^3$, it is a smooth function that satisfies in $\mathbb{R}^3$ the Helmholtz equation, and the same is true for $u_1(x, \alpha_0, k)$. Therefore, the scattered parts v_1 and v_2 of the scattering solutions u_1 and u_2 satisfy the Helmholtz equation (2.1.23) in $\mathbb{R}^3$ and the radiation condition. A function satisfying the radiation condition and the Helmholtz equation in $\mathbb{R}^3$ is equal to zero in $\mathbb{R}^3$. Therefore, $v_1 = v_2 = 0$ and $u_1 = u_2 = e^{ik\alpha_0 \cdot x}$ in $\mathbb{R}^3$. This is impossible since $u_m = 0$ on S_m, $m = 1, 2$, while $e^{ik\alpha_0 \cdot x} \neq 0$ on S_m.

Theorem 2.1.4 is proved. □

The second possibility is $D_1 \neq D_2$ and $D_1 \subset D_2$.

Proof of Theorem 2.1.5. One proves Theorem 2.1.5 using Lemma 2.1.9. By Lemma 2.1.9, one has

$$G_1(x,y_0) = G_2(x,y_0), \quad \forall\, x \in D_2', \quad y_0 = -\tau\alpha_0 + \eta, \quad y_0 \in D_2', \\ \eta \cdot \alpha_0 = 0, \quad \tau \in (0,\infty). \tag{2.1.50}$$

Note that

$$\lim_{x \to y_0} |G_1(x,y_0)| = \infty \tag{2.1.51}$$

since both x and y_0 belong to D_1' and are away from S_1 if $D_1 \neq D_2$. On the other hand, if $y_0 \in S_2$, then $G_2(x,y_0) = 0$ for all $x \in D_2$, $x \neq y_0$ and

$$\lim_{x \to y_0} |G_2(x,y_0)| = 0. \tag{2.1.52}$$

This is a contradiction unless $D_1 = D_2$.

Theorem 2.1.5 is proved. □

It follows from Theorems 2.1.3–2.1.5 that Corollary holds: the solution to problem (2.1.23)–(2.1.24) (the scattering solution) cannot have a closed surface of zeros except the surface S, the boundary of the obstacle.

Remark 2.1.4. For the transmission problem, the uniqueness theorem with the data $f(n_0,\nu,k_0)$ does not hold: one can have the same scattering amplitude $f(n_0,\nu,k)$ and various transmission constants and k_1^2.

Example. Let $D = \{x : |x| \leq 1\}$,

$$(\nabla^2 + 1)u = 0 \text{ in } \Omega, \quad (\nabla^2 + k_1^2)u = 0 \quad \text{in } D, \\ u_+ = u_-, \quad \alpha\frac{\partial u}{\partial N_+} = \frac{\partial u}{\partial N_-} \tag{2.1.53}$$

$$u = f\,\exp(ir)r^{-1}, \quad r \geq 1, \quad f = \text{const}, \\ u = cr^{-1}\sin(k_1 r), \quad r \leq 1, \quad c = \text{const}. \tag{2.1.54}$$

Equation (2.1.53) is satisfied if

$$f \exp i = c \sin k_1, \quad f(i-1) \exp i = \alpha c(k_1 \cos k_1 - \sin k_1). \tag{2.1.55}$$

Equation (2.1.55) is satisfied if

$$k_1 \cot k_1 = 1 + \alpha^{-1}(i-1). \tag{2.1.56}$$

For a fixed α, equation (2.1.56) can have several roots $k_{1m}, m = 1, 2, \ldots$. Thus, various triples $(D, \alpha, k_{1m}), m = 1, 2, \ldots$ produce the same field in Ω and the same scattering amplitude d.

Remark 2.1.5. The uniqueness theorem does not hold for the inverse source problem. This problem consists in finding the source $h(x)$ from the scattering amplitude $f(\nu, k_0)$. Here

$$(\nabla^2 + k_0^2)v = h(x), \quad h = 0 \quad \text{in } \Omega_R, \tag{2.1.57}$$

$$v \sim \gamma(r) f(\nu, k_0), \quad r \to \infty, \quad \nu = xr^{-1}. \tag{2.1.58}$$

To prove that $f(\nu, k_0)$ does not determine h uniquely, consider $h(x) \neq 0$ which produces $v = 0$ in Ω_R and therefore $f = 0$. Such h are called nonradiating sources.

Take any $v \in C_0^2(B_R)$ and set $h = (\nabla^2 + k_0^2)v$. Then $v = 0$ in Ω_R and $f = 0$.

Remark 2.1.6. Consider the inverse problem of potential scattering. Let $q \in C_0^\infty(\mathbb{R}^3)$

$$(\nabla^2 + k_0^2 - q(x))u = 0 \text{ in } \mathbb{R}^3, \quad k_0 > 0, \tag{2.1.59}$$

$$u = u_0 + v, \quad v \sim r^{-1} \exp(ik_0 r) f(n_0, \nu, k_0). \tag{2.1.60}$$

Do the data $f(n_0, \nu, k_0)$ determine $q(x)$ uniquely? The answer is no. To prove this, consider $v = r^{-1} \exp(ik_0 r)$ in Ω_R, where $R > 0$ is a fixed number, continue v in B_R so that $v \in C^2(\mathbb{R}^3)$ and $v + u_0 \neq 0$ in B_R. There are infinitely many continuations v which satisfy these requirements. Define $q = ((\nabla^2 + k_0^2)v)/(v + u_0)$. Then $q \in C_0^\infty(B_R)$, $q \neq 0$, and the corresponding scattering amplitude is $f = 1$ for any of the potential q.

One can prove, however, that the data $f(n_0, \nu, k), 0 < a \le k \le b$, $b > a$, $\forall \nu \in S^2$, determine the potential $q(x) \in C_0^\infty(\mathbb{R}^3)$ uniquely. Actually this can be proved for $q \in C_0(\mathbb{R}^3)$ (see Part II, Scattering by Potentials).

Remark 2.1.7. If one knows $f(n, \nu, k)$ for small k, $0 < k < k_0$, $n, \nu \in S^2$, then, by analyticity, $f(n, \nu, k)$ is uniquely determined for all k, and the obstacle is determined uniquely. But practically only the first one or two coefficients in the expansion of $f(n, \nu, k)$ in the power series about $k = 0$ can be measured. Practically, in general, any body scatters as an ellipsoid if $ka \ll 1$ (see Chapter 4.4.6), so that the scattering data allows one to determine the axes of this ellipsoid, i.e., three scalar parameters. This is the reason why most of the applied work on inverse scattering problem is done in the high-frequency region.

Remark 2.1.8. Let us give a new proof of Theorems 2.1.3 and 2.1.5. Assume first that $D_1 \subset D_2$, so that assumptions of Theorem 2.1.5 hold. Denote by u the analytic continuation of u_2 into $D_2 \setminus D_1$. This u is equal to u_1 in D_1'. Green's formula yields the following relations:

$$u(x) = \int_{S_2} g(x, s) u_N(s) ds - \int_{S_1} g(x, s) u_N(s) ds, \quad x \in D_2 \setminus D_1, \tag{2.1.61}$$

$$u = u_0 - \int_{S_2} g(x, s) u_N(s) ds, \quad x \in D_2', \tag{2.1.62}$$

$$u = u_0 - \int_{S_1} g(x, s) u_N(s) ds, \quad x \in D_1'. \tag{2.1.63}$$

From (2.1.61) and (2.1.63) one derives

$$0 = u_0 - \int_{S_2} g(x, s) u_N(s) ds, \quad x \in D_2 \setminus D_1. \tag{2.1.64}$$

Denote $I^+ := \int_{S_2} g(x, s) u_N(s) ds, \quad x \in D_2 \setminus D_1$ and $I^- := \int_{S_2} g(x, s) u_N(s) ds, \quad x \in D_2'$. The functions u and u_0 are uniquely extended from D_2' into $D_2 \setminus D_1$ and integral I^- is extended uniquely from D_2' into $D_2 \setminus D_1$ as integral I^+. Equations (2.1.64) can be written as $I^+ = u_0$ in $D_2 \setminus D_1$. Equation (2.1.62) can be uniquely extended into $D_2 \setminus D_1$ and written there as $u = u_0 - I^+ = 0$ (by (2.1.64)). Therefore $u = 0$ in $D_2 \setminus D_1$. This is a contradiction since u is the scattering solution in D_1', it solves the elliptic Helmholtz equation in D_1' and if $u = 0$ in $D_2 \setminus D_1$ then $u = 0$ everywhere in D_1', which is impossible since $|u| \to 1$ as $|x| \to \infty$.

If the assumptions of Theorem 2.1.3 hold, then the argument is similar. The roles of S_2 and S_1 are played respectively by S_{12}, the boundary of D_{12}, and the boundary of the intersection $D^{12} = D_1 \cap D_2$. □

2.2. Construction of Obstacles from the Scattering Data at High Frequencies

2.2.1. Let D be a convex bounded domain with a smooth connected, boundary Γ. Let $\psi : \Gamma \to S^2$ be such a map (spherical map of the surface) that $\psi(s)$ is the unit exterior (positive) normal to Γ at the point s. This unit vector $\psi(s)$ is identified with a point on the unit sphere S^2, namely with the end point of the vector $\psi(s)$ the initial point of which is placed at the center of the unit ball with the surface S^2. Let $K_1(s)$, $K_2(s)$ and $K(s) = K_1K_2$ be the principal curvatures and the Gaussian curvature of Γ at the point s, $H = (K_1 + K_2)/2$ be the mean curvature. Let 0 be the origin of the coordinate system, Γ_+ be the part of Γ which is seen from 0, the illuminated part of Γ, and Γ_- is the shadowed part of Γ. Let l be a unit vector. If $0 \in D$, then $\Gamma_+ = \Gamma$. Define the support function of the convex surface Γ with respect to the origin 0 by the formulas

$$a(l) = \max_{s\in\Gamma}(l, s), \quad 0 \in D, \tag{2.2.1}$$

$$a(l) = -\max_{s\in\Gamma_+}(l, s), \quad 0 \notin D. \tag{2.2.2}$$

Note that if v runs through a subset of negative numbers then $-\max v = \min |v|$. The support function is the distance from the origin to the tangent plane to Γ perpendicular to the vector l. If $0 \notin D$, $s \in \Gamma_+$, and $l = \psi(s)$ is the unit positive normal to Γ at the point s, then $(l, s) < 0$. The support function defined by formula (2.2.2) is the distance from the origin 0 to the tangent plane to Γ_+ perpendicular to l.

Let us derive the equation of the surface Γ from the support function. It is clear from the definition that $a(l)$ can be extended from S^2 to $\mathbb{R}^3$ as a homogeneous function of order 1 so that

$$a(tl) = ta(l), \quad t > 0. \tag{2.2.3}$$

Since the surface is smooth, $a(l)$ is smooth. It follows from (2.2.3) that

$$\sum_{j=1}^{3} \frac{\partial a}{\partial \alpha_j} \alpha_j = a(l), \quad l = (\alpha_1, \alpha_2, \alpha_3). \tag{2.2.4}$$

If $0 \in D$ so that formula (2.2.1) holds, then the equation of the surface Γ is of the form

$$x_j = \partial a / \partial \alpha_j, \quad 1 \leq j \leq 3. \tag{2.2.5}$$

This is a parametric equation of the surface, and the role of the parameters is played by the unit vector $l = (\alpha_1, \alpha_2, \alpha_3)$. In order to derive (2.2.5), let us write the equation of the tangent plane to Γ perpendicular to l at the point at which $(l, s) = a(l)$:

$$\sum_{j=1}^{3} x_j \alpha_j = a(l). \tag{2.2.6}$$

The surface Γ is the envelope of the family of the tangent planes (2.2.6). Therefore, its equation can be obtained by the formula

$$\frac{\partial}{\partial \alpha_j} \left(\sum_{j=1}^{3} x_j \alpha_j - a(\alpha_1, \alpha_2, \alpha_3) \right) = 0. \tag{2.2.7}$$

Equation (2.2.7) is equivalent to (2.2.5). If $0 \notin D$ and formula (2.2.2) holds, then the same argument shows that the equation of the tangent plane is

$$\sum_{j=1}^{3} x_j \alpha_j = -a(l), \tag{2.2.8}$$

and the equation of the surface is

$$x_j = -\partial a/\partial \alpha_j, \quad 1 \le j \le 3. \tag{2.2.9}$$

Example. Consider a sphere $\sum_{j=1}^{3}(x_j - b_j)^2 = R^2$. The tangent planes to the sphere orthogonal to the unit vector $\sum_{j=1}^{3} \alpha_j e_j = l$, where e_j are the orthogonal unit vectors of the coordinate system, are $\sum_{j=1}^{3}(x_j - b_j \pm R\alpha_j)\alpha_j = 0$. Assuming that the origin is outside of the sphere, i.e., $|b|^2 = \sum_{j=1}^{3} |b_j|^2 > R^2$, one has $a(l) = -(b, l) - R$ on Γ_+. Therefore, equation (2.2.9) is the parametric equation of Γ_+:

$$x_j = b_j + R\alpha_j, \quad 1 \le j \le 3. \tag{2.2.10}$$

One should have in mind that the constant $R = R(\alpha_1^2 + \alpha_2^2 + \alpha_3^2)^{1/2}$ should be considered as a restriction on S^2 of the homogeneous function of order 1 defined on $\mathbb{R}^3$ and equal to constant R on S^2 so that $\partial R/\partial \alpha_j = R\alpha_j$.

If the support function can be obtained from the scattering data, then the surface can be easily recovered by formula (2.2.9). Usually, it is convenient to place the origin at some known point and in this case $0 \notin D$, where D is the scatterer. This is the reason why formula (2.2.9) rather than (2.2.5) is used.

2.2.2. Let us derive the high-frequency asymptotics of the scattering amplitude assuming that the scatterer is convex. The idea is to use the Kirchhoff approximation which gives the correct high-frequency asymptotics.

The starting point is the formula (1.4.2) for the scattering amplitude. The function $h = \partial u/\partial N$ in this formula satisfies the equation analogous to (1.3.5)

$$h + Ah = 2\frac{\partial \exp(-ik\nu \cdot s)}{\partial N}, \tag{2.2.11}$$

where A is defined in (1.1.4).

In what follows, it is convenient to use $-\nu$ in place of ν.

The Kirchhoff approximation is

$$h = 0 \text{ on } \Gamma_-, \quad h = 2ik(N_s, \nu)\exp(ik\nu \cdot s) \quad \text{on } \Gamma_+, \tag{2.2.12}$$

where $\Gamma_+(\Gamma_-)$ is the illuminated (shadowed) part of Γ, i.e., $(N_s, \nu) < 0$ on Γ_+, $(N_s, \nu) > 0$ on Γ_-. Substitute (2.2.12) into (1.4.2) to obtain

$$f = -\frac{ik}{2\pi}\int_{\Gamma_+} \exp\{ik(\nu - n)\cdot s\}(N_s, \nu)ds. \tag{2.2.13}$$

Here $f = f(-n, -\nu, k) = f(\nu, n, k)$.

Take the origin outside of D and define the support function by formula (2.2.2). Compute integral (2.2.13) asymptotically as $k \to +\infty$ by the stationary phase method. The following lemma is well known (see [63]).

Lemma 2.2.1. *Let $f \in C_0^\infty(D), S(x) \in C^\infty(D)$, and $D \subset \mathbb{R}^r$ be a finite domain. Let $S(x)$ be a real-valued function which has only one stationary (critical) point x_0 in D, and assume that this point is nondegenerate (i.e., $dS(x_0) = 0, d^2S(x_0) \neq 0$). Then*

$$\int_D f(x)\exp(ikS(x))dx = \left(\frac{2\pi}{k}\right)^{r/2}\exp\left\{ikS(x_0) + \frac{i\pi}{4}\text{sgn}\, S''(x_0)\right\} \times |\det S''(x_0)|^{-1/2}\left[f(x_0) + O\left(\frac{1}{k}\right)\right], \tag{2.2.14}$$

where

$$S''(x_0) = \frac{\partial^2 S(x_0)}{\partial x_i \partial x_j}, \quad \text{sgn}\, S''(x_0) = n_+ - n_-, \tag{2.2.15}$$

and $n_+(n_-)$ is the number of positive (negative) eigenvalues of the matrix $S''(x_0)$.

In order to apply this lemma, let us calculate the term $S(s) = (\nu - n) \cdot s$ in (2.2.13) around the stationary point s. One has

$$(\nu - n) \cdot s = -|n - \nu| l \cdot s, \quad \text{where } l = \frac{n - \nu}{|n - \nu|}. \tag{2.2.16}$$

The stationary point of the function (2.2.16) is the point $s_0 \in \Gamma_+$ at which $l = N_{s_0}$. Since Γ is convex, there is only one stationary point on Γ_+, and

$$l \cdot s = l \cdot s_0 + l \cdot (s - s_0) = l \cdot s_0 + \frac{1}{2} b_{ij} u^i u^j + o(|u|^2), \tag{2.2.17}$$

where one should sum over the repeated indices, b_{ij} is the matrix of the second differential form of the surface Γ at the point s_0, $u^i, i = 1, 2$, are the local coordinates in the tangential plane to Γ, $s_0 = (0, 0)$ in these coordinates and the third axis is directed along $l = N_{s_0}$, the unit positive normal to Γ at the point s_0. Therefore,

$$S''(s_0) = -|n - \nu| b_{ij}, \quad r = 2, \tag{2.2.18}$$

$$\det(-|n - \nu| b_{ij}) = |n - \nu|^2 \det b_{ij} = |n - \nu|^2 K(s_0), \tag{2.2.19}$$

where $K = K_1 K_2 = \det b_{ij}$ is the Gaussian curvature, and the principal curvatures $K_j, j = 1, 2$, are the eigenvalues of the matrix b_{ij}. If the surface is convex, then the matrix b_{ij} is positive definite, $K_j > 0$, $j = 1, 2$, $n_+ = 2$, $n_- = 0$, and sgn $b_{ij} = 2$. Thus, applying formula (2.2.14) to integral (2.2.13) yields

$$f = -\frac{ik}{2\pi} \cdot \frac{2\pi}{k} \exp\left\{-ik|n - \nu| l \cdot s_0 + \frac{i\pi}{4} \cdot 2\right\}. \tag{2.2.20}$$

Since $|n - \nu| = 2n \cdot l, l \cdot \nu = -l \cdot n$ and $l \cdot s_0 = -a(l)$, one can write (2.2.20) as

$$f = f(\nu, n, k) = -\frac{\exp\{i2kn \cdot la(l)\}}{2K^{1/2}(s_0)} \times \left[1 + O\left(\frac{1}{k|l \cdot \nu|}\right)\right], \quad k \to \infty. \tag{2.2.21}$$

Let us formulate the result.

Theorem 2.2.2. *If Γ is a smooth convex surface, then the asymptotic behavior of the scattering amplitude as $k \to +\infty$ is given by formula* (2.2.21) *in which s_0 is the specular point; that is, the point at which the unit positive normal to Γ_+ is the vector $l = (n-\nu)/|n-\nu|$, $K(s_0)$ is the Gaussian curvature of Γ at the specular point s_0, $a(l)$ is the support function of Γ_+ defined by formula* (2.2.21), *$n \cdot l > 0$ and $\nu \cdot l = -n \cdot l < 0$.*

From formula (2.2.21), it follows that

$$|f(\nu, n, k)| = \frac{1}{2K^{1/2}(s_0)} \quad \text{as } k \to \infty. \tag{2.2.22}$$

Actually the Gaussian curvature in (2.2.22) is known as a function of the unit normal $l = (n-\nu)/|n-\nu|$ to Γ as follows from the derivation of formula (2.2.21). The problem of recovering a convex compact closed surface from its Gaussian curvature given as a function of the exterior normal to the surface (Minkowski's problem) has been studied extensively (see [82, 167] where further references are given). It is known (see [82]), that this problem has at most one solution apart from the translations. Existence of a convex surface with its Gaussian curvature equal to a given function $K(l)$ has also been studied (see [167]). However, there is no constructive way to recover the surface from its Gaussian curvature: the problem was reduced to a nonlinear elliptic problem. Therefore, it is important to use the phase information in the scattering amplitude. The knowledge of the phase of the scattering amplitude (2.2.21) allows one to compute the support function $a(l)$ and to recover the surface effectively by formula (2.2.9). Actually one needs only the knowledge of the high-frequency scattering amplitude for a set of n and ν such that the vector $l = (n-\nu)/|n-\nu|$ runs through all of S^2. For example, the backscattering amplitude $f(n, -\nu, k)$, $n \in S^2$ is sufficient for the effective recovery of the surface. If the high-frequency scattering amplitude is known for a set of n and ν such that the vector l runs through a subset $\tilde{S}^2$, then formula (2.2.9) gives the corresponding part of the surface Γ: the recovery of a part of $\tilde{\Gamma}$ of the surface by formula (2.2.9) does not require the knowledge of $a(l)$ on all of S^2

but only on $\tilde{S}^2 \subset S^2$, where $\tilde{S}^2 = \psi(\tilde{\Gamma})$ and ψ is the spherical map of the surface defined in Section 2.1.

One sees from (2.2.21) that the knowledge of the scattering amplitude at one sufficiently high frequency and n and ν running through such a subset of $S^2 \times S^2$ that $l = (n - \nu)/|n - \nu|$ runs through all of S^2 is sufficient for recovering the support function $a(l)$ and, therefore, the surface Γ itself.

2.2.3. The knowledge of the mean curvature of a convex closed smooth surface determines the surface uniquely up to translations. Existence of the surface with the mean curvature equal to a given function was studied in the literature (see [82, 167]). The problem was reduced to a nonlinear elliptic problem and there is no constructive way to recover the surface from its mean curvature.

2.2.4. However, there is a simple and constructive way to recover the surface from its harmonic curvature which is defined as

$$h = K_1^{-1} + K_2^{-1}. \tag{2.2.23}$$

The function $h = h(l)$ is assumed to be known as a function of the unit normal l to the surface. Let us derive the equation of the surface from the knowledge of $h(l)$. Actually, the support function $a(l)$ will be recovered from $h(l)$. If $a(l)$ is found, then the equation of the surface is equation (2.2.9).

Let us extend $a(l)$ and $h(l)$ from S^2 to $\mathbb{R}^3$ and set $y = rl$

$$A(y) = ra(l), \quad h(y) = r^{-1}h(l). \tag{2.2.24}$$

Lemma 2.2.3. *The following equation holds*:

$$\nabla_y^2 A = h(y). \tag{2.2.25}$$

Proof. Let $l = (\alpha_1, \alpha_2, \alpha_3)$ and $x = x(\alpha_1, \alpha_2, \alpha_3)$ be the radius vector of a point of the surface Γ. Let R be the radius of the curvature of the normal cut of Γ in the principal direction $(d\alpha_1, d\alpha_2, d\alpha_3)$. From the Rodrigues's formula

$$dx_j - Rd\alpha_j = 0, \quad 0 \le j \le 3, \tag{2.2.26}$$

and formula (2.2.9) it follows that

$$0 = \frac{\partial^2 a}{\partial\alpha_j \partial\alpha_i} d\alpha_i - Rd\alpha_j = \left(\frac{\partial^2 a}{\partial\alpha_i \partial\alpha_j} - R\delta_{ij} \right) d\alpha_i. \tag{2.2.27}$$

Here and below one sums over the repeated indices, $\delta_{ij} = \begin{cases} 0 & i \ne j \\ 1 & i - j \end{cases}$.
Since the vector $(d\alpha_1, d\alpha_2, d\alpha_3) \ne 0$, one concludes from (2.2.27) that

$$\det \left(\frac{\partial^2 a}{\partial\alpha_i \partial\alpha_j} - R\delta_{ij} \right) = 0. \tag{2.2.28}$$

Furthermore,

$$\det \frac{\partial^2 a}{\partial\alpha_i \partial\alpha_j} = 0. \tag{2.2.29}$$

Indeed, differentiate (2.2.4) in α_i to obtain

$$\frac{\partial^2 a}{\partial\alpha_i \partial\alpha_j} + \frac{\partial a}{\partial\alpha_j} \delta_{ij} = \frac{\partial a}{\partial\alpha_i}, \quad 1 \le i \le 3 \tag{2.2.30}$$

which is equivalent to

$$\frac{\partial^2 a}{\partial\alpha_i \partial\alpha_j} \alpha_j = 0. \tag{2.2.31}$$

Since $(\alpha_1, \alpha_2, \alpha_3) = l \ne 0$, one concludes from (2.2.31) that (2.2.29) holds.

It follows from (2.2.29) that equation (2.2.28) has a root $R = 0$. Its two other roots R_1 and R_2 are the principal radii of the surface, $R_i = K_i^{-1}$. Since the surface is convex, $R_i > 0, i = 1, 2$. The trace of the matrix $\partial^2 a/(\partial\alpha_i \partial\alpha_j)$ is equal to $R_1 + R_2 + 0 = h(\alpha_1, \alpha_2, \alpha_3) = h(l)$. Since the trace $\text{tr}(\partial^2 a/(\partial\alpha_i \partial\alpha_j)) = \nabla_\alpha^2 a$, one has

$$\nabla_\alpha^2 a = h. \tag{2.2.32}$$

Because of the properties (2.2.24) equation (2.2.32), which is equation (2.2.25) for $|y| = 1$, implies (2.2.25).

Let us give a simple effective algorithm for solving equation (2.2.25). Let (θ, ϕ) be the spherical coordinates of the end point of vector $l = (\alpha_1, \alpha_2, \alpha_3)$ and Y_{nm} be the spherical harmonics normalized in $L^2(S^2)$:

$$Y_{nm}(\theta, \phi) = \left(\frac{(2n+1)(n-m)!}{4\pi(n+m)!} \right)^{1/2} P_{n,m}(\cos\theta) \exp(\mathrm{im}\,\varphi), \tag{2.2.33}$$

$n = 0, 1, 2, \ldots,\ -n \leq m \leq n$. The known function $H(l) = h(\theta, \phi)$ can be expanded in the Fourier series

$$h = \sum_{n=0}^{\infty} \sum_{m=-n}^{n} h_{nm} Y_{nm}, \tag{2.2.34}$$

where

$$h_{nm} = \int_{S^2} h(\theta, \phi) \bar{Y}_{nm}(\theta, \phi) d\omega, \quad d\omega = \sin\theta d\theta d\varphi, \tag{2.2.35}$$

and the bar denotes the complex conjugate. From (2.2.34) and (2.2.24), it follows that

$$h(y) = \frac{1}{r} \sum_{n=0}^{\infty} \sum_{m=-n}^{\infty} h_{nm} Y_{nm}(\theta, \phi), \tag{2.2.34$'$}$$

where (r, θ, ϕ) are the spherial coordinates of the point $y = rl$. A simple calculation shows that

$$\begin{aligned} \nabla_y^2 \left(r Y_{nm}(\theta, \phi) \right) &= \left(\frac{\partial^2}{\partial y_1^2} + \frac{\partial^2}{\partial y_2^2} + \frac{\partial^2}{\partial y_3^2} \right) (r Y_{nm}) \\ &= -\frac{(n-1)(n+2)}{r} Y_{nm}. \end{aligned} \tag{2.2.36}$$

Therefore, equation (2.2.25) can be written as

$$-\sum_{n=0}^{\infty} \sum_{m=-n}^{n} \frac{(n-1)(n+2)}{r} A_{nm} Y_{nm} = \frac{1}{r} \sum_{h=0}^{\infty} \sum_{m=-n}^{n} h_{nm} Y_{nm}, \tag{2.2.37}$$

where A_{nm} are the coefficients in the expansion

$$A(y) = ra(\theta, \phi) = r \sum_{n=0}^{\infty} \sum_{m=-n}^{n} A_{nm} Y_{nm}. \tag{2.2.38}$$

From (2.2.37), it follows that

$$A_{nm} = -\frac{h_{nm}}{(n-1)(n+2)}, \quad n \neq 1. \tag{2.2.39}$$

We will prove shortly that a necessary condition for a function $h(\theta, \phi)$ to be equal to $R_1 + R_2$, where R_i, $i = 1, 2$ are the principal radii of a closed smooth surface, is

$$h_{1m} = 0, \quad m = 0, \pm 1. \tag{2.2.40}$$

□

Let us summarize the result as follows.

Theorem 2.2.4. *If $h(\theta, \phi)$ is the sum of the principal radii of a smooth closed convex surface, then the support function $a(l)$ of this surface is given by the formula*

$$a(\theta, \phi) = - \sum_{n=0, n=1}^{\infty} \sum_{n=-m}^{m} \frac{h_{nm}}{(n-1)(n+2)} Y_{nm}(\theta, \phi), \tag{2.2.41}$$

where h_{nm} are given by formula (2.2.35). The equation of the surface is given by formula (2.2.9) in which $(\alpha_1, \alpha_2, \alpha_3)$ are the cartesian coordinates of the point $(\theta, \phi) \in S^2$.

It remains to be proved that (2.2.40) holds.

Lemma 2.2.5. *If Γ is a closed compact surface (smooth enough for the Gauss formula to hold), then equation (2.2.40) holds.*

Proof. Let ds be the element of the surface area of Γ, $ds = R_1 ds_1 ds_2$, where $ds_1 ds_2 = d\omega$ is the element of the surface area of S^2, $|R_j ds_j|, j = 1, 2$, are the elements of the lengths along the lines of principal curvatures on Γ, and $ds_1 ds_2$ is the inner product of the vectors ds_1 and ds_2. Consider a family Γ_2 of parallel surfaces for which the normals are the same as for $\Gamma = \Gamma_0$. Then

$$ds(z) = (R_1 + z)(R_2 + z)ds_1 ds_2 = ds + z(R_1 + R_2)d\omega = z^2 d\omega, \tag{2.2.42}$$

where $ds(z)$ is the element of the surface area of the surface Γ_2. From the Gauss formula, it follows that

$$\int_{\Gamma_2} N_j ds(z) = 0, \quad j = 1, 2, 3, \quad 0 \leq z \leq z_0, \tag{2.2.43}$$

where N_j is the jth component of the exterior unit normal to Γ_z, and z_0 is so small that N does not depend on z. From (2.2.42) and (2.2.43), one obtains

$$\int_{S^2} (R_1 + R_2) N_j d\omega = 0, \quad 1 \leq j \leq 3. \tag{2.2.44}$$

Since $R_1 + R_2 = h, N_1 = \sin\theta\cos\varphi = \frac{1}{2}(Y_{11} + Y_{1-1}), N_2 = \sin\theta\sin\varphi = (1/2i)(Y_{11} - Y_{1-1})$ and $N_3 = \cos\theta = Y_{10}$, it is clear that (2.2.44) is equivalent to (2.2.40). □

Example. If Γ is a sphere, then $R_1 = R_2 = R = \text{const}$, $h_{nm} = 0$ for $n > 0$. Therefore, $A_{nm} = 0$ for $n > 0$, $a(\theta, \phi) = A_0 Y_0 = A_0/\sqrt{4\pi}, A_0 = h_0/2, h_0 = 2R\sqrt{4\pi}$. Thus, $a(l) = R$ and the sphere is recovered from the knowledge of the function $h = 2R = \text{const}$.

We proved that (2.2.40) is a necessary condition for $h(\theta, \phi)$ to be the sum of the principal radii of a closed smooth surface Γ. Is this condition also sufficient?

By the algorithm given in Theorem 2.2.4 one can construct $a(\theta, \phi)$ (formula (2.2.41)) and then the surface by formula (2.2.9). It is not guaranteed, however, that this algorithm produces a convex surface.

Lemma 2.2.6. *A necessary and sufficient condition for h to be the sum of the principal radii of a closed convex surface is*

$$(2.2.40) \textit{ holds}, \tag{2.2.45}$$

and

$$\int_{S^2} \frac{(l \cdot l_2)(\nabla h(l) \cdot l_2}{1-(l \cdot l_1)} d\omega \leq 0, \quad l_2 \cdot l_1 = 0 \tag{2.2.46}$$

for all $l_2, l_1 \in S^2$ *such that* $l_2 \cdot l_1 = 0$, *with strict inequality in* (2.2.46) *for some* l_2 *and* l_1. *The symbol* $\nabla h(l)$ *means* $\nabla_y h(y)|_{y=l}$.

A proof of this result can be found in [66]. Since the result is not used in the sequel, its proof is not given here.

One can give another necessary and sufficient condition for h to be the sum of the principal radii of a convex closed smooth surface: (2.2.45) and

$$\text{the matrix } a_{ij}(\theta,\phi) \text{ is semi-positive definite for all } (\theta,\phi), \tag{2.2.47}$$

where

$$a_{ij}(\theta,\phi) = \frac{\partial^2 a(l)}{\partial\alpha_i \partial\alpha_j} = -\sum_{n=0,n\neq 1}^{\infty} \sum_{m=-n}^{n} \frac{h_{nm}}{(n-1)(n+2)} \frac{\partial^2 r Y_{nm}(\theta,\phi)}{\partial y_i \partial y_j}\Bigg|_{r=1}. \tag{2.2.48}$$

A matrix a_{ij} is semi-positive definite if and only if $\sum_{i,j=1}^{n} a_{ij} t_i \bar{t}_j \geq 0$ for all t_i. Condition (2.2.47) follows from (2.2.41) and from the following lemma.

Lemma 2.2.7. *Let* $x(l), l = (\alpha_1, \alpha_2, \alpha_3)$ *be the normal of the surface* Γ, *that is* $x(l)$ *is the point of* Γ *at which the exterior unit normal to* Γ *is* l. *A continuously differentiable function* $x(l)$ *is the normal representation of a convex surface* Γ *if and only if the matrix* $\partial x_i/\partial\alpha_j$ *is a semi-positive definite matrix for all* $l \in S^2$.

Proof. (i) If Γ is convex, then $a(l) = l \cdot x(l), x_j = \partial a(l)/\partial\alpha_j$, $\partial x_i/\partial\alpha_j = \partial^2 a(l)/(\partial\alpha_i\partial\alpha_j) = a_{ij}$, and this matrix is semi-positive definite if Γ is convex. This proves the necessity. (ii) Suppose now that the matrix $\partial x_i/\partial\alpha_j$ is semi-positive definite. Then so is the matrix

a_{ij}. But then the support function $a(l)$ of Γ is convex and therefore Γ is convex. □

2.3. Stability of the Surface with Respect to Small Perturbations of the Data

2.3.1. If $k = k_0$ is so large that the remainder term in formula (2.2.21) satisfies the estimate $O(1/(k_0|l \cdot \nu|)) < 1$, then formula (2.2.21) shows that $f(\nu, n, k_0)/|f(n, \nu, k_0)| = -\exp\{i2k_0 n \cdot l a(l)\}(1 + O(1/(k_0 b))$ where $|l \cdot \nu| \geq b > 0$.

Therefore,

$$a(l) = \frac{1}{2ik_0 n \cdot l} \ln\left\{-\frac{f(\nu, n, k_0)}{|f(\nu, n, k_0)|}\right\} + O\left(\frac{1}{(k_0 b)^2}\right). \tag{2.3.1}$$

Let us formulate the algorithm for recovering the surface from the high-frequency scattering data based on formula (2.3.1).

Theorem 2.3.1. *If the surface is smooth, convex and closed and k_0 is sufficiently large, then the support function of the surface is given by formula* (2.3.1) *and the equation of the surface is given by formula* (2.2.9).

2.3.2. Suppose that the scattering data are measured with some error $\delta > 0$, so that f_δ is measured, and $|f_\delta - f| \leq \delta$. Assume that

$$\sup_{s \in \Gamma} K(s) \leq d^2, \quad |l \cdot \nu| \geq b > 0. \tag{2.3.2}$$

Then

$$f_\delta = -\frac{\exp\{i2kn \cdot la(l)\}}{2K^{1/2}(s_0)}\left[1 + O\left(\frac{1}{kb}\right) + O(\delta d)\right]. \tag{2.3.3}$$

Therefore,

$$a(l) = \frac{1}{2ik_0 n \cdot l} \ln\left\{-\frac{f_\delta(\nu, n, k_0)}{|f_\delta(\nu, n, k_0)|}\right\} + O\left(\frac{1}{(k_0 b)^2} + \frac{\delta d}{k_0 b}\right). \tag{2.3.4}$$

The remainder in formula (2.3.4) depends on the error δ. Therefore, one sees that the error in the measurements is added to the error of the asymptotic formula. The reconstruction algorithm for noisy data is the same as in Theorem 2.3.1, but one uses formula (2.3.4) for $a(l)$ instead of formula (2.3.1). The basic point to discuss is the ill-posedness of the inverse diffraction problem, which comes from the differentiation of the data given with some error.

Let us denote

$$a_\beta(l) = \frac{1}{2ik_0 n \cdot l} \ln \left\{ -\frac{f_\delta(\nu, n, k_0)}{|f_\delta(\nu, n, k_0)|} \right\}. \tag{2.3.5}$$

Then

$$|a_\beta(l) - a(l)| < \beta, \quad \beta = O\left(\frac{1}{(k_0 b)^2} + \frac{\delta d}{k_0 b} \right). \tag{2.3.6}$$

We want to estimate the derivatives $\partial a(l)/\partial \alpha_j$, $l = (\alpha_1, \alpha_2, \alpha_3)$ from the knowledge of $a_\beta(l)$ and β. It is impossible to estimate $\partial a/\partial \alpha_j$ stably without additional information about $a(l)$. This additional information can be taken in the form of the *a priori* assumption

$$|a''(l)| \leq M. \tag{2.3.7}$$

Here, a'' denotes any second derivative of $a(l)$.

The stable estimate $\partial \hat{a}/\partial \alpha_j$ is such estimate that

$$\epsilon(\beta) = \left| \frac{\partial \hat{a}}{\partial \alpha_j} - \frac{\partial a}{\partial \alpha_j} \right| \to 0 \quad \text{as } \beta \to 0. \tag{2.3.8}$$

Under the assumption (2.3.7), one calculates the stable estimate of the derivative $\partial \hat{a}/\partial \alpha_j$ by the formula in which $j = 1$

$$\frac{\partial \hat{a}}{\partial \alpha_j} = \frac{a(\alpha_1 + h(\beta), \alpha_2, \alpha_3) - a(\alpha_1 - h(\beta), \alpha_2, \alpha_3)}{2h(\beta)}. \tag{2.3.9}$$

Here

$$h(\beta) = \left(\frac{2\beta}{M} \right)^{1/2} \tag{2.3.10}$$

and the error of the estimate (2.3.9) is

$$\epsilon(\beta) \leq (2M\beta)^{1/2}. \tag{2.3.11}$$

These results are proved in Appendix D, where it is also proved that the estimate (2.3.9) is the best possible estimate (in a certain sense) among all linear and nonlinear estimates on the set of data specified by the data number (β, M).

To conclude this section, let us summarize the basic result:

An algorithm is given for recovery of the surface from the scattering amplitude measured at one sufficiently high frequency and for the set of n and ν such that the vector $l = (n - \nu)/|n - \nu|$ runs through all of S^2. The measurements are noisy. The algorithm is stable with respect to the noise and to the error due to the asymptotic nature of the high frequency approximation of the scattering amplitude.

Remark. One can solve the inverse diffraction problem with the impedance boundary condition (2.1.6). The argument, based on the Kirchhoff approximation, leads to the following formula for the leading term of the scattering amplitude for the convex impedance obstacle:

$$f_\eta = Bf, \quad \text{where } B = \frac{ik\nu \cdot l + \eta}{-ik\nu \cdot l + \eta}, \tag{2.3.12}$$

where $\eta = \eta(s)$ is the function in the boundary condition (2.1.6), and $f = f(\nu, n, k)$ is given by (2.2.21). Notice that if $\eta = \infty$ (the Dirichlet boundary condition) then $B = 1$; if $\eta = 0$ (the Neumann boundary condition) then $B = -1$. If $|k\nu \cdot l| \gg |\eta|$, then $B = -1$, that is the body scatters as if it were acoustically hard $(\eta = 0)$. Formulas (2.3.12) and (2.2.21) allow one to find both the shape of Γ and the impedance $\eta(s)$ from the scattering data $f(\nu, n, k)$ at a fixed large k and $\nu, n \in S^2$.

Chapter 3

Time-Dependent Problems

3.1. Statement of the Problems

3.1.1. Consider the problem

$$u_{tt} = \nabla^2 u \quad \text{in } \Omega, \quad t > 0, \tag{3.1.1}$$

$$u = 0 \quad \text{on } \Gamma, \tag{3.1.2}$$

$$u(x,0) = 0, \quad u_t(x,0) = f(x), \tag{3.1.3}$$

where $f \in C_0^\infty(\Omega)$.
Define

$$v(x,k) = \int_0^\infty \exp(ikt)u(x,t)dt. \tag{3.1.4}$$

Then

$$\nabla^2 v + k^v = -f \quad \text{in } \Omega, \tag{3.1.5}$$

$$v = 0 \quad \text{on } \Gamma, \tag{3.1.6}$$

$$\int_{S_r} \left| \frac{\partial v}{\partial r} - ikv \right|^2 ds \to 0, \quad r \to \infty, \tag{3.1.7}$$

and

$$u(x,t) = \frac{1}{2\pi} \int_{-\infty}^\infty \exp(-ikt)v(x,k)dk. \tag{3.1.8}$$

The solution to (3.1.5)–(3.1.7) can be written as

$$v(x,k) = \int_\Omega G(x,y,k) f(y)dy, \tag{3.1.9}$$

where $G(x,y,k)$ is the Green function for problem (3.1.5)–(3.1.7). The Green function satisfies the equation

$$G = g - \int_\Gamma ghds, \quad h = \frac{\partial G(s,y,k)}{\partial N_s} \tag{3.1.10}$$

and h solves the equation

$$(I + A(k))h = 2\frac{\partial g}{\partial N_s}, \tag{3.1.11}$$

where $A(k)$ is defined by formula (1.1.4). Since the operator function $A(k)$ is compact and analytic in k on the whole complex plane k, and since $I + A(k)$ is invertible for some k (e.g., for $\operatorname{Im} k > 0$, or for $k > 0$ such that k^2 is not an eigenvalue of the interior Neumann Laplacian), it follows from Lemma 1.4.2 that h is a meromorphic function of k on the whole k-plane and the Laurent coefficients of h are finite rank (degenerate) kernels. The Green function G has the same properties. This follows from (3.1.10) since g is analytic in k. The functions G and h are analytic in k in the half-plane $\operatorname{Im} k \geq 0$ since for $\operatorname{Im} k > 0$, the operator $(I + A(k))^{-1}$ is invertible and on the real line $\operatorname{Im} k = 0$, the Green function has bounded limit values as follows from the limiting absorption principle (Theorem 1.2.3). Therefore, G is analytic in a neighborhood of the real axis. The poles of G in the half-plane $\operatorname{Im} k < 0$ are called complex poles, resonances, and natural frequencies by mathematicians, physicists, and engineers, respectively. These poles play an important role in the study of the behavior of the solution to problem (3.1.1)–(3.1.3) for large t. In order to explain this, let us assume that function (3.1.9) satisfies the estimate

$$\begin{gathered} |v(x,k)| \leq c(b)(1+|k|)^{-a}, \quad a > \tfrac{1}{2}, \\ |\operatorname{Re} k| \to \infty, \quad \operatorname{Im} k = b, \quad -\infty < b < \infty, \end{gathered} \tag{3.1.12}$$

uniformly in x in compact domains, where b is an arbitrary constant. Estimate (3.1.12) and the meromorphic nature of v imply that

$$|\operatorname{Im} k_j| \to \infty \quad \text{as } j \to \infty, \tag{3.1.13}$$

where k_j are the poles of v ordered so that

$$|\operatorname{Im} k_1| \le |\operatorname{Im} k_2| \le \cdots . \tag{3.1.14}$$

The poles of v are among the poles of G, but due to some special choice of f, it may lead to a situation where some of the poles of G are not poles of v. If (3.1.12) holds, then one can move contour in (3.1.8) down and obtain the asymptotic expansion of $u(x,t)$ for large t:

$$\begin{aligned} u(x,t) = &\sum_{j=1}^{N} c_j(x,t)\exp(-ik_j t) \\ &+ O(\exp\{-|\operatorname{Im} k_{N+1}|t\}), \quad t \to +\infty. \end{aligned} \tag{3.1.15}$$

This argument is known as the asymptotic *singularity expansion method* (SEM). The coefficients $c_j(x,t)$ can be written as

$$c_j(x,t) = \operatorname{Res}_{k=k_j}\{v(x,k)\exp(-ikt)\}. \tag{3.1.16}$$

If m_j is the order of the pole k_j, then it follows from (3.1.16) that

$$c_j(x,t) = O(t^{m_j-1}), \quad t \to +\infty. \tag{3.1.17}$$

If the pole k_j is simple, $m_j = 1$, then

$$c_j(x,t) = c_j(x). \tag{3.1.18}$$

In the literature, one can see the formula

$$u(x,t) = \sum_{j=1}^{\infty} c_j(x,t)\exp(-ik_j t). \tag{3.1.19}$$

This formula is not justified and does not hold in general. It is an open problem to find out the cases when (3.1.19) holds. Convergence of the series in (3.1.19) can be understood in various ways. For example, it can be uniform convergence on compacts in $\mathbb{R}^3 \times \mathbb{R}_+$, $x \in \mathbb{R}^3$, $t \in \mathbb{R}_+$.

It is interesting to study the properties of the complex poles. Let us mention some questions of interest: (1) existence of infinitely many poles, (2) behavior of the pole k_j as $j \to \infty$, e.g., property (3.1.13), (3) estimates from below and from above of $\operatorname{Im} k_1$ on the class of obstacles $D \subset B_R$, where R is a fixed number; it is possible that a bound on the curvatures of the surface Γ will enter the estimates, (4) in which cases does (3.1.13) not hold and there exists a sequence of poles k_j such that $\operatorname{Im} k_j \to 0$, $j \to \infty$? in which cases does (3.1.13) not hold but there exists a constant $b > 0$ such that the strip $-b < \operatorname{Im}\, j < 0$ is free from the poles? (5) when does (3.1.12) hold and when does it not hold?

These questions are essentially open although partial answers are known and will be discussed later.

3.1.2. Let us write the integral equation similar to (3.1.10) in the time domain. Applying to (3.1.10), the inverse Fourier transform, (3.1.8), one obtains

$$\mathcal{G}(x, y, t) = \mathcal{G}_0(x, y, t) - \int_\Gamma \frac{ds}{4\pi r_{xs}} \frac{\partial \mathcal{G}(s, y, t - r_{xs})}{\partial N_s}, \tag{3.1.20}$$

where

$$\mathcal{G}_0(x, y, t) = \frac{\delta(r_{xy} - t)}{4\pi r_{xy}}, \tag{3.1.21}$$

$\delta(x)$ is the delta function, and $\mathcal{G} = 0$ for $t < 0$.

The scattering problem in the time domain is obtained by taking the inverse Fourier transform (3.1.8) of the problem (0.0.1)–(0.0.5):

$$\nabla^2 U - U_{tt} = 0 \quad \text{in } \Omega \times \mathbb{R}_+ \tag{3.1.22}$$

$$U(x, t) = 0, \quad x \in \Gamma, \tag{3.1.23}$$

$$U(x, t) = \delta(t - n \cdot x) + V(x, t), \tag{3.1.24}$$

$$r\left(\frac{\partial V}{\partial r} + \frac{\partial V}{\partial t}\right) \to 0, \quad r \to \infty. \tag{3.1.25}$$

Note that this is a problem without initial data. The integral equation (3.1.20) was not used for deriving any qualitative properties of Green's function such as its existence, uniqueness, and behavior at large t.

3.1.3. Consider the problem ($f \in C_0^\infty(\Omega)$):

$$u_{tt} = \nabla^2 u + f(x) \exp(-ik_0 t) \quad \text{in } \Omega, \quad k_0 > 0, \tag{3.1.26}$$

$$u(x, 0) = u_t(x, 0) = 0, \tag{3.1.27}$$

$$u(s, t) = 0 \quad \text{on } \Gamma. \tag{3.1.28}$$

The limiting amplitude principle says that the limit

$$v(x, k_0) = \lim_{T \to \infty} T^{-1} \int_0^T u(x, t) \exp(ik_0 t) dt \tag{3.1.29}$$

exists in an appropriate topology (e.g., uniformly in $x \in B_R$, for any fixed $R > 0$, or in $L^2(B_R)$) and solves the steady state problem (3.1.5)–(3.1.7). The validity of this principle can be established for problem (3.1.26)–(3.1.28). This is done in Section 3.2 in an abstract setting. Of some interest is to give conditions under which the limit (3.1.29) can be substituted by the limit

$$v(x, k_0) = \lim_{t \to \infty} u(x, t) \exp(ik_0 t). \tag{3.1.30}$$

It is interesting to estimate the rate at which $u(x, t)$ approaches the steady state $v(x, k_0) \exp(-ik_0 t)$, that is to estimate

$$u(x, t) - v(x, k_0) \exp(-ik_0 t).$$

If one uses the analytic properties of $v(x, k)$ in the half-plane Im $k < 0$, then this question is a part of the singularity expansion method, and the result is of the type (3.1.15). We also give another method which is based on the spectral theorem for self-adjoint operators.

Some of the basic results and definitions from the spectral theory of self-adjoint operators are given in Appendix B. These results and definitions are used in the rest of this chapter.

3.2. The Limiting Amplitude Principle (Abstract Results)

3.2.1. Let us develop an abstract theory. Let H be a Hilbert space, L be a linear, densely defined in H, closed, not necessarily self-adjoint operator. Let P be an orthoprojection in H onto a certain subspace $F \subset H$. In applications $F = L^2(B_R \setminus D)$, so that F is a subspace of functions with support in a certain compact domain. Let

$$u_{tt} + Lu = f \exp(-ik_0 t), \quad k_0 > 0, \quad f \in H, \tag{3.2.1}$$

$$u(0) = u_t(0) = 0. \tag{3.2.2}$$

The limiting amplitude principle says that the limit in H

$$Pv = \lim_{T \to \infty} T^{-1} \int_0^T Pu(t) \exp(ik_0 t) dt \tag{3.2.3}$$

exists, and v solves the problem

$$Lv - k_0^2 v = f. \tag{3.2.4}$$

In the problems for differential equations, equation (3.2.4) has many solutions, and the requirement that its solution is the limit (3.2.3) of the uniquely solvable problem (3.2.1)–(3.2.2) is one of the ways to select a unique solution to (3.2.4). The other way, which is called the limiting absorption principle, is to solve the problem

$$(L - (k_0 + i\epsilon)^2) v_\epsilon = f, \quad \epsilon > 0 \tag{3.2.5}$$

and to pass to the limit $\epsilon \to +0$

$$Pv = \lim Pv(k_0 + i\epsilon), \quad \epsilon \to +0. \tag{3.2.6}$$

If this limit exists and the element v solves in some sense equation (3.2.4), then the limiting absorption principle holds. In the problems for differential equations, the element v often does not belong to H, but it solves equation (3.2.4) in the sense of the theory of distributions and in H^s_{loc}. The classical way to select a unique solution to (3.2.4) in the case when L is the Dirichlet or Neumann Laplacian in Ω is to use the radiation condition. The limiting amplitude and limiting absorption principles in this case will give the solution which satisfies the radiation condition. The limiting absorption and limiting amplitude principles are more general than the radiation condition. For example, in the cylindrical domain, the radiation condition (3.1.7) is not applicable.

Let us introduce the class of linear operators which we study. Applying the Laplace transform to (3.2.1) yields

$$\begin{aligned} &(L + p^2 I)\tilde{u} = (p + ik_0)^{-1} f, \\ &\qquad \tilde{u} = \int_0^\infty \exp(-pt)u(t)dt, \quad p = \sigma + i\tau. \end{aligned} \tag{3.2.7}$$

Let us formulate the following basic assumptions:

A. The operator $R(-p^2) = (L + p^2 I)^{-1}$ exists, is bounded, and is defined on all of H for all p with $\operatorname{Re} p = \sigma > 0$ except, possibly, for a finite number of points $p_0, \ldots, p_M$, $\operatorname{Re} p_0 > \operatorname{Re} p_1 > \ldots$, which are poles of the operators $R(-p^2)$, and the following estimates hold:

$$\begin{aligned} &||PR(-p^2)f|| \le \frac{c}{1 + |p|^\gamma}, \quad \gamma > 0, \\ &\sigma > 0, \quad |\operatorname{Im} p| = |\tau| > N > 0, \quad c = c(f), \end{aligned} \tag{3.2.8}$$

where N is sufficiently large, γ is a positive number, and the constant c does not depend on p. If $|\tau| \le N$ and $\sigma = 0$, then $PR(-p^2)f$ may have a finite number of points $p = i\omega_m$, $\omega_m \ne 0$ in a neighborhood of which

$$PR(-p^2)f = \frac{q(p, f)}{(p - i\omega_m)^\delta},$$

$$0 \le \delta < 2, \quad 1 \le m \le m_0, \quad \omega_m \ne 0, \tag{3.2.9}$$

where $||q(p, f)|| < c$ if $|p - i\omega_m| \le \epsilon, \sigma > 0, \epsilon > 0$ is a small number, c denotes various constants which do not depend on p. Without loss of generality, it is assumed that δ does not depend on m, and $||PR(-p^2)f|| < c$ if $|p - i\omega_m| > \epsilon$, $\sigma > 0$, $m = 1, 2, \ldots, m_0$, $|\tau| < N$ and $c = c(\epsilon, N)$ does not depend on $\sigma > 0$. If there are no poles of $R(-p^2)$ in the half-plane $\sigma > 0$ and A

holds, then we say that A_0 holds. Assume that $\{\delta = 0$ and $A_0\}$ implies (3.2.11) for $-\infty < \tau < \infty$ and Hölder continuity of $W(\tau)$ in (3.2.11).

B_τ. There exists the following limit in H:

$$\lim_{p\to 0} PR(-(p-i\tau)^2)f = W(\tau), \quad |\arg p| < \frac{\pi}{2}, \tag{3.2.10}$$

where τ is a real number. If this limit exists for all $-\infty < \tau < \infty$, then we say that condition B holds.

C. Suppose that B_τ holds and

$$\begin{aligned} &||PR(-(p-i\tau)^2)f - W(\tau)|| \le c(\tau)|p|^\alpha, \\ &0 < \alpha \le 1, \quad |\arg p| < \frac{\pi}{2}. \end{aligned} \tag{3.2.11}$$

Then we say that condition C_τ holds. If (3.2.11) holds for any $\tau \neq 0$ and $c(\tau) \le c(a)$ if $|\tau| \ge a > 0$, where $a > 0$ is an arbitrary number, then we say that condition C holds.

Remark 3.2.1. The above assumptions and most of our arguments and results below remain valid for linear operators in a Banach space.

3.2.2. In this section, a lemma will be proved. This lemma is used extensively in what follows.

Lemma 3.2.1. *Let $F(p)$ be a function with values in H,*

$$F(p) = \int_0^\infty \exp(-pt)f(t)dt, \quad \sigma > \sigma_0 \ge 0. \tag{3.2.12}$$

Assume that

$$F(p) \text{ is analytic in the half-plane } \sigma > 0, \tag{3.2.13}$$

$$\begin{aligned} &F(p) \text{ satisfies conditions (3.2.8) and (3.2.9)} \\ &\text{with } F(p) \text{ in the place of } PR(-p^2)f, \end{aligned} \tag{3.2.14}$$

and

$$\lim_{p\to 0} p^a F(p) = b, \quad a > 0, 1 + a > \delta, \text{ where } \delta \text{ is defined in (3.2.9).} \tag{3.2.15}$$

Then

$$\lim_{T\to\infty} T^{-a}\int_0^T f(t)dt = \frac{b}{\Gamma(1+a)}, \tag{3.2.16}$$

where $\Gamma(t)$ *is the Gamma function. If*

$$\lim_{p\to 0}\sup ||p^a F(p)|| \le c, \tag{3.2.17}$$

then

$$\lim_{T\to\infty}\sup \left\| T^{-a}\int_0^T f(t)dt\right\| \le c_1. \tag{3.2.18}$$

Remark 3.2.2. Lemma 3.2.1 is a Tauberian lemma in the sense that some conclusion about $f(t)$ is obtained from the assumption about its Laplace's transform. But it is an unusual Tauberian lemma since the known Tauberian theorems have different logical structure: one assumes something about the integral transform of $f(t)$ and about $f(t)$ itself. These additional assumptions on f are called Tauberian conditions. For example, one assumes often that $f(t) > -c$ where c is a constant (see, e.g., [80]). In Lemma 3.2.1, there are no assumptions on $f(t)$. All the assumptions are on $F(p)$. This is the unusual feature of Lemma 3.2.1.

Proof of Lemma 3.2.1. If $F(p)$ satisfies conditions (3.2.13) and (3.2.8), then $F(p)$ is a Laplace transform of a function $f(t) \in L^1(0,\infty)$, and

$$\int_0^T f dt = \frac{1}{2\pi i}\int_{c-i\infty}^{c+i\infty} \frac{\exp(pT)-1}{p}F(p)dp, \quad c>0. \tag{3.2.19}$$

Let us set $c = T^{-1}$ and $y = pT$. Then

$$\begin{aligned} I(T) &= \frac{1}{T^a}\int_0^T f(t)dt \\ &= \frac{1}{2\pi i T^a}\int_{T^{-1}-i\infty}^{T^{-1}+i\infty} \frac{\exp(pT)-1}{p}F(p)dp \\ &= \frac{1}{2\pi i}\int_{t-i\infty}^{t+i\infty} \frac{\exp(y)-1}{y^{1+a}}\left(\frac{y}{T}\right)^a F\left(\frac{y}{T}\right)dy. \end{aligned} \tag{3.2.20}$$

Let $T \to +\infty$. Condition (3.2.15) says that for any fixed y, the integrand in (3.2.20) tends to $((\exp(y)-1)/y^{1+a})b$. We want to apply Lebesgue's dominated convergence theorem in order to pass to the limit $T \to \infty$ in (3.2.20). To check the basic condition of this theorem, assume first that $\delta = 0$ in the estimate (3.2.9) for $F(p)$, and consider the inequality,

$$\left\| \frac{\exp(y)-1}{y^{1+a}} \left(\frac{y}{T}\right)^a F\left(\frac{y}{T}\right) \right\| = \left\| \frac{\exp(y)-1}{y^{1+\gamma_1}} \frac{1}{T^{a-\gamma_1}} \left(\frac{y}{T}\right)^{\gamma_1} F\left(\frac{y}{T}\right) \right\| \leq \frac{c}{|y|^{1+\gamma_1}}, \tag{3.2.21}$$

where estimate (3.2.8) for $F(p)$ was used in the form

$$||p^{\gamma_1} F(p)|| \leq c, \tag{3.2.22}$$

and γ_1 is an arbitrary number in the interval $0 < \gamma_1 < \min(\gamma, a)$. It follows from (3.2.21) that the Lebesgue's theorem is applicable to the integral (3.2.20). This theorem yields

$$\lim_{T\to\infty} I(T) = \frac{b}{2\pi i} \int_{1-i\infty}^{1+i\infty} \frac{\exp(y)-1}{y^{1+a}} dy = \frac{b}{\Gamma(1+a)}, \tag{3.2.23}$$

where $\Gamma(a)$ is the Gamma function, and the integral is computed by the residue theorem. Thus, Lemma 3.2.1 is proved under the assumption that $\delta = 0$, where δ is defined in (3.2.9). If $1 < \delta < 2$, then the estimate (3.2.22) does not hold in neighborhoods of the points $y_m = 1 + i\omega_m T$. In this case, we write

$$I(T) = I_0(T, \epsilon) + \sum_{m=1}^{m_0} I_m(T, \epsilon), \tag{3.2.24}$$

where I_m is the part of the integral (3.2.20) taken over the contour $|T^{-1}y - i\omega_m| < \epsilon$, $\operatorname{Re} y = 1$, $\epsilon > 0$ is a fixed, sufficiently small number, $\epsilon < \min_{m \neq m'} |\omega_m - \omega_{m'}|$, and $I_0(T, \epsilon)$ is defined by (3.2.24). The integrand in $I_0(T)$ can be majorized as in (3.2.21). To estimate

$I_m(T, \epsilon)$, one writes

$$||I_m|| = \left\| \frac{1}{2\pi i T^a} \int_{|T^{-1}+i\tau-i\omega_m|<\epsilon} \frac{\exp(pt)-1}{p} F(p)dp \right\|$$
$$\leq \frac{c}{T^a} \int_{-\epsilon}^{\epsilon} \frac{dx}{(T^{-2}+x^2)^{\delta/2}} \leq \frac{c(\delta)}{T^{a+1-\delta}},$$
$$1 < \delta < \min(1+a, 2), \tag{3.2.25}$$

where $c(\delta)$ is a constant which does not depend on ϵ and T. Therefore, if $1 < \delta < \min(1+a, 2)$, then

$$\lim_{T\to\infty} I(T) = \lim_{T\to\infty} I_0(T,\epsilon) + \sum_{m=1}^{m_0} \lim_{T\to\infty} I_m(T,\epsilon) = \frac{b}{\Gamma(1+a)}. \tag{3.2.26}$$

Note that although I_0 and I_m depend on ϵ, their limits, as $T \to \infty$, do not.

Consider now the case of $0 < \delta < 1$. In this case,

$$||I_m|| < \frac{c}{T^a} \int_{-\epsilon}^{\epsilon} \frac{dx}{(T^{-2}+x^2)^{\delta/2}} \leq \frac{c(\delta)}{T^a}, \quad a > 0. \tag{3.2.27}$$

Finally, if $\delta = 1$, then

$$||I_m|| \leq \frac{c \ln T}{T^a}, \quad a > 0. \tag{3.2.28}$$

Given (3.2.27) or (3.2.28), one obtains (3.2.26) as before. The same argument shows that (3.2.17) implies (3.2.18). □

3.2.3. Let us formulate the main results which establish the relationship between the spectral properties of the operator L and the behavior of the solution to the nonstationary problem as $t \to +\infty$. In this section, we drop the projection P in front of $w(t)$ and $\bar{w}(p, h)$ in the proof of Theorem 3.2.2 below. This is possible if the analytic properties of $\bar{w}(p, h)$ and $P\bar{w}(p, h)$ are the same. In the problems for differential equations, the operator P is the operator of multiplication by a characteristic function of a compact set, $P\bar{w} = \eta\bar{w}, \eta = 1$ if $|x| \leq R, \eta = 0$ if $|x| > R$. Therefore, $\bar{w}(p)$ and $\eta\bar{w}(p)$ have the same analytic properties as functions of p.

Theorem 3.2.2. *Let the assumption A hold. The operator L does not have eigenvalues in the half-plane* Re $\lambda < 0$ *if and only if*

$$\left\| \int_0^t w(t)dt \right\| = O(\exp(\epsilon t)) \quad \text{as } t \to +\infty, \quad \epsilon > 0, \tag{3.2.29}$$

where $w(t)$ solves the problem

$$w_{tt} + Lw = 0, \quad t > 0, \quad w(0) = 0, \quad w_t(0) = h, \tag{3.2.30}$$

h runs through a dense set S in H, and $\epsilon > 0$ is an arbitrarily small number.

If $L = L^*$ is self-adjoint and $L \geq 0$, then a sufficient condition for L not to have an eigenvalue at the point k_0^2, $k_0 > 0$, is

$$\left\| \int_0^t w(t) \exp(ik_0 t)dt \right\| = O(t), \quad \text{as } t \to +\infty. \tag{3.2.31}$$

A necessary condition for $L \geq 0$ to have no positive eigenvalues is

$$\left\| \int_0^t w(t)dt \right\| = O(1), \quad \text{as } t \to +\infty. \tag{3.2.32}$$

Proof. (1) If h runs through a dense set of H, then the element $R(-p^2)h$ and the operator function $R(-p^2)$ have the same analytic properties as functions of p. The Laplace transform of (3.2.30) yields

$$(L + p^2 I)\bar{w} = h, \quad \bar{w} = \bar{w}(p, h) = (L + p^2 I)^{-1}h. \tag{3.2.33}$$

If λ_0 is an eigenvalue of L, then the resolvent $(L - \lambda I)^{-1}$ has a pole at $\lambda = \lambda_0$. Set $\lambda = -p^2$, $p = -i\sqrt{\lambda}$, where $\sqrt{\lambda} = |\lambda|^{1/2}$ $\exp(i \arg\lambda/2), 0 < \arg\lambda < 2\pi$. Suppose λ_0 is an eigenvalue of L, Re $\lambda_0 < 0$. Then Re $p = \text{Re}(-i\sqrt{\lambda_0}) > 0$, and $\bar{w}(p, h)$ has a pole at $p = p_0$ for some $h \in S$. Suppose for simplicity that p_0 is the pole with maximal real part. Then, if $\delta >$Re p_0 and Re $p_1 <$Re p_0, one has

$$\begin{aligned}\left\|\int_0^t w(t)dt\right\| &= \left\|\frac{1}{2\pi i}\int_{c_0-i\infty}^{c_0+i\infty}\frac{\exp(pt)-1}{p}\bar{w}(p,h)dp\right\|\\ &= \left\|\frac{1}{2\pi i}\int_{c_1-i\infty}^{c_1+i\infty}\frac{\exp(pt)-1}{p}\bar{w}(p,h)dp\right.\\ &\quad\left.+\operatorname{Res}_{p=p_0}\frac{\exp(pt)-1}{p}\bar{w}(p,h)\right\|\\ &\geq \exp(t\operatorname{Re}p_0)d_0 + O(e^{t\operatorname{Re}p_1}) \text{ as } t\to+\infty, \qquad (3.2.34)\end{aligned}$$

where

$$d_0 = \operatorname{Res}_{p=p_0}\frac{\bar{w}(p,h)}{p},$$
$$d_0 \neq 0.$$

Therefore, condition (3.2.29) does not hold if $\operatorname{Re}p_0 > 0$. This proves that condition (3.2.29) is sufficient for the absence of the eigenvalues of L in the half-plane $\operatorname{Re}\lambda < 0$. To prove the necessity of (3.2.29), let us assume that L has no eigenvalues in the half-plane $\operatorname{Re}\lambda < 0$. This is equivalent to the assumption that $\bar{w}(p,h)$ has no poles in the half-plane $\operatorname{Re}p > 0$. Therefore, the number c_0 in (3.2.34) can be an arbitrarily small positive number, and, using (3.2.8), one obtains (3.2.29).

(2) If L is self-adjoint, then its eigenvalues are real numbers. Suppose $\lambda_0 = k_0^2, k_0 > 0$, is an eigenvalue of L. Then $p = -ik_0$ is a pole of $\bar{w}(p,h)$ for some h, and the function $\overline{w(t)\exp(ik_0t)} = \bar{w}(p-ik_0,h)$ has a pole at $p=0$.

Since L is self-adjoint, the pole $p=0$ is simple. Therefore,

$$\lim_{p\to0} pF(p) = b, \quad F(p) = \bar{w}(p-ik_0,h), \qquad (3.2.35)$$

where $b \neq 0, b \notin H$. The assumption of Lemma 3.2.1 holds for $\bar{w}(p-ik_0,h)$ with $\sigma_0 = 0$ (since $L \geq 0$), $\delta = 1$, and $a = 1$ (see (3.2.35)). Therefore,

$$\lim_{t\to\infty}\frac{1}{t}\int_0^t w(t)\exp(ik_0t)dt = b. \qquad (3.2.36)$$

Thus, if (3.2.31) holds, then $b = 0$ and $\lambda_0 = k_0$ is not a pole of $(L + p^2 I)^{-1}$. Therefore, $\lambda_0 = k_0$ is not an eigenvalue of L.

Finally, assume that $L \geq 0$ has no positive eigenvalues. Then $\delta = 0$ in (3.2.9) and one has

$$\begin{aligned}\left\| \int_0^t w(t)dt \right\| &= \left\| \frac{1}{2\pi i} \int_{-i\infty}^{+i\infty} \frac{\exp(pt) - 1}{p} \bar{w}(p,h)dp \right\| \\ &\leq c \int_{|\tau|>N} \frac{|\exp(pt) - 1|}{|p|(1+|p|)^\gamma} |dp| \\ &\quad + \frac{1}{2\pi} \left\| \int_{|\tau|<N} \frac{\exp(i\tau t) - 1}{\tau} W(\tau)d\tau \right\| < \infty. \end{aligned} \tag{3.2.37}$$

Thus, (3.2.32) holds. □

3.2.4. *Here, we prove that the limiting absorption principle is equivalent to the limiting amplitude principle defined in* (3.2.3). We also give an estimate of the remainder of the form

$$Pv = \frac{1}{T} \int_0^T \exp(ik_0 t) Pu(t)dt + O\left(\frac{1}{T^\alpha}\right), \tag{3.2.38}$$

where α is defined in (3.2.11).

Theorem 3.2.3. *Let the assumption A hold. Then for the limiting amplitude principle* (3.2.3) *to hold, it is necessary and sufficient that assumptions A_0 and B_{k_0} hold. If in addition assumption C_{k_0} holds and $\delta < 2 - \alpha$, where δ is defined in* (3.2.9) *and α in* (3.2.11), *then* (3.2.38) *holds.*

Remark 3.2.3. If one defines the limiting amplitude principle by the formula

$$Pv = \exp(ik_0 t) Pu(t) + o(1), \quad \text{as } t \to +\infty, \tag{3.2.39}$$

then the limiting absorption principle is not equivalent to the limiting amplitude principle. This will be clear from the proof of Theorem 3.2.3, and can be illustrated as follows. Let $u =$

$\exp(-it) + \exp(-ik_0t), k_0 \neq 1$. Then the function $\exp(ik_0t)u = 1+\exp\{i(k_0-1)t\}$ does not have a limit as $t \to +\infty$, but the function $T^{-1}\int_0^T \exp(ik_0t)u(t)dt \to 1$ as $T \to +\infty$.

To prove Theorem 3.2.3, we need the following well-known (Abelian type) lemma.

Lemma 3.2.4. *If $f \in L^1_{\text{loc}}$, and*

$$\lim_{T\to+\infty} T^{-1}\int_0^T f(t)dt = b, \tag{3.2.40}$$

then

$$\lim_{p\to 0} p\int_0^\infty \exp(-pt)f(t)dt = b, \quad |\arg p| < \frac{\pi}{2}. \tag{3.2.41}$$

Proof. One has

$$\begin{aligned} p\int_0^\infty \exp(-pt)f(t)dt &= p\exp(-pt)\int_0^t f dt\Big|_0^\infty \\ &\quad + p^2\int_0^\infty \exp(-pt)tt^{-1}\int_0^t f dt\, dt \\ &= p^2\int_0^\infty \exp(-pt)t\left(t^{-1}\int_0^t f dt\right)dt, \end{aligned}$$

$$|\arg p| < \frac{\pi}{2}, \quad p \neq 0. \tag{3.2.42}$$

Furthermore, if $b = \text{const}$, then

$$\lim_{p\to 0} p^2\int_0^\infty \exp(-pt)tbdt = b. \tag{3.2.43}$$

Let $h(t) = t^{-1}\int_0^t f(t)dt - b$. Then $\lim_{t\to+\infty} h(t) = 0$, and $h \in L^1_{\text{loc}}$.

The desired conclusion (3.2.41) follows from (3.2.42), (3.2.43), and (3.2.40) if one shows that

$$\lim_{t\to+\infty} h(t) = 0 \tag{3.2.44}$$

implies

$$\lim_{p\to 0} p^2 \int_0^\infty \exp(-pt)th(t)dt = 0, \quad |\arg p| < \frac{\pi}{2}. \tag{3.2.45}$$

Let us prove (3.2.45). Fix $t(\epsilon)$ such that $|h(t)| < \epsilon$ if $t \geq t(\epsilon)$. Clearly,

$$\lim_{p\to 0} p^2 \int_0^{t(\epsilon)} \exp(-pt)th(t)dt = 0. \tag{3.2.46}$$

Moreover, if Re $p = \sigma = |p|\cos\beta$, $|\beta| < \pi/2$, then

$$\begin{aligned}\left| p^2 \int_{t(\epsilon)}^\infty \exp(-pt)th(t)dt \right| &\leq \epsilon|p|^2 \int_{t(\epsilon)}^\infty \exp(-\sigma t)dt \\ &\leq \epsilon|p|^2\sigma^{-1} \leq \frac{\epsilon}{\cos^2\beta}.\end{aligned} \tag{3.2.47}$$

Since $\epsilon > 0$ is arbitrary, (3.2.46) and (3.2.47) imply (3.2.45). □

Proof of Theorem 3.2.3. *Necessity.* Assume (3.2.3). Then, by Lemma 3.2.4, the following limit exists:

$$\begin{aligned}Pv &= \lim_{p\to 0} p \int_0^\infty \exp(-pt)Pu(t)\ \exp(ik_0 t)dt \\ &= \lim_{p\to 0} Pp\frac{v(p-ik_0)}{p} = \lim_{p\to 0} Pv(p-ik_0).\end{aligned} \tag{3.2.48}$$

Therefore, condition B_{k_0} holds. Condition A_0 holds as well. Indeed, if (3.2.3) holds, then

$$\left\| \int_0^t Pu(t)\ \exp(ik_0 t)dt \right\| = O(t), \quad \text{as } t \to +\infty. \tag{3.2.49}$$

Moreover,

$$\overline{Pu(t)\ \exp(ik_0 t)} = \frac{1}{p} PR(-(p-ik_0)^2)f. \tag{3.2.50}$$

If condition A_0 does not hold, then there exists at least one pole p_0, $\mathrm{Re}\, p_0 > 0$, of the resolvent $R(-p^2)$. Therefore, as in the proof of Theorem 3.2.2, one can show that for some f,

$$\left\| \int_0^t Pu(t) \exp(ik_0 t) dt \right\| \geq d \, \exp(t \, \mathrm{Re}\, p_0), \quad d > 0,$$

which contradicts (3.2.49). This contradiction proves that A_0 holds.

Sufficiency. Assume A_0 and B_{k_0}. Consider (3.2.50). Condition B_{k_0} and (3.2.50) imply that the following limit exists:

$$\lim_{p \to 0} p\overline{Pu(t) \exp(ik_0 t)} = \lim_{p \to 0} PR(-(p - ik_0)^2) f, \quad |\arg p| < \frac{\pi}{2}. \tag{3.2.51}$$

Equation (3.2.51) and condition A_0 show that the assumptions of Lemma 3.2.1 hold.

It follows from (3.2.51) and Lemma 3.2.1 that (3.2.3) holds.

(3) Let us prove the last conclusion of Theorem 3.2.3. Suppose that A_0, B_{k_0} and C_{k_0} hold, and $\delta < 2 - \alpha$. Then, from (3.2.50) and (3.2.11), one obtains

$$\begin{aligned} \|\overline{Pu(t) \exp(ik_0 t) - W(k_0)}\| &= \left\| \frac{PR(-(p - ik_0)^2) f - W(k_0)}{p} \right\| \\ &\leq c(k_0) |p|^{\alpha - 1}. \end{aligned} \tag{3.2.52}$$

Therefore,

$$\lim_{p \to 0} p^{1-\alpha} \|\overline{Pu(t) \exp(ik_0 t) - W(k_0)}\| \leq c. \tag{3.2.53}$$

Inequality (3.2.53) implies condition (3.2.17) of Lemma 3.2.1 with $a = 1 - \alpha$. Inequality $\delta < 2 - \alpha$ says that condition $1 + a > \delta$ in (3.2.15) is satisfied. Thus, inequality (3.2.53) and Lemma 3.2.1 yield

$$\overline{\lim_{p \to 0}} \left\| \frac{1}{T^{1-\alpha}} \int_0^T [Pu(t) \exp(ik_0 t) - W(k_0)] dt \right\| \leq c. \tag{3.2.54}$$

This inequality is equivalent to (3.2.38) since

$$W(k_0) = \lim_{p\to 0} Pv(p - ik_0) = Pv, \quad |\arg p| < \frac{\pi}{2}. \tag{3.2.55}$$

This completes the proof. □

3.2.5. The limiting amplitude principle for some self-adjoint operators in a Hilbert space H can be justified with the help of the spectral theorem. Let $L \geq 0$ be a self-adjoint operator in H, and E_λ be its resolution of identity. The solution to problem (3.2.1)–(3.2.2) can be written as

$$u = \int_0^\infty a(t,\lambda)(\lambda - k_0^2)^{-1} dE_\lambda f, \quad k_0 > 0, \tag{3.2.56}$$

where the integral is understood as the principal value and

$$\begin{aligned} a(t,\lambda) &= \exp(-ik_0 t) - \cos(t\lambda^{1/2}) + ik_0\lambda^{-1/2}\sin(t\lambda^{1/2}) \\ &= \exp(-ik_0 t) + b(t,\lambda). \end{aligned} \tag{3.2.57}$$

It is clear that (3.2.56) satisfies the initial condition (3.2.2) and equation (3.2.1). When checking that (3.2.56) solves (3.2.1), one uses the identity $L\int_0^\infty h(\lambda)dE_\lambda f = \int_0^\infty \lambda h(\lambda)dE_\lambda f$. Formula (3.2.56) is the solution to (3.2.1)–(3.2.2) by Fourier's method. From the spectral theory, one has $dE_\lambda f = \operatorname{Im} \pi^{-1} R(\lambda + i0) f d\lambda$. Therefore,

$$\begin{aligned} PdE_\lambda f &= \operatorname{Im} \frac{1}{\pi} PR(\lambda + i0) f d\lambda = h(\lambda, f) d\lambda, \\ R(\lambda + i0) f &= (L - \lambda - i0)^{-1} f, \end{aligned} \tag{3.2.58}$$

where P is the orthoprojection on a subspace F of H. It has the same meaning as in Section 2.3. The function $h(\lambda, f), a < \lambda < b$, is defined almost everywhere if $||PR(\lambda \pm i\epsilon)f|| \leq c$ where $c =$ const does not depend on ϵ, and $\lambda\epsilon(a,b)$. Indeed, in this case, the weak limits $v_\pm = PR(\lambda \pm i0)f$ exist almost everywhere in (a,b).

Assume that

$$\int_0^\infty (1+\lambda)^{-1} ||h(\lambda, f)|| (1 + \lambda^{1/2})\lambda^{-1/2} d\lambda < \infty. \tag{3.2.59}$$

This assumption means that $||h(\lambda)||\lambda^{-1}$ is integrable near infinity, and $||h(\lambda)||\lambda^{-1/2}$ is integrable near zero. This allows one to use the Riemann–Lebesgue lemma which yields:

$$\lim_{t\to\infty}\left\|\int_{|\lambda-k_0^2|>\epsilon} b(t,\lambda)\frac{dE_\lambda f}{\lambda-k_0^2}\right\| = 0, \tag{3.2.60}$$

where $b(t,\lambda)$ is defined in (3.2.57).

Let us assume that

$$||h(\lambda)-h(k_0^2)|| \le c|\lambda-k_0^2|^\alpha, \quad 0<\alpha\le 1. \tag{3.2.61}$$

Then

$$\begin{aligned}\int_{|\lambda-k_0^2|\le\epsilon} & b(t,\lambda)\frac{dE_\lambda f}{\lambda-k_0^2} \\ &= \int_{|\lambda-k_0^2|\le\epsilon} b(t,\lambda)\frac{h(\lambda)-h(k_0^2)}{\lambda-k_0^2}d\lambda + h(k_0^2) \\ &\quad\times\int_{|\lambda-k_0^2|\le\epsilon}\frac{b(t,\lambda)d\lambda}{\lambda-k_0^2} = J_1+J_2.\end{aligned} \tag{3.2.62}$$

It follows from (3.2.61) and the Riemann–Lebesgue lemma that

$$||J_1|| \to 0 \quad \text{as } t\to+\infty. \tag{3.2.63}$$

One can check that

$$\lim_{t\to+\infty}\int_a^b \frac{\exp(ixt)}{x}dx = i\pi, \quad a<0, \quad b>0. \tag{3.2.64}$$

Therefore, if $||h(x)-h(\lambda)|| \le c|\lambda-x|^\alpha$, $0<\alpha\le 1$, then

$$\lim_{t\to+\infty}\int_a^b \frac{\exp\{i(\lambda-x)t\}}{\lambda-x}h(x)dx = i\pi h(\lambda), \quad a<\lambda<b. \tag{3.2.65}$$

Consider the integral

$$J = \int_{|\lambda-k_0^2|\le\epsilon}\frac{b(t,\lambda)d\lambda}{\lambda-k_0^2}$$

$$= \int_{|\lambda-k_0^2|\le\epsilon} \left[-\frac{\exp(it\lambda^{1/2}) + \exp(-it\lambda^{1/2})}{2} \right.$$

$$\left. + \frac{ik_0\lambda^{-1/2}}{2i}(\exp(it\lambda^{1/2}) - \exp(-it\lambda^{1/2})) \right] \frac{d\lambda}{\lambda - k_0^2}$$

$$= \int_{|\lambda-k_0^2|\le\epsilon} \frac{d\lambda}{\lambda - k_0^2} \left[\frac{\exp(it\lambda^{1/2})}{2}(-1 + k_0\lambda^{-1/2}) \right.$$

$$\left. - \frac{\exp(-it\lambda^{1/2})}{2}(1 + k_0\lambda^{-1/2}) \right] d\lambda$$

$$= \exp(-ik_0t) \int_{|\lambda-k_0^2|\le\epsilon} \frac{d\lambda}{\lambda - k_0^2} \left[-\frac{\exp\{it(\lambda^{1/2} + k_0)\}}{2} \right.$$

$$\left. \times (1 - k_0\lambda^{-1/2}) - \frac{\exp\{i(k_0 - \lambda^{1/2})t\}}{2}(1 + k_0\lambda^{1/2}) \right]. \tag{3.2.66}$$

It follows from (3.2.65) and (3.2.66) that

$$\lim_{t\to+\infty} \{J \ \exp(ik_0t)\} = \frac{1}{2k_0} \lim_{t\to+\infty} \int_{|\lambda-k_0^2|\le\epsilon} \frac{d\lambda \ \exp\{i(k_0 - \lambda^{1/2})t\}}{k_0 - \lambda^{1/2}} = i\pi. \tag{3.2.67}$$

Combining (3.2.56), (3.2.57), (3.2.59)–(3.2.63), (3.2.65) and (3.2.67), one obtains

$$\lim_{t\to+\infty} \exp(ik_0t)Pu = \int_0^\infty \frac{h(\lambda, f)d\lambda}{\lambda - k_0^2} + i\pi h(k_0^2, f)$$

$$= \int_0^\infty \frac{PdE_\lambda f}{\lambda - k_0^2 - i0}, \tag{3.2.68}$$

where $h(\lambda, f)$ is defined in (3.2.58), the well-known formula

$$(x - i0)^{-1} = x^{-1} + i\pi\delta(x) \tag{3.2.69}$$

is used, the distribution x^{-1} is understood as Cauchy's principal value, i.e., $\int_{-\infty}^{\infty} x^{-1}h(x)dx = \lim_{\epsilon\to 0}\int_{|x|>\epsilon} x^{-1}h(x)dx$, and $h(\lambda) = h(\lambda, f)$ is assumed to be Hölder-continuous at $\lambda = k_0^2$ (see (3.2.61)).

It follows from the spectral theorem that

$$\int_0^\infty \frac{PdE_\lambda f}{\lambda - k_0^2 - i0} = \lim_{\epsilon\to+0} PR(k_0^2 + i\epsilon)f = PR(k_0^2 + i0)f. \tag{3.2.70}$$

Therefore, formula (3.2.68) can be written as

$$Pu(t) = \exp(-ik_0 t)v_+ + o(1), \quad t \to +\infty, \tag{3.2.71}$$

where

$$v_+ = PR(k_0^2 + i0)f \tag{3.2.72}$$

is the solution to the stationary problem

$$(K - k_0^2)v_+ = f \tag{3.2.73}$$

obtained by the limiting absorption principle. This means that we call v_+ the solution to (3.2.73) if and only if

$$v_+ = \lim_{\epsilon\to+0} PR(k_0^2 + i\epsilon)f. \tag{3.2.74}$$

Let us study the dependence of the asymptotics of $Pu(t)$ as $t \to +\infty$ on the singularity of $h(\lambda)$ at $\lambda = 0$. Condition (3.2.59) allows $||h(\lambda)||$ to be $O(\lambda^{-a}), a < \frac{1}{2}$.

Let us assume that in a neighborhood of $\lambda = 0$ the function $h(\lambda)$ behaves as follows:

$$h(\lambda) = h_0\lambda^{-a} + h_1(\lambda), \quad 1 > a \geq \tfrac{1}{2}, \quad 0 < \lambda \leq \lambda_0, \tag{3.2.75}$$

where $h_1(\lambda)$ satisfies inequality (3.2.59), λ_0 is a small number, $0 < \lambda_0 < k_0^2, h_0 \in H$ is a constant, i.e., an element which does not depend on λ. The arguments in Section 3.2.4 remain valid except for the conclusion (3.2.60). This conclusion is no longer valid because inequality (3.2.59) is not valid now. The term $h_0\lambda^{-a}$ produces the new expression for the limiting behavior of the integral

$$I(t,a) = \int_{|\lambda-k_0^2|>\epsilon} b(t,\lambda)h(\lambda)d\lambda$$
$$= h_0 \int_0^{\lambda_0} ik_0\lambda^{-1/2-a}\sin(\lambda^{1/2}t)d\lambda + o(1) \quad \text{as } t \to +\infty. \tag{3.2.76}$$

The term $o(1)$ in (3.2.76) is obtained as before from the Riemann–Lebesgue lemma. But, if $\frac{1}{2} \le a < 1$, the integral

$$I_0(t,a) = \int_0^{\lambda_0} \lambda^{-1/2-a}\sin(\lambda^{1/2}t)d\lambda \tag{3.2.77}$$

does not tend to zero as $t \to +\infty$. If $a = \frac{1}{2}$, then

$$I_0\left(t,\frac{1}{2}\right) = \int_0^{\lambda_0} \frac{\sin(\lambda^{1/2}t)}{\lambda}d\lambda = 2\int_0^{\lambda_0^{1/2}t} \frac{\sin x}{x}dx \xrightarrow[t\to\infty]{} \pi. \tag{3.2.78}$$

From (3.2.76) and (3.2.78), it follows that

$$I(t,\tfrac{1}{2}) = ik_0\pi h_0 + o(1), \quad t \to +\infty. \tag{3.2.79}$$

If $\frac{1}{2} < a < 1$, one has

$$I_0(t,a) = 2t^{2a-1}\int_0^{\lambda_0^{1/2}t} \frac{\sin x}{x^{2a}}dx \sim I_0 t^{2a-1}, \text{ as } t \to +\infty, \tag{3.2.80}$$

where (see [75, formula 3.761.4])

$$I_0 = 2\int_0^{\infty} \frac{\sin x}{x^{2a}}dx = 2\Gamma(1-2a)\cos\pi a, \quad \frac{1}{2} < a < 1. \tag{3.2.81}$$

Therefore,

$$I(t,a) = ik_0 I_0 h_0 t^{2a-1} + o(1), \text{ as } t \to +\infty, \quad \tfrac{1}{2} < a < 1, \tag{3.2.82}$$

where the constant I_0 is given by (3.2.81).

It follows from (3.2.79) and (3.2.81) that

$$Pu(t) = \exp(-ik_0 t)v_+ + I(t,a), \quad \tfrac{1}{2} \le a < 1, \tag{3.2.83}$$

where $I(t,a)$ is given by formula (3.2.76)–(3.2.82), and the assumption (3.2.75) holds.

Let us summarize our results.

Theorem 3.2.5. *Let $L \ge 0$ be a self-adjoint operator on a Hilbert space, and P be the orthoprojection on a subspace F. If assumptions* (3.2.59) *and* (3.2.61) *hold, then the solution $u(t)$ to problem* (3.2.1)–(3.2.2) *has asymptotics* (3.2.71) *where v_+ is defined by* (3.2.72). *If assumptions* (3.2.75) *and* (3.2.61) *hold, then $u(t)$ is given by* (3.2.83), *where the asymptotics of $I(t,a)$ is given by* (3.2.76)–(3.2.82).

3.2.6. In the proof of Theorems 3.2.3 and 3.2.5, we used conditions (3.2.8) and (3.2.59) which give some decay as $|p| \to \infty$ and $\lambda \to +\infty$. If L is self-adjoint and the absolutely continuous subspace H_{ac} of L is the whole space $H = L^2(\Omega)$, then one can drop condition (3.2.59). It is difficult to prove an estimate of the form

$$||h(\lambda, f)|| \le c(1+|\lambda|)^{-\gamma}, \quad \gamma > 0, \tag{3.2.84}$$

which is sufficient for (3.2.59) to hold. Since it is possible to establish absolute continuity of the spectrum of the Dirichlet or Neumann Laplacian in the exterior domain, it is important to be able to drop assumption (3.2.59) and substitute it by absolute continuity of the spectrum and the equality $H_{\text{ac}} = H = L^2(\Omega)$. Let us show that the following theorem holds.

Theorem 3.2.6. *Let $L \ge 0$ be a self-adjoint operator, $H_{\text{ac}} = H$ where H_{ac} is the absolutely continuous subspace of H, and assumption* (3.2.61) *holds, then the conclusion Theorem* 3.2.5 *holds.*

Proof. Condition (3.2.59) was used when we referred to the Riemann–Lebesgue lemma. The conclusion of this lemma holds for the integrals of the type (3.2.60) and (3.2.76). For the Riemann–Lebesgue lemma to hold it is sufficient, for example, that

$$\int_0^\infty ||h(\lambda, f)||^2 d\lambda < \infty. \tag{3.2.85}$$

If $H_{\text{ac}} = H$, then

$$\int_0^\infty ||h(\lambda, f)||^2 d\lambda = \int_0^\infty (dE_\lambda f, f) < \infty. \tag{3.2.86}$$

Therefore, the Riemann–Lebesgue lemma is applicable. The rest of the proof is the same as in the proof of Theorem 3.2.5. □

3.3. The Limiting Amplitude Principle for the Laplacian in Exterior Domains

3.3.1. Let us apply the result in Section 3.2 to the Dirichlet Laplacian in Ω. The assumption about Ω and notations are the same as in Chapter 1. Thus, consider the problem

$$\begin{aligned} u_{tt} + Lu &= u_{tt} - \nabla^2 u \\ &= f(x)\ \exp(-ik_0 t) \quad \text{in } \Omega \subset \mathbb{R}^3, \quad t > 0, k_0 > 0, \end{aligned} \tag{3.3.1}$$

$$u = 0 \quad \text{on } \Gamma, \tag{3.3.2}$$

$$u(x, 0) = u_t(x, 0) = 0, \quad x \in \Omega, \tag{3.3.3}$$

where L stands for the self-adjoint realization of the Dirichlet Laplacian in $L^2(\Omega)$.

Let us assume that $f(x) \in C_0^\infty(\Omega)$. A much weaker assumption on f would be sufficient, e.g.,

$$f(x) \in L^2_{\text{loc}} \text{ and } |f(x)| \le c(1 + |x|)^{-a}, \quad a > 2, c = \text{const} > 0. \tag{3.3.4}$$

In Chapter 1, it was proved that the resolvent kernel $G(x, y, k)$ of the operator $L \ge 0$ is meromorphic in k and analytic in a neighborhood of the real axis on the complex k-plane. The resolvent kernel $G(x, y, k)$ is the kernel of the operator $(L - k^2 I)^{-1}$. If $\text{Im}\, k^2 > 0$, this kernel is uniquely defined by the requirement $G(x, y, k) \in L^2(\Omega_x)$, that is, for any fixed $y \in \Omega$ the function $G(x, y, k)$ belongs to $L^2(\Omega)$. For $k > 0$, the resolvent kernel is defined as the limit $G(x, y, k + i0) = G(x, y, k)$. It was proved in Chapter 1 that this limit satisfies the radiation condition, and the estimate

$$|G(x, y, k)| \leq c|x - y|^{-1}, \quad x, y \in \Omega, \quad 0 \leq a \leq k \leq b, \tag{3.3.5}$$

where $c = \text{const} > 0$ does not depend on x, y, and $k, a \leq k \leq b$.

Therefore, the function

$$h(\lambda, f) = \frac{1}{\pi} \operatorname{Im} \int_\Omega G(x, y, \sqrt{\lambda + i0}) f dy, \quad f \in C_0^\infty(\Omega), \tag{3.3.6}$$

is not only Hölder-continuous in $\lambda, \lambda > 0$, but analytic in $k = \sqrt{\lambda}$ in a neighborhood of any point $\lambda > 0$. Therefore, condition (3.2.61) is satisfied for $k_0 \geq 0$, as well as condition (3.2.11) with $\alpha = 1$ and P being an orthoprojection in $L^2(\Omega)$ onto $L^2(B_R \setminus D)$, where $D \subset B_R$ and $R > 0$ is an arbitrarily large fixed number.

Conditions (3.2.8) and (3.2.59) are satisfied as follows from Theorem 1.6.1. Conditions (3.2.9) and (3.2.10) are obviously satisfied. Therefore, Theorems 3.2.3 and 3.2.5 in Section 3.2 yield the following result.

Theorem 3.3.1. *The limiting amplitude principle is valid for the solution to problem* (3.3.1)–(3.3.3) *under the assumptions of Theorem* 1.6.1 *about D. More precisely, the following formulas hold:*

$$Pv = \exp(ik_0 t) Pu(x, t) + o(1), \quad t \to +\infty, \tag{3.3.7}$$

and

$$Pv = \frac{1}{T} \int_0^T \exp(ik_0 t) Pu(x, t) dt + O\left(\frac{1}{T}\right), \quad T \to \infty, \tag{3.3.8}$$

where Pv is defined in (3.2.6), *and v is given by* (3.1.9).

Actually, under the assumptions of Theorem 1.6.1 a stronger form of the limiting amplitude principle holds

$$u(x, t) = v(x) \exp(-ik_0 t) + O(\exp(-at)), \quad a > 0, \quad t \to +\infty, \tag{3.3.9}$$

where O is uniform in $x \subset B_R \setminus D$ for any fixed R, and v solves the stationary problem (3.1.5)–(3.1.7). The proof of (3.3.9) is much more difficult than the proof of (3.3.7) and (3.3.8). The abstract result does not help. One should use analytical estimates.

Let us give a sketch of the proof. One starts with the formula

$$u(x,t) = (2\pi)^{-1} \int_{-\infty}^{\infty} \exp(-ikt)V(x,k)dk, \tag{3.3.10}$$

where $V(x,k)$ solves the stationary problem

$$-\nabla^2 V - k^2 V = f(x)i(k - k_0 + i0)^{-1}, \tag{3.3.11}$$

$$V = 0 \quad \text{on } \Gamma. \tag{3.3.12}$$

Therefore

$$\begin{aligned} V &= i(k - k_0 + i0)^{-1} \int_{\Omega} G(x,y,k) f(y) dy \\ &= i(k - k_0 + i0)^{-1} v(x,k), \end{aligned} \tag{3.3.13}$$

and

$$u(x,t) = i(2\pi)^{-1} \int_{-\infty}^{\infty} \exp(-ikt)v(x,k)(k - k_0 + i0)^{-1} dk. \tag{3.3.14}$$

If the assumption (3.1.12) holds, then one can move the contour of integration in (3.3.14) down and obtain

$$\begin{aligned} u(x,t) &= i(2\pi)^{-1} \int_{-ib-\infty}^{-ib+\infty} \exp(-ikt)v(x,k)(k - k_0 + i0)^{-1} dk \\ &\quad + \exp(-ik_0 t)v(x,k_0) \\ &= \exp(-ik_0 t)v(x,k_0) + O(\exp(-bt)), \quad 0 < b < |\text{Im } k_1|, \end{aligned} \tag{3.3.15}$$

where k_j are the poles of $v(x,k)$, $|\text{Im } k_1| \le |\text{Im } k_2| \le \cdots$. The argument is complete if (3.1.12) is established. Another approach is given in Section 3.4.

In the proof of the limiting amplitude principle and in the study of the asymptotic behavior of solutions to time-dependent problems for large t we use three different types of assumptions. The first deals with the local behavior of the limiting values of the resolvent on the continuous spectrum of L (assumptions (3.2.10), (3.2.11), and (3.2.61)). The second deals with the behavior of the resolvent on the continuous spectrum for large λ (assumptions (3.2.8), (3.2.59), (3.2.84), and (3.2.85)). The third deals with the behavior of the resolvent near its singular points (assumptions (3.2.9), (3.2.15), and (3.2.75)).

In scattering problems, assumptions (3.2.8) and (3.2.84) are often difficult to check. For example, these estimates are not obtained for the Robin and Neumann boundary conditions. If they are available, then one can usually obtain the exponential decay of the remainder in the limiting amplitude principle (see (3.3.15)).

But for a self-adjoint operator with absolutely continuous spectrum the limiting amplitude principle holds in the form (3.2.71) without assumptions for (3.2.8) or (3.2.84) (see Theorem 3.2.5). Therefore, we obtain the following theorem.

Theorem 3.3.2. *For the solution to problem* (3.3.1)–(3.3.3), *the limiting amplitude principle holds in the form*

$$u(x,t) = \exp(-ik_0t)v + o(1), \quad t \to +\infty, \quad k_0 > 0, \tag{3.3.16}$$

where v solves the stationary problem (3.1.5)–(3.1.7), *and $o(1)$ means that for any fixed $R > 0$, such that the ball B_R contains the obstacle D, one has*

$$||u(x,t) - \exp(-ik_0t)v||_{L^2(B_R\setminus D)} \to 0 \text{ as } t \to +\infty. \tag{3.3.17}$$

If the boundary condition (3.3.2) *is replaced by the Neumann or Robin boundary condition, the limiting amplitude principle* (3.3.16) *holds provided that the function $\eta(s)$ in the Robin boundary condition*

$$\frac{\partial u}{\partial N} + \eta(s)u = 0 \tag{3.3.18}$$

is real-valued and the spectrum of the exterior Robin Laplacian is absolutely continuous (which is the case, e.g., if $\eta(s) < 0$).

Proof. The conclusion of Theorem 3.3.2 follows from Theorem 3.2.5. The assumptions of Theorem 3.2.5 are satisfied for the operator L

corresponding to the problem (3.3.1)–(3.3.3), that is for the exterior Dirichlet Laplacian in $H = L^2(\Omega)$.

The same argument holds for the Neumann boundary condition. If $\operatorname{Im}\eta(s) = 0$, and $\eta(s) \in C(\Gamma)$ in the Robin boundary condition, then the corresponding operator L is self-adjoint and its positive spectrum is absolutely continuous. The resolvent kernel of the Robin Laplacian in Ω solves the problem

$$(-\nabla^2 - k^2)\mathcal{G}(x, y, k) = \delta(x - y) \quad \text{in } \Omega, \tag{3.3.19}$$

$$\frac{\partial \mathcal{G}}{\partial N} + \eta(s)\mathcal{G} = 0 \quad \text{on } \Gamma, \quad \operatorname{Im} \eta = 0, \tag{3.3.20}$$

$$r\left(\frac{\partial \mathcal{G}}{\partial r} - ik\mathcal{G}\right) \to 0, \quad r = |x| \to \infty. \tag{3.3.21}$$

This kernel may have a pole $k = 0$ if $\eta > 0$. For example, let $\Gamma = S^2 = \{s : |s| = 1\}$ be a sphere, and $\eta = 1$. Then the homogeneous problem (3.3.19)–(3.3.21) with $k = 0$ has a nontrivial solution $u = r^{-1}$:

$$\nabla^2 v = 0 \quad \text{if } r > 1, \quad \frac{\partial v}{\partial r} + v = 0, \quad \text{if } r = 1. \tag{3.3.22}$$

Thus, the point $k = 0$ is a pole of $\mathcal{G}$ if $\eta = 1$ and $\Gamma = S^2$. Similarly one can check that $\mathcal{G}$ may have poles $k = ib, b > 0$, if $\eta > 0$. Indeed, if $\Gamma = S^2$ and $v = r^{-1}\exp(-br), b > 0$, then

$$\begin{aligned} &-\nabla^2 v + b^2 v = 0 \quad \text{if } r > 1, \\ &\frac{\partial v}{\partial r} + \eta v = 0 \quad \text{if } r = 1 \quad \text{and} \quad \eta = b + 1. \end{aligned} \tag{3.3.23}$$

Thus, $\mathcal{G}$ has a pole $ib, b > 0$, if $\eta = b + 1$ and $\Gamma = S^2$. In these two examples, the assumptions of Theorem 3.3.2 do not hold, and the limiting amplitude principle does not hold. If $\eta \leq 0$, then the Robin Laplacian is a nonnegative self-adjoint operator L in $H = L^2(\Omega)$, and $H_{\text{ac}} = H$. In this case, the limiting amplitude principle for L holds as follows from Theorem 3.2.5. Since the positive spectrum of L is absolutely continuous, we need only to show that the condition $\eta \leq 0$ guarantees that $L \geq 0$ and $k = 0$ belongs to the absolutely continuous spectrum. We have, for any v which satisfies (3.3.18) and decays rapidly at infinity,

$$\begin{aligned}(Lv, v) &= \int_{\Omega} -\nabla^2 v\bar{v}dx \\ &= \int_{\Omega} |\nabla v|^2 dx + \int_{\Gamma} v_N \bar{v} ds - \lim_{R\to\infty} \int_{S_R} v_r \bar{v} ds \\ &= \int_{\Omega} |\nabla v|^2 dx - \int_{\Gamma} \eta(s)|v|^2 ds > 0 \quad \text{if } \eta(s) \leq 0.\end{aligned} \tag{3.3.24}$$

Thus $L > 0$, and $k = 0$ is not a pole of $\mathcal{G}$. Therefore, we proved (under the assumptions $\eta \leq 0$, $\eta \in C(\Gamma)$) the uniqueness theorem

$$\{Lv = 0, v = O(r^{-1}) \text{ as } r \to \infty\} \Rightarrow v = 0. \tag{3.3.25}$$

From (3.3.25), as in Section 1.2, one can derive that the point $\lambda = 0$ belongs to the absolutely continuous spectrum of L.

It is sufficient to prove that for any $f \in C_0^\infty(\Omega)$ the limit

$$\lim_{\epsilon\to+0} \int_{\Omega} \mathcal{G}(x, y, i\epsilon) f dy = v \tag{3.3.26}$$

exists in L^2_{loc}. This is done as in Theorem 1.2.3. □

Remark 3.3.1. The limiting amplitude principle holds for the problem of the type

$$u_{tt} = \nabla^2 u \quad \text{in } \Omega, \tag{3.3.27}$$

$$u(x, 0) = u_t(x, 0) = 0, \tag{3.3.28}$$

$$u(s, t) = f(s)\exp(-ik_0 t) \quad \text{on } \Gamma, \quad k_0 > 0. \tag{3.3.29}$$

In this case, the function v in (3.3.16) solves the problem

$$\nabla^2 v + k_0^2 v = 0 \quad \text{in } \Omega, \tag{3.3.30}$$

$$v = f(s) \quad \text{on } \Gamma, \tag{3.3.31}$$

v satisfies the radiation condition (3.1.7). (3.3.32)

3.4. Decay of Energy

3.4.1. Consider the problem (3.1.1)–(3.1.3). The solution to this problem can be written as

$$u = L^{-1/2}\sin(tL^{1.2})f. \tag{3.4.1}$$

This is a particular case of the formula (3.2.56), and L is the Dirichlet Laplacian in $H = L^2(\Omega)$. Define the energy norm by the formula

$$\begin{aligned} &||u||_1 = \{(u,u) + (Lu,u)\}^{1/2}, \quad u \in D(L), \\ &(u,u) = (u,u)_{L^2(\Omega)} = ||u||_0^2. \end{aligned} \tag{3.4.2}$$

The closure of $D(L_0) = \mathring{H}^2(\Omega)$ in the norm (3.4.2) is the Sobolev space $\mathring{H}^1(\Omega)$. The norm of this space is the graph norm of the operator $L^{1/2}$, $||\cdot||_1 = ||\cdot||_{\mathring{H}^1(\Omega)}$. Let E_λ be the resolution of the identity for L. Since the embedding $\mathring{H}^1(\Omega) \to L^2(B_R\backslash D)$ is compact, it is sufficient to prove that

$$(u(t),h)_1 \to 0 \quad \text{as } t \to +\infty, \quad h \in \mathring{H}^1(\Omega) \tag{3.4.3}$$

in order to obtain that

$$||u(t)||_{L^2(B_R\backslash D)} \to 0 \quad \text{as } t \to +\infty. \tag{3.4.4}$$

In this section, $R > 0$ denotes an arbitrarily large fixed number, $D \subset B_R$. From (3.4.2), it follows that

$$(u(t),h)_1 = \int_0^\infty [\lambda^{-1/2}\sin(t\lambda^{1/2}) + \lambda^{1/2}\sin(t\lambda^{1/2})]d(E_\lambda f, h). \tag{3.4.5}$$

Since $f \in D(L^{1/2})$ one has

$$\int_0^\infty \lambda d(E_\lambda f, f) < \infty, \tag{3.4.6}$$

and

$$\begin{aligned} \int_0^\infty \lambda^{1/2}|d(E_\lambda f,h)| &\le \left\{\int_0^\infty \lambda d(E_\lambda f,f)\right\}^{1/2}\left\{\int_0^\infty d(E_\lambda h,h)\right\}^{1/2} \\ &= ||L^{1/2}f||_0 \cdot ||h||_0 < \infty. \end{aligned} \tag{3.4.7}$$

Moreover,

$$\int_0^{\epsilon} \lambda^{-1/2}|d(E_\lambda f, h)| < \infty \quad \text{for } f, g \in H_0^1(\Omega).$$

The spectrum of L is absolutely continuous. Therefore, the Riemann–Lebesgue lemma is applicable to the integral (3.4.5) and this lemma yields (3.4.3). Thus, we proved (3.4.4). Let us summarize the result.

Lemma 3.4.1. *If $f \in \mathring{H}^1(\Omega)$, then the solution to problem (3.1.1)–(3.1.3) satisfies equation (3.4.4).*

One can obtain a stronger decay:

$$||u(t)||_{1,R} + ||\dot{u}(t)||_{1,R} \to 0 \quad \textit{as } t \to +\infty, \quad \dot{u} = \frac{du}{dt}, \tag{3.4.8}$$

where $||\cdot||_{1,R}$ is the $\mathring{H}^1(B_R \setminus D)$ norm, and $f \in \mathring{H}^1(\Omega)$.

Theorem 3.4.2. *If $f \in D(L) = \mathring{H}^2(\Omega)$ then (3.4.8) holds.*

Proof. Let us prove that

$$||\dot{u}||_{1,R} \to 0 \quad \text{as} \quad t \to +\infty. \tag{3.4.9}$$

It follows from (3.4.1) that $\dot{u} = \cos(tL^{1/2})f$. As in the proof of Lemma 3.4.1, one obtains

$$(\dot{u}, h)_1 = \int_0^{\infty} [\cos(t\lambda^{1/2}) + \lambda\cos(t\lambda^{1/2})] d(E_\lambda f, h). \tag{3.4.10}$$

If $f, h \in \mathring{H}^1(\Omega) = D(L^{1/2})$, then

$$\int_0^{\infty} \lambda|d(E_\lambda f, h)| \le \left\{\int_0^{\infty} \lambda d(E_\lambda f, f)\right\}^{1/2} \left\{\int_0^{\infty} \lambda d(E_\lambda h, h)\right\}^{1/2} < \infty. \tag{3.4.11}$$

Therefore, the Riemann–Lebesgue lemma applied to (3.4.10) yields

$$\dot{u} \to 0 \quad \text{weakly in } \mathring{H}^1, \text{ as } t \to +\infty. \tag{3.4.12}$$

As in the proof of Lemma 3.4.1, this and the embedding theorem imply that

$$||\dot{u}||_{0,R} \to 0, \quad t \to +\infty. \tag{3.4.13}$$

Furthermore $L\dot{u} = L\cos(tL^{1/2})f$, and

$$(\dot{u},h)_L = (\dot{u},h_0) + (L\dot{u},Lh)_0$$
$$= \int_0^\infty [\cos(t\lambda^{1/2}) + \lambda^2 \cos(t\lambda)^{1/2}] d(E_\lambda f,h), \quad h \in D(L). \tag{3.4.14}$$

Since $f \in D(L)$ one has $\int_0^\infty \lambda d(E_\lambda f,f) < \infty$, and the Riemann–Lebesgue lemma applied to (3.4.14) yields

$$(\dot{u},h)_L \to 0 \quad \text{as } t \to +\infty, \quad \forall h \in \mathring{H}^2(\Omega) = D(L). \tag{3.4.15}$$

From (3.4.15) and the embedding theorem equation (3.4.9) follows. Equation (3.4.8) can be proved similarly. □

Remark 3.4.1. If $f \in D(L^n) = H^{2n}(\Omega), L^n u|_\Gamma = 0, 0 \le m \le n$, then

$$((L+I)^n\dot{u},(L+I)^n h)$$
$$= (\dot{u},h)_{H^{2n}(\Omega)} \to 0 \quad \text{as } t \to +\infty, \quad h \in \mathring{H}^{2n}(\Omega). \tag{3.4.16}$$

From (3.4.16) and the embedding theorem, it follows that

$$||\dot{u}||_{\mathring{H}^{2n-1}(B_R\backslash D)} \to 0 \quad \text{as } t \to +\infty. \tag{3.4.17}$$

3.4.2. If the obstacle D satisfies a nontrapping condition, e.g., D is convex or star-shaped (or even more sophisticated nontrapping conditions like in [464] or [144]) then function (3.1.4) satisfies inequality (3.1.12) and the solution to problem (3.1.1)–(3.1.3) decays exponentially

$$|u(x,t)| \le c(R,D)\exp(-b_0 t), \quad b_0 > 0, \quad |x| \le R. \tag{3.4.18}$$

In [144], a similar result is mentioned for the problem with the Neumann boundary condition. We discuss the literature on this question in the Bibliographical notes. In the literature, the estimate (3.4.18) is obtained in the course of a study of the time-dependent problem. If (3.4.18) is established, then the function (3.4.4) is seen to be analytic in the half-plane $\operatorname{Im} k > -a$, and by Parseval's equality,

$$\int_{\operatorname{Im} k=b} |v(x,k)|^2 |dk| < c(b,r), \quad b > -b_0, \quad |x| \leq R. \tag{3.4.19}$$

This estimate and the analyticity of v imply estimate (3.4.12) for $b > -b_0$. Conversely, analyticity of $v(x,k)$ in the half-plane $b > -b_0$ and the estimate (3.4.19) imply (3.4.18). A direct proof of the estimates (3.1.12) or (3.4.19), which does not use the time-dependent problem, is not yet found. For the Schrödinger operator in $\mathbb{R}^3$, a proof of such estimates was given in [194, 199], and, under weaker assumptions on the potential, in [437].

3.5. Singularity and Eigenmode Expansion Methods

The singularity expansion method (SEM) has been explained in Section 3.1. The eigenmode expansion method (EEM) is a method for solving linear equations using expansions in root vectors of B. Consider an example. Look for the solution to the problem

$$(\nabla^2 + k^2)w = 0 \text{ in } \Omega, \quad w = f \text{ on } \Gamma \tag{3.5.1}$$

w satisfies the radiation condition, in the form

$$w = \int_\Gamma g\sigma ds, \quad g = \frac{\exp(ik|x-s|)}{4\pi|x-s|}. \tag{3.5.2}$$

Then

$$Q\sigma = \int_\Gamma g(s,t)\sigma(t)dt = f(s) \tag{3.5.3}$$

is an equation in $H = L^2(\Omega)$ with a linear nonself-adjoint compact operator Q. If B is a compact self-adjoint operator in H, then the equation

$$B\sigma = f \tag{3.5.4}$$

is solvable in H if and only if

$$\sum_{n=1}^{\infty} |f_n|^2 \lambda_n^{-2} < \infty, \quad B\phi_n = \lambda_n \phi_n, \tag{3.5.5}$$

where $f_n = (f, \phi_n)$, and the system $\{\phi_n\}$ of the eigenvectors of B forms an orthonormal basis of H. If (3.5.5) holds, then the solution to (3.5.4) is of the form

$$\sigma = \sum_{n=1}^{\infty} \lambda_n^{-1} f_n \phi_n. \tag{3.5.6}$$

If $\lambda_j = 0$ and $f_j = 0$, then the solution to (3.5.4) exists but is not unique:

$$\sigma = \sum_{n \neq j} \lambda_n^{-1} f_n \phi_n + c_j \phi_j, \tag{3.5.7}$$

where c_j are arbitrary constants.

If B is nonself-adjoint, then it may have no eigenvectors (e.g., $Bh = \int_0^x b(x,y)h(y)dy, B : L^2[0,1] \to L^2[0,1]$). A nonself-adjoint operator may have root vectors (generalized eigenvectors), its eigenvectors are not necessarily orthogonal and their linear span may not be the whole space. Root vectors of a compact linear operator B corresponding to an eigenvalue λ are the vectors which satisfy the equation

$$(B - \lambda I)^j \phi = 0, \quad \phi \neq 0 \tag{3.5.8}$$

with some integer j. If $j = 1$, then ϕ is an eigenvector. If B is compact, then one knows that $N\{(B - \lambda I)^j\} = N\{(B - \lambda I)^{j+1}\}$ if $j \geq r$, where $r = r(\lambda)$ is an integer which is called the Riesz number corresponding to the eigenvalue λ. The elements of $N\{(B - \lambda I)^r\}$ are called root vectors of B corresponding to the eigenvalue λ. Let us define the Jordan chain associated with a pair (λ, ϕ) where $B\phi = \lambda\phi$. Consider the equation $B\phi^{(1)} - \lambda\phi^{(1)} = \phi$. If it is not solvable, then we say that the Jordan chain corresponding to the pair (λ, ϵ) consists of one vector ϕ and has length 1. If it is solvable, consider the equation $B\phi^{(2)} - \lambda\phi^{(2)} = \phi^{(1)}$. Continue this process. Since $r(\lambda) < \infty$, there is a number $r = r(\lambda, \phi)$ such that the equations

$$(B - \lambda I)\phi^{(j)} = \phi^{(j-1)}, \quad j = 1, 2, \dots, r(\lambda, \phi), \quad \phi^{(0)} = \phi, \tag{3.5.9}$$

are solvable but the equation with $j = r(\lambda, \phi)+1$ is not solvable. Then we say that $\{\phi^{(j)}\}, 0 \le j \le r(\lambda, \phi)$ is the Jordan chain associated with the pair (λ, ϕ) and $r(\lambda, \phi)+1$ is the length of this chain. Clearly $r(\lambda) = \max r(\lambda, \phi)$, where the maximum is taken over $\phi \in N(B-\lambda I)$. The basic question concerning EEM is:

When does B have a root system which forms a basis of H? (3.5.10)

It is well known (see [52]) that if B is normal, i.e., $B^*B = BB^*$, then its eigenvectors form an orthonormal basis of H. One can check (see [226]) that the operator (3.5.3) is normal if Γ is a sphere.

Before answering the basic question (3.5.10) for the operator $B = Q$, one can discuss a much simpler question:

Is the root system of Q complete in H? (3.5.11)

The answer is positive. An abstract theorem which is sufficient for our purposes can be formulated.

Theorem 3.5.1. *Let Q be a compact linear operator on a Hilbert space H, $Q = Q_0 + T, Q_0 \ge 0, T_j = \text{Im } T = (T - T^*)/2i \ge 0$, and T belongs to the trace class. Then the root system of Q is complete in H.*

Proof. If $Q_0 \ge 0$ and $S = S^*$ are compact then

$$N_\lambda^-(Q_0 + S) \le N_\lambda^-(S), \quad \text{where } N_\lambda^-(S) = \sum_{\lambda_n(S) \le -\lambda} 1,$$

and $\lambda_n = \lambda_n(S)$ are the eigenvalues of S, $|\lambda_1| \ge |\lambda_2| \ge \cdots$.

Take $S = T_R = (T + T^*)/2$. Then

$$0 \le \lim_{\lambda \to 0} \lambda N_\lambda^-(Q_0 + T_R) \le \lim_{\lambda \to 0} \lambda N_\lambda^-(T_R) = 0. \tag{3.5.12}$$

The last equation holds because T and, therefore, T_R are trace-class. One can now use the following proposition (see [71, p. 292]).

Proposition 3.5.2. *The root system of a compact operator Q such that $Im Q \geq 0$ and $\lim_{\lambda\to 0} \lambda N_\lambda^-(Q_R) = 0$, is complete in H.*

Since Im $Q = T_j \geq 0$ by assumption, $Q_R = Q_0 + T_R$, and (3.5.12) holds, the conclusion of Theorem 3.5.1 follows from Proposition 3.5.2. □

Theorem 3.5.3. *The root system of the operator* (3.5.3) *is complete in $H = L^2(\Gamma)$.*

Proof. Let $Q = Q(k) = Q_0 + T$, where $Q_0 = Q(0) \geq 0$ and $T = Q(k) - Q(0)$ is trace-class.

The conclusion of Theorem 3.5.3 follows from Theorem 3.5.1 if one notices that the operator (3.5.3) is compact, and

$$\begin{aligned}(\operatorname{Im} Tf, f) &= (\operatorname{Im} Qf, f)\\ &= \int_\Gamma \int_\Gamma \frac{\sin kr_{st}}{4\pi r_{st}} f(t)\overline{f(s)}dsdt \geq 0, \quad r_{st} = |s-t|.\end{aligned} \tag{3.5.13}$$

This equality shows that $\operatorname{Im} Q \geq 0$. The inequality (3.5.13) follows from the identity

$$\frac{\sin kr_{st}}{4\pi r_{st}} = \frac{k}{(4\pi)^2} \int_{S^2} \exp\{ik\omega \cdot (s-t)\}d\omega.$$

This identity implies that

$$\begin{aligned}&\int_\Gamma \int_\Gamma \frac{\sin kr_{st}}{4\pi r_{st}} f(t)\overline{f(s)}dsdt\\ &\quad = \frac{k}{(4\pi)^2} \int_{S^2} \left| \int_\Gamma f(s) \exp(ik\omega \cdot s)ds \right|^2 d\omega \geq 0.\end{aligned} \tag{3.5.14}$$

Theorem 3.5.3 is proved. □

We prove that the root system of the operator (3.5.3) is not only complete in H but forms a Bari basis with brackets of H. The definition of this notion is given in Appendix C. The basic idea of the proof is as follows: $Q = Q_R + iQ_J$. The operator Q_R is a self-adjoint compact operator of order -1, i.e.,

$$||Q_R f||_{\tilde{H}^{q+1}} \le c||f||_{\tilde{H}^q}, \quad c > 0. \tag{3.5.15}$$

If $k^2 \notin \sigma(-\Delta_D)$, i.e., k^2 is not an eigenvalue of the Dirichlet Laplacian in D, then Q is an isomorphism of $\tilde{H}^q$ onto $\tilde{H}^{q+1}$ (Lemma 1.7.4), and so is Q_R. The operator Q_J has an analytic kernel $(4\pi r_{st})^{-1}\sin(kr_{st})$, so that $Q_J : \tilde{H}^2 \to \tilde{H}^\infty$. From the abstract results in Appendix C, the following theorem follows.

Theorem 3.5.4. *The root system of the operator* (3.5.3) *forms a Bari basis with brackets of* $L^2_{\cdot}(\Gamma)$.

Proof. We assume first that $0 < k^2 \notin \sigma(-\Delta_D)$. Then Q and Q_R are invertible as operators from $\tilde{H}^0$ onto $\tilde{H}^1$, and

$$Q = Q_R(I + iQ_R^{-1}Q_J), \;\text{ ord } Q_R = -1, \;\text{ ord } Q_R^{-1}Q_J = -\infty, \tag{3.5.16}$$

$$\lambda_n(Q_R) \sim cn^{-1/2}, \quad c = \text{ const}, n \to \infty. \tag{3.5.17}$$

Formula (3.5.17) is a particular case of the formula

$$\lambda_n(L) \sim cn^{ld^{-1}}, \quad c = \text{ const}, \quad n \to \infty,$$

where $\lambda_n(L)$ are the eigenvalues (counted by their multiplicities) of a pseudo-differential elliptic operator L of order l on a compact manifold of dimension d. In our case, $l = -1$ and $d = 2$. The conclusion of Theorem 3.5.4 follows from (3.5.16), (3.5.17) and Theorem C.0.2 in Appendix C. In our case $p = \frac{1}{2}$, and a can be taken arbitrary negative, so that $p(1-a)$ can be made as large as one wishes. Theorem C.0.5 in Appendix C gives additional information: the eigenvector expansion for Q_R and root vector expansion for Q are equiconvergent uniformly on bounded sets in $\tilde{H}^0$. If $\Gamma \in C^\infty$, the same is true in $\tilde{H}^q$ for any real q. □

Chapter 4

The T-matrix Scheme

4.1. Statement of the Problem

4.1.1. The T-matrix scheme for solving exterior boundary value problems and scattering problems has been popular for many decades. Although many papers have been published on this method and its applications to acoustical electromagnetic and elastic wave scattering, the theoretical foundations of the method were given only in 1982 [108, 244, 245, 247–248].

These foundations are given in this chapter. They include analysis of the convergence of the method, stability of the numerical scheme towards small perturbations of the data and round-off errors, and estimates of the rate of convergence. Numerous practical applications of the method are not discussed. References concerning these applications can be found in [466]. Only scalar wave scattering is discussed in detail since the basic principles and the results of the theory are fully represented in this case.

4.1.2. Let us describe the T-matrix scheme in a general formulation. Let $D \subset \mathbb{R}^3$ be a bounded obstacle with boundary Γ and Ω be the exterior domain. Let N be the exterior unit normal to Γ, i.e., the normal pointing into Ω. Consider the problem

$$(\nabla^2 + k^2)u = 0 \quad \text{in } \Omega, \quad k > 0, \tag{4.1.1}$$

$$u = f \quad \text{on } \Gamma, \tag{4.1.2}$$

$$r\left(\frac{\partial u}{\partial r} - iku\right) \to 0, \quad r \to \infty. \tag{4.1.3}$$

We are not concerned here with determining the minimal smoothness requirements on the surface Γ under which the theory is valid. It is certainly sufficient to assume that Γ is a Lyapunov surface but this requirement can be considerably relaxed, e.g., edges and conical points are admissible.

Let $\{\psi_n\}_{\lambda=1}^{\infty}$ be a system of outgoing waves, i.e.,

$$(\nabla^2 + k^2)\psi_n = 0 \quad \text{in } \Omega, \tag{4.1.4}$$

$$r\left(\frac{\partial \psi_n}{\partial r} - ik\psi_n\right) \to 0, \quad r \to \infty. \tag{4.1.5}$$

From Green's formula, it follows that

$$\int_\Gamma \psi_n \frac{\partial u}{\partial N} ds = \int_\Gamma u \frac{\partial \psi_n}{\partial N} ds, \quad \forall n, \tag{4.1.6}$$

where u solves (4.1.1)–(4.1.3). Using the boundary condition (4.1.2), one obtains

$$\int_\Gamma \psi_n h ds = f_n, \quad \forall n, \tag{4.1.7}$$

where

$$f_n = \int_\Gamma f \frac{\partial \psi_n}{\partial N} ds, \tag{4.1.8}$$

$$h = \left.\frac{\partial u}{\partial N}\right|_\Gamma. \tag{4.1.9}$$

The T-matrix scheme consists of the following. Let $\{\phi_j\}$ be a linearly independent, complete system of functions in $H_0 = L^2(\Gamma)$. Let

$$h_m = \sum_{j=1}^{m} c_j^{(m)} \phi_j, \tag{4.1.10}$$

where $c_j^{(m)}, 1 \le j \le m$, are constant coefficients which should be determined from the linear algebraic system

$$\sum_{j=1}^{m} a_{nj} c_j^{(m)} = f_n, \quad 1 \le n \le m, \tag{4.1.11}$$

$$a_{nj} = \int_\Gamma \psi_n \phi_j ds = (\phi_j, \bar{\psi}_n). \tag{4.1.12}$$

One obtains this system if (4.1.10) is substituted in (4.1.7) and only the first m equalities (4.1.7) are used.

4.1.3. Justification of the T-matrix scheme requires positive answers to the following questions:

Q.1. If (4.1.11) solvable for sufficiently large m? Is the solution of (4.1.11) unique?

Q.2. Does $h_m \underset{H_0}{\longrightarrow} h$ as $m \to \infty$? Here, h is defined in (4.1.9).

Q.3. Does $c_j^{(m)} \to c_j$ as $m \to \infty$? Is the convergence uniform in $j, 1 \le j < \infty$?

Q.4. Does the equality $h = \sum_{j=1}^{\infty} c_j \phi_j$ hold, where c_j are defined in Q.3 and it is assumed that the limit c_j exist?

Q.5. How does the rate of convergence depend on the choice of the systems $\{\phi_j\}$ and $\{\psi_n\}$?

Q.6. Is the numerical scheme based on equation (4.1.11) stable with respect to small perturbations of f_n and the matrix a_{nj}?

Remark 4.1.1. The following questions are easy to answer. Is the set of equations (4.1.7) solvable? Is the solution to (4.1.7) unique? The answer to the first question is yes. Indeed, one knows that problem (4.1.1)–(4.1.3) is solvable. If u solves (4.1.1)–(4.1.3), then apply Green's formula to u and ψ_n and obtain (4.1.7).

The answer to the second question is yes if and only if the system $\{\psi_n\}$ is closed in $H_0 = L^2(\Gamma)$. This means that

$$\left\{ \int_\Gamma h \psi_n ds = 0, \forall n \right\} \Rightarrow h = 0. \tag{4.1.13}$$

This requirement is equivalent to completeness of the system $\{\psi_n\}$ in H_0. The system $\{\psi_n\}$ is called complete in H_0 if and only if for any $\epsilon > 0$ and $h \in H_0$ there exists $m(\epsilon)$ and coefficients $c_j = c_j(\epsilon, h)$ such that $\|h - \sum_{j=1}^{m(\epsilon)} c_j\psi_j\| < \epsilon$. We assume below that the system $\{\psi_j\}$ is closed in H_0.

4.1.4. If one takes as ψ_1 in (4.1.7), the function $g = \exp(ik|s - y|/(4\pi|s-y|), y \in D$, and does not use equation (4.1.7) for $n > 1$, then one obtains the integral equation

$$\int_\Gamma g(s,y)h(s)ds = \int_\Gamma f(s)\frac{\partial g(s,y)}{\partial N}ds = F(y), \quad y \in D. \tag{4.1.14}$$

If this equation holds for $y \in B \subset D$, where B is any ball lying strictly in D, then (4.1.14) holds in D, because both sides in (4.1.14) are solutions to Helmholtz's equation and, by the unique continuation property, they are identical in D if they are identical in B. If $y \to s' \in \Gamma$, then one obtains from (4.1.14) the following equation

$$Qh = \int_\Gamma g(s,s')h(s)ds = b(s'), \tag{4.1.15}$$

where

$$b(s') = \lim F(y), \quad y \in D, \quad y \to s'. \tag{4.1.16}$$

If one looks for a solution to (4.1.15) of the form (4.1.10) and uses a projection method for finding the coefficients $c_j^{(m)}$, then one obtains the system

$$\sum_{j=1}^{m} A_{nj}c_j^{(m)} = b_n, \quad 1 \le n \le m, \tag{4.1.17}$$

where

$$A_{nj} = (Q\phi_j, \eta_n), \quad b_n = (b, \eta_n), \tag{4.1.18}$$

$$(f,h) = \int_\Gamma f\bar{h}ds, \tag{4.1.19}$$

$\{\eta_n\}$ is a system of linearly independent functions in H_0 and the bar denotes complex conjugation. The same questions Q.1–Q.6 as in

Section 4.3 can be studied for the system (4.1.17). The η_n plays the role of ψ_n, but now there is no need to assume anything about the properties of η_n in Ω. The functions η_n are defined only on Γ. One can go back from the integral equation (4.1.14) to the system (4.1.7) assuming that

$$g(s, y) = \sum_{j=1}^{\infty} \phi_j(y)\psi_j(s), \quad |y| < |s|, \tag{4.1.20}$$

substituting (4.1.20) into (4.1.14) and equating coefficients in front of ψ_j. In the literature, expansion (4.1.20) is used with ψ_j being the outgoing spherical waves and ϕ_j being the regular (i.e., finite at the origin) solutions to Helmholtz's equation. Matrix (4.1.18) is identical to matrix (4.1.12) if ψ_n in (4.1.12) are chosen so that $Q\bar{\eta}_n = \psi_n$. This corresponds to a specific choice of the outgoing waves, since for any linearly independent system of functions η_n in $L^2(\Gamma)$, the system $\psi_n = Q\bar{\eta}_n$ is a system of outgoing waves whose boundary values on Γ form a linearly independent system in H_0 provided that $Q : H_0 \to H_0$ is injective, i.e., $Q\eta = 0$ implies $\eta = 0$. This will be the case if and only if k^2 is not an eigenvalue of the Dirichlet Laplacian in D. If k^2 is an eigenvalue of the Dirichlet Laplacian in D, then one can use g_ϵ instead of g and the corresponding operator Q_ϵ has no zeros. Here, g_ϵ is defined in Section 1.1.3.

Discussion of questions Q.1–Q.6 is given in Section 4.2.

4.2. Justification of the *T*-matrix Scheme

4.2.1. The strategy is to treat equation (4.1.11) as an operator equation obtained by applying a projection method to equation (4.1.7) and to use some general results about the convergence of projection methods.

Therefore, some background material is needed. First, we give some definitions.

A system of elements of Hilbert space H is called *minimal* if none of the elements belongs to the closure of the linear span of the others. A minimal system $\{\phi_j\}$ is called *strongly minimal* if $\lim_{m\to\infty} \lambda_m = \lambda > 0$, where λ_m is the minimal eigenvalue of the Gram matrix

$\phi_{ij} = (\phi_i, \phi_j), 1 \le i, j \le m$. A system $\{\tilde{\phi}_j\}$ is called biorthogonal to the system $\{\phi_j\}$ if $(\tilde{\phi}_j, \phi_i) = \delta_{ij}$. The biorthogonal system $\{\tilde{\phi}_j\}$ is uniquely defined if and only if the system $\{\phi_j\}$ is minimal and closed in H.

A minimal system is called a Schauder basis of a Banach space X if any element $f \in X$ can be uniquely represented as $f = \sum_{j=1}^{\infty} c_j \phi_j$, where the series converges in the norm of X. Uniqueness means that if $f = \sum_{j=1}^{\infty} c_j \phi_j$ and $f = \sum_{j=1}^{\infty} c'_j \phi_j$, then $c_j = c'_j, \forall j$.

A minimal system $\{\phi_j\}$ in a Hilbert space H is called a *Riesz basis* of H if $\phi_j = Th_j$, where $\{h_j\}$ is an orthonormal basis of H and T is an isomorphism on H, i.e., T is a linear bounded injective and surjective operator defined on all of H. In this case, T^{-1} has the same properties. A system $\{\tilde{\phi}_j\}$ biorthogonal to a Riesz basis is also a Riesz basis (see Appendix C).

A complete system $\{\phi_j\}$ in H is a Riesz basis if and only if the matrix $\phi_{ij} = (\phi_i, \phi_j)$ generates an isomorphism on l^2. Another useful criterion for a complete system $\{\phi_j\}$ to form a Riesz basis of H is the inequality

$$a_1^2 \sum_{j=1}^{m} |c_j|^2 \le \left\| \sum_{j=1}^{m} c_j \phi_j \right\|^2 \le a_2^2 \sum_{j=1}^{m} |c_j|^2, \quad a_1 > 0,$$

which should hold for any constants c_j and for any m, and the positive constants a_1 and a_2 should not depend on m and $c_j, 1 \le j \le m$. If $\{\phi_j\}$ is a Riesz basis, then $\sup_j \|\phi_j\|^2 < a_2^2, \inf_j \|\phi_j\|^2 > a_1^2$ and similar inequalities hold for $\tilde{\phi}_j$. In particular, $\sup_j \|\tilde{\phi}_j\| < \infty$. Proofs of the above statements one can find in [71, 482].

Consider an interpolation problem in the Hilbert space H. The problem consists of finding an element $h \in H$ such that

$$(h, \psi_n) = f_n, \quad n = 1, 2, \ldots, \tag{4.2.1}$$

where $\{f_n\}$ is a given sequence of numbers and $\{\psi_n\}$ is a system of linearly independent elements of H, i.e., any finite number of elements ψ_n forms a linearly independent system of elements in H.

Definition 4.2.1. A system $\{\psi_n\}$ is called a Bessel system if $\sum_{n=1}^{\infty} |(h, \psi_n)|^2 < \infty$ whenever $h \in H$.

Definition 4.2.2. A system $\{\psi_n\}$ is called a Riesz–Fischer system if for any sequence $\{f_n\} \in l^2$ there exists an element $f \in H$ such that $(f, \psi_n) = f_n, \forall n$. The set of sequences $\{(h, \psi_n)\}, h \in H$, is called the moment space M for the system $\{\psi_n\}$. We are interested in the questions:

(1) When does a sequence $\{f_n\} \in M$. In other words, when is (4.2.1) solvable?
(2) Is the solution to (4.2.1) unique?
(3) How does one construct the solution to (4.2.1)?

The answer to question (4.2.1) is as follows:

For (4.2.1) *to be solvable in H, it is necessary and sufficient that for any m and any scalars c_j the inequality*

$$\left| \sum_{j=1}^{m} c_j \tilde{f}_j \right| \leq c \left\| \sum_{j=1}^{m} c_j \psi_j \right\|, \quad c > 0, \quad f_j = (f, \psi_j) \tag{4.2.2}$$

be valid, where c does not depend on m and c_j; if (4.2.2) *holds, then* $\|h\| \leq c$.

The necessity follows from the estimate

$$\left| \sum_{j=1}^{m} c_j \tilde{f}_j \right| = \left| \left(h, \sum_{j=1}^{m} c_j \psi_j \right) \right| \leq \|h\| \left\| \sum_{j=1}^{m} c_j \psi_j \right\|.$$

To prove the sufficiency, notice that (4.2.2) and the inequality $\|h\| \leq c$ imply that the functional defined on the elements ψ_n by formula (4.2.1) can be extended to the linear space of these elements as a linear bounded functional by the formula $(\sum_{j=1}^{m} c_j \psi_j, h) = \sum_{j=1}^{m} c_j \bar{f}_j$. By continuity, it can be uniquely extended on the closure Ψ of the linear span of the elements $\{\psi_j\}$ and then by the Haln–Banach theorem it can be extended on all of H with the same norm.

The answer to question (2) is obvious.

Equation (4.2.1) *has no more than one solution if and only if the system* $\{\psi_j\}$ *is closed.*

One can prove that if (4.2.1) is solvable then the solution to (4.2.1) of minimal norm exists and is unique. This solution $h \in \Psi$. Indeed, let h_0, $\|h_0\| \leq c$, be a solution to (4.2.1). Consider the set S of all solutions to (4.2.1) such that $\|h\| \leq c$. Let $h \in S$. One can write $h = h_1 + h_2, h_1 \in \Psi, h_2 \perp \Psi$. Since $(h_2, \psi_n) = 0, \forall n$, the element h_1 solves (4.2.1) and $\|h_1\|^2 = \|h\|^2 - \|h_2\|^2 \leq \|h\|^2$. Let us show that h_1 is the solution to (4.2.1) with minimal norm and that this solution is unique. Clearly it is sufficient to prove that if $h_1', h_1 \in \Psi$ and solve (4.2.1), then $h_1' = h_1$. But if h_1' and h_1 solve (4.2.1), then $(h_1' - h_1, \psi_n) = 0, \forall n$. Since $h_1' - h_1 \in \Psi$, one concludes that $h_1' = h_1$. For any fixed m, one can find the solution of minimal norm to the first m equations (4.2.1) by the formula

$$h = h_m = -[\det(\psi_i, \psi_j)]^{-1} \begin{pmatrix} 0 & \psi_1 & \psi_m \\ f_1 & (\psi_1, \psi_1) & (\psi_1, \psi_m) \\ \vdots & \vdots & \vdots \\ f_m & (\psi_m, \psi_1) & (\psi_m, \psi_m) \end{pmatrix}. \tag{4.2.3}$$

Moreover, if (4.2.1) is solvable, then its unique solution of minimal norm can be obtained by the formula

$$h = \lim h_m, \quad m \to \infty, \tag{4.2.4}$$

where h_m is defined in (4.2.3).

4.2.2. Let $(a_{ij}), 1 \leq i, j < \infty$ be a matrix of a linear operator on l^2, i.e., $(Ax)_i = a_{ij}x_j$. Here and below, one should sum over the repeated indices from 1 to ∞. When is A bounded and when is it compact in l^2?

One has

$$\|Ax\|^2 = a_{ij}x_j\bar{a}_{ik}\bar{x}_k \leq |a_{ij}\bar{a}_{ik}|\frac{|x_k|^2 + |x_j|^2}{2} \leq \sup_{1\leq k<\infty} \sum_{j=1}^{\infty} |a_{ij}a_{ik}| \, \|x\|^2.$$

Therefore, A is bounded if and only if

$$\|A\| \leq \left(\sup_k \sum_{j=1}^{\infty} \sum_{i=1}^{\infty} |a_{ij} a_{ik}| \right)^{1/2} < \infty. \tag{4.2.5}$$

The operator A is compact if (4.2.5) holds and

$$\sup_{1 \leq k < \infty} \sum_{j=1}^{\infty} \sum_{i=N}^{\infty} |a_{ij} a_{ik}| \to 0 \quad \text{as } N \to \infty. \tag{4.2.6}$$

A useful criterion for boundedness of A gives the following lemma.

Lemma 4.2.1 (Schur). *Assume that* $a_{ij} = \bar{a}_{ji}$ *and* $\sup_i \sum_{j=1}^{\infty} |a_{ij}| \leq c$. *Then* $\|A\| \leq c$, $A : l^2 \to l^2$.

Proof. If A is symmetric, densely defined and $|(Ax, x)| \leq c\|x\|^2$, then $\|A\| \leq c$. One has

$$|(Ax, x)| = \left| \sum_{i,j=1}^{\infty} a_{ij} x_j \bar{x}_i \right| \leq \sum_{i,j=1}^{\infty} |a_{ij}| \frac{|x_i|^2 + |x_j|^2}{2}$$

$$\leq \sum_{i,j=1}^{\infty} |a_{ij}||x_i|^2 \leq \sup_i \sum_{j=1}^{\infty} |a_{ij}| \|x\|^2 \leq c\|x\|^2.$$

Therefore, $\|A\| \leq c$. □

4.2.3. Let us discuss convergence of projection methods . Let A be a linear bounded and boundedly invertible operator from a Hilbert space H onto a Hilbert space G. Let $L_m \subset L_{m+1}$, $\dim L_m = m$, be a sequence of subspaces *limit dense* in H, i.e., $\rho(h, L_m) \to 0$ as $m \to \infty$ for any $h \in H$, where $\rho(h, L)$ is the distance from the element h to the subspace L. Let P_m be the orthoprojection onto L_m. Let $M_m \subset M_{m+1}$ be a limit dense sequence of subspaces in G, $\dim M_m = m$, and Q_m be the orthoprojection onto M_m in G.

Consider the equation

$$Ah = f \tag{4.2.7}$$

and the projection method for its approximate solution:

$$Q_m A P_m h_m = Q_m f. \tag{4.2.8}$$

Consider the following statement:

$$\text{Equation (4.2.8) is uniquely solvable for all sufficiently large } m \text{ and } \|h_m - h\| \to 0 \text{ as } m \to \infty, \tag{4.2.9}$$

where h_m is the solution to (4.2.8). We want to give conditions for (4.2.9) to be true.

Equation (4.1.11) is of the form (4.2.8). Indeed, in equation (4.1.11) $H = L^2(\Gamma), G = l^2$, L_m is the linear span of $(\psi_1, \ldots, \psi_m)$, M_m is the subspace of the sequences in l^2 whose components $f = 0$ for $j > m$, the elements h_m are given by (4.1.10) and $Q_m f = (f_1, \ldots, f_m)$. Therefore, the abstract results will be immediately applicable to equation (4.1.11).

Theorem 4.2.2. *For (4.2.9) to hold, it is necessary and sufficient that*

$$\|Q_m A P_m h\| \geq c\|P_m h\|, \quad m > m_0, \quad h \in H, \quad c > 0, \tag{4.2.10}$$

where m_0 is some integer and c does not depend on m and h.

Proof. Assume (4.2.9). Then (4.2.8) is uniquely solvable for $m > m_0$ and $h_m = (Q_m A P_m)^{-1} Q_m f \to h = A^{-1} f, m \to \infty, f \in G$. Therefore, $\sup_m \|(Q_m A P_m)^{-1} Q_m\| \leq c < \infty$. Here and below c denotes various positive constants which do not depend on m. Thus, $\|P_m h\| = \|(Q_m A P_m)^{-1} Q_m Q_m A P_m h\| \leq c\|Q_m A P_m h\|$, i.e., (4.2.10) holds. Note that $(Q_m A P_m)^{-1} Q_m A P_m = I_m$, where I_m is the identity in $P_m H$ but not in all of H. The necessity is proved.

Assume (4.2.10). The operator $Q_m A P_m : P_m H \to Q_m G, m > m_0$, is injective because of (4.2.10). Therefore, it is surjective since $\dim P_m H = \dim Q_m G = m < \infty$. Therefore, equation (4.2.8) is uniquely solvable for $m > m_0$. To prove that $\|h_m - h\| \to 0, m \to \infty$, consider the equations

$$Q_m A[P_m h + (I - P_m)h] = Q_m f,$$
$$Q_m A P_m h_m = Q_m f,$$

which imply that

$$Q_m A P_m (h_m - P_m h) = Q_m A(I - P_m)h. \tag{4.2.11}$$

Since the sequence of the subspaces L_m is limit dense in H, one has $\|(I - P_m)h\| \to 0, m \to \infty, h \in H$. Therefore, (4.2.11) and (4.2.10) imply that

$$\begin{aligned} \|h_m - P_m h\| &= \|P_m(h_m - P_m h)\| \\ &\le c^{-1}\|Q_m A(I - P_m)h\| \to 0 \text{ as } m \to \infty. \end{aligned} \tag{4.2.12}$$

Thus,

$$\|h - h_m\| \le \|h - P_m h\| + \|P_m h - h_m\| \to 0, m \to \infty. \tag{4.2.13}$$

The proof is complete. □

Remark 4.2.1. One can give the same proof without the assumption $\dim L_m = \dim M_m = m < \infty$. In this case, one should assume in addition to (4.2.10) that

$$Q_m A P_m H = Q_m G, \quad m > m_0. \tag{4.2.14}$$

If $m < \infty$, then as was shown in the proof of Theorem 4.2.2, assumption (4.2.10) implies assumption (4.2.14).

Let us study the *stability of the projection method*. Suppose that (4.2.9) holds for the operator A in (4.2.7). These are the questions we are interested in:

(1) Will (4.2.9) hold for $A + B$ if $\|B\| < \delta$ and $\delta > 0$ is sufficiently small?
(2) Will (4.2.9) hold for $A+B$ if B is compact and $A+B$ is boundedly invertible?

Theorem 4.2.3. *Under the above assumptions, the answers to questions* (1) *and* (2) *are yes.*

Proof. (1) Let $\delta = c - \epsilon$, where c is the constant in (4.2.10) and ϵ is an arbitrary number in the interval $0 < \epsilon < c$. Let $\|B\| < \delta$. Then

$$\|Q_m(A+B)P_m h\| \geq c\|P_m h\| - \delta\|P_m h\| = \epsilon\|P_m h\|,$$

and condition (4.2.10) holds with the constant $\epsilon > 0$.

(2) If B is compact, then $\|(Q_m A P_m)^{-1} Q_m B - A^{-1}B\| \to 0, m \to \infty$, because $(Q_m A P_m)^{-1} Q_m - A^{-1} \to 0, m \to \infty$ strongly. Here we use the following simple statement: if $A_m \to 0$ strongly and B is compact then $\|A_m B\| \to 0, m \to \infty$. If A and $A+B$ are invertible, then $I + A^{-1}B$ is invertible. Thus, $\|P_m h + A^{-1}BP_m h\| \geq c_1\|P_m h\|, c_1 \to 0$. Therefore,

$$\begin{aligned} \|Q_m(A+B)P_m h\| &= \|Q_m A P_m[P_m h + (Q_m A P_m)^{-1} Q_m B P_m]\| \\ &\geq c\|P_m h + (Q_m A P_m)^{-1} Q_m B P_m h\| \\ &\geq c\|P_m h + A^{-1}BP_m h\| - c\|[(Q_m A P_m)^{-1} Q_m B \\ &\quad - A^{-1}B]P_m h\| \geq cc_1\|P_m h\| - c\alpha_m\|P_m h\|, \\ a_m &\to 0, \quad m \to \infty. \end{aligned}$$

Therefore, (4.2.10) holds for sufficiently large m with, e.g., the constant $0.5cc_1$. □

Example 4.2.1. As an application of Theorems 4.2.2 and 4.2.3, let us study the projection method for the equation

$$(I+T)h = f, \tag{4.2.15}$$

where T is a compact linear operator on a Hilbert space H. Many integral equations in Chapter are of this type. Assume that $N(I + T) = \{0\}$, where $N(A) = \{h : Ah = 0\}$. Then, by Fredholm's alternative, the operator $I + T$ is an isomorphism of H onto H. Let $\{\phi_j\}$ be a complete linearly independent system of elements in H, $L_m = \text{span}(\phi_1, \ldots, \phi_m)$, P_m be the orthoprojection onto L_m in

H. The projection method with $H = G$, $L_m = M_m$, leads to the equation

$$h_m + P_m T h_m = P_m f. \tag{4.2.16}$$

One can apply Theorem 4.2.3 with $A = I$ and $B = T$. Since T is compact, $I + T$ is an isomorphism, and for $A = I$ condition (4.2.10) is obviously satisfied with $c = 1$, one concludes that equation (4.2.16) are uniquely solvable for $m > m_0$ and $\|h_m - h\| \to 0, m \to \infty$, where h is the solution to equation (4.2.15).

4.2.4. Let us turn to the analysis of the T-matrix scheme, i.e., let us answer questions Q.1–Q.6 in Section 4.1.

Let us write equation (4.1.7) as an operator equation

$$ah = p, \quad p = (f_1, f_2, \ldots), \tag{4.2.17}$$

where $a : H_0 \to l^2, H_0 = L^2(\Gamma), p \in l^2$.

The operator a is bounded if and only if

$$\sum_{j=1}^{\infty} |(h, \bar{\psi}_j)|^2 \le c^2 \|h\|_{H_0}^2, \quad c > 0, \quad h \in H_0. \tag{4.2.18}$$

If (4.2.18) does not hold but the system $\{\psi_j\}$ forms a Schauder basis of H_0, then the operator a is densely defined on H_0 (i.e., $D(a)$, the domain of the definition of a, is a dense set in H_0). Indeed, if the system $\{\psi_j\}$ is a basis, then the biorthogonal system $\{\tilde{\psi}_j\}$ is also a basis of H_0 and any linear combination $h_m = \sum_{j=1}^{m} c_j \tilde{\psi}_j$ belongs to $D(a)$, $ah_m = (c_1, c_2, \ldots, c_m, 0, 0, \ldots)$. The operator a transforms functions $h \in H_0$ into sequences $\{(h, \bar{\psi}_n\}$. The range of a, $\operatorname{Ran} a$, is dense in l^2 if the system $\{\psi_j\}$ is a basis of H_0. Indeed, suppose $(ah, p) = 0, \forall h \in H_0$ and some $p \in l^2$. Take $h = \tilde{\psi}_j$. Then $a\tilde{\psi}_i = \delta_{ij}, (ah, p) = p_i = 0$, and, since i is arbitrary, one concludes that $p = 0$. The operator a^{-1} is bounded and defined on all of l^2 if and only if $\|ah\| \ge \|h\|, c > 0, h \in D(a)$. The operator a is an isomorphism of

H_0 onto l^2 if

$$c_1\|h\| \le \|ah\| \le c_2\|h\|, \quad c_1 > 0, \quad h \in H_0 \tag{4.2.19}$$

and they hold if and only if the system $\{\psi_j\}$ is a Riesz basis of H_0. Consider the projection method for solving equation (4.2.17). Let $Q_m p = (f_1, \dots, f_m, 0, \dots)$ be an orthoprojection in l^2 and P_m be an orthoprojection in H_0 onto the linear span of $\phi_1, \phi_2, \dots, \phi_m$. Here $\{\phi_j\}$ is a closed linearly independent system of elements of H_0. Equation (4.1.11) can be written as

$$Q_m a P_m h_m = Q_m p, \tag{4.2.20}$$

where

$$h_m = P_m h_m = \sum_{j=1}^{m} c_j^{(m)} \phi_j. \tag{4.2.21}$$

Let us write condition (4.2.10) for equation (4.2.20):

$$\begin{aligned} c^2 \left\| \sum_{j=1}^{m} c_j \phi_j \right\|_{H_0}^2 &\le \left\| Q_m a \left(\sum_{j=1}^{m} c_j \phi_j \right) \right\|_{l^2}^2 \\ &= \sum_{i=1}^{m} \left| \sum_{j=1}^{m} a_{ij} c_j \right|^2, m > m_0, \end{aligned} \tag{4.2.22}$$

where $c_1, \dots, c_m$ are arbitrary constants, $c > 0$,

$$a_{ij} = (\phi_j, \bar{\psi}_i). \tag{4.2.23}$$

One can write (4.2.22) as

$$c^2 \sum_{j,j'=1}^{m} (\phi_j, \phi_{j'}) c_j \bar{c}_{j'} \le \sum_{j,j'=1}^{m} \sum_{i=1}^{m} a_{ij} \bar{a}_{ij'} c_j \bar{c}_{j'}, \quad m > m_0. \tag{4.2.24}$$

Let us denote by $(a)_m$ the truncated matrix:

$$(a)_m = a_{jj'}, \quad 1 \le j, j' \le m. \tag{4.2.25}$$

Let Φ denote the Gram matrix

$$\Phi_{jj'} = (\phi_j, \phi_{j'}). \tag{4.2.26}$$

Then (4.2.24) can be rewritten as the matrix inequality

$$c^2(\Phi)_m \le (a^*a)_m, \quad c > 0, \quad \forall m > m_0, \tag{4.2.27}$$

where a^* is the matrix $a^*_{jj'} = \bar{a}_{j'j}$,

$$(a^*a)_m = (a^*)_m(a)_m \tag{4.2.28}$$

and the inequality $A \le B$ means that $(Ap,p) \le (Bp,p), \forall p$.

Theorem 4.2.2 implies the following result.

Theorem 4.2.4. *Assume that* $\{\psi_j\}$ *is a Riesz basis of* H_0 *and* $\{\phi_j\}$ *is a closed nearly independent system in* H_0. *The inequality* (4.2.27) *is a necessary and sufficient condition for the system* (4.1.11) *to be uniquely solvable for all* $m > m_0$ *and for the equation* $\|h_m - h\| \to 0$, $m \to \infty$ *to hold, where* h_m *is defined by* (4.1.10), $\{c_j^{(m)}\}, 1 \le j \le m$, *is the solution to* (4.1.11), *the sequence* $\{f_n\} \in l^2$, *and* h *solves equation* (4.1.7).

Proof. It was already established that the operator a is an isomorphism of H_0 onto l^2 if the system $\{\psi_j\}$ is a Riesz basis of H_0. Therefore, equation (4.1.7) is uniquely solvable in H_0 and inequality (4.2.27) implies that condition (4.2.10) holds. The conclusion of Theorem 4.2.4 follows from Theorem 4.2.2. □

Let us derive some sufficient conditions for the conclusion of Theorem 4.2.4 to hold. These sufficient conditions will be easy to verify in practice. Let $\lambda(A)(\Lambda(A))$ denote the minimal (maximal) eigenvalue of a self-adjoint matrix A.

Proposition 4.2.5. *If*

$$\inf_m \lambda((a^*a)_m) \ge \lambda > 0 \quad \textit{and} \quad \sup_m \Lambda((\Phi)_m) \le \Lambda < \infty, \tag{4.2.29}$$

then (4.2.27) *holds and therefore the conclusion of Theorem* 4.2.4 *holds.*

Proof. Indeed, $(a^*a)_m \geq \lambda I_m$, where I_m is the identity operator in l^2, and $c^2(\Phi)_m \leq c^2 \Lambda I_m$. Therefore, inequality (4.2.27) holds with $c^2 = \Lambda^{-1}$. □

Condition (4.2.29) is a spectral condition in which one can verify directly if the system $\{\phi_j\}$ and $\{\psi_j\}$ are given. The practical conclusion is that the T-matrix scheme converges if the system $\{\psi_j\}$ is a Riesz basis of H_0, the system $\{\phi_j\}$ is a closed linearly independent system in H_0, and (4.2.29) holds.

If one takes $\phi_j = \bar{\psi}_j$ then $a_{ij} = \Phi_{ij}, a = a^*$, and inequality (4.2.27) takes the form

$$c^2(a)_m \leq (a)^2_m, \quad c > 0. \tag{4.2.30}$$

Inequality (4.2.30) holds for the self-adjoint positive definite matrix $a_{ij} = (\phi_i, \phi_j)$ if and only if the first inequality (4.2.29) holds. To see this, one can use the spectral theorem for a self-adjoint operator $A \geq 0$ and write

$$((A^2 - c^2 A)f, f) = \int_0^\infty (t^2 - c^2 t) d(E_t f, f), \tag{4.2.31}$$

where E_t is the resolution of the identity for A. Since $\min_{t \geq 0}(t^2 - c^2 t) = -c^4/4 < 0$, one sees that a necessary and sufficient condition for the inequality $A^2 \geq c^2 A$ to hold with some positive c^2 is the validity of the inequality $A \geq bI$ with some $b > 0$. If $A \geq bI, b > 0$, then, e.g., $A^2 - bA \geq 0$. Let us formulate the result.

Proposition 4.2.6. *If* $\phi_j = \bar{\psi}_j$ *and* $\inf_m \lambda(a)_m \geq \lambda > 0$, *then* (4.2.27) *holds and therefore the conclusion of Theorem* 4.2.4 *holds.*

4.2.5. Theorem 4.2.4 and Propositions 4.2.5 and 4.2.6 answer questions Q.1 and Q.2 from Section 4.1. The rate of convergence of the T-matrix scheme with the fixed systems $\{\psi_j\}$ and $\{\phi_j\}$ depends on the rate of convergence of the quantity $\|h - P_m h\|$, in other words, it depends on the rate of approximation of the function $h = a^{-1}p$ by linear combinations of function $\sum_{j=1}^m c_j \phi_j$. It follows from (4.2.12) and (4.2.13) that

$$\|h - h_m\| \leq \|h - P_m h\|(1 + c^{-1}\|a\|). \tag{4.2.32}$$

This inequality gives the rate of convergence of the T-matrix scheme. It is a problem of approximation theory to estimate

$$\|h - P_m h\| = \inf_{c_1,\ldots,c_m} \left\| h - \sum_{j=1}^{m} c_j \phi_j \right\|. \tag{4.2.33}$$

Estimate (4.2.32) gives an answer to question Q.5 from Section 4.1.

Let us assume that

$$\{\phi_j\} \text{ is a complete minimal system in } H_0. \tag{4.2.34}$$

Then there exists a unique biorthogonal system $\{\tilde{\phi}_j\}, (\phi_i, \tilde{\phi}_j) = \delta_{ji}$. The system (4.1.11) can be written as $(a)_m c^{(m)} = p^{(m)}$, where $c^{(m)} = (c_1^{(m)}, \ldots, c_m^{(m)}), p^{(m)} = (f_1, \ldots, f_m)$, and the solution to (4.1.11) gives

$$h_m = \sum_{j=1}^{m} c_j^{(m)} \phi_j(x).$$

Therefore,

$$c_j^{(m)} = (h_m, \tilde{\phi}_j). \tag{4.2.35}$$

If $\|h_m - h\| \to 0, m \to \infty$, then equation (4.2.35) shows that

$$|c_j^{(m)} - c_j| \to 0, \quad m \to \infty, \quad c_j = (h, \tilde{\phi}_j). \tag{4.2.36}$$

The rate of convergence can be estimated:

$$|c_j^{(m)} - c_j| \leq \|h_m - h\| \, \|\tilde{\phi}_j\|. \tag{4.2.37}$$

Therefore, if

$$\sup_{1 \leq j < \infty} \|\tilde{\phi}_j\| \leq c < \infty \tag{4.2.38}$$

then

$$\sup_{1 \leq j < \infty} |c_j^{(m)} - c_j| \leq c\|h_m - h\|. \tag{4.2.39}$$

Inequality (4.2.39) implies uniform (in j) convergence of the sequence $c_j^{(m)}$ provided that $\|h_m - h\| \to 0, m \to \infty$.

Condition (4.2.38) holds, e.g., if the system $\{\phi_j\}$ is a Riesz basis of H_0. Let us summarize the result.

Proposition 4.2.7. *Assume that the systems* $\{\psi_j\}$ *and* $\{\phi_j\}$ *are Riesz bases of* H_0 *and the first inequality* (4.2.29) *holds. Then the* T*-matrix scheme converges, i.e., equation* (4.1.11) *is uniquely solvable for all* $m > m_0$ *and* $\|h_m - h\| \to 0, m \to \infty$, *where* h_m *is defined by* (4.1.10), *and* (4.2.36) *holds uniformly in* $j, 1 \leq j < \infty$. *Moreover, the series* $h = \sum_{j=1}^{\infty} c_j \phi_j$ *converges in* H_0 *to the solution of* (4.1.7).

Proof. All the statements of this proposition were proved except for the very last. To prove the last statement, let us write the series for the solution to the problem (4.1.7): $h = \sum_{j=1}^{\infty} a_j \phi_j$. This series converges in H_0 because the system $\{\phi_j\}$ is a basis of H_0. The coefficients a_j are uniquely determined by h: apply the functional $\tilde{\phi}_j$ and use the biorthogonality $(\phi_j, \tilde{\phi}_j) = \delta_{ij}$ to obtain $(h, \tilde{\phi}_j) = a_j$. Therefore, $a_j = c_j$ (see (4.2.36)).

The questions Q.1–Q.5 from Section 4.1 are answered. Let us discuss the last question Q.6. □

4.2.6. Consider the perturbed system (4.1.11):

$$\sum_{j=1}^{m} (a_{nj}^{(m)} + b_{nj}) \tilde{c}_j^{(m)} = \tilde{f}_n, \quad 1 \leq n \leq m, \quad |\tilde{f}_n - f_n| \leq \epsilon_n, \tag{4.2.40}$$

where $\tilde{f}_n$ and b_{nj} are small perturbations of the right side and of the matrix $a_{nj}^{(m)}$, respectively.

Let us assume that: the operator b in l^2, generated by the matrix b_{nj}, has the norm

$$\|b\| \leq \delta \tag{4.2.41}$$

and

$$\|\tilde{f} - f\| \leq \|\{\epsilon_n\}\| < \epsilon, \quad f = \{f_n\}, \quad \tilde{f} = \{\tilde{f}_n\}, \tag{4.2.42}$$

where $\{\epsilon_n\}$ is an element of l^2. The matrix b_{nj} can be considered as the perturbation due to round-off errors and to perturbations of the basis system $\{\psi_j\}$ and $\{\phi_j\}$.

Proposition 4.2.8. *If the assumptions in Proposition* (4.2.7) *hold and* $\delta > 0$ *is sufficiently small, then equation* (4.2.40) *is uniquely solvable for all* $m > m_0$*, the function* $\tilde{h}_m = \sum_{j=1}^m \tilde{c}_j^{(m)} \phi_j$ *converges in* H_0 *to a function* $\tilde{h}$ *such that* $\|\tilde{h} - h\| \le c(\epsilon + \delta)$*, where* $c > 0$ *is a constant which does not depend on* ϵ *and* δ*.*

Proof. The unique solvability of system (4.2.40) for all sufficiently large m follows from Theorem 4.2.3 provided that $\|b\| < \delta$, where $\delta > 0$ is sufficiently small.

Convergence in H_0 of $\tilde{h}_m$ to an element $\tilde{h} \in H_0$ follows from the same theorem. In order to estimate $\|h - \tilde{h}\|$, one writes

$$\begin{aligned}
\|h - \tilde{h}\| &= \|a^{-1}f - (a+b)^{-1}\tilde{f}\| \le \|a^{-1}\|\epsilon + \|[a^{-1} - (a+b)^{-1}]\tilde{f}\| \\
&= \|a^{-1}\|\epsilon + \|(a+b)^{-1}ba^{-1}\tilde{f}\| \le \|a^{-1}\|\epsilon \\
&\quad + \|(a+b)^{-1}\|\,\|a^{-1}\| \cdot \|\tilde{f}\| \cdot \delta \le c(\epsilon + \delta), \\
c &= \|a^{-1}\| + \|\tilde{f}\|\,\|a^{-1}\|^2(1 - \delta\|a^{-1}\|)^{-1}.
\end{aligned} \tag{4.2.43}$$

Thus, if (4.2.41) and (4.2.42) hold and δ is sufficiently small, the answer to the question Q.6 is yes.

Assumptions (4.2.41) and (4.2.42) mean that the perturbations decay as $n \to \infty$. Let us consider the case when the perturbations do not depend on n:

$$|f_n - \tilde{f}_n| \le \epsilon \quad \forall n, \tag{4.2.44}$$

$$|b_{nj}| < \delta \qquad \forall n, j. \tag{4.2.45}$$

In this case, the number m which determines the size of the truncated matrix should be chosen as a function of ϵ and δ, $m = m(\epsilon, \delta)$, so that it will not be too large or too small. The problem of how to choose this $m(\epsilon, \delta)$ is discussed in [235]. □

4.2.7. Equation (4.1.7) was obtained for the Dirichlet boundary condition (4.1.2), but similar equations can be obtained in the same way for other boundary conditions, e.g.,

$$\frac{\partial u}{\partial N} = f \quad \text{on } \Gamma, \tag{4.2.46}$$

or

$$\frac{\partial u}{\partial N} + \eta(s)u = f \quad \text{on } \Gamma, \quad \operatorname{Im} \eta \geq 0, \tag{4.2.47}$$

where $\eta(s) \in C(\Gamma)$ is a given function.

For example, equations (4.1.6) and (4.2.47) imply

$$\int_\Gamma u \left\{ \frac{\partial \psi_n}{\partial N} + \eta(s)\psi_n \right\} ds = \int_\Gamma f \psi_n ds. \tag{4.2.48}$$

This equation is of the form (4.1.7) with $\int_\Gamma f\psi_n ds, u(s)$, and $\partial\psi_n/\partial N + \eta(s)\psi_n$ playing the roles of f_m, h, and ψ_n, respectively.

If $\{\psi_n\}$ is the system of the outgoing spherical waves, $\psi_n = h_n^{(1)}(kr)Y_n(\omega)$, where $h_n^{(1)}(kr) = (\pi/(2kr))^{1/2}H_{n+1/2}^{(1)}(kr)$ is the spherical Hankel function, $H_{n+1/2}^{(1)}(kr)$ is the Hankel function of the first kind, and $Y_n(\omega)$ are the spherical harmonics normalized in $L^2(S^2), \omega = (\theta, \phi)$, then this system on Γ as well as the systems $\{\partial\psi_n/\partial N + \eta(s)\psi_n\}$ and $\{\partial\psi_n/\partial N\}$ on Γ are closed (total, complete) in $H_0 = L^2(\Gamma)$. In order to prove this, assume, e.g., that

$$\int_\Gamma f\psi_n ds = 0, \quad \forall n. \tag{4.2.49}$$

The Green function g can be expanded as

$$\begin{aligned} g &= \frac{\exp(ik|x-y|)}{4\pi|x-y|} \\ &= \sum_n ik\psi_n^{re}(x)\psi_n(y), \quad |x| < \min_{s\in\Gamma} |s| = a, \end{aligned} \tag{4.2.50}$$

where $\{\psi_n^{re}\}$ are regular solutions of the Helmholtz equation, $\psi_n^{re} = j_n(kr)Y_n(\omega), j_n(kr) = (\pi/(2kr))^{1/2}J_{n+1/2}(kr)$ is the spherical Bessel function, and the origin is inside D. From (4.2.49) and (4.2.50), one obtains

$$v(x) = \int_\Gamma g(x,s) f ds = 0, \quad |x| < a. \tag{4.2.51}$$

The function v is a solution of Helmholtz's equation in D and in Ω, $v = 0$ in D by unique continuation theorem. Therefore, $v = 0$ on Γ, and $v = 0$ in Ω by Lemma 1.2.2. Therefore, $f = 0$ by the jump relation (1.1.3). This argument proves the closedness of the two other systems. For example, if

$$\int_\Gamma f \left(\frac{\partial \psi_n}{\partial N} + \eta(s)\psi_n \right) ds = 0, \tag{4.2.52}$$

then

$$u(x) = \int_\Gamma \left\{ \frac{\partial g(x,s)}{\partial N} + \eta(s) g(x,s) \right\} f ds = 0, \quad |x| < a. \tag{4.2.53}$$

Therefore,

$$(\nabla^2 + k^2) u(x) = 0 \text{ in } D \cup \Omega, \quad u = 0 \text{ in } D. \tag{4.2.54}$$

Thus,

$$u_+ = \frac{\partial u}{\partial N_+} = 0. \tag{4.2.55}$$

From (1.1.3), it follows that

$$u_- - u_+ = f, \quad \frac{\partial u}{\partial N_+} - \frac{\partial u}{\partial N_-} = \eta f. \tag{4.2.56}$$

From (4.2.55) and (4.2.56), one obtains

$$u_- = f, \quad \frac{\partial u}{\partial N_-} = -\eta f, \quad \frac{\partial u}{\partial N_-} + \eta u_- = 0. \tag{4.2.57}$$

Therefore, $u(x)$ solves the problem

$$(\nabla^2 + k^2) u = 0 \text{ in } \Omega,\ k > 0,\ \frac{\partial u}{\partial N_-} + \eta(s) u = 0 \text{ on } \Gamma \tag{4.2.58}$$

$$u \text{ satisfies the radiation condition (0.0.5).} \tag{4.2.59}$$

The argument in Lemma 1.2.2 shows that $u = 0$ in Ω. Therefore, $f = u_- = 0$. Let us summarize this result.

Lemma 4.2.9. *Let D be a bounded domain with a Lyapunov boundary Γ. Then each of the systems*

$$\{\psi_n\}, \quad \left\{\frac{\partial \psi_n}{\partial N}\right\} \text{ and } \left\{\frac{\partial \psi_n}{\partial N} + \eta(s)\psi_n\right\}, \quad \text{where } \psi_n = h_n^{(1)}(kr)Y_n(\omega), \tag{4.2.60}$$

is closed in $L^2(\Gamma)$. Here, $\operatorname{Im}\eta \geq 0, \eta \in C(\Gamma)$.

Proof. A proof of this lemma was given above for the first and third systems. The proof for the second system is essentially the same as for the first. □

Our argument holds for domains which are not necessarily connected. Actually the following stronger lemma is valid.

Lemma 4.2.10. *Let Γ be an open set on a Lyapunov surface. Then each of the systems* (4.2.60) *is closed in $L^2(\tilde{\Gamma})$.*

Proof. Assume that, e.g.,

$$\int_\Gamma f\tilde{\psi}_n ds = 0 \quad \forall n. \tag{4.2.61}$$

Then as in the proof of Lemma 4.2.9, one obtains

$$v(x) = \int_\Gamma fg(x,s)ds = 0, \quad |x| < a. \tag{4.2.62}$$

By the unique continuation theorem, one has

$$v(x) = 0 \text{ outside } \tilde{\Gamma}. \tag{4.2.63}$$

From the jump relation (1.1.3), it follows that $f = 0$. Similar argument holds for the other two systems (4.2.60). □

Remark 4.2.2. The system $\{\psi_n^{\text{re}}\}$ may not be closed in $L^2(\Gamma)$. For example, if $\Gamma = \{s : |s| = 1\}$ and $k > 0$ is a zero of the Bessel function: $J_{n+1/2}(k) = 0$ for some $n = n_0$, then the function $Y_{n_0}(\omega)$ is orthogonal in $L^2(\Gamma)$ to all of the functions $\psi_n^{\rm re}$ but $Y_{n_0}(\omega) \not\equiv 0$.

Lemma 4.2.11. *The system $\{\psi_n^{\rm re}\}$ is closed in $L^2(\Gamma)$ if and only if k^2 is not an eigenvalue of the Dirichlet Laplacian $-\Delta_D$ in D. If k^2 is an eigenvalue of $-\Delta_D$, then the system $\{\psi_n^{\rm re}\}$ is not closed in $L^2(\Gamma)$ but it has a finite defect m, where m is the multiplicity of the eigenvalue k^2. This means that there exists m functions $\mu_1, \dots, \mu_m$ such that the system*

$$\{\mu_j\} \cup \{\psi_n^{\rm re}\} \textit{ is complete in } L^2(\Gamma).$$

Proof. Let $\int_\Gamma f\psi_n^{re} ds = 0, \forall n$. Then $v(x) = \int_\Gamma f g(x,s) ds = 0$, $x \in \Omega$. Therefore,

$$(\nabla^2 + k^2)v = 0 \text{ in } D, \quad v = 0 \text{ on } \Gamma. \tag{4.2.64}$$

The jump relation (1.1.3) shows that $f = \partial u/\partial N_+$. If there are m linearly independent in $L^2(D)$ eigenfunctions $v_1, \dots, v_m$, then the functions $\mu_j = \partial v_j/\partial N_+$ are linearly independent in $L^2(\Gamma)$. Indeed, suppose $\sum_{j=1}^m c_j\mu_j = 0$. Then $v(x) = \sum_{j=1}^m c_j v_j(x)$ solves (4.2.64) and $\partial v/\partial N_+ = 0$ on Γ. This implies that $v = 0$ in D. Since the system $\{v_j\}, 1 \le j \le m$, is linearly independent in $L^2(D)$, it follows that $c_j = 0, 1 \le j \le m$. Thus, the system $\{\mu_j\}, 1 \le j \le m$, is linearly independent in $L^2(\Gamma)$. Therefore, if $\int_\Gamma f\psi_n^{re} ds = 0 \,\forall n$ and $\int_\Gamma f\mu_j ds = 0, 1 \le j \le m$, then $f = 0$. □

Remark 4.2.3. The system (4.2.60), although closed in $L^2(\Gamma)$, do not form a Riesz basis on Γ unless Γ is a sphere. This conclusion follows from the fact that the condition number of the truncated Gramians of systems (4.2.60) grows to infinity as the truncation size grows to infinity. This is illustrated by the numerical results given in Section 4.3 below. In general, these systems do not form a Schauder basis either. This question is closely connected with the so-called

Rayleigh conjecture (see Part III). This conjecture says that the series

$$u(x) = \sum_{n=0}^{\infty} c_n \psi_n(x), \quad |x| > R, \tag{4.2.65}$$

where u solves problem (4.1.1)–(4.1.3) and ψ_n are given in (4.2.60), converges on the surface Γ. This hypothesis is not true in general: although the solution to problem (4.1.1)–(4.1.3) for any f can be represented by the series (4.2.65) if R is so large that $D \subset B_R = \{x : |x| \leq R\}$, this series diverges on Γ for certain shapes of Γ. A discussion of this question and further references can be found in [13, 21, 148]. Examples of finite scatterers and periodic surfaces for which the Rayleigh hypothesis does not hold are given in these references.

Remark 4.2.4. The systems (4.2.60) are complete in $L^p(\Gamma), p > 1$. The proof is the same as the proof of Lemma 4.2.9. The jump relation for the L^p densities, $p > 1$, is given in [61].

Suppose that

$$\{(f, \phi_n) = 0 \ \forall n \,\&\, f \in M\} \Rightarrow f = 0, \tag{*}$$

where M is a linear dense set in the Hilbert space H, $\{\phi_n\}$ is a linearly independent system in H, and (f, ϕ) is the inner product in H. Given (*), is it true that

$$\{(f, \phi_n) = 0 \ \forall n \,\&\, f \in H\} \Rightarrow f = 0? \tag{**}$$

This question is of interest in the following situation.

Suppose that $H = L^2(\Gamma)$, ψ_n is the system (4.2.60), and $M = C(\Gamma)$. For the densities $f \in C(\Gamma)$, the classical potential theory is well known while for the densities $f \in L^p(\Gamma)$ the theory is more complicated. Since $C(\Gamma)$ is dense in $H = L^2(\Gamma)$, one could avoid the reference to this more complicated theory if one could show that (*) implies (**). However, *this implication does not hold in general.*

Example. Let $H = l^2$, $\phi_1 = (1, -2, 0, 0, \dots)$, $\phi_{n+1} = S\phi_n, n \geq 1$, where S is the shift operator defined by $Sf = (0, f_1, f_2, \dots), f =$

$(f_1, f_2, \dots)$, and M be the set of all compactly supported sequences $f \in l^2$, i.e., $f \in M$ if and only if $f \in l^2$ and $f_j = 0$ if $j > N(f)$. Suppose that $(f, \phi_n) = 0 \forall n$. Then $f_n - 2f_{n+1} = 0, f_{n+1} = f_n/2, f_{n+1} = f_1 2^{-n}$. If $f \in M$, then $f_1 2^{-n} = 0$ for $n > N(f)$. Therefore, $f_1 = 0$, and $f = 0$. Thus, (*) holds. But (**) does not hold: take $f = (1, 2^{-1}, 2^{-2}, \dots)$ and check that $(f, \phi_n) = 0 \, \forall n, f \in l^2$, and $f \neq 0$.

If $M = C[0, 1]$, then (*) does not imply (**). Indeed, choose a set $S = \bigcup_{j=1}^{\infty} I_j$, where I_j is an open interval of length $|I_j| \leq \epsilon 2^{-(j+1)}$ centered at r_j, where $\{r_j\} \subset I = [0, 1]$ is the countable set of all rationals in the segment $[0, 1]$. The set S is dense in $I, |S| \leq \epsilon, |T| = |I \backslash S| \geq 1 - \epsilon$, where $|S|$ is the Lebesgue measure of the set S. Let ϕ_j be a nonnegative function with suppose I_j, $\int_0^1 \phi_j dx = 1$. One can assume that $\phi_j(r_j) \geq \phi_j(x)$, that $\phi_j(x)$ is monotone on the intervals $[0, r_j], [r_j, 1]$, and that the system $\{\phi_n\}, n = 1, 2, 3, \dots$ is linearly independent in $L^2[0, 1]$. If $M = C[0, 1]$, then (*) holds for the system $\{\phi_j\}$. Indeed, suppose $f \in C[0, 1]$ and $(f, \phi_n) = 0 \forall n$. Let us show that $f(r_1) = 0 \forall j$. This and the continuity if f imply that $f = 0$. Assume that $f(r_m) \neq 0$ at some rational number r_m. Without loss of generality, assume $f(r_m) \geq \delta > 0$. Then there exists an interval $I_{j(m)}, r_m \in I_{j(m)}$ such that $f(x) \geq \delta/2$ if $x \in I_{j(m)}$. Therefore, $\int_0^1 f(x)\phi_{j(m)}(x)dx \geq \delta/2 > 0$, which is a contradicton. Thus, (*) holds. Let us show that (**) does not hold. Let $\kappa_T(x)$ be the characteristic function of the set $T = I \backslash S, \kappa_T(x) = 1$ if $x \in T$ and 0 otherwise. Then $0 \neq \kappa_T \in L^2[0, 1]$, and $\int_0^1 \kappa_T \phi_n dx = 0 \forall n$. Thus, (**) does not hold.

4.2.8. One can use the T-matrix method to solve problems of potential scattering theory.

Let

$$(-\nabla^2 + q(x) - k^2)u = 0 \text{ in } \mathbb{R}^3, \quad k > 0, \tag{4.2.66}$$

$$q(x) = 0, \quad |x| \geq R, \quad q \in L^2(B_R), \quad \operatorname{Im} q(x) = 0, \tag{4.2.67}$$

$$\begin{gathered} u = u_0 + v, \quad u_0 = \exp(-ikn \cdot x), \\ v \text{ satisfies the radiation condition.} \end{gathered} \tag{4.2.68}$$

The integral equation for u is

$$u = u_0 - \int_{B_R} gqudy = u_0 - Tu, \quad g = \frac{\exp(ik|x-y|}{4\pi|x-y|}. \tag{4.2.69}$$

Let $\{f_i\}, 1 \le j < \infty$, be an orthonormal basis of $L^2(B_R), u_{0R} = u_0$ in $B_R, u_{0R} = 0$ in $\Omega_R = \mathbb{R}^3 \backslash B_R$,

$$u_N = \sum_{j=1}^{N} c_j f_j, \tag{4.2.70}$$

$L_N = \text{span}\{f_1, \dots, f_N\}$, and P_N is the orthoprojection in $L^2(B_R) = H$ onto L_N. Consider the equation

$$u_N + P_N T u_N = P_N u_{0R}. \tag{4.2.71}$$

The problem is to prove that (4.2.71) is uniquely solvable for all sufficiently large N and $u_N \to u$ as $N \to \infty$, where u solves (4.2.69) with $u_0 = u_{0R}$.

Clearly

$$\|P_N T - T\| \to 0, \quad N \to \infty, \tag{4.2.72}$$

because T is compact in H and $P_N \to I$ strongly in H, where I is the identity in H. Assume for a moment that $I + T$ is invertible. (This assumption is proved later). Then (4.2.71) is identical with (4.2.16) and the result of Example 4.2.1 allows one to show that the following proposition holds.

Proposition 4.2.12. *Under the assumption* (4.2.67), *equation* (4.2.71) *is uniquely solvable for all sufficiently large* N *and* $\|u_N - u_R\| \to 0, N \to \infty$, *where* u_R *solves equation* (4.2.69) *with* $u_0 = u_{0R}$. *The solution* u_R *determines uniquely the solution to equation* (4.2.69) *in* $\mathbb{R}^3 : u = u_R$ *in* $B_R, u = u_0 - Tu_R$ *in* Ω_R.

Proof. The result of Example 4.2.1 shows that equation (4.2.71) is uniquely solvable for all sufficiently large N and $\|u_N - u_R\| \to 0, N \to \infty$, where u_R solves equation (4.2.69) with $u_0 = u_{0R}$. This is true under the assumption that $I + T$ is invertible. The last statement of the proposition can be verified immediately. It remains to be proved that $I + T$ is invertible. This follows from compactness of T and Lemma 4.2.13. □

Lemma 4.2.13. *If $(I + T)w = 0$, then $w = 0$.*

Proof. If $(I+T)w = 0$, then w solves equation (4.2.66) and satisfies the radiation condition. Therefore, $w = 0$. Indeed the radiation condition can be written as

$$\begin{aligned} 0 &= \lim_{r\to\infty} \int_{S_r} |w_r - ikw|^2 ds \\ &= \lim_{r\to\infty} \left\{ \int_{S_r} [|w_r|^2 + k^2|w|^2] ds + ik \int_{S_r} (\bar{w} w_r - w\bar{w}_r) ds \right\}. \end{aligned} \tag{4.2.73}$$

One can show that (4.2.67) implies

$$\lim_{r\to\infty} \int_{S_r} (\bar{w} w_r - w\bar{w}_r) ds = 0. \tag{4.2.74}$$

Therefore,

$$\lim_{r\to\infty} \int_{S_r} (|w_r|^2 + k^2|w|^2) ds = 0. \tag{4.2.75}$$

This and the remark at the end of the proof of Lemma 1.2.1 imply that $w = 0$ in Ω_R, and, by the unique continuation property of solutions to elliptic equation, $w = 0$ in $\mathbb{R}^3$. Let us finally check (4.2.74). Multiply equation (4.2.66) by $\bar{w}$ and equation (4.2.66) for $\bar{w}$ by w, subtract and use the Green's formula. As a result, one obtains (4.2.74). □

Remark 4.2.5. If $\operatorname{Im} q \neq 0$, then Lemma 4.2.13 is not valid: there exists a potential $q(x) \in C_0^\infty$ such that equation (4.2.66) has a nontrivial solution satisfying the radiation condition.

Example. Take $w = r^{-1}\exp(ikr)$ if $r = |x| \geq 1$ and continue w in B_1 as a function $w \in C^\infty$ such that $|w| \geq \epsilon > 0$. This is possible since $|w| = 1$ on ∂B_1. Define

$$q(x) = 0, \quad r \geq 1, \tag{4.2.76}$$

$$q(x) = w^{-1}(x)(\nabla^2 + k^2)w \text{ in } B_1. \tag{4.2.77}$$

Then $q \in C_0^\infty, w \neq 0$ satisfies equation (4.2.66) with this q and the radiation condition.

4.2.9. Consider the transmission problem (scattering by a penetrable body) (1.1.40)–(1.1.44). In Section 1.2.6, Theorem 1.2.9, the existence and uniqueness of the solution to this problem is established. Notations introduced in Section 1.2.6 are used below. Let $\{\phi_j\}$ be a complete linearly independent system of functions in H^0 and $\{\psi_j\}$ be a complete linearly independent system in $H^{-1/2}(\Gamma)$. The union of the systems $\begin{Bmatrix} \phi_j \\ 0 \end{Bmatrix}$ and $\begin{Bmatrix} 0 \\ \psi_j \end{Bmatrix}$ is a complete linearly independent system in $H_0 = H^0(D) \oplus H^{-1/2}(\Gamma)$. Let $\psi_j = \partial\phi_j/\partial N_+$ and $\{\phi_j\}$ be the orthonormal (in $L^2(\Delta)$) system of eigenfunctions of the Dirichlet Laplacian, where $\Delta \supset D$ is a box or a ball, so that the system, $\{\phi_j\}$ can be given explicitly. The system $\{\phi_j\}$ is complete in H^0 while the system $\{\partial\phi_j/\partial N\}$ is complete in $H^{-1/2}(\Gamma)$. Indeed, let

$$\int_\Gamma f \overline{\frac{\partial \phi_j}{\partial N}} ds = 0, \quad \forall j. \tag{4.2.78}$$

Multiply (4.2.78) by $\phi_j(x)/k_j^2$, where $(\nabla^2 + k_j^2)\phi_j = 0$ in Δ, $\phi_j = 0$ on $\partial\Delta$, and sum over j. Since

$$\sum_{j=1}^{\infty} \phi_j(x)\overline{\phi_j(s)}k_j^{-2} = G(x,s),$$

where

$$-\nabla^2 G = \delta(x-y) \text{ in } \Delta, \quad G = 0 \text{ on } \partial\Delta,$$

one obtains

$$\int_\Gamma f(s) \frac{\partial G(x,s)}{\partial N} ds = 0, \quad x \in \Delta. \tag{4.2.79}$$

From (4.2.79) and the jump relations for the potential of the double layer, one concludes that $f = 0$.

Let

$$w_m = \begin{pmatrix} \sum_{j=1}^{m} c_j^{(m)} \phi_j \\ \sum_{j=1}^{m} d_j^{(m)} \frac{\partial \phi_j}{\partial N} \end{pmatrix}. \tag{4.2.80}$$

The projection method leads to the system

$$(w_m - Bw_m - h, \eta_j)_{H^0} = 0, \quad 1 \le j \le 2m, \tag{4.2.81}$$

for finding $c_j^{(m)}$ and $d_j^{(m)}$. Here, B and h are given in (1.2.35) and

$$\eta_j = \begin{pmatrix} \phi_j \\ 0 \end{pmatrix}, \quad 1 \le j \le m,$$
$$\eta_j = \begin{pmatrix} 0 \\ \dfrac{\partial \phi_j}{\partial N} \end{pmatrix}, \quad m+1 \le j \le 2m. \tag{4.2.82}$$

System (4.2.81) is a linear system of $2m$ equations for $2m$ unknowns $c_j^{(m)}$ and $d_j^{(m)}$. The result in Example 4.2.1 shows that system (4.2.81) is uniquely solvable for all sufficiently large m and $w_m \to w$ in H^0 when $m \to \infty$, where w solves equation (1.2.34).

One can reduce by half the number of unknowns if one uses equation (1.2.32) (which is the second equation in the vector equation (1.2.34)) in the form (1.2.29) and set

$$d_j^{(m)} = c_j^{(m)}. \tag{4.2.83}$$

In this case, system (4.2.81) takes the form

$$\sum_{j=1}^{m} a_{nj} c_j^{(m)} = u_{0n}, \quad 1 \le n \le m, \tag{4.2.84}$$

where

$$u_{0n} = (u_0, \phi_n), \quad (f, g) = (f, g)_{H_0}, \tag{4.2.85}$$

$$a_{nj} = (\phi_j, \phi_n) + \kappa(T\phi_j, \phi_n) - (1-\rho)\left(Q\frac{\partial \phi_j}{\partial N}, \phi_n\right) \tag{4.2.86}$$

and κ, ρ, T and Q are defined by formulas (1.2.26), (1.1.43), (1.1.20), and (1.1.6), respectively. One can also obtain the system (4.2.84) by applying the projection method to the equation

$$u = F - \kappa T y + (1-\rho) Q \frac{\partial u}{\partial N_+} \tag{4.2.87}$$

which is equation (1.2.28) with σ given by (1.2.29).

Convergence of the scheme based on equation (4.2.84) does not follow from the result in Example 4.2.1 because the operator $Q\partial/\partial N_+$ in (4.2.87) is not compact in H^0. It is not known if this scheme converges. It does converge if $\rho = 1$.

4.2.10. One can suggest a numerical solution to exterior boundary value problems based on the completeness of the systems (4.2.60) in $L^2(\Gamma)$. Consider, for example, problem (4.1.1)–(4.1.3). If $D \subset B_R$, then the solution to problem (4.1.1)–(4.1.3) in Ω_R can be represented by the series

$$u = \sum_{n=0}^{\infty} c_n \psi_n(x), \quad x \in \Omega_R \tag{4.2.88}$$

which converges uniformly in $x \in K$, where $K \subset \Omega_R$ is compact. Define

$$u_m(x) = \sum_{n=0}^{m} c_n(m) \psi_n(x), \quad x \in \bar{\Omega}, \tag{4.2.89}$$

where the numbers $c_n(m)$ are chosen so that

$$\left\| f - \sum_{n=0}^{m} c_n(m) \psi_n(s) \right\|_{L^2(\Gamma)} = \epsilon(m) \to 0 \quad \text{as } m \to \infty. \tag{4.2.90}$$

For example, one can choose $c_n(m)$ by solving the minimization problem

$$\epsilon(m) = \min . \tag{4.2.91}$$

This is equivalent to finding the distance from f to the subspace span $\{\psi_0(s) \dots \psi_m(s)\}$. Because of the completeness of the system $\{\psi_n(s)\}$ in $L^2(\Gamma)$, condition (4.2.90) can be satisfied. It will be satisfied if the numbers $c_n(m)$ are chosen so that (4.2.91) holds. Problem (4.2.91) determines $c_n(m)$ uniquely because the system $\{\psi_0 \dots \psi_m\}$ is linearly independent in $L^2(\Gamma)$. If (4.2.90) holds, then

$$|c_n - c_n(m)| \to 0 \quad \text{as } m \to \infty, \tag{4.2.92}$$

uniformly in $n, 0 \le n \le n_0$, where n_0 is an arbitrary large fixed number.

To prove (4.2.92), let us use the formulas

$$u(x) = \int_\Gamma u(s) \frac{\partial G(x,s)}{\partial N} ds, \quad u_m(s) = \int_\Gamma u_m(s) \frac{\partial G(x,s)}{\partial N} ds, \tag{4.2.93}$$

where G is the Green function corresponding to problem (4.1.1)–(4.1.3):

$$(\nabla^2 + k^2)G = -\delta(x-y) \text{ in } \Omega, \quad G = 0 \text{ on } \Gamma,$$

and G satisfies the radiation condition.

It follows from (4.2.93) that

$$|u(x) - u_m(x)| \le \left(\int_\Gamma |u - u_m|^2 ds \right)^{1/2} \times \left(\int_\Gamma \left| \frac{\partial G(x,s)}{\partial N_s} \right|^2 ds \right)^{1/2}, \quad x \in \Omega. \tag{4.2.94}$$

Therefore,

$$\begin{aligned} &\int_{|x|=R} |u(x) - u_m(x)|^2 ds \le \epsilon^2(m) c(R), \\ &c(R) = \int_{|x|=R} ds \int_\Gamma \left| \frac{\partial G(x,s')}{\partial N} \right|^2 ds'. \end{aligned} \tag{4.2.95}$$

From (4.2.88), (4.2.89), and (4.2.95), it follows that

$$\sum_{n=0}^{m} |c_n - c_n(m)|^2 + \sum_{n>m} |c_n|^2 \to 0 \text{ as } m \to \infty. \tag{4.2.96}$$

The relation (4.2.92) follows from (4.2.96) immediately.

Let us summarize the results.

Lemma 4.2.14. *If the numbers $c_n(m)$ are chosen so that* (4.2.90) *holds, then* (4.2.92) *holds uniformly in $n, 0 \le n \le n_0$, where n_0 is an arbitrary large fixed number. The solution to problem* (4.1.1)–(4.1.3) *can be computed by formula* (4.2.88) *in which c_n are defined by* (4.2.91)*. The function* (4.2.89) *is an approximate solution to problem* (4.1.1)–(4.1.3) *in the sense that*

$$\sup_{r\ge R} \|u(x) - u_m(x)\|_{L^2(S_r)} \to 0 \quad \text{as } m \to \infty, \tag{4.2.97}$$

$$\max_{x\in\tilde{\Omega}} |D^j u(x) - D^j u_m(x)| \to 0 \quad \text{as } m \to \infty, \quad j = 0, 1, 2, \dots, \tag{4.2.97$'$}$$

where $\tilde{\Omega} \subset \Omega$ is an arbitrary compact subdomain.

Proof. Equation (4.2.97) follows from (1.2.72). Equation (4.2.97′) for $j = 0$ follows from inequality (4.2.94), and for $j > 0$ it follows in a similar way. The rest has already been proved. □

Remark 4.2.6. If f can be represented by the convergent in $L^2(\Gamma)$ series

$$f = \sum_{n=0}^{\infty} c'_n \psi_n(s), \tag{4.2.90$'$}$$

then, as (4.2.96) shows, $c'_n = c_n$, where c_n are the coefficients in the series (4.2.88). In this case, problem (4.1.1)–(4.1.3) is solved by formula (4.2.88) if the coefficients in the series (4.2.90′) are known.

4.2.11. Let us discuss briefly the completeness properties of solutions to equation (4.1.1) in a compact subdomain $\tilde{\Omega} \subset \Omega$. Let u be an arbitrary solution to equation (4.1.1) in Ω. Consider u in $\tilde{\Omega}$. It is well known that u in $\tilde{\Omega}$ is a C^∞ function. Let us prove that the system $\{u_0(x,n,k)\}, n \in S^2, u_0 = \exp(-ikn \cdot x)$ is complete in the linear manifold of solutions to equation (4.1.1). The number $k > 0$ is fixed.

Lemma 4.2.15. *If u solves equation* (4.1.1) *and*

$$\int_{\tilde{\Omega}} u(x)u_0(x,n,k)dx = 0, \quad \forall n \in S^2, \tag{4.2.98}$$

then $u = 0$ in $\tilde{\Omega}$.

Proof. Let

$$v(x) = \int_{\tilde{\Omega}} g(x,y)u(y)dy. \tag{4.2.99}$$

Then $v(x)$ solves equation (4.1.1) in Ω_R (where R is sufficiently large so that $\tilde{\Omega} \subset B_R$) and satisfies the radiation condition. It follows from (4.2.98) that

$$v(x) = O(|x|^{-2}) \quad \text{as } |x| \to \infty. \tag{4.2.100}$$

Thus, by Lemma 1.2.1, $v = 0$ in Ω_R. By the unique continuation, property of solutions to Helmholtz's equation

$$v = \frac{\partial v}{\partial N} = 0 \quad \text{on } \partial\tilde{\Omega}. \tag{4.2.101}$$

Moreover,

$$(\nabla^2 + k^2)v = u(x) \quad \text{in } \tilde{\Omega}. \tag{4.2.102}$$

From (4.2.101), (4.2.102), and equation (4.1.1) for u, it follows that $(\nabla^2 + k^2)v = 0$ in $\tilde{\Omega}$, and, therefore, $u = 0$ in $\tilde{\Omega}$. Indeed, from (4.1.1) and (4.2.102), it follows that

$$(\nabla^2 + k^2)^2 v = 0 \quad \text{in } \tilde{\Omega}. \tag{4.2.103}$$

Multiply (4.2.103) by $\bar{v}$ and integrate over $\tilde{\Omega}$. Apply Green's formula to the identity and use (4.2.101) to obtain

$$0 = \int_{\tilde{\Omega}} (\nabla^2 + k^2)^2 v\bar{v} dx = \int_{\tilde{\Omega}} |(\nabla^2 + k^2)v|^2 dx. \tag{4.2.104}$$

Thus, $(\nabla^2 + k^2)v = 0$, and, by (4.2.102), $u = 0$ in $\tilde{\Omega}$. $\square$

Remark 4.2.7. Similar result holds for solutions to the Schrödinger equation

$$(-\nabla^2 + q(x) - k^2)u = 0, \quad k > 0, \tag{4.2.105}$$

where

$$|q(x)| = o(|x|^{-1}) \quad \text{as } |x| \to \infty. \tag{4.2.106}$$

The role of u_0 in this case is played by the solutions to the scattering problem:

$$u_0(x, n, k) = \exp(-ikn \cdot x) + v, \tag{4.2.107}$$

where u_0 solves equation (4.2.105) and v satisfies the radiation condition.

4.3. Numerical Results

4.3.1. The purpose of the numerical results presented here is to illustrate the properties of the commonly used complete sets of functions in $L^2(\Gamma)$ as basis systems in $H_0 = L^2(\Gamma)$. Do they form a Riesz basis of H_0? Are they good for expanding of functions $f \in H_0$? Consider first a simple example of a complete system in $L^2[0, \pi]$:

$$\phi_n(x) = (2\pi^{-1})^{1/2} \sin nx \exp(-qnx), \quad n = 1, 2, \ldots. \tag{4.3.1}$$

This set is obtained from an orthonormal basis of $L^2[0, \pi]$ consisting of the functions $\phi_{0n} = (2\pi^{-1})^{1/2} \sin nx, n = 1, 2, \ldots,$ by multiplying ϕ_{0n} by the function $\exp(-qnx)$. Qualitatively, the set $\{\phi_n\}$ is similar to the set (4.2.60). The constant $q > 0$ will serve in the case of the system (4.2.60) as a measure of the eccentricity of the obstacle D.

The Gram matrix (ϕ_m, ϕ_n) can be calculated analytically:

$$\int_0^\pi \phi_m \bar{\phi}_n dx = \pi^{-1} q(m+n)[1 - (-1)^{m+n} \exp(-q\pi(n+m))] \times \left\{ \frac{1}{q^2(m+n)^2 + (m-n)^2} - \frac{1}{q^2(m+n)^2 + (m+n)^2} \right\}. \tag{4.3.2}$$

Let $\kappa = \|A\| \, \|A^{-1}\|$ be the *condition number* of a matrix A. The Gram matrix is always self-adjoint nonnegative definite, and if the system $\{\phi_j\}$ is linearly independent, then the matrix $(\phi_i, \phi_j), 1 \leq i, j \leq m$ is positive definite. For positive definite matrices A, the condition number $\kappa(A) = \Lambda\lambda^{-1}$ where $\lambda(\Lambda)$ is the minimal (maximal) eigenvalue of A. In Table 4.1, the condition number is given for the matrix (ϕ_i, ϕ_j) and (in parentheses) for the matrix (ϕ_i^0, ϕ_j^0) where $\{\phi_j^0\} = \{\phi_j/\|\phi_j\|\}$ is a normalized system. It is seen that: (1) $\kappa = \kappa(m, q)$ grows rapidly as m grows, where m is the size of the matrix $(\phi_i, \phi_j), 1 \leq i, j \leq m$; (2) $\kappa(m, q)$ increases as q increases; (3) the Gram matrix for the normalized system has a smaller condition number than that for the corresponding nonnormalized system.

The system $\{\phi_j\}$ and the biorthogonal system $\{\tilde{\phi}_j\}$ do not form a Riesz basis since the necessary conditions $\inf_m \|\phi_m\| > 0$, $\sup_m \|\tilde{\phi}_m\| < \infty$ are violated as Table 4.1 shows.

Table 4.1. The condition number κ as a function of different truncation sizes and eccentricities for the model problem (4.3.2).

	Truncation size							
Eccentricity	5	5	10	10	20	20	40	40
$q = 0.1$	6	(4)	60	(40)	8×10^3	(6×10^3)	2×10^8	(1×10^8)
$q = 1/3$	1×10^2	(80)	7×10^4	(4×10^4)	3×10^{10}	(2×10^{10})	$>10^{16}$	$(>10^{16})$
$q = 1$	8×10^3	(5×10^3)	1×10^9	(6×10^8)	$>10^{16}$	$(>10^{16})$	$>10^{16}$	$(>10^{16})$

Note: The corresponding condition number κ for the normalized functions $\phi_m \cdot (\phi_m, \phi_m)^{-1/2}$ is given in parentheses. This table is taken from [108].

4.3.2. The following systems have similar properties: $\{\psi_n\}$, $\{\psi_n^{\text{re}}\}$, and $\{\chi_n\}$, where $\chi_n = (kr)^{-l-1}Y_n(\omega), n = (l, m)$,

$$Y_n = \left\{ \frac{(2l+1)(l-m)!}{4\pi(l+m)!} \right\}^{1/2} P_{l,m}(\cos\theta)\exp(\mathrm{i}m\phi), \quad l = 0, 1, 2, \ldots,$$

$-l \leq m \leq l, P_{l,m}(\cos\theta)$ are the associated Legendre functions, $\psi_n(x) = h_n^{(1)}(kr)Y_n(\omega)$, $\psi_n^{\text{re}}(x) = j_n(kr)Y_n(\omega), h_n^{(1)}(kr)$ is the spherical Hankel function of the first kind and $j_n(kr)$ is the spherical Bessel function. The system $\{Y_n(\omega)\}, n = 0, 1, 2, \ldots$ forms an orthonormal basis of $L^2(S^2)$. If $n \gg kr$, then

$$\psi_n \sim -i(2n-1)!!(kr)^{-n-1}Y_n(\omega), \tag{4.3.3}$$

$$\psi_n^{\text{re}} \sim [(2n+1)!!]^{-1}(kr)^n Y_n(\omega). \tag{4.3.4}$$

Therefore, ψ_n and ψ_n^{re} for large n and $r = r(\omega)$, where $r = r(\omega)$ is the equation of the surface Γ, the boundary of an obstacle D, have a structure similar to that of functions (4.3.1): both systems are obtained from orthonormal systems by multiplication by functions of the form $\exp(n\rho(\omega))$, where $\rho = -qx$ in (4.3.1) and $\rho = \ln\{kr(\omega)\}$ in (4.3.3) and (4.3.4). This similarity motivated us to consider the model example (4.3.1).

Let us consider a spheroidal obstacle. The equation for its surface Γ is

$$r(\theta) = (a^{-2}\sin^2\theta + b^{-2}\cos^2\theta)^{-1/2}, \tag{4.3.5}$$

where a and b are the semi-axes, the spheroid is axially symmetric and the semi-axis corresponding to b is the axis of symmetry. Since $r(\theta) = r(\pi - \theta)$, the elements of the Gram matrix are zeros if $n + n'$ is odd. One can change the numeration of the columns and rows in the matrix so that it becomes a block diagonal matrix with two blocks. The size of the first block, which corresponds to the rows with even numbers, is $(n_{\max} + 1)/2$ if $n_{\max}$ is odd, and $n_{\max}/2 + 1$ if $n_{\max}$ is even. In Table 4.2, the condition numbers of the truncated Gram matrices for the systems $\{\psi_n\}, \{\psi_n^{\text{re}}\}, \{\chi_n\}$ and $\{Y_n\}$ on Γ are given. The numerical results in Table 4.2 indicate that the systems $\{\psi_n\}, \{\psi_n^{\text{re}}\}$ and $\{\chi_n\}$ do not form a Riesz basis of $L^2(\Gamma)$. However,

Table 4.2. The condition number κ as a function of truncation sizes and eccentricities for three different spherical waves $\{\psi_n\}$, $\{\mathrm{Re}\,\psi_n\}$, $\{\chi_n\}$, and the spherical harmonics $\{Y_n\}$.

	5 even	5 odd	9 even	9 odd	19 even	19 odd
$ka = 4$						
$kb = 2$						
ψ_n	30 (3)	400 (4)	$2 \times 10^7 (30)$	$1 \times 10^9 (50)$	$> 10^{16} (3 \times 10^4)$	$>10^{16} (6 \times 10^4)$
$\mathrm{Re}\,\psi_n$	30 (20)	400 (10)	$5 \times 10^6 (200)$	$2 \times 10^8 (400)$	$>10^{16} (1 \times 10^6)$	$>10^{16} (3 \times 10^6)$
χ_n	$1 \times 10^3 (5)$	$1 \times 10^3 (8)$	$2 \times 10^6 (50)$	$2 \times 10^6 (100)$	$1 \times 10^{14} (4 \times 10^4)$	$1 \times 10^{14} (1 \times 10^5)$
Y_n	3 (3)	3 (3)	4 (4)	4 (4)	4 (4)	4 (4)
$ka = 6$						
$kb = 2$						
ψ_n	30 (3)	40 (5)	$2 \times 10^7 (40)$	$1 \times 10^9 (90)$	$>10^{16} (1 \times 10^5)$	$>10^{16} (3 \times 10^5)$
$\mathrm{Re}\,\psi_n$	400 (400)	20 (80)	$4 \times 10^4 (500)$	$8 \times 10^5 (1 \times 10^3)$	$>10^{16} (1 \times 10^8)$	$>10^{16} (3 \times 10^8)$
χ_n	$2 \times 10^4 (6)$	$2 \times 10^4 (10)$	$3 \times 10^6 (100)$	$3 \times 10^6 (200)$	$5 \times 10^{14} (2 \times 10^5)$	$4 \times 10^{14} (4 \times 10^5)$
Y_n	7 (7)	6 (6)	9 (9)	9 (9)	10 (10)	10 (10)
$ka = 10$						
$kb = 2$						
ψ_n	30 (3)	300 (5)	$2 \times 10^7 (50)$	$1 \times 10^9 (100)$	$>10^{16} (2 \times 10^5)$	$>10^{16} (4 \times 10^5)$
$\mathrm{Re}\,\psi_n$	30 (30)	3 (2)	60 (30)	600 (200)	$5 \times 10^{14} (8 \times 10^8)$	$7 \times 10^{15} (2 \times 10^9)$
χ_n	$3 \times 10^3 (6)$	$2 \times 10^3 (10)$	$5 \times 10^6 (100)$	$4 \times 10^6 (300)$	$1 \times 10^{14} (3 \times 10^5)$	$7 \times 10^{14} (8 \times 10^5)$
Y_n	20 (20)	10 (10)	30 (30)	30 (30)	40 (40)	40 (40)

Note: The corresponding condition number for the normalized functions is given in parentheses, $m = 0$ in all cases. This table is taken from [108].

the system $\{Y_n(\omega)\} = Y_n\{p^{-1}(s)\}$ on Γ does form a Riesz basis of $L^2(\Gamma)$. Here, $s = p(\omega)$ is the equation of Γ and $p : S_2^2 \to \Gamma$ is assumed to be a diffeomorphism. The three properties of the system $\{\tilde{\phi}_j\}$ formulated in Section 4.1 hold for the system $\{\psi_n\}$, $\{\psi_n^{\text{re}}\}$ and $\{\chi_n\}$. One sees that the Gram matrices of the systems $\{\psi_n\}$ and $\{\chi_n\}$ have smaller condition numbers than that for the system $\{\psi_n^{\text{re}}\}$. The numerical results indicate that the normalized functions should be used for expansions of the field on the surface since their Gram matrix has a smaller condition number. A large condition number means that the corresponding Gram matrix is difficult to invert numerically. It means numerical instability, ill-posedness, i.e., strong dependence of the numerical results on round-off errors and errors in the data.

Let us prove that the system $Y_n(p^{-1}(s))$ forms a Riesz basis of $L^2(\Gamma)$.

Lemma 4.3.1. *Assume that the surface Γ can be defined by the equation $s = p(\omega)$ where $p : S^2 \to \Gamma$ is a diffeomorphism. Then the system $\{Y_n(p^{-1}(s))\}, n = 0, 1, 2, \ldots$ forms a Riesz basis of $L^2(\Gamma)$.*

Proof. If ds is the element of the area of the surface Γ and $d\omega$ is the corresponding element of the area of S^2, then our assumption implies that $0 < c_1 \leq ds/d\omega \leq c_2$. Let $q(s) = (ds/d\omega)^{-1/2}$ and

$$\phi_n(s) = Y_n(p^{-1}(s))q(s). \tag{4.3.6}$$

Then

$$\begin{aligned}\int_\Gamma \phi_n(s)\overline{\phi_m(s)}ds &= \int_{S^2} Y_n(\omega)\overline{Y_n(\omega)}q^2\frac{ds}{d\omega}d\omega \\ &= \int_{S^2} Y_n(\omega)\overline{Y_n(\omega)}d\omega = \delta_{nm}.\end{aligned} \tag{4.3.7}$$

The operator $T : L^2(\Gamma) \to L^2(\Gamma)$ defined by the formula

$$Tf = q(s)f \tag{4.3.8}$$

is an isomorphism of $L^2(\Gamma)$ onto $L^2(\Gamma)$ because

$$c_2^{-1/2} \leq q(s) \leq c_1^{-1/2}. \tag{4.3.9}$$

From (4.3.6) and (4.3.7), it follows that the system $\{Y_n(p^{-1}(s))\}$ forms a Riesz basis of $L^2(\Gamma)$. □

Remark 4.3.1. If $p: S^2 \to \Gamma$ is a C^{l+1} diffeomorphism, then $Y_n(p^{-1}(s))$ is a Riesz basis of $H^l(\Gamma), l \geq 0$. The proof is similar to that given above.

4.4. Other Schemes

4.4.1. Problem (4.1.1)–(4.1.3) can be reduced in many ways to various boundary integral equations. These equations were discussed in Section 1.1. The classical approach is to look for the solution in the form of a double layer potential. This leads to equation (1.1.12) which is not solvable for all $k > 0$, and therefore is not equivalent to the problem (4.1.1)–(4.1.3) at some discrete set of $k_j > 0$, namely at those values k_j^2 which are eigenvalues of the interior Neumann Laplacian. The problem (4.1.1)–(4.1.3) is uniquely solvable for all $k > 0$. Therefore, equation (1.1.12) is not convenient for numerical solution especially for large k, because for large k_j the distance between k_j^2 becomes very small: $k_j^2 \sim j^{2/3}$ as $j \to \infty$. Thus, every large k lies in a neighborhood of an exceptional value k_j at which equation (1.1.2) is not solvable for some f. Numerically this means that the inverse of the operator $I + A'(k)$, if it exists, has a large norm.

4.4.2. Another integral equation, equation (1.1.7), was derived in Section 1.1 from Green's representation (1.1.5). This equation is solvable for all $k > 0$ and all f. But for the exceptional values k_j (which are sometimes called irregular frequencies) this equation has more than one solution if no other conditions on the solution are imposed. Uniqueness can be guaranteed, for example, if the solution satisfies both equations (1.1.6) and (1.1.7). Therefore, again one has difficulties in the numerical treatment.

4.4.3. Equation (1.1.16) is uniquely solvable for all $k > 0$ and is therefore more convenient for the numerical treatment.

But the representation (1.1.15) is not very convenient for the Neumann and Robin boundary conditions since the derivative of the double layer potential leads to a strong singularity in the kernel of the boundary integral equation.

4.4.4. It is not quite convenient numerically to use the so-called modified Green functions of the type (1.1.26). First, if one takes a finite number of terms in the sum of the right-hand side of formula (1.1.26), then only a finite interval $[0, k_0]$ is free from the irregular frequencies. Moreover, the modified Green functions are more complicated than the free space Green function.

4.4.5. One can use an equation of the first kind which is convenient from the following points of view: (1) it is solvable for any $k > 0$; (2) at the values of k_j at which it has more than one solution all of these solutions will produce the same (unique) solution to problem (4.1.1)–(4.1.3). Let us discuss this in more detail. Consider equation (1.1.8) which is an integral equation of the first kind with respect to h. If h solves this equation, then the unique solution to problem (4.1.1)–(4.1.3) is given by the formula

$$u(x) = Df - Sh. \tag{4.4.1}$$

Suppose that equation (1.1.8):

$$Qh = \frac{A'f - f}{2} = (Df)_+ \tag{4.4.2}$$

has more than one solution. Let

$$h = h_p + h_0 \tag{4.4.3}$$

be the general solution to (4.4.2), where h_0 is the general solution to the homogeneous equation (4.4.2) and h_p is a particular solution to (4.4.2).

Lemma 4.4.1. *The solution $u(x)$ to problem* (4.1.1)–(4.1.3) *given by formula* (4.4.1) *does not depend on the choice of the solution to equation* (4.4.2).

Proof. It is sufficient to prove that

$$Sh_0 = 0 \text{ in } \Omega \tag{4.4.4}$$

provided that $Qh_0 = 0$. But this follows from Lemma 1.2.2 since Sh_0 solves the homogeneous problem (4.1.1)–(4.1.3) and this problem has only the trivial solution. □

Lemma 4.4.2. *Equation* (4.4.2) *is solvable for any* $k > 0$.

Proof. This follows from the fact that problem (4.1.1)–(4.1.3) is solvable (see Section 1.2) and the solution can be represented by Green's formula (1.1.5).

Equation (4.4.2) is therefore convenient for numerical purposes. One can look for any solution to this equation. Practically it is useful to look for the solution with minimal norm. This solution is stable towards perturbations of f and round-off errors.

A numerical scheme for solving equation (4.4.2) can be formulated: take a complete linearly independent system of elements $\{\phi_j\} \in H_0 = L^2(\Gamma)$, let

$$h_m = \sum_{j=1}^{m} c_j\phi_j, \quad Q\phi_j = \psi_j, \quad c_j = \text{ const.}$$

Let

$$\left\| \sum_{j=1}^{m} c_j\psi_j - F \right\| = \min, \quad F = (Df)_+, \tag{4.4.5}$$

where $\|\cdot\|$ is the norm in H_0. Variational problem (4.4.5) leads to a linear system for the coefficients c_j. This system is uniquely solvable if the Gramian $\det(\psi_i, \psi_j) \neq 0, 1 \leq i, j \leq m$. If $\det(\psi_i, \psi_j) = 0$, then (4.4.5) is still solvable because geometrically (4.4.5) is the problem of finding the distance from F to the finite dimensional space $\text{span}(\psi_1, \dots, \psi_m)$. It will have many solutions in the case $\det(\psi_i, \psi_j) = 0$, and one should find among them the solution with minimal norm. Such solution exists and is unique. Indeed, for a linear equation $Ah = f$ finding the solution with minimal norm among the set of all solutions $h_p + h_0$ is equivalent to solving the problem of finding the distance from h_p to the null space $N(A)$ of the operator A. If $\dim N(A) < \infty$, this problem is uniquely solvable. In the case of the finite dimensional equation (4.4.5), this problem is always uniquely solvable. Notice that the operator Q has a finite dimensional null space, $\dim N(Q) = q < \infty$, and therefore for any m the null-space of the Gramian has dimension $\leq q$ and the rank of the Gramian is $\leq m - q$ for all m. □

4.4.6. We will not discuss here finite differences and finite element methods (see [72]).

Chapter 5

Scattering by Small Bodies

5.1. Scattering by a Single Small Body

5.1.1. If the body $\mathcal{D}$ is small in comparison with the wavelength, i.e., $ka \ll 1$, where a is the characteristic dimension of $\mathcal{D}$ and k is the wave number, then one does not have to solve integral equations in order to find the scattering amplitude. The basic idea in this chapter is as follows: for small bodies, the scattering amplitude depends on some functionals of the field: on capacitances or polarizability tensors as we will see shortly. If $ka \ll 1$, then one can find linear algebraic equations for these quantities. Let us illustrate this by an example. Consider problem (0.0.1)–(0.0.5). For the scattered field v, one has

$$(\nabla^2 + k^2)v = 0 \quad \text{in } \Omega, \quad k > 0, \quad \Omega \subset \mathbb{R}^3, \tag{5.1.1}$$

$$v = -u_0 = -\exp(ikn \cdot s) \quad \text{on } \Gamma \tag{5.1.2}$$

$$\begin{aligned} &v \sim \gamma(r) f(n, \nu, k) \quad \text{as} \quad |x| = r \to \infty, \quad xr^{-1} = \nu, \\ &\gamma(r) = r^{-1}\exp(ikr). \end{aligned} \tag{5.1.3}$$

Let

$$v = \int_\Gamma g(x,t)\sigma(t)dt, \quad g = \frac{\exp(ik|x-y|)}{4\pi|x-y|}. \tag{5.1.4}$$

Then equation (5.1.1) is satisfied, the radiation condition (5.1.3) is satisfied, and

$$f(n, \nu, k) = \frac{1}{4\pi}\int_\Gamma \exp(-ik\nu \cdot t)\sigma(t)dt. \tag{5.1.5}$$

The boundary condition (5.1.2) leads to the equation for σ:

$$Q\sigma = \int_\Gamma g(s,t)\sigma(t)dt = -\exp(ikn \cdot s). \tag{5.1.6}$$

So far the assumption $ka \ll 1$ was not used. For an arbitrary ka, one should solve equation (5.1.6) for σ and then compute the scattering amplitude by formula (5.1.5). If $ka \ll 1$, then equation (5.1.5) can be written as

$$f(n,\nu,k) = \frac{1}{4\pi}\int_\Gamma \sigma(t)dt(1+O(ka)) \quad \text{if} \int_\Gamma \sigma dt \neq 0. \tag{5.1.7}$$

Equation (5.1.6) can be written as

$$\int_\Gamma \frac{\sigma(t)dt}{4\pi|s-t|} = -1 + O(ka), \quad ka \ll 1. \tag{5.1.8}$$

If one neglects the term $O(ka)$, then equation (5.1.8) becomes

$$\int_\Gamma \frac{\sigma(t)dt}{4\pi|s-t|} = -1. \tag{5.1.9}$$

This equation can be interpreted as the equation for the equilibrium charge distribution on the surface Γ of the perfect conductor $\mathcal{D}$ charged to the potential $U = -1$. Formula (5.1.7) shows that the scattering amplitude is defined by one number $q = \int_\Gamma \sigma dt$, which has the physical meaning of the total charge on the conductor. If C is the electrical capacitance of the conductor $\mathcal{D}$, then $q = CU = -C$. Thus, it follows from (5.1.9) that

$$\int_\Gamma \sigma dt = -C \tag{5.1.10}$$

and

$$f(n,\nu,k) = -\frac{C}{4\pi}. \tag{5.1.11}$$

The assumption $\int_\Gamma \sigma dt \neq 0$ is satisfied. One can see that there is no need to solve equation (5.1.9) for σ since all one needs is the number $\int_\Gamma \sigma dt$. Of course, there is a question of how to compute C. This question we answer later.

5.1.2. Let us pass to the general method for calculating the scattering amplitude by a small body of an arbitrary shape.

Let

$$(\nabla^2 + k^2)v = 0 \quad \text{in } \Omega, \quad k > 0, \tag{5.1.12}$$

$$\frac{\partial v}{\partial N_-} + \eta v = -\left(\frac{\partial u_0}{\partial N} + \eta u_0\right) \quad \text{on } \Gamma, \tag{5.1.13}$$

v satisfies (5.1.3).

Substitute (5.1.4) into (5.1.13) to obtain

$$\sigma = A(k)\sigma + 2\eta Q(k)\sigma + 2\eta u_0 + 2\frac{\partial u_0}{\partial N}, \tag{5.1.14}$$

where A and Q are defined in (1.1.4) and (1.1.6). Let us expand $\sigma = \sigma(k), A(k), 2Q(k)$, and u_0 in powers of ik:

$$\sigma = \sigma_0 + ik\sigma_1 + \frac{(ik)^2}{2}\sigma_2 + \cdots, \tag{5.1.15}$$

$$A(k) = A_0 + ikA_1 + \frac{(ik)^2}{2}A_2 + \cdots, \tag{5.1.16}$$

$$2Q(k) = Q_0 + ikQ_1 + \frac{(ik)^2}{2}Q_2 + \cdots, \tag{5.1.17}$$

$$u_0 = u_{00} + iku_{01} + \frac{(ik)^2}{2}u_{02} + \cdots. \tag{5.1.18}$$

Convergence of the expansions (5.1.16)–(5.1.18) is obvious. Convergence of the expansion (5.1.15) follows from analyticity of v in k in a neighborhood of $k = 0$ and the fact that the operator $Q(k)$ in (5.1.6) is invertible and analytic in k in a neighborhood of $k = 0$.

From (5.1.15)–(5.1.18) and (5.1.14), it follows that

$$\sigma_0 = A_0\sigma_0 + \eta Q_0\sigma_0 + 2\eta u_{00} + 2\frac{\partial u_{00}}{\partial N}, \tag{5.1.19}$$

$$\sigma_1 = A_0\sigma_1 + \eta Q_0\sigma_1 + A_1\sigma_0 + \eta Q_1\sigma_0 + 2\eta u_{01} + 2\frac{\partial u_{01}}{\partial N}, \tag{5.1.20}$$

$$\sigma_2 = A_0\sigma_2 + \eta Q_0\sigma_2 + A_2\sigma_0 + 2A_1\sigma_1 + \eta Q_2\sigma_0 + 2\eta Q_1\sigma_1 + 2\eta u_{02} + 2\frac{\partial u_{02}}{\partial N}. \tag{5.1.21}$$

From (5.1.5) and (5.1.15), it follows that

$$\begin{aligned} 4\pi f &= \int_\Gamma \left[1 - ik\nu \cdot t + \frac{(ik)^2}{2}(\nu \cdot t)^2 + \cdots\right] \\ &\quad \times \left[\sigma_0 + ik\sigma_1 + \frac{(ik)^2}{2}\sigma_2 + \cdots\right] dt \\ &= \int_\Gamma \sigma_0 dt + ik\left[\int_\Gamma \sigma_1 dt - \int_\Gamma \nu \cdot t\sigma_0 dt\right] \\ &\quad + \frac{(ik)^2}{2}\left[\int_\Gamma \sigma_2 dt - 2\int_\Gamma \nu \cdot t\sigma_1 dt\right. \\ &\quad \left. + \int_\Gamma \sigma_0(\nu \cdot t)^2 dt\right] + \cdots . \end{aligned} \tag{5.1.22}$$

We assume that $u_0 = \exp(ikn \cdot x)$ and that the origin is inside $\mathcal{D}$. Then

$$u_{00} = 1, u_{01} = n \cdot s, \quad u_{02} = (n \cdot s)^2, \tag{5.1.23}$$

$$\frac{\partial u_{00}}{\partial N} = 0, \quad \frac{\partial u_{01}}{\partial N} = n \cdot N, \quad \frac{\partial u_{02}}{\partial N} = 2(n \cdot s)(n \cdot N), \tag{5.1.24}$$

where $s \in \Gamma$. We cannot take $s = 0$ in (5.1.23) and (5.1.24) because we are interested in the dependence of the scattering amplitude on the characteristic size a of the body. As one can see later, the formulas for the scattering amplitude will depend on the parameter ka. Let us note that

$$\begin{aligned} A_0\sigma &= \int_\Gamma dt\sigma(t)\frac{\partial}{\partial N_s}\frac{1}{2\pi r_{st}}, \\ A_1\sigma &= 0, \quad r_{st} = |s - t|, \end{aligned} \tag{5.1.25}$$

$$-\int_\Gamma A_0\sigma dt = \int_\Gamma \sigma dt, \tag{5.1.26}$$

$$Q_0\sigma = \int_\Gamma \frac{\sigma dt}{2\pi r_{st}},$$
$$Q_1\sigma = \frac{1}{2\pi}\int_\Gamma \sigma dt. \qquad (5.1.27)$$

Let us integrate (5.1.19) over Γ and use (5.1.23)–(5.1.27). This yields

$$\int_\Gamma \sigma_0 dt = \int_\Gamma \eta dt + \int_\Gamma\int_\Gamma \frac{\eta(s)\sigma_0(t)dsdt}{4\pi r_{st}}. \qquad (5.1.28)$$

If one assumes that $\eta =$ const and uses the approximation formula

$$\int_\Gamma \frac{ds}{r_{st}} \approx \frac{1}{S}\int_\Gamma dt\int_\Gamma \frac{ds}{r_{st}} = JS^{-1}, \quad J = \int_\Gamma\int_\Gamma \frac{dsdt}{r_{st}}, \qquad (5.1.29)$$

where S is the area of the surface Γ, then (5.1.28) can be written as

$$\int_\Gamma \sigma_0 dt \approx \eta S + \frac{\eta J}{4\pi S}\int_\Gamma \sigma_0 dt. \qquad (5.1.30)$$

Therefore,

$$\int_\Gamma \sigma_0 dt \approx \frac{\eta S}{1-\eta J(4\pi S)^{-1}}, \quad \eta J(4\pi S)^{-1} \ll 1. \qquad (5.1.31)$$

If one uses the approximate formula for the capacitance ([239, p. 88])

$$C \approx 4\pi S^2 J^{-1},$$

then (5.1.31) can be written as

$$\int_\Gamma \sigma_0 dt \approx \frac{\eta S}{1-\eta S C^{-1}}. \qquad (5.1.32)$$

From (5.1.22) and (5.1.32), it follows that

$$f = \frac{\eta S}{4\pi(1-\eta S C^{-1})}, \quad \eta S C^{-1} \ll 1, \qquad (5.1.33)$$

where C is the electrical capacitance of the perfect conductor with the shape and size of the scatterer $\mathcal{D}$. If $\eta \to \infty$, then the boundary condition (5.1.13) becomes condition (5.1.2), and formula (5.1.33) becomes formula (5.1.11). If $\eta \to 0$, then, up to the terms $O(ka)$ the scattering amplitude vanishes. Let us calculate the scattering amplitude for the limiting case $\eta = 0$ corresponding to the acoustically hard obstacle. Note that for the acoustically soft obstacle, that is, for the boundary condition (5.1.2), the main term in the scattering amplitude for a small body does not depend on n, and k, the scattering is isotropic, and $f \sim a$ since $C \sim a$.

In the next section, we will see that for the acoustically hard small obstacle, the main term in the scattering amplitude depends on n, ν, and k, the scattering is anisotropic, and $f \sim k^2a^3$.

5.1.3. Assume $\eta = 0$. Then (5.1.19) takes the form

$$\sigma_0 = A_0\sigma_0 + 2\frac{\partial u_{00}}{\partial N} = A_0\sigma_0. \tag{5.1.34}$$

Since 1 is not an eigenvalue of A_0 [79], equation (5.1.34) implies that $\sigma_0 = 0$. Let us prove that 1 is not an eigenvalue of A_0. Consider the potential of the single layer

$$w = \int_\Gamma \frac{1}{4\pi r_{st}}\sigma_0 dt.$$

Equation (5.1.34) and formula (1.1.3) imply that $\partial w/\partial N_- = 0$. Therefore, $w = 0$ in Ω and $w_- = 0$. Since $w_+ = w_-$, one has $w_+ = 0, \nabla^2 w = 0$ in $\mathcal{D}$. Thus, $w = 0$ in $\mathcal{D}$, and $\sigma_0 = \partial w/\partial N_+ - \partial w/\partial N_- = 0$.

Consider equation (5.1.20) for $\eta = 0$ and $\sigma_0 = 0$, and use (5.1.24) to obtain:

$$\sigma_1 = A_0\sigma_1 + 2\frac{\partial u_{01}}{\partial N} = A_0\sigma_1 + 2n \cdot N. \tag{5.1.35}$$

Integrating (5.1.35) over Γ and using (5.1.26), one obtains

$$2\int_\Gamma \sigma_1 dt = 2\int_\Gamma \frac{\partial u_{01}}{\partial N} dt = 2n \cdot \int_\Gamma N_t dt = 0. \tag{5.1.36}$$

Therefore, formula (5.1.22) when $\eta = 0$ takes the form

$$4\pi f = -\frac{k^2}{2}\int_\Gamma \sigma_2 dt + k^2 \int_\Gamma \nu \cdot t\sigma_1 dt. \tag{5.1.37}$$

If $\sigma_q = A\sigma_q - 2N_q$, then the tensor

$$\beta_{pq} = V^{-1} \int_\Gamma t_p \sigma_q dt \tag{5.1.38}$$

is the magnetic polarizability tensor of the body $\mathcal{D}$ [239, formula (5.1.19), p. 58], and V is the volume of the body $\mathcal{D}$. From (5.1.35) and (5.1.38), it follows that

$$k^2 \nu \int_\Gamma t\sigma_1 dt = -k^2 V \beta_{pq} n_p \nu_q. \tag{5.1.39}$$

Here and below, one should sum over the repeated indices.

In order to calculate the first integral in (5.1.37), one uses equation (5.1.21) and takes into account that $\eta = 0, \sigma_0 = 0$, and $A_1\sigma = 0$:

$$\sigma_2 = A_0\sigma_2 + 2\frac{\partial u_{02}}{\partial N} = A_0\sigma_2 + 4(n \cdot s)(n \cdot N). \tag{5.1.40}$$

Here, formula (5.1.24) was used.

Integrating (5.1.40) over Γ and using (5.1.26), one obtains

$$\begin{aligned} \int_\Gamma \sigma_2 dt &= 2\int_\Gamma (n \cdot t)(n \cdot N_t) dt = 2n \int_\Gamma N_t (n \cdot t) dt \\ &= 2n \int_\mathcal{D} \nabla(n \cdot x) dx = 2n \cdot nV = 2V. \end{aligned} \tag{5.1.41}$$

From (5.1.37), (5.1.39), and (5.1.41), it follows that

$$f(n, \nu, k) = -\frac{k^2 V}{4\pi}\beta_{pq} n_p \nu_q - \frac{k^2 V}{4\pi}, \quad f \sim k^2 a^3. \tag{5.1.42}$$

This formula justifies the statement at the end of Section 5.2 since $k^2 V \sim k^2 a^3$.

5.1.4. If the initial field u_0 is not necessarily a plane wave but any solution to equation (5.1.1), then, for $ka \ll 1$ and $\eta = 0$, the main term in the scattering amplitude can be written as (see [239, formula (39), p. 90]):

$$f(\nu, k) = \frac{ikV}{4\pi}\beta_{pq}\frac{\partial u_0(0)}{\partial x_p}\nu_q + \frac{V\nabla^2 u_0(0)}{4\pi}, \tag{5.1.43}$$

where $u_0(0)$ is the value of the incident field at the point 0 where the small body is placed and $\partial u_0/\partial x_p$ should be calculated at the same point, and $\nabla^2 u_0 = -k^2 u_0$. For $u_0 = \exp(ikn \cdot x)$, one has $\partial u_0(0)/\partial x_p = ikn_p, u_0(0) = 1$, and (5.1.43) reduces to (5.1.42).

5.2. Scattering by Many Small Bodies

5.2.1. Consider scattering by r small bodies. There are two aspects of this problem. The first aspect is the scattering by $r \sim 10$ bodies. In this problem, one is interested in mutual interference of waves. This is a many-body problem, in which the scattering amplitude again will be defined by a finite number of functionals of the scattered field. The second aspect of the problem is the scattering by $r \sim 10^{23}$ small bodies. In this problem, one describes the scattering as the scattering in a continuous medium with effective parameters to be calculated from the knowledge of the distribution of the bodies (particles) and the scattering properties of a single particle [410].

Let us start with the statement of the scattering problem. Let

$$\begin{gathered} \mathcal{D} = \bigcup_{j=1}^{r} \mathcal{D}_j, \quad \Gamma = \bigcup_{j=1}^{r} \Gamma_j, \quad \mathcal{D}_i \cup \mathcal{D}_j = \emptyset \\ \text{if } i \neq j, \quad \Omega = \mathbb{R}^3 \setminus \mathcal{D}, \end{gathered} \tag{5.2.1}$$

where $\emptyset$ is the empty set and Ω is the complement to $\mathcal{D}$ in $\mathbb{R}^3$. Let

$$\eta = \eta_j \quad \text{on } \Gamma_j, \quad 1 \leq j \leq r, \tag{5.2.2}$$

$$a = \max_{1 \leq j \leq r} a_j, \quad a_j = \operatorname{diam} D_j, \tag{5.2.3}$$

$$d = \min d_{ij}, \quad i \neq j, 1 \leq i, j \leq r, \; d_{ij} = \operatorname{dist}(D_i, D_j), \tag{5.2.4}$$

$$l = \max d_{ij}, \quad 1 \leq i, j \leq r, \quad \int_{\Gamma} = \sum_{j=1}^{r} \int_{\Gamma_j}. \tag{5.2.5}$$

The scattering problem consists of finding the solution to problem (5.1.12), (5.1.13), (5.1.3). Let $s_j \in \mathcal{D}_j, 1 \leq j \leq r$, denote the positions of the particles and $\sigma = (\sigma_1, \ldots, \sigma_r)$. We look for the solution of the form (5.1.4) and the scattering amplitude is given as in Secion 5.1 by formula (5.1.5). The basic difference between the one- and many-body problems is that in the one-body problem $|t| \sim a$ while in the many-body problem $|t|$ can be large. Therefore, let us rewrite formula (5.1.5) as

$$f = \frac{1}{4\pi} \sum_{j=1}^{r} \exp(-ik\nu \cdot s_j) \int_{\Gamma_j} \exp\{-ik\nu \cdot (t - s_j)\} \sigma_j dt. \tag{5.2.6}$$

Note that $|t - s_j| \sim a, 1 \leq j \leq r$, and the phase shifts between the bodies are described by the factors $\exp(-ik\nu \cdot s_j)$. One obtains an integral equation for σ_j by substituting (5.1.4) into the boundary condition (5.1.13). This yields

$$\begin{aligned} \sigma_j = {} & A_j(k)\sigma_j + 2\eta_j Q_j(k)\sigma_j + \sum{}' A_{jp}(k)\sigma_p + 2\eta_j \sum{}' Q_{jp}(k)\sigma_p \\ & + 2\frac{\partial u_0}{\partial N(t_j)} + 2\eta_j u_0(t_j), \quad 1 \leq j \leq r, \end{aligned} \tag{5.2.7}$$

where $\sum' = \sum_{p=1, p\neq j}^{t}$,

$$A_{jp}\sigma_p = 2\int_{\Gamma_p} \frac{\partial g(t_j, t_p)}{\partial N(t_j)} \sigma_p(t_p) dt_p, \tag{5.2.8}$$

$$Q_{jp}\sigma_p = \int_{\Gamma_p} g(t_j, t_p)\sigma_p(t_p) dt_p. \tag{5.2.9}$$

Suppose that

$$d \gg a, \quad \lambda \gg a, \tag{5.2.10}$$

where λ is the wavelength, $\lambda = 2\pi k^{-1}$.

This assumption does not exclude the case when there are many particles on the wavelength: the inequalities $\lambda \gg d \gg a$ are compatible. The second inequality (5.2.10) is equivalent to the basic assumption $ka \ll 1$. If (5.2.10) holds and $\eta_j = \infty, 1 \leq j \leq r$, then equation (5.2.7) becomes

$$Q_j(k)\sigma_j = -\sum{}'Q_{jp}\sigma_p - u_0(t_j), \quad 1 \leq j \leq r. \tag{5.2.11}$$

This equation can be written as

$$\int_{\Gamma_j} \frac{\sigma_j(t)dt}{4\pi|t_j - t|} = -\exp(ikn \cdot s_j) - \sum_{p=1, j\neq p}^{r} \int_{\Gamma_p} g(s_j, t)\sigma_p(t)dt + O(ka), \quad 1 \leq j \leq r. \tag{5.2.12}$$

It follows from (5.2.12) that

$$q_j = \int_{\Gamma_j} \sigma_j dt = -C_j \left\{ \exp(ikn \cdot s_j) + \sum_{p=1, j\neq p}^{r} g(s_j, s_p)q_p \right\}, \quad 1 \leq j \leq r, \tag{5.2.13}$$

where C_j is the electrical capacitance of the perfect conductor $\mathcal{D}_j$. Equation (5.2.13) is a linear system of algebraic equations for $q_j, 1 \leq j \leq r$. The matrix of this system is $\delta_{jp} + C_j g(s_j, s_p)(1 - \delta_{jp})$. The system (5.2.13) can be solved by iterations if

$$\max_{1\leq j\leq r} \sum_{p=1, p\neq j}^{r} C_j \frac{1}{4\pi|s_j - s_p|} < 1. \tag{5.2.14}$$

Condition (5.2.14) will be satisfied if the much stronger condition

$$(4\pi)^{-1}Crd^{-1} < 1, \quad C = \max_{1\leq j\leq r} C_j \tag{5.2.15}$$

holds. If the system (5.2.13) is solved and the numbers q_j are found, then it follows (5.2.6) that

$$f = \frac{1}{4\pi} \sum_{j=1}^{r} \exp(-ik\nu \cdot s_j)q_j + O(rka). \tag{5.2.16}$$

The idea of this derivation is exactly the same as in Section 5.1.1.

5.2.2. Consider now the case $\eta_j \neq 0, 1 \le j \le r$. Let us assume that (5.2.10) holds and define

$$U_j = u_0(x) + \sum_{p \neq j}{}' \int_{\Gamma_p} g(x,t)\sigma_p(t)dt. \qquad (5.2.17)$$

Then equation (5.2.7) can be written as

$$\sigma_j(k) = A_j(k)\sigma_j + 2\eta_j Q_j(k)\sigma_j + 2\frac{\partial U_j}{\partial N(t_j)} + 2\eta_j U_j(t_j). \qquad (5.2.18)$$

This equation is the same as (5.1.14) if one substitutes U_j in the place of u_0 in (5.1.14). Therefore, U_j defined in (5.2.17) plays the role of the effective field. One can repeat the arguments given after equation (5.1.14) and obtain for the scattering amplitude (5.2.6) the following approximate expression:

$$f = \frac{1}{4\pi}\sum_{j=1}^{r} \exp(-ik\nu \cdot s_j)\frac{\eta_j S_j U_j}{1 - \eta_j S_j C_j^{-1}}, \quad \eta_j S_j C_j^{-1} \ll 1, \quad (5.2.19)$$

where S_j is the area of the surface Γ_j, C_j is the capacitance of $\mathcal{D}_j$ if it were a perfect conductor, and it is assumed that η_j do not vanish; and if $\eta_j = 0, 1 \le j \le r$, then

$$f = \frac{1}{4\pi}\sum_{j=1}^{r} \exp(-ik\nu \cdot s_j)\left(ikV_j\beta_{pq}^{(j)}\frac{\partial U^{(j)}(s_j)}{\partial x_p}\nu_q + V_j\nabla^2 U_j(s_j)\right), \qquad (5.2.20)$$

where $V_j = |\mathcal{D}_j|$ is the volume of $\mathcal{D}_|$, $\beta_{pq}^{(j)}$ is the magnetic polarizability tensor of $\mathcal{D}_j$.

If $\eta_j = \infty, 1 \le j \le r$, and $r \gg 1$, then (5.2.16) (or (5.2.19)) can be written as

$$f = -\frac{1}{4\pi}\sum_{j=1}^{r} \exp(-ik\nu \cdot s_j)C_j U_j. \qquad (5.2.21)$$

If the medium consisting of many small particles is rarefied, then one can use formulas (5.2.19)–(5.2.20) with $U_j = u_0(s_j)$. In general, in order to use formulas (5.2.19)–(5.2.21), one should calculate the effective field U.

Let us derive the equations for the effective field U. Let $N(x)dx$ be the number of the particles in the volume dx around the point x, $C(x)$ and $\beta_{pq}(x)$ be the average capacitance and magnetic polarizability tensor of the particles in the volume dx. If the particles are acoustically soft, i.e., $\eta = \infty, 1 \leq j \leq r$, then the field in the medium is

$$U(s_j) = u_0(s_j) - \sum_{\substack{p \neq j, \\ p=1}}^{r}{}' g(s_j, t_p) C_p U(t_p) V_p. \tag{5.2.22}$$

This follows from (5.2.17) if one takes into account that, for $j \neq p$,

$$\int_{\Gamma_p} g(s_j, t)\sigma_p dt = g(s_j, t_p) q_p = -g(s_j, t_p) C(t_p) U(t_p) V_p. \tag{5.2.23}$$

Indeed, the total field is $u = U + u_p$, where u_p is the field due to the p-th body $\mathcal{D}_p$. Although away from $\mathcal{D}_p$ the field u_p is negligible, but on Γ_p this field should enforce the boundary condition $U(t_p)+u_p = 0$. Thus, $U(t_p) = -u_p$ on Γ_p, and the total charge on Γ_p, i.e., q_p, is given by $q_p = -C(t_p)U(t_p)V_p$. Passing to the limit $r \to \infty$ in (5.2.22) yields

$$U(x) = u_0(x) - \int g(x, y) q(y) U(y) dy, \tag{5.2.24}$$

where

$$q(y) = N(y) C(y). \tag{5.2.25}$$

If equation (5.2.24) is solved, then the scattering amplitude is given by the formula

$$f = -\frac{1}{4\pi} \int \exp(-ik\nu \cdot y) q(y) U(y) dy. \tag{5.2.26}$$

A similar argument gives the same equations (5.2.24) and (5.2.25) in the case when $\eta_j, 1 \leq j \leq r$, are finite and do not vanish.

In this case,

$$q(y) = N(y)a(y), \quad \text{where} \quad a(y) = \overline{\left(\frac{\eta S}{1 - \eta S C^{-1}}\right)}, \qquad (5.2.27)$$

and the bar denotes the mean value.

If $\eta_j = 0, 1 \le j \le r$, then the equation for the effective field is

$$U(x) = u_0(x) + \int g(x,y)\left\{ik\beta_{pq}(y)\frac{\partial u(y)}{\partial y_q}\frac{x_p - y_p}{|x-y|} + b(y)\nabla^2 u(y)\right\} dy, \qquad (5.2.28)$$

where

$$b(y) = \lim_{r \to 0} \frac{\sum V_j}{|B_r(y)|}, \quad B_{pq}(y) = \lim_{r \to 0} \frac{\sum V_j \beta_{pq}^{(j)}}{|B_r(y)|}, \qquad (5.2.29)$$

$|B_r(y)|$ is the volume of the ball of radius r centered at the point y, the sum is taken over the particles situated in this ball, and it is assumed that the limits (5.2.29) exist. The function $b(y)$ is the average volume accupied by the particles around the point y, and $B_{pq}(y)$ is the average polarizability tensor around point y.

5.3. Electromagnetic Wave Scattering by Small Particles

5.3.1. The theory is developed in [239] and here we just formulate one of the basic results. Let $\mathcal{D}$ be a homogeneous body with characteristic dimension a. Let ϵ, μ and σ be its dielectric permeability, magnetic permeability, and conductivity, ϵ_0, μ_0 and $\sigma_0 = 0$ be the corresponding parameters of the exterior medium, $\epsilon' = \epsilon + i\sigma\omega^{-1}$, ω be the frequency of the initial field, λ_0 be its wavelength and $k_0 = 2\pi\lambda_0^{-1}$. Let $\delta = 1/(|\text{Im}(\epsilon'\mu)^{1/2}|\omega)$ be the depth of the skin layer in $\mathcal{D}$. If $\epsilon \ll \sigma\omega^{-1}$, then $\delta = (2/(\sigma\mu\omega))^{1/2}$. We consider scattering under the following assumptions:

$$|\epsilon'| \gg 1, \quad \delta \gg a \quad \text{and} \quad k_0 a \ll 1, \tag{5.3.1}$$

$$|\epsilon'| \gg 1, \quad \delta \ll a \quad \text{and} \quad k_0 a \ll 1, \tag{5.3.2}$$

$$\{|(\epsilon' - \epsilon_0)\epsilon_0^{-1}| + |(\mu - \mu_0)\mu_0^{-1}|\}(1 + k_0^2 a^2) \ll 1. \tag{5.3.3}$$

Assumption (5.3.1) means that the body is dielectric and ϵ' is not very large; assumption (5.3.2) means that the body is a perfect conductor or that $|\epsilon'|$ is very large; and assumption (5.3.3) means that the body can be of arbitrary size, not necessarily small, but it should be a small perturbation of the medium. In fact, assumption (5.3.3) is a sufficient condition for convergence of the Born series for the corresponding integro-differential equation for the electromagnetic field in the presence of the body (see [239]).

Let us choose a Cartesian coordinate system with the origin inside $\mathcal{D}$. Let the initial field be a plane wave propagating in the positive direction e_3 of the x_3-axis, ν be the unit vector, the direction of propagation of the scattered field, θ be the angle between e_3 and ν, the angle of scattering. Let E_1 and E_2 be the components of the incident field along the x_1 and x_2-axes. The scattered field is considered in the (x_1', x_2', x_3') coordinate system. The x_3'-axis of this system is directed along the vector ν, the x_1'-axis is the same as the x_1-axis, and the planes (x_2, x_3) and (x_2', x_3') are the same. Thus, the x'-coordinate system is obtained from the X-coordinate system by a rotation around the x_1-axis over the angle θ. Let f_1 and f_2 be the components of the scattered electric field onto the axes x_1' and x_2', respectively. The (x_2, x_3) plane is called the plane of scattering.

The scattering matrix, S-matrix, is defined by the formula

$$f = SE, \quad f = \begin{pmatrix} f_2 \\ f_1 \end{pmatrix}, \quad E = \begin{pmatrix} E_2 \\ E_1 \end{pmatrix}, \quad S = \begin{pmatrix} S_2 S_3 \\ S_4 S_1 \end{pmatrix}, \tag{5.3.4}$$

or

$$\begin{pmatrix} f_2 \\ f_1 \end{pmatrix} = \begin{pmatrix} S_2 S_3 \\ S_4 S_1 \end{pmatrix} \begin{pmatrix} E_2 \\ E_1 \end{pmatrix}. \tag{5.3.5}$$

Our basic results are the following theorems.

Theorem 5.3.1. *Under the assumptions (5.3.1) or (5.3.2), the following formula holds if $\mu = \mu_0$*

$$S = \frac{k_0^2 V}{4\pi} \times \begin{pmatrix} \mu_0\beta_{11} + \alpha_{22}\cos\theta - \alpha_{32}\sin\theta & \alpha_{21}\cos\theta - \alpha_{31}\sin\theta - \mu_0\beta_{12} \\ \alpha_{12} - \mu_0\beta_{21}\cos\theta + \mu_0\beta_{31}\sin\theta & \alpha_{11} + \mu_0\beta_{22}\cos\theta - \mu_0\beta_{32}\sin\theta \end{pmatrix}, \tag{5.3.6}$$

where α_{ij} and β_{ij} are the polarizability tensors.

If $\mu \neq \mu_0$, then (5.3.6) holds with $\tilde{\beta}_{ij} := \beta_{ij} + \alpha_{ij}(\gamma_\mu)$ in place of β_{ij}, where $\gamma_\mu = (\mu - \mu_0)/(\mu + \mu_0)$.

This theorem is proved in [239], where the polarizability tensors $\alpha_{ij} = \alpha_{ij}(\gamma), \gamma = (\epsilon' - \epsilon_0)/(\epsilon' + \epsilon_0)$, and $\beta_{ij} = \alpha_{ij}(-1)$ are defined and the following formula is proved:

$$|\alpha_{ij}(\gamma) - \alpha_{ij}^{(n)}(\gamma)| \leq cq^n, \quad 0 < q < 1, \tag{5.3.7}$$

where q and c are constants which depend on the geometry of Γ,

$$\alpha_{ij}^{(n)}(\gamma) = \frac{2}{V}\sum_{m=0}^{n}\frac{(-1)^m}{(2\pi)^m}\frac{\gamma^{n+2} - \gamma^{m+1}}{\gamma - 1}b_{ij}^{(m)}, \quad n \geq 1, \tag{5.3.8}$$

$$b_{ij}^{(0)} = V\delta_{ij}, \quad b_{ij}^{(1)} = \int_\Gamma\int_\Gamma \frac{N_i(t)N_j(s)}{r_{st}}dsdt, \tag{5.3.9}$$

$N_i(t)$ is the i-th Cartesian component of the unit normal to Γ at the point t, N is directed into Ω, $r_{st} = |s - t|$, ds and dt are the elements of the area of the surface Γ,

$$b_{ij}^{(m)} = \int_\Gamma\int_\Gamma dsdtN_i(t)N_j(s)\int_\Gamma \cdots \int_\Gamma \frac{1}{r_{st_{m-1}}}\psi(t_1, t) \cdot \psi(t_2, t_1)\ldots\psi(t_{m-1}, t_{m-2})dt_1\ldots dt_{m-1}, \tag{5.3.10}$$

and

$$\psi(t, s) = \frac{\partial}{\partial N}\frac{1}{r_{ts}}. \tag{5.3.11}$$

In particular,

$$\alpha_{ij}^{(1)}(\gamma) = 2(\gamma + \gamma^2)\delta_{ij} - \frac{\gamma^2}{\pi V} b_{ij}^{(1)}, \tag{5.3.12}$$

$$\beta_{ij}^{(1)} = \alpha_{ij}^{(1)}(-1) = -\frac{1}{\pi V} b_{ij}^{(1)}. \tag{5.3.13}$$

Theorem 5.3.2. *If (5.3.3) holds, then the scattering amplitude for the electric field is given by*

$$f_E = -\frac{\epsilon' - \epsilon_0}{4\pi\epsilon_0} k_0^2 \nu \times [\nu \times A_E(\nu)] - \frac{k_0\omega(\mu - \mu_0)}{4\pi} \nu \times A_H(\nu), \tag{5.3.14}$$

where

$$\begin{aligned} A_E(\omega) &= \int_{\mathcal{D}} \exp(-ik_0\nu \cdot y) E_0(y) dy, \\ A_H(\nu) &= \int_{\mathcal{D}} \exp(-ik_0\nu \cdot y) H_0(y) dy \end{aligned} \tag{5.3.15}$$

and (E_0, H_0) *is the initial field.*

Sketch of the proof (see [239]). The integral equations for the total field (E, H) are

$$\begin{aligned} E(x) = E_0(x) &+ \frac{\epsilon' - \epsilon_0}{\epsilon_0} \nabla \times \nabla \times \int_{\mathcal{D}} g(x, y) E(y) dy \\ &+ i\omega(\mu - \mu_0) \nabla \times \int_{\mathcal{D}} g(x, y) H(y) dy - \frac{\epsilon' - \epsilon_0}{\epsilon_0} \chi_{\mathcal{D}} E(x), \end{aligned} \tag{5.3.16}$$

$$\begin{aligned} H(x) = H_0(x) &+ \frac{\mu - \mu_0}{\mu_0} \nabla \times \nabla \int_{\mathcal{D}} g(x, y) H dy \\ &- i\omega(\epsilon' - \epsilon_0) \nabla \times \int_{\mathcal{D}} g(x, y) E(y) dy - \frac{\mu - \mu_0}{\mu_0} \chi_{\mathcal{D}} H(x), \end{aligned} \tag{5.3.17}$$

where $\chi_{\mathcal{D}} = 1$ if $x \in \mathcal{D}$, $\chi_{\mathcal{D}} = 0$ if $x \in \Omega$ (see Lemma 5.3.3 below). Under the assumption (5.3.3), one can prove [239, p. 101], that the system (5.3.16)–(5.3.17) is uniquely solvable by iterations. If one iterates once and then computes the scattering amplitude f_E, which is defined by the formula

$$E - E_0 = \frac{\exp(ik_0 r)}{r} f_E + o\left(\frac{1}{r}\right), \quad r = |x| \to \infty, \quad \frac{x}{r} = \nu, \tag{5.3.18}$$

then one obtains formula (5.3.14).

Scattering by thin wires and by small holds in a metallic screen is studied in [239]. □

Lemma 5.3.3. *Equations* (5.3.16)–(5.3.17) *hold.*

Proof. The Maxwell equations are

$$\nabla \times E = i\omega\mu_0 H + i\omega(\mu - \mu_0)\chi_{\mathcal{D}} H, \tag{5.3.19}$$

$$\nabla \times H = -i\omega\epsilon_0 E - i\omega(\epsilon' - \epsilon)\chi_{\mathcal{D}} E, \tag{5.3.20}$$

$$E = E_0 + E_1, \quad H = H_0 + H_1, \tag{5.3.21}$$

where (E_0, H_0) is the incident field which satisfies equations (5.3.19)–(5.3.20) with $\mu = \mu_0$ and $\epsilon' = \epsilon_0$. Therefore,

$$\nabla \times E_1 = i\omega\mu_0 H_1 + i\omega(\mu - \mu_0)\chi_{\mathcal{D}} H, \tag{5.3.22}$$

$$\nabla \times H_1 = -i\omega\epsilon_0 E_1 - i\omega(\epsilon' - \epsilon_0)\chi_{\mathcal{D}} E. \tag{5.3.23}$$

Let us denote

$$j_e = -i\omega(\epsilon' - \epsilon)\chi_{\mathcal{D}} E, \quad j_m = i\omega(\mu - \mu_0)\chi_{\mathcal{D}} H, \tag{5.3.24}$$

$$A_e = \int g(x,y) j_e(y) dy, \quad A_m = \int g(x,y) j_m(y) dy, \quad \int = \int_{\mathcal{D}}. \tag{5.3.25}$$

Then

$$E_1 = -(i\omega\epsilon_0)^{-1}(\nabla \times \nabla \times A_e - j_e) - \nabla \times A_m \tag{5.3.26}$$

$$H_1 = -(i\omega\mu_0)^{-1}(\nabla \times \nabla \times A_m - j_m) + \nabla \times A_e. \tag{5.3.27}$$

These formulas are equivalent to (5.3.16)–(5.3.17). □

5.4. Behavior of the Solutions to the Exterior Boundary Value Problems at Low Frequencies

5.4.1. In $\mathbb{R}^3$, the Green functions of exterior Dirichlet and Neumann boundary value problems are analytic in k near any real point, in particular, near $k = 0$. In $\mathbb{R}^2$, the free space Green function $(i/4)H_0^{(1)}(kr_{xy}), r_{xy} = |x - y|$, does not have a finite limit as $k \to 0$, which is clear from the formula

$$\begin{aligned} g &= \frac{i}{4} H_0^{(1)}(kr_{xy}) \\ &= \alpha(k) + g_0 + O\left((kr_{xy})^2 \ln \frac{1}{kr_{xy}}\right), \quad kr_{xy} \to 0, \quad (5.4.1) \end{aligned}$$

where $g_0 = (2\pi)^{-1} \ln r_{xy}^{-1}$, and $\alpha(k) = \frac{1}{2\pi} \ln \frac{2}{k} - \frac{C}{2\pi} + \frac{i}{4}, C = 0.5572\ldots$ is Euler's constant. Therefore, it is of interest to study the behavior of solutions to the exterior boundary value problems in $\mathbb{R}^2$ as $k \to 0$. This was done in the literature (see [62, 134, 199, 477], and references therein), where integral equations of the second kind were used. In [62, 134, 199], the case of the boundary with one connected component was considered. In [477], the general case of the boundary with m connected components was treated. In [199], not only the Helmholtz equation but also the Schrödinger equation was treated. The basic result in [134], is that Green's function of the exterior Dirichlet problem has a finite limit as $k \to 0$, while Green's function of the exterior Neumann problem does not. The aim of this section is to obtain this result for the more general case of the boundary with many components by new methods based on integral equations of the first kind and *a priori* estimates. These methods lead to the results in an easier and shorter way than the previously used methods, and the results are more complete. Below $\mathcal{D}$ is a bounded domain with boundary $\Gamma \in C^\infty$, Ω is the exterior domain, $\partial/\partial N_\pm$ are the limit values of outer normal derivatives on Γ from inside $(+)$ and outside $(-)$. The boundary function f (e.g., in (5.4.2a) below) is assumed to belong to $H^p(\Gamma)$ with $p = 0, 1$ or $3/2$. These assumptions can be relaxed. However, if $p < 3/2$, then the solution is not in H^2 in a neighborhood of Γ; if $p < 1$, then the boundary integral equation, e.g., (5.4.5) should be studied in a class of distributions.

5.4.2. The Dirichlet problem.

Let

$$(\nabla^2 + k^2)u = 0 \quad \text{in } \Omega \subset \mathbb{R}^2, k > 0, u = f(s) \text{ on } \Gamma, \tag{5.4.2a}$$

$$r^{1/2}(u_r - iku) = o(1), \quad r = |x| \to \infty. \tag{5.4.2b}$$

Theorem 5.4.1. *For any $f \in H^1(\Gamma)$, there exists*

$$\lim_{k \to 0} u(x, k) = u_0(x), \tag{5.4.3}$$

where the limit is attained uniformly on compacts, and the function $u_0(x)$ solves the problem

$$\nabla^2 u = 0 \text{ in } \Omega, \quad u_0 = f \text{ on } \Gamma, \quad u_0 = O(1), \quad r \to \infty. \tag{5.4.4}$$

We need some lemmas for the proof of Theorem 5.4.1. These lemmas are of independent interest.

Lemma 5.4.2. *If the equation*

$$Q_0\sigma = \int_\Gamma g_0(s,t)\sigma(t)dt = f(s) \tag{5.4.5}$$

has at most one solution, then $Q_0 : \tilde{H}^q \to \tilde{H}^{q+1}$ is an isomorphism. Here, $\tilde{H}^q = H^q(\Gamma) = W^{2,q}(\Gamma)$ is the Sobolev space, and q is a real number.

A proof of this lemma is given in Section 1.7, Lemma 1.7.2.

Lemma 5.4.3. *If Γ is connected and equation $Q_0\sigma = 0$ has a nontrivial solution σ, then $\dim N(Q_0) = 1$ and $\int_\Gamma \sigma dt \neq 0$. If Γ consists of m connected components $\Gamma_j, 1 \leq j \leq m$, where Γ_j is a closed smooth curve, then $\dim N(Q_0) \leq 1$. If $d > \operatorname{diam}\Gamma := \max_{s,s' \in \Gamma} |s - s'|$ then the operator Q_d with the kernel $(2\pi)^{-1}\ln(d/r_{xy})$ is an isomorphism between $\tilde{H}^q$ and $\tilde{H}^{q+1}$.*

Proof. (1) Let $m = 1$. If $Q_0\sigma = 0, \sigma \neq 0$, then $\int_\Gamma \sigma dt \neq 0$. Indeed, if $Q_0\sigma = 0$ and $\int_\Gamma \sigma dt = 0$, then $\sigma = 0$. In order to prove this, consider the single-layer potential

$$w(x) = S\sigma = \frac{1}{2\pi}\int_\Gamma \ln\frac{1}{|x-t|}\sigma(t)dt$$
$$= \frac{1}{2\pi}\ln\frac{1}{|x|}\int_\Gamma \sigma dt + w_1, \tag{5.4.6}$$

where the origin is placed inside $\mathcal{D}$,

$$\nabla^2 w_1 = 0 \text{ in } \Omega, \quad w_1 = O(|x|^{-1}) \text{ as } |x| \to \infty. \tag{5.4.7}$$

If $\int_\Gamma \sigma dt = 0$, then $w = w_1 = O(|x|^{-1})$. If $Q_0\sigma = 0$, then $w = 0$ on Γ. Thus,

$$\nabla^2 w = 0 \text{ in } \Omega, \quad w = 0 \text{ on } \Gamma, \quad w = O(|x|^{-1}). \tag{5.4.8}$$

By the maximum principle, $w = 0$ in Ω. Moreover,

$$\{\nabla^2 w = 0 \text{ in } \mathcal{D},\ w = 0 \text{ on } \Gamma\} \Rightarrow w = 0 \text{ in } \mathcal{D}. \tag{5.4.9}$$

Therefore, $\sigma = (\partial w/\partial N_+) - (\partial w/\partial N_-) = 0$ by the jump relation. It is clear now that $\dim N(Q_0) = 1$ if $N(Q_0) \neq \{0\}$. Indeed, suppose $\sigma_j \in N(Q_0), j = 1, 2$. Since $\int_\Gamma \sigma_j dt \neq 0$, one can find a constant c such that $\int_\Gamma(\sigma_1 - c\sigma_2)dt = 0$. This and the equation $Q_0(\sigma_1 - c\sigma_2) = 0$ imply that $\sigma_1 = c\sigma_2$, i.e., $\dim N(Q_0) = 1$.

Let us prove that $Q_d\sigma = 0$ implies $\sigma = 0$. If $Q_d\sigma = 0$, then

$$Q_0\sigma = -\frac{\ln d}{2\pi}\int_\Gamma \sigma dt. \tag{5.4.10}$$

The operator $Q_0 : \tilde{H}^0 \to \tilde{H}^0$ is symmetric. Therefore, equation (5.4.10) is solvable only if σ_0, satisfying the conditions $Q_0\sigma_0 = 0, \int_\Gamma \sigma_0 dt \neq 0$, is orthogonal to the right-hand side of (5.4.10) which is a constant. If $\ln d \neq 0$, then this orthogonality condition implies that $\int_\Gamma \sigma dt = 0$, $Q_0\sigma = 0$, and therefore, $\sigma = 0$. If $d = 1$, then $Q_d = Q_0, \ln(1/r_{st}) > 0$ and equation $Q_0\sigma = 0$ implies $\sigma = 0$.

Indeed, one can assume without loss of generality that $\int_\Gamma \sigma dt > 0$ and $\sigma = 1 + \sigma_1$, where $\int_\Gamma \sigma_1 dt = 0$. Then

$$Q_0\sigma_1 = -Q_0 1 < 0. \tag{5.4.11}$$

A necessary condition for (5.4.11) to have a solution is the orthogonality of the right-hand side of (5.4.11) to the solution σ of the homogeneous equation (5.4.11). This condition can be written as:

$$0 = \int_\Gamma ds\sigma(s)Q_0 1. \tag{5.4.12}$$

One can show that

$$\sigma(s) \geq 0. \tag{5.4.13}$$

Therefore, (5.4.12) and (5.4.11) imply that $\sigma = 0$.

Let us prove (5.4.13). If $Q_0\sigma = 0$, then the function $w(x)$, defined by (5.4.6), solves the problem

$$\begin{aligned} &\nabla^2 w = 0 \text{ in } \Omega, \quad w = 0 \text{ in } \Gamma, \\ &w = O\left(\int_\Gamma \sigma dt \ln \frac{1}{|x|}\right), \quad |x| \to \infty. \end{aligned} \tag{5.4.14}$$

Thus, $w \to -\infty$ as $|x| \to \infty$, and by the maximum principle, one concludes that $w < 0$ inside Ω. Therefore, $\partial w/\partial N_- \leq 0$ on Γ. Since $w = 0$ in $\mathcal{D}$, one has $\partial w/\partial N_+ = 0$. Thus, $\sigma = \partial w/\partial N_+ - \partial w/\partial N_- = -\partial w/\partial N_- \geq 0$. The first part of Lemma 5.4.3 is proved.

(2) If $m > 1$, then $\dim N(Q_0) \leq 1$ as in the case $m = 1$. Indeed, assume that

$$0 = Q_0\sigma = \sum_{j=1}^m \frac{1}{2\pi} \int_{\Gamma_j} \ln \frac{1}{|s_p - t|}\sigma_j(t)dt, \, 1 \leq p \leq m, \quad s_p \in \Gamma_p.$$

Define $w(x)$ by formula (5.4.6) where now

$$\Gamma = \bigcup_{j=1}^m \Gamma_j, \quad \int_\Gamma \sigma dt = \sum_{j=1}^m \int_{\Gamma_j} \sigma_j(t)dt. \tag{5.4.15}$$

The argument after formula (5.4.6) shows that

$$\int_\Gamma \sigma dt = 0 \Rightarrow \sigma = 0, \quad \text{i.e., } \sigma_j = 0,\ 1 \le j \le m. \tag{5.4.16}$$

Therefore, if $\sigma \in N(Q_0)$ then $\int_\Gamma \sigma dt \neq 0$. If $0 \neq \sigma \in N(Q_0)$ and $0 \neq \tilde{\sigma} \in N(Q_0)$, then one can find a constant $c \neq 0$ such that $\int_\Gamma (\sigma - c\tilde{\sigma})dt = 0, \sigma - c\tilde{\sigma} \in N(Q_0)$. This implies that $\sigma = c\tilde{\sigma}$, and $\dim N(Q_0) = 1$. It is worthwhile to mention that in the traditional approach, when one uses the double-layer potential for solving the exterior Dirichlet problem, the corresponding integral equation has the null space of dimension m, where m is the number of the connected components of Γ (see [79]).

The rest of the argument is the same as in the first part of the proof. □

Proof of Theorem 5.4.1. Let

$$u(x) = \int_\Gamma g(x,t)\sigma dt = \sum_{j=1}^m \int_{\Gamma_j} g(x,t)\sigma_j dt, \tag{5.4.17}$$

then u solves (5.4.2b) if and only if

$$\int_\Gamma g(s,t)\sigma dt = f, \quad s \in \Gamma. \tag{5.4.18}$$

From (5.4.1) and (5.4.18), it follows that

$$\begin{aligned} &\sigma(k)\sum_{j=1}^m \int_\Gamma \sigma_j dt + \int_\Gamma g_0 \sigma dt + \epsilon(k)\sigma = f, \\ &g_0 = \frac{1}{2\pi}\ln\frac{1}{r_{st}}, \end{aligned} \tag{5.4.19}$$

where $\epsilon(k)$ is a linear operator from $\tilde{H}^q$ into $\tilde{H}^{q+1}$ and

$$\|\epsilon(k)\|_{\tilde{H}^q \to \tilde{H}^{q+1}} \to 0 \quad \text{as } k \to 0. \tag{5.4.20}$$

Let $d > \operatorname{diam}\Gamma$. Then

$$\frac{1}{2\pi}\ln\frac{d}{r_{st}} > 0. \tag{5.4.21}$$

Let us write equation (5.4.19) as

$$\alpha(k,d)c_\sigma + Q(k,d)\sigma = d, \quad c_\sigma = \sum_{j=1}^{m} \int_{\Gamma_j} \sigma_j dt, \tag{5.4.22}$$

where

$$\begin{aligned} Q(k,d)\sigma &= Q(0,d)\sigma + \epsilon(k)\sigma = \frac{1}{2\pi}\int_\Gamma \ln\frac{d}{r_{st}}\sigma dt + \epsilon(k)\sigma, \\ \alpha(k,d) &= \alpha(k) - (2\pi)^{-1}\ln d. \end{aligned} \tag{5.4.23}$$

From Lemma 5.4.3 and (5.4.20), it follows that

$$\|Q^{-1}(k,d) - Q^{-1}(0,d)\|_{\tilde{H}^{q+1}\to\tilde{H}^q} \to 0, \quad k \to 0. \tag{5.4.24}$$

Let us write equation (5.4.22) as

$$\sigma + \alpha(k,d)c_\sigma Q^{-1}(k,d)1 = Q^{-1}(k,d)f, \tag{5.4.25}$$

where 1 is the function equal to 1 on $\Gamma_j, 1 \le j \le m$. Integrate (5.4.25) over Γ to obtain

$$\begin{aligned} c_\sigma &\left(1 + \alpha(k,d)\sum_{j=1}^{m}\int_{\Gamma_j} Q^{-1}(k,d)1dt\right) \\ &= \sum_{j=1}^{m}\int_{\Gamma_j} Q^{-1}(k,d)fdt. \end{aligned} \tag{5.4.26}$$

It follows from (5.4.21) that

$$\int_{\Gamma_j} Q^{-1}(k,d)1dt \ne 0, \quad k \to 0. \tag{5.4.27}$$

Therefore, one can solve (5.4.26) for c_σ

$$\begin{aligned} c_\sigma &= \sum_{j=1}^{m}\int_{\Gamma_j} Q^{-1}(k,d)fdt \\ &\times \left\{1 + \alpha(k,d)\sum_{j=1}^{m}\int_{\Gamma_j} Q^{-1}(k,d)1dt\right\}^{-1}. \end{aligned} \tag{5.4.28}$$

Since $|\alpha(k,d)| \to +\infty$ as $k \to +0$, it follows from (5.4.28), (5.4.24), and (5.4.23) that

$$\lim_{k\to+0} c_\sigma \alpha(k,d) = \sum_{j=1}^{m} \int_{\Gamma_j} Q^{-1}(0,d) f\,dt \times \left\{ \sum_{j=1}^{m} \int_{\Gamma_j} Q^{-1}(0,d) 1 dt \right\}^{-1} = \beta. \quad (5.4.29)$$

Equations (5.4.29), (5.4.25), and (5.4.24) imply that

$$\|\sigma(k) - \sigma(0)\|_{\tilde{H}^0} \to 0 \quad \text{as} \quad k \to 0, \quad (5.4.30)$$

where

$$\sigma(0) = \sigma_0 = Q^{-1}(0,d) f - \beta Q^{-1}(0,d) 1. \quad (5.4.31)$$

It follows from (5.4.17), (5.4.29), and (5.4.30) that

$$u(x,k) \to u_0(x), \quad k \to 0, \quad (5.4.32)$$

where

$$u_0(x) = \beta + \int_\Gamma g_0(x,t)\sigma_0(t)dt, \quad g_0 = \frac{1}{2\pi} \ln \frac{1}{r_{xt}}. \quad (5.4.33)$$

Here, $\sigma_0(t)$ is given by (5.4.31). If $x \in \Gamma$, then $u_0 = f$ on Γ and convergence in (5.4.32) is in $\tilde{H}^1$. If $x \in \tilde{\Omega}$, where $\tilde{\Omega} \subset \Omega$ is a compact which has no common points with Γ, then convergence in (5.4.32) is in $C^\infty(\tilde{\Omega})$. Finally, $\lim_{|x|\to\infty} u_0(x) = \beta$ since $\int_\Gamma \sigma_0 dt = \lim_{k\to 0} c_\sigma = 0$. □

5.4.3. The Neumann problem. Let

$$(\nabla^2 + k^2)u = 0 \text{ in } \Omega \subset \mathbb{R}^2,\ k > 0,\ \frac{\partial u}{\partial N} = f \text{ on } \Gamma, \quad (5.4.34)$$

and u satisfies the radiation condition (5.4.2b).

Theorem 5.4.4. *The limit (5.4.3) exists if and only if*

$$\sum_{j=1}^{m} \int_{\Gamma_j} f dt = 0. \tag{5.4.35}$$

Proof. (1) We start with Green's formula

$$u(x,k) = \sum_{j=1}^{k} \left(\int_{\Gamma_j} u(t) \frac{\partial g(x,t)}{\partial N_t} - \int_{\Gamma_j} g f dt \right), \quad x \in \Omega. \tag{5.4.36}$$

Denote

$$\sigma = u \quad \text{on} \quad \Gamma. \tag{5.4.37}$$

It follows from (5.4.36) that

$$\sigma_p(s) = \frac{A'_p \sigma_p + \sigma_p}{2} + \sum_{j=1, j\neq p}^{m} A_{pj}\sigma_j - F(s), \quad s \in \Gamma_p, \tag{5.4.38}$$

where

$$A_{pj}\sigma_j = \int_{\Gamma_j} \sigma_j(t) \frac{\partial g(s,t)}{\partial N_t} dt, \quad s \in \Gamma_p, \tag{5.4.39}$$

$$A'_p \sigma_p = \int_{\Gamma_p} \sigma_p(t) \frac{\partial g(s,t)}{\partial N_t} dt, \quad s \in \Gamma_p, \tag{5.4.40}$$

and

$$F(s) = \sum_{j=1}^{m} \int_{\Gamma_j} g(s,t) f_j(t) dt, \quad s \in \Gamma. \tag{5.4.41}$$

Consider first the simpler case, $m = 1$. Then (5.4.38) can be written as

$$\frac{A'(k)\sigma - \sigma}{2} = F. \tag{5.4.42}$$

One has

$$F = \alpha(k) \int_{\Gamma} f dt + \int_{\Gamma} g_0 f dt + o(1), \quad k \to 0, \tag{5.4.43}$$

and

$$\|A'(k) - A'\|_{\tilde{H}^0} \to 0, \quad k \to 0, \tag{5.4.44}$$

where $A' = A'(0)$. The operator $A' - I$ is an isomorphism of $\tilde{H}^0$ onto $\tilde{H}^0$ since 1 is not an eigenvalue of A'. This is a well-known fact [79] but we give a short proof for the convenience of the reader: if $A'\mu = \mu$, then the potential of the double layer $v = D\mu = \int_\Gamma dt\mu(t)\partial g_0(x,t)/\partial N_t, v_\pm = (A' \mp I)\mu/2, \partial v/\partial N_+ = \partial v/\partial N_-$, satisfies the condition $v_+ = 0$ on Γ. By the uniqueness of the solution to the interior Dirichlet problem for Laplace's equation, one has $v = 0$ in $\mathcal{D}$. Therefore, $0 = \partial v/\partial N_+ = \partial v/\partial N_-$, and, by the uniqueness of the solution to the exterior Neumann problem for Laplace's equation, $v = 0$ in Ω. Note that $v(\infty) = 0$. Thus, $\mu = v_- - v_+ = 0$.

If $A' - I$ is invertible and (5.4.44) holds, then

$$\|(A'(k) - I)^{-1} - (A' - I)^{-1}\|_{\tilde{H}^0} \to 0, \quad k \to 0. \tag{5.4.45}$$

It follows from (5.4.42), (5.4.43), and (5.4.45) that

$$\sigma(k) = 2\alpha(k)\int_\Gamma f dt(A' - I)^{-1}1 + a(k) \quad \text{as} \quad k \to 0, \tag{5.4.46}$$

where $a(k)$ is continuous at $k = 0$. Clearly $(A' - I)^{-1}1 \neq 0$. Therefore, $\sigma(k)$ has a finite limit as $k \to +0$ if and only if

$$\int_\Gamma f dt = 0. \tag{5.4.47}$$

It follows from (5.4.36) and (5.4.37) that if and only if $\sigma(k)$ has a finite limit, $u(x,k)$ has a finite limit as $k \to 0$. Theorem 5.4.4 is proved for $m = 1$.

(2) If $m > 1$, then the argument is practically the same. The role of (5.4.47) is played by (5.4.35). □

5.4.4. The Robin problem. Consider the Robin problem

$$\begin{aligned} &(\nabla^2 + k^2)u = 0 \quad \text{in} \quad \Omega \subset \mathbb{R}^2, \quad k > 0, \\ &\frac{\partial u}{\partial N} + \eta(s)u = f \quad \text{on} \quad \Gamma, \quad f \in \tilde{H}^0, \end{aligned} \tag{5.4.48}$$

$$\begin{aligned} &u \text{ satisfies (5.4.2b)}, \quad \eta(s) \in C(\Gamma), \\ &\text{and } \operatorname{Im} \eta(s) \geq 0. \end{aligned} \tag{5.4.49}$$

This problem is uniquely solvable for all sufficiently small $k > 0$. Assume for simplicity that $m = 1$ and look for the solution of (5.4.48) of the form (5.4.17). Then the boundary condition (5.4.48) leads to the equation

$$\frac{A(k)\sigma - \sigma}{2} + \eta(s)Q(k)\sigma = f. \tag{5.4.50}$$

This is a Fredholm integral equation in $\tilde{H}^0$. This equation is uniquely solvable for all sufficiently small $k > 0$. Indeed, it is sufficient to prove that the homogeneous equation (5.4.50) has only the trivial solution. If ν is a nontrivial solution of the homogeneous equation (5.4.50), then the single-layer potential w with density ν solves problem (5.4.48)–(5.4.49) with $f = 0$. Therefore, $w = 0$ in Ω, and $w = 0$ on Γ. Thus,

$$(\nabla^2 + k^2)w = 0 \text{ in } \mathcal{D}, \quad w = 0 \text{ on } \Gamma. \tag{5.4.51}$$

Since $\mathcal{D}$ is a finite domain, the interval $0 \leq k \leq a$ is free from the eigenvalues of problem (5.4.51) if a is sufficiently small. For these k, one has $w = 0$ in $\mathcal{D}$ and $\nu = \partial w/\partial N_+ - \partial w/\partial N_- = 0$. Equation (5.4.50) can be written as

$$\frac{A_0\sigma - \sigma}{2} + \eta(s)\alpha(k) \int_\Gamma \sigma ds + \eta Q_0\sigma + \epsilon(k)\sigma = f, \tag{5.4.52}$$

where $\|\epsilon(k)\|_{\tilde{H}^q \to \tilde{H}^{q+1}} \to 0$ as $k \to 0$. Let $B\sigma = (A_0\sigma - \sigma)/2 + \eta Q_0\sigma$. If $N(B) = \{0\}$, then B is an isomorphism of $\tilde{H}^0$ onto $\tilde{H}^0$, and (5.4.52) can be written as

$$\sigma + \alpha(k)c_\sigma B^{-1}\eta + B^{-1}\epsilon(k)\sigma = B^{-1}f, \; c_\sigma = \int_\Gamma \sigma ds. \tag{5.4.53}$$

Integrate (5.4.53) over Γ to obtain

$$c_\sigma \left(1 + \alpha(k)\int_\Gamma B^{-1}\eta ds\right) + \int_\Gamma B^{-1}\epsilon(k)\sigma ds = \int_\Gamma B^{-1}f ds. \tag{5.4.54}$$

If

$$\int_\Gamma B^{-1}\eta ds \neq 0 \tag{5.4.55}$$

then, as in the proof of Theorem 5.4.1, we conclude from (5.4.54) that $\lim_{k\to 0} c_\sigma \alpha(k) = \beta$ exists and is finite, and then it follows from (5.4.53) that $\lim_{k\to 0}\sigma = \sigma_0$ does exist and is finite. Existence of these two limits implies that the solution u in the form (5.4.17) has a finite limit as $k \to 0$. This limit $u_0 = \beta + \int_\Gamma g_0(x,t)\sigma_0(t)dt$ solves the equation $\nabla^2 u_0 = 0$ in Ω. It satisfies the boundary condition (5.4.48) since (5.4.53) implies that

$$B\sigma_0 + \beta\eta = f. \tag{5.4.56}$$

Therefore, if $N(B) = \{0\}$ and (5.4.55) holds then the solution to (5.4.48)–(5.4.49) exists, is unique, can be found in the form (5.4.17), and has a finite limit $u_0(x)$ as $k \to 0$ for any $f \in H^0$. This limit solves problem (5.4.48) with $k = 0$ and satisfies the condition $\lim_{|x|\to\infty} u_0(x) = \beta$. This argument is the same as in the proof of Theorem 5.4.1. If $N(B) = \{0\}$ but $\int_\Gamma B^{-1}\eta ds = 0$, then (5.4.53) and (5.4.54) imply that $\lim_{k\to 0} c_\sigma = \int_\Gamma B^{-1}f ds$. Thus, if $\int_\Gamma B^{-1}f ds \neq 0$, it follows from (5.4.53) that $\sigma = O(\alpha(k))$ as $k \to 0$, so that $\lim_{k\to 0}\sigma$ does not exist.

Lemma 5.4.5. *If $N(B) = \{0\}$, then condition (5.4.55) is necessary and sufficient for the existence of the limit $\lim_{k\to 0} u(x,k)$ for all $f \in \tilde{H}^0$.*

Assume now that $\dim N(B) > 0$. Let us show that this assumption implies $\dim N(B) = 1$, provided that

$$\eta \leq 0 \quad \text{if } \operatorname{Im} \eta = 0. \tag{5.4.49$'$}$$

Suppose that $h \in N(B)$. If $\int_\Gamma h ds \neq 0$ for any $h \in N(B), h \neq 0$, then $\dim N(B) = 1$. Indeed, if $h_1 \in N(B), h_2 \in N(B), \int_\Gamma h_j ds \neq 0, j = 1, 2$, then $h_2 = ch_1, c = \int_\Gamma h_2 ds (\int_\Gamma h_1 ds)^{-1}$. If there exists an $h \in N(B)$, such that $\int_\Gamma h ds = 0$, then $h = 0$. Indeed, let

$$\frac{A_0 h - h}{2} + \eta Q_0 h = 0 \tag{5.4.57}$$

and $\int_\Gamma h ds = 0$. Let $w = Sh = \int_\Gamma g_0(x, s) h ds$ be the single-layer potential with density h. Then (5.4.57) says that

$$\begin{aligned} &\nabla^2 w = 0 \quad \text{in } \Omega, \quad \frac{\partial w}{\partial N_-} + \eta w = 0 \quad \text{on } \Gamma, \\ &w = O(|x|^{-1}) \quad \text{as } |x| \to \infty. \end{aligned} \tag{5.4.58}$$

Assumptions (5.4.49)–(5.4.49$'$) imply that $w = 0$ in Ω. Therefore, $w = 0$ on Γ, $\nabla^2 w = 0$ in $\mathcal{D}$, and so $w = 0$ in $\mathcal{D}$. From the jump formula, $h = \partial w / \partial N_+ - \partial w / \partial N_- = 0$. Thus, $\int_\Gamma h ds \neq 0$ for all $h \in N(B), h \neq 0$, and $\dim N(B) = 1$.

Let us derive a necessary condition for $h \in N(B)$. One has $\int_\Gamma A_0 h ds = (A_0 h, 1) = (h, A_0' 1) = -(h, 1)$ where $(h, f) = \int_\Gamma h \bar{f} ds$, and the well-known equality $A_0' 1 = -1$ was used. Thus, integrating (5.4.57) over Γ yields a necessary condition for $h \in N(B)$:

$$\int_\Gamma \eta Q_0 h ds = \int_\Gamma h ds. \tag{5.4.59}$$

Let us note that

$$A_0' Q_0 = Q_0 A_0. \tag{5.4.60}$$

Formula (5.4.60) follows from the identity

$$\int_\Gamma ds' g(s, s') \frac{\partial g(s', t)}{\partial N_{s'}} = \int_\Gamma ds' g(t, s') \frac{\partial g(s, s')}{\partial N_{s'}} \tag{5.4.61}$$

which is a consequence of Green's formula.

Apply Q_0 to (5.4.57) and use (5.4.60) and (5.4.59) to obtain

$$\frac{A_0' - I}{2}p + Q_0\eta p = 0, \quad p = Q_0 h, \quad \int_\Gamma \eta p ds \neq 0. \tag{5.4.62}$$

Let us summarize what we have proved.

Lemma 5.4.6. *If* $\dim N(B) > 0$ *and* η *satisfies assumptions* (5.4.49)–(5.4.49$'$) *then* $\dim N(B) = 1$. *If* $h \in N(B)$ *and* $h \neq 0$, *then*

$$\int_\Gamma \eta p ds \neq 0, \quad \textit{where } p = Q_0 h, \quad B'p = 0. \tag{5.4.63}$$

To complete our analysis, let us prove Lemma 5.4.7.

Lemma 5.4.7. *Let* $\dim N(B) > 0$ *and* η *satisfy assumptions* (5.4.49)–(5.4.49$'$). *For the limit* $\lim_{k\to 0} u(x,k)$ *to exist for any* $f \in \tilde{H}^0$, *it is necessary and sufficient that condition* (5.4.63) *holds.*

Proof. *Necessity.* Suppose $\lim_{k\to 0} u(x,k)$ exists for any $f \in \tilde{H}^0$. Without loss of generality, one can assume that the operator Q_0 in equation (5.4.52) is injective. If it is not, then, as in Lemma 5.4.3, one can use an injective operator Q_d and the constant $\alpha(k,d) = \alpha(k) - (1/2\pi)\ln d$ in place of $\alpha(k)$.

If Q_0 is injective, then it is an isomorphism of $\tilde{H}^0$ onto $\tilde{H}^1$. Therefore, if $\lim_{k\to 0} u(x,k) = \lim_{k\to 0} Q\sigma$ exists, then the limits (5.4.29) and (5.4.30) exist, and the limit of equation (5.4.52) is

$$B\sigma_0 + \beta\eta = \frac{A_0\sigma_0 - \sigma_0}{2} + \eta Q_0\sigma_0 + \beta\eta = f. \tag{5.4.64}$$

This equation is solvable if and only if

$$\int_\Gamma f p ds = \beta \int_\Gamma \eta p ds, \quad p \in N(B'). \tag{5.4.65}$$

If $\beta \neq 0$ and condition (5.4.63) does not hold, then $\int_\Gamma f p ds = 0$. This cannot hold for all $f \in \tilde{H}^0$ since $p \neq 0$. Therefore, if $\beta \neq 0$, then condition (5.4.63) is necessary for the limit $\lim_{k\to 0} u(x,k)$ to exist for all $f \in \tilde{H}^0$. If $\beta = 0$, then the same argument shows that the limit $\lim_{k\to 0} u(x,k)$ does not exist for some f (e.g., for $f = \bar{p}$). In the proof of sufficiency, we will show that if (5.4.63) holds then $\beta \neq 0$ for some f.

Sufficiency. Suppose (5.4.63) holds, $h \in N(B), p \in N(B'), (p,p) = 1$. The $\lim_{k\to 0} u(x,k)$ exists if $\lim_{k\to 0}\sigma = \sigma_0$ and $\lim_{k\to 0}\alpha(k)c_\sigma$ exist. In order to establish that these limits exist provided that (5.4.63) holds, let us write equation (5.4.52) as

$$\begin{aligned} &B_1\sigma + \alpha(k)c_\sigma\eta + \epsilon(\kappa)\sigma - P\sigma = f, \\ &B_1 = B + P, \quad P\sigma = (\sigma, h)\eta. \end{aligned} \tag{5.4.66}$$

The operator B_1 is invertible because it is Fredholm and injective. Indeed, if $B_1\sigma = 0$ then $B\sigma = -(\sigma,h)\eta$. Therefore, $(\eta,\bar{p})(\sigma,h) = 0$. Thus, $(\sigma,h) = 0, B\sigma = 0$, and $\sigma \in N(B)$. Since $\dim N(B) = 1$ and $h \in N(B)$, one concludes that $\sigma = 0$. If B_1 is invertible, equation (5.4.66) can be written as

$$\begin{aligned} &\sigma + \alpha(k)c_\sigma B_1^{-1}(k)\eta - (\sigma,h)B_1^{-1}(k)\eta = B_1^{-1}(k)f, \\ &B_1(k) = B_1 + \epsilon(k). \end{aligned} \tag{5.4.67}$$

Multiply (5.4.67) by $\bar{h}$ and integrate over Γ to obtain

$$\begin{aligned} &\sigma_h + \alpha(k)c_\sigma(B_1^{-1}(k)\eta, h) - \sigma_h(B_1^{-1}(k)\eta, h) = (B_1^{-1}(k)f, h), \\ &\sigma_h = (\sigma, h), \quad \|B_1^{-1}(k) - B_1^{-1}\|_{\tilde{H}^0} \to 0, \quad k \to 0. \end{aligned} \tag{5.4.68}$$

Integrate (5.4.67) over Γ to obtain

$$c_\sigma + \alpha(k)c_\sigma(B_1^{-1}(k)\eta, 1) - \sigma_h(B_1^{-1}(k)\eta, 1) = (B_1^{-1}(k)f, 1). \tag{5.4.69}$$

Denote

$$(B_1^{-1}(k)\eta, h) = a, \quad (B_1^{-1}(k)\eta, 1) = b. \tag{5.4.70}$$

Then (5.4.68) and (5.4.69) can be written as a linear system with respect to c_σ and σ_h:

$$\begin{aligned} &(1-a)\sigma_h + \alpha(k)ac_\sigma = (B_1^{-1}(k)f, h), \\ &-b\sigma_h + (1+\alpha(k)b)c_\sigma = (B_1^{-1}(k)f, 1). \end{aligned} \tag{5.4.71}$$

The determinant of this system is $1 + ab - a$. It is not zero since $\lim_{k\to 0} b = b_0 \neq 0$. Indeed, $a \to (B_1^{-1}\eta, h) = 1$, and $a(k) \to \infty$.

The term $(B_1^{-1}\eta, h)$ can be computed as follows. Let $\eta = B_1 q$. Then $(\eta, \bar{p}) = (Bq, \bar{p}) + (q, h)(\eta, \bar{p}) = (q, \overline{B'p}) + (q, h)(\eta, \bar{p}) = (q, h)(\eta, \bar{p})$. Since $(\eta, \bar{p}) \neq 0$, one sees that $(q, h) = 1$. But $(q, h) = (B_1^{-1}\eta, h)$. Thus, $(B_1^{-1}\eta, h) = 1$. Furthermore, $b_0 = (B_1^{-1}\eta, 1) = (q, 1) \neq 0$. One has $\eta = B_1 q = Bq + (q, h)\eta = Bq + \eta$. Thus, $Bq = 0, q \neq 0$, and equations (5.4.59) and (5.4.63) imply that $(q, 1) \neq 0$. Therefore, the system (5.4.71) is uniquely solvable for all sufficiently small k. It is an easy matter to solve this system and see that the following limits exist:

$$\begin{aligned} &\lim_{k\to 0} \alpha(k) c_\sigma = \beta = (B_1^{-1} f, h), \\ &\lim_{k\to 0} \alpha_h = (B_1^{-1} f, h) - b_0^{-1}(B_1^{-1} f, 1) = \gamma, \end{aligned} \tag{5.4.72}$$

where

$$b_0 = (B_1^{-1}\eta, 1). \tag{5.4.73}$$

From (5.4.72) and (5.4.67), it follows that

$$\lim_{k\to 0} \sigma(k) = (\gamma - \beta) B_1^{-1}\eta + B_1^{-1} f = \sigma_0. \tag{5.4.74}$$

From (5.4.17), (5.4.72), and (5.4.74), it follows that

$$\lim_{k\to 0} u(x, k) = \beta + \int_\Gamma g_0(x, s)\sigma_0(s) ds. \tag{5.4.75}$$

Lemma 5.4.7 is proved. □

Let us summarize our results.

Theorem 5.4.8. *The solution to problem* (5.4.48)–(5.4.49′) (*with* $m = 1$) *has a finite limit as* $k \to 0$ *for any* $f \in \tilde{H}^0$ *if and only if either*

$$N(B) = \{0\} \quad \text{and} \quad (5.4.55) \text{ hold}, \tag{5.4.76}$$

or $\dim N(B) > 0$.

$$\begin{aligned}&\textit{If } \dim N(B) > 0, \textit{ then } \dim N(B) = 1 \textit{ and } (5.4.63) \textit{ hold,}\\ &\textit{where } h \in N(B) \textit{ and } B\sigma = (A_0\sigma - \sigma)/2 + \eta(s)Q_0\sigma. \end{aligned} \tag{5.4.77}$$

Example. If condition (5.4.49′) does not hold, then the conclusion of Theorem 5.4.8 may not hold. Indeed, consider a circle of radius 1 centered at the origin. Green's function of the problem (5.4.48) is

$$G(x, x'k) = \sum_{m=-\infty}^{\infty} \frac{\exp\{\mathrm{im}(\phi - \phi')\}}{2\pi} G_m(r, r'),$$

where, if $\eta = \text{const}$,

$$G_m(r, r') = \frac{i\pi}{2}[J_m(kr) + b_m H_m^{(1)}(kr)]H_m^{(1)}(kr'), \quad r < r'$$

and

$$b_m = -\frac{kJ_m'(k) + \eta J_m(k)}{kH_m^{(1)'}(k) + \eta H_m^{(1)}(k)}.$$

The solution to (5.4.48) has a finite limit as $k \to 0$ for any $f \in \tilde{H}^0$ if and only if $G(x, x', k)$ has a finite limit as $k \to 0$ for all $x \neq x'$, and this is so if and only if $b_m H_m^{(1)}(kr')H_m^{(1)}(kr)$ has a finite limit as $k \to 0$ for all $m = 0, \pm 1, \pm 2, \ldots$. A simple computation shows that $b_m H_m^{(1)}(kr')H_m^{(1)}(kr)$ has a finite limit as $k \to 0$ if and only if $\eta \neq m, m = 0, 1, 2, \ldots$. Assumption (5.4.49′) guarantees that the last condition is satisfied.

We assumed (5.4.49) and (5.4.49′). Let us drop these assumptions. Now problem (5.4.48) may have no solutions for some $k > 0$ and some f. However, the following result holds.

Theorem 5.4.9. *The solution to problem* (5.4.48) *has a finite limit as* $k \to 0$ *for any* $f \in \tilde{H}^0$ *if and only if either* (5.4.76) *holds or*

$$\dim N(B) = 1 \quad \textit{and} \quad (5.4.63) \textit{ hold.} \tag{5.4.78}$$

If $\dim N(B) > 1$, *then there exists* f *such that the solution to* (5.4.48) *does not have a finite limit as* $k \to 0$.

Proof. (1) If $u(x,k) \to u_0(k)$ as $k \to 0$, where $u(x,k)$ solves (5.4.48), and u_0 solves the limit problem

$$\nabla^2 u_0 = 0 \text{ in } \Omega, u_{0N} + \eta u_0 = f \quad \text{on } \Gamma, |u(\infty)| < \infty, \tag{5.4.79}$$

then the formula ($\epsilon\sigma = Q(k)\sigma - \alpha(k)c_\sigma - Q_0\sigma, c_\sigma = \int_\Gamma \sigma dt$)

$$\begin{aligned} u(s,k) &= \alpha(k)c_\sigma + (Q_0 + \epsilon)\sigma, \\ \sigma + \alpha(k)c_\sigma(Q_0+\epsilon)^{-1}1 &= (Q_0+\epsilon)^{-1}u \end{aligned} \tag{5.4.80}$$

shows that

$$\begin{aligned} c_\sigma &= O(\alpha^{-1}), \quad \alpha c_\sigma \to \beta, \\ \sigma &\to \sigma_0 = Q_0^{-1}u - \beta Q_0^{-1}1, \\ u &\to u_0 = \beta + \int_\Gamma g_0\sigma_0 dt, \end{aligned} \tag{5.4.81}$$

provided that $N(Q_0) = \{0\}$, so that $\int Q_0^{-1}1dt \neq 0$. Otherwise, write (5.4.80) as

$$\begin{aligned} u(s,k) &= (\alpha(k) - c)c_\sigma + (Q_1+\epsilon)\sigma, \\ Q\sigma &= Q_0\sigma + cc_\sigma, \end{aligned} \tag{5.4.82}$$

where $c = \text{const} \neq 0$, and check that $N(Q_1) = \{0\}$, so that $\int_\Gamma Q_1^{-1}1dt \neq 0$. To see that $N(Q_1) = \{0\}$, one writes $Q_0\sigma + cc_\sigma = 0$. This equation is solvable only if

$$cc_\sigma \int_\Gamma \sigma_0 dt = 0, \quad \text{where } \ Q_0\sigma_0 = 0. \tag{5.4.83}$$

Remember that $\dim N(Q_0) = 1$ and $\int_\Gamma \sigma_0 dt \neq 0$. Thus, $c_\sigma = \int_\Gamma \sigma dt = 0, \sigma \in N(Q_0)$, and therefore $\sigma = 0$. So $N(Q_1) = \{0\}$.

If (5.4.81) holds, then

$$B\sigma_0 + \beta\eta = f \tag{5.4.84}$$

and

$$\beta \int_\Gamma B^{-1}\eta dt = \int_\Gamma B^{-1} f dt. \tag{5.4.85}$$

Choose f such that $\int_\Gamma B^{-1} f dt \neq 0$ and obtain (5.4.55). This proves necessity.

To prove sufficiency, assume (5.4.76) and repeat the argument after formula (5.4.55).

(2) Assume now that $\dim N(B) = 1$. If $u \to u_0$ and (5.4.79) holds then, as above, we derive equation (5.4.84). This equation is solvable only if

$$\int_\Gamma fpdt = \beta \int_\Gamma \eta pdt, \quad p \in N(B'). \tag{5.4.86}$$

Choose f such that $\int_\Gamma fpdt \neq 0$ and obtain (5.4.63). This proves necessity.

To prove sufficiency, assume that (5.4.78) holds. Write equation

$$B_\epsilon \sigma + \alpha c_\sigma \eta = f, \quad B_\epsilon = B + \epsilon(k),$$

where $\epsilon = \epsilon(k)$ is defined in (5.4.52), as

$$B_1(k)\sigma + \alpha c_\sigma \eta - \sigma_h \eta = f, \quad \sigma_h = \int \sigma h dt, \quad h \in N(B), \tag{5.4.87}$$

where

$$B_1(k) = B_1 + \epsilon, \quad B_1\sigma = B\sigma + \sigma_h \eta. \tag{5.4.88}$$

Check that $N(B_1) = \{0\}$, so that B_1 is an isomorphism. Indeed, if $0 = B_1\sigma = B\sigma + \sigma_h\eta$, then $\sigma_h \int_\Gamma \eta pdt = 0$. This and (5.4.78) imply that $\sigma_h = 0$. Thus, $B\sigma = 0, \sigma = ch, c = \text{const}$. Since $\sigma_h = 0$ we conclude that $\sigma = 0$. So $N(B_1) = \{0\}$.

Write (5.4.87) as

$$\sigma + \alpha c_\sigma B_1^{-1}(k)\eta - \sigma_h B_1^{-1}(k)\eta = B_1^{-1}(k) f. \tag{5.4.89}$$

The further argument is the same as after formula (5.4.67) and it proves sufficiency.

(3) Assume that $\dim N(B) > 1$ and prove that there exists f such that the solution to (5.4.48) does not have a finite limit as $k \to 0$. Indeed, if this limit does exist then, as above, one obtains (5.4.86) as a necessary condition for the existence of the limit. Since $\dim N(B') = \dim N(B) > 1$, there exist at least two linearly independent elements $p_j \in N(B'), j = 1, 2$. If $\int_\Gamma \eta p_j dt = 0$ for $j = 1$ or 2, then choose

f such that $\int_\Gamma f p_j dt \neq 0$ for this j, and condition (5.4.86) cannot hold for this f. If $\int_\Gamma \eta p_j dt \neq 0, j = 1, 2$, then take f such that $\int_\Gamma f p_1 dt = 1, \int_\Gamma f p_2 dt = 0$. Then $1 = \beta \int_\Gamma \eta p_1 dt$ and $0 = \beta \int_\Gamma \eta p_2 dt$, which is a contradiction if $\int_\Gamma \eta p_j dt \neq 0, j = 1, 2$. This contradiction proves our claim.

Theorem 5.4.9 is proved. □

5.4.5. In this section, we give an alternative approach to the study of the behavior of the solutions to the problem

$$Lu + k^2 u = 0 \text{ in } \Omega, \quad k > 0, \tag{5.4.90}$$

$$\left.\begin{aligned} &u = f \quad (a), \\ &u_N = f \quad (b), \\ &u_N + \eta(s)u = f \quad (c) \end{aligned}\right\} \text{ on } \Gamma, \tag{5.4.91}$$

$$r^{1/2}(u_r - iku) \to 0, \quad r \to \infty, \tag{5.4.92}$$

with one of the boundary conditions (5.4.91), where Ω is the exterior domain with a smooth boundary Γ, L is a general second order strongly elliptic self-adjoint differential expression which coincides with ∇^2 outside of a sphere of radius R. Let us assume that $f \in H^{3/2}(\Gamma)$ so that $u \in H^2_{\text{loc}}(\Omega)$. We wish to show that the following result holds. Our argument is valid (with obvious modifications) in $\mathbb{R}^n, n > 2$.

Theorem 5.4.10. *The limit*

$$\lim_{k \to 0} u(x, k) = u_0(x) \quad \text{in } H^2_{\text{loc}}(\Omega) \tag{5.4.93}$$

exists for any $f \in H^{3/2}(\Gamma)$ *and solves the limiting problem*

$$Lu_0 = 0 \text{ in } \Omega, \ u_0 \text{ satisfies } (5.4.91), \tag{5.4.94}$$

$$u_0 = O(1) \quad \text{as} \quad |x| \to \infty, \quad \text{if } \Omega \subset \mathbb{R}^2, \tag{5.4.95}$$

if and only if

the problem (5.4.94)–(5.4.95) *has at most one solution.* (5.4.96)

This theorem reduces the problem of existence of the limit (5.4.93)–(5.4.95) *to a proof of a uniqueness theorem.*

Proof. (1) Assume that (5.4.96) holds and derive (5.4.94)–(5.4.95). Suppose that the solution to problem (5.4.90)–(5.4.92) satisfies the estimate

$$\sup_{0<k<k_0} \|u(x,k)\| \le c, \qquad \|u\|^2 = \int_\Omega |u|^2(1+|x|)^{-a}dx, \quad a>2, \tag{5.4.97}$$

where c is a constant which does not depend on $k, r > R$ is an arbitrary large fixed number, $k_0 > 0$ is an arbitrary small fixed number. From (5.4.97) and the known estimate (see [20])

$$\|u\|_{H^2(\Omega'_r)} \le c(\|Lu\|_{L^2(\Omega'_{r+1})} + \|u\|_{L^2(\Omega'_{r+1})} + \|u\|_{H^{3/2}(\Gamma)}, \; \Omega'_r = \Omega \cap B_r \tag{5.4.98}$$

it follows that

$$\|u(x,k)\|_{H^2(\Omega'_r)} \le c(r). \tag{5.4.99}$$

This equation and the embedding theorem imply that there exists a subsequence

$$u_n = u(x,k_n) \to u_0(x) \text{ in } H^2_{\text{loc}}(\Omega), \quad \text{and (5.4.94) holds.} \tag{5.4.100}$$

In $\Omega_r = \mathbb{R}^2 \setminus B_r$, one writes $(g_{0\epsilon} = g(k) - \alpha(k))$

$$\begin{aligned} u_n &= \int_{S_r} (g_r u_n - u_{nr} g)dt \\ &= \int_{S_r} (g_r u_n - u_{nr} g_{0\epsilon})dt - \alpha(k_n)\int_{S_r} u_{nr}dt. \end{aligned} \tag{5.4.101}$$

If (5.4.100) holds on S_r, then u_n and the first integral on the right of (5.4.101) tend to a limit as $k \to 0$. Therefore,

$$\int_{S_r} u_{0r}dt = 0, \quad \lim_{k\to 0}\alpha(k)\int_{S_r} u_{nr}dt = \beta. \tag{5.4.102}$$

Thus,

$$u_0(x) = \int_{S_r} (g_{0r}u_0 - u_{0r}g_0)dt - \beta \tag{5.4.103}$$

and (5.4.94)–(5.4.95) hold.

We prove sufficiency of (5.4.96) if we show that (5.4.96) implies (5.4.97). If (5.4.97) does not hold, then there exists a sequence $u(x, k_n) = u_n, k_n \to 0$, such that

$$\|u_n\| \geq n. \tag{5.4.104}$$

Let $w_n = u_n/\|u_n\|, \|w_n\| = 1$, and w_n solves (5.4.90)–(5.4.92) with $f_n = f\|u_n\|^{-1}$ in place of f. The above argument shows that $w_n \to w_0$ in $H^2_{\text{loc}}(\Omega)$ and w_0 solves the homogeneous problem (5.4.94)–(5.4.95). By (5.4.96), we conclude that $w_0 = 0$. On the other hand, one can pass to the limit in the equation

$$1 = \|w_n\| \to \|w_0\| = 0, \quad n \to \infty. \tag{5.4.105}$$

This is possible by the Lebesgue dominated convergence theorem since $w_n \to w_0$ in $H^2_{\text{loc}}(\Omega)$, $|w_n| \leq c(1+|\ln|x||)$, and from the formula for w_0 similar to (5.4.103) one obtains

$$|w_0(x)| \leq c, \quad x \in \Omega_r = \Omega \setminus B_r, \tag{5.4.106}$$

where c does not depend on x. Convergence $u_n \to u_0$ is not uniform in Ω.

Example. Let $\Omega = \{x : |x| \geq 1\}$, $u|_{|x|=1} = 1$, $u = \frac{H_0^1(k|x|)}{H_0^1(k)}$. Then $u \to 1$ as $k \to 0$, but not uniformly in Ω. The contradiction (5.4.105) proves that (5.4.104) is wrong, and (5.4.97) holds. We proved that (5.4.96) implies (5.4.93)–(5.4.95).

(2) Assume that (5.4.93)–(5.4.95) hold for any $f \in H^{3/2}(\Gamma)$. Then the problem (5.4.94)–(5.4.95) is solvable for any $f \in H^{3/2}(\Gamma)$. This problem is of Fredholm's type. Therefore, the homogeneous problem has only the trivial solution, so that (5.4.96) holds. To see that problem (5.4.94)–(5.4.95) is of Fredholm's type, one noticed that any solution to equation (5.4.94) which satisfies (5.4.95) can be represented as $\beta + \int_\Gamma G(x,t)\sigma(t)dt$, where $\beta = \text{const}$ and G is a fundamental solution for L. The boundary condition (5.4.94) is equivalent to a Fredholm-type equation for σ. This proves that problem (5.4.94)–(5.4.95) is of Fredholm's type.

Theorem 5.4.10 is proved. □

Chapter 6

Some Inverse Scattering Problems of Geophysics

There are many problems in geophysics, optics, nondestructive testing, and elsewhere in which a finite inhomogeneity in a homogeneous space is to be recovered from the measurements of a field scattered by the inhomogeneity. The field can be measured far away from the inhomogeneity or in some region which is not too far from the inhomogeneity. Several problems of this type are solved in this chapter. The problems are formulated in Sections 6.1–6.4. Section 6.1 deals with recovery of the refraction index (velocity profile) from the knowledge of the field on the plane for all positions of the source and receiver. Other problems of similar nature are also considered. Section 6.2 deals with the two-parameter inversion: both the density and velocity are recovered from the knowledge of the field for all positions of the source and receiver on the plane and for two distinct frequencies. In Section 6.1, the theory is exact, in Section 6.2 it is based on the first Born approximation. In Section 6.3, the three-dimensional inverse problem of the scattering theory by a potential is solved in the Born approximation. Section 6.4 deals with a model problem of induction logging theory.

6.1. Inverse Scattering for Geophysical Problems

6.1.1. Let

$$\nabla^2 u + \omega^2 n(x) u = -\delta(x-y) \quad \text{in } \mathbb{R}^3, \tag{6.1.1}$$

$$n(x) = 1 \text{ if } |x| \geq R, \quad n(x) = 1 + v(x) \text{ if } |x| \leq R, \quad (6.1.2)$$

$$n(x) > 0, \quad v(x) = 0 \quad \text{if} \quad |x| \geq R \quad \text{or} \quad x_3 > 0. \quad (6.1.3)$$

The function u is the outgoing (i.e., u satisfies the radiation condition) scalar, e.g., acoustic, field generated by a point source located at the point y in the homogeneous space with a compactly supported inhomogeneity $v(x)$ located in the lower half-space $\mathbb{R}^3_- = \{x : x_3 < 0\}$. Let $P = \{x : x_3 = 0\}, B_R = \{x : |x| \leq R\}$. The inhomogeneity is not assumed small in any sense. No symmetry assumptions of the type $v = v(x_3)$ or $v = v(\rho, x_3)$, where $\rho = (x_1^2 + x_2^2)^{1/2}$, are made. The only assumptions about $v(x)$ are:

$$v(x) = 0, \quad |x| > R \text{ or } x_3 \geq 0, \quad v(x) \in L^2(\mathbb{R}^3). \quad (6.1.4)$$

The inverse problem we are concerned with consists in finding $v(x)$ from measurements of the field $u(x, y, \omega)$ for all positions $x, y \in P$ of the receiver and source and for small ω. The notion of the smallness of ω will be specified later. This is a model *seismic exploration problem* in which the refraction coefficient $n(x) = c^{-2}(x)$, where $c(x)$ is the wave velocity, and it is assumed that the measured field is the acoustic scalar field. One could assume that $c = c(x, \omega)$ but in this case only $c(x, 0)$ is recovered. Our method is exact and analytic. The Born approximation is not used. The assumption $v(x) = 0$ if $|x| \geq R$ can be relaxed but fast decay of $v(x)$ at infinity is necessary for our method. For example, one can assume that

$$|v(x)| \leq c(1 + |x_1|^2 + |x_2|^2)^{-a}(1 + |x_3|)^{-a}, \quad a > 1, \quad (6.1.5)$$

but further relaxation of the assumptions on the rate of decay of v would require some new ideas.

The requirement that $u(x, y, \omega)$ is measured for $\omega \to 0$ does not restrict the resolution ability of our method as will be clear later. Therefore, the usual concept, according to which the size of the details that can be resolved by testing an object with waves of length λ, is of order λ and is not applicable to our scheme.

There are quite a few practical examples in which the condition $\omega \to 0$ is satisfied. Currently the wavelength λ used in seismic prospecting is in the range from 50 m to 400 m. If the characteristic dimension a of the inhomogeneity $v(x)$ satisfies the condition $a \ll \lambda$, then the condition $\omega \to 0$ is satisfied.

Practical examples include: (1) metallic ore lodes ($a \sim 30$ m), (2) caves ($a \sim 20$–80 m), (3) old mine structures ($a \sim 50$ m), (4) submarines ($a \sim 50$–100 m), and (5) pipelines ($a \sim 3$ m). The last problem requires some modifications in the theory because the pipeline is infinite in one direction.

6.1.2. The method we use for solving the inverse problem is simple. One starts with the integral equation

$$\begin{aligned} u(x, y, \omega) &= g(x, y, \omega) + \omega^2 \int g(x, z, \omega) v(z) u(z, y, \omega) dz \\ &= g + \omega^2 Bu, \end{aligned} \tag{6.1.6}$$

where $g = \exp(i\omega|x-y|)/(4\pi|x-y|)$ and the integral is taken actually over $\mathbb{R}^3_-$ because of (6.1.4). Equation (6.1.6) is equivalent to (6.1.1) which can be checked by differentiation, and the solution to (6.1.6) is an outgoing wave.

Under assumption (6.1.4), the linear integral operator B in (6.1.6) is compact in the space C_y of functions with a fixed singularity at the point y. The functions are of the form $u = C/|x-y| + h(x, y)$ where $h(x, y)$ is continuous, and the norm in this space is given by the equation $\|u\| = |C| + \max_{x \in \mathbb{R}^3} |h(x, y)|$.

Lemma 6.1.1. *For sufficiently small ω, equation (6.1.6) is uniquely solvable in C_y, its solution is analytic in ω in a neighborhood of the point $\omega = 0$ and can be obtained by iterations*

$$u_{n+1} = g + \omega^2 B u_n, \quad u_0 = g. \tag{6.1.7}$$

Proof. The conclusion of Lemma 6.1.1 is an immediate consequence of the Banach fixed point theorem. □

Let us rewrite (6.1.6) as

$$\frac{u - g}{\omega^2} = Bg + O(\omega^2) \tag{6.1.8}$$

and, using Lemma 6.1.1, pass to the limit $\omega \to 0$ in (6.1.8). The result is

$$Mv = \int \frac{v(z)dz}{|x-z||z-y|} = f(x,y), \tag{6.1.9}$$

where

$$f(x,y) = 16\pi^2 \lim_{\omega\to 0} \frac{u-g}{\omega^2}. \tag{6.1.10}$$

Let

$$x^1 = (x_1, x_2, 0), \quad y^1 = (y_1, y_2, 0). \tag{6.1.11}$$

Setting $x = x^1$ and $y = y^1$ in (6.1.9) yields

$$M_P v = \int \frac{v(z)dz}{|x^1-z||y^1-z|} = f(x^1,y^1), \quad x^1, y^1 \in P. \tag{6.1.12}$$

This is our basic equation for finding $v(z)$ from the data $f(x^1, y^1)$.

Remark 6.1.1. One can measure the data on surfaces other than P. For example, on a sphere S_{R_1} with radius $R_1 > R$, or on a cylinder $C_{R_1} = \{x : x_1^2 + x_2^2 = R_1^2\}, R_1 > R$. The corresponding inverse problems can be useful also. See Section 6.1.5.

6.1.3. Let us solve equation (6.1.12). Taking the Fourier transform in x^1 and y^1 and using the formula

$$\begin{aligned}
&\frac{1}{(2\pi)^2}\int_{-\infty}^{\infty}\int \frac{\exp(i\lambda\cdot x^1)}{|x^1-z|}dx_1dx_2\\
&\quad = \frac{\exp(i\lambda\cdot z^1)}{(2\pi)^2}\int_0^\infty \frac{d\rho\rho}{(\rho^2+z_3^2)^{1/2}}\int_0^{2\pi}\exp(i|\lambda|\rho\cos\phi)d\phi\\
&\quad = \frac{\exp(i\lambda\cdot z^1)}{2\pi}\int_0^\infty \frac{d\rho\rho J_0(|\lambda|\rho)}{(\rho^2+z_3^2)^{1/2}}\\
&\quad = (2\pi)^{-1}|\lambda|^{-1}\exp(i\lambda\cdot z^1)\exp(-|\lambda||z_3|),
\end{aligned} \tag{6.1.13}$$

one obtains,

$$\frac{1}{(2\pi)^2}\int dzv(x)\exp\{i(\lambda+\mu)\cdot z^1\}\exp\{-(|\lambda|+|\mu|)|z_3|\}$$
$$=|\lambda||\mu|\tilde{f}(\lambda,\mu), \tag{6.1.14}$$

where

$$\begin{aligned}\lambda&=(\lambda_1,\lambda_2),\quad |\lambda|=(|\lambda_1|^2+|\lambda_2|^2)^{1/2},\\ \mu&=(\mu_1,\mu_2),\quad |\mu|=(|\mu_1|^2+|\mu_2|^2)^{1/2},\end{aligned} \tag{6.1.15}$$

and

$$\tilde{f}(\lambda,\mu)=\frac{1}{(2\pi)^4}\int\int_{-\infty}^{\infty}\int\int f(x^1,y^1)$$
$$\times\exp\{i(\lambda\cdot x^1+\mu\cdot y^1)\}dx_1dx_2dy_1dy_2. \tag{6.1.16}$$

Let

$$\begin{aligned}&p=\lambda+\mu,\quad p_1=\lambda_1+\mu_1,\quad p_2=\lambda_2+\mu_2,\\ &p_3=|\lambda|,\quad p_4=|\mu|.\end{aligned} \tag{6.1.17}$$

Then (6.1.14) can be written as

$$\int dzv(z)\exp(ip\cdot z^1)\exp\{-(p_3+p_4)|z_3|\}$$
$$=(2\pi)^2p_3p_4F(p_1,p_2,p_3,p_4), \tag{6.1.18}$$

where $F(p_1,p_2,p_3,p_4)$ is the function $\tilde{f}(\lambda,\mu)$ in the variables p_1,p_2,p_3,p_4. The Jacobian of the mapping $(\lambda_1,\lambda_2,\lambda_3,\lambda_4)\to(p_1,p_2,p_3,p_4)$ is

$$J=\begin{vmatrix}1&0&1&0\\0&1&0&1\\ \lambda_1|\lambda|^{-1}&\lambda_2|\lambda|^{-1}&0&0\\0&0&\mu_1|\mu|^{-1}&\mu_2|\mu|^{-1}\end{vmatrix}=\frac{\mu_1\lambda_2-\mu_2\lambda_1}{|\mu||\lambda|}, \tag{6.1.19}$$

so that $J \neq 0$ if μ and λ are linearly independent. Let

$$p_3 = p_4 = \frac{q}{2}. \tag{6.1.20}$$

Equation (6.1.20) determines a three-dimensional manifold in the four-dimensional space of variables p_1, p_2, p_3, p_4. Equation (6.1.18) reduces to

$$\int_0^\infty \exp(-q\xi) \left\{ \int_{-\infty}^{\infty} \int dz_1 dz_2 \exp(ip \cdot z^1) v(z_1, z_2, -\xi) \right\} d\xi$$
$$= \pi^2 \phi(p_1, p_2, q), \tag{6.1.21}$$

where

$$\phi(p_1, p_2, q) = q^2 F\left(p_1, p_2, \frac{q}{2}, \frac{q}{2}\right), \tag{6.1.22}$$

and $\xi = |z_3| = -z_3$.

Equation (6.1.21) can be solved analytically by taking the two-dimensional inverse Fourier transform and one-dimensional inverse Laplace transform:

$$v(z_1, z_2, z_3) = \frac{1}{4} \int_{-\infty}^{\infty} \int dp_1 dp_2 \exp(-ip \cdot z^1) \frac{1}{2\pi i}$$
$$\times \int_{c-i\infty}^{c+i\infty} dq \phi(p_1, p_2, q) \exp(-q z_3), \tag{6.1.23}$$

where $p \cdot z^1 = p_1 z_1 + p_2 z_2$ and $c > 0$ is an arbitrary constant. The Mellin integral, which gives the inversion of the Laplace transform, does not depend on $c > 0$. Computational aspects of the problem will be discussed in Section 5.1.6. Let us summarize the result that we have just proved.

Theorem 6.1.2. *Equation* (6.1.12) *is uniquely solvable by formula* (6.1.23) *in which* ϕ *is given in* (6.1.22) *and* $F(p_1, p_2, p_3, p_4)$ *is the function* (6.1.16) *in the variables* (6.1.17).

Remark 6.1.2. The above argument does not give an explicit characterization of the function $f(x^1, y^1)$ for which equation (6.1.12) is solvable. Equation (6.1.12) is an integral equation of the first kind since the support of $v(z)$ is a compact set. Therefore, the range of the operator $M_p : L^2(B_R) \to C(P \times P)$ is not closed, small perturbations of $f(x^1, y^1)$ in the norm of $C(P \times P)$ may lead to an equation which has no solution in $L^2(B_R)$, and therefore the problem of solving equation (6.1.12) is ill-posed. An implicit characterization of admissible data, that is of the functions $f(x^1, y^1)$ for which equation (6.1.12) is solvable in $L^2(B_R)$, can be given as follows: $f(x^1, y^1)$ *are admissible data if and only if formula* (6.1.23) *gives a function* v *which satisfies conditions* (6.1.4).

If the data are not measured on all of P but only on a finite domain, e.g., on a disc $\mathcal{D}_{R_1}$ of radius R_1, then one can still apply the inversion formula (6.1.23), $\tilde{f}(\lambda, \mu)$ will be the Fourier transform of the data collected on $\mathcal{D}_{R_1}$, but the integral (6.1.23) will not necessarily vanish outside of B_R.

One could try a numerical method to solve equation (6.1.12). Since this is an equation of the first kind, some regularization is needed.

6.1.4. If the data are measured with a random error η uniformly distributed on the interval $(-\delta, \delta), \delta > 0$, then one can define the right-hand side of equation (6.1.12) as the solution of a linear regression problem. One measures $u = u(x, y, \omega)_j$ at several small frequencies ω_j,

$$u_j = g_j + \omega_j^2 b + \eta_j, \quad 1 \le j \le n \tag{6.1.24}$$

and estimates b from these measurements. This b is then taken as the right-hand side of equation (6.1.12). The least squares method gives

$$b = \left(\sum_{j=1}^{n} \omega_j^4 \right)^{-1} \sum_{j=1}^{n} (u_j - g_j) \omega_j^2, \tag{6.1.25}$$

where $g_j = g(x, y, \omega_j)$, n is the number of various ω_j used, and $b = b(x^1, y^1)$.

6.1.5. Consider the case when the lower half-space which contains the inhomogeneity, has different properties from the upper half-space.

Let ρ_1, α_1 and ρ_2, $\alpha_2(x)$ be density and velocity in the upper $(x_3 > 0)$ and lower $(x_3 < 0)$ half-spaces, respectively. Let us assume that there is a compactly supported inhomogeneity in the lower half-space $\mathbb{R}^3_-$, that $\rho_1 = \text{const}$, $\alpha_1 = \text{const}$, $\rho_2 = \text{const}$, $\alpha_2(x) = \alpha_2 + v(x)$, where $\alpha_2 = \text{const}$ and $v(x) = 0$ if $|x| > R$ or $x_3 \geq 0$. The Green function for the problem without an inhomogeneity, i.e., for $v = 0$, satisfies the equations

$$\nabla^2\phi_i + \frac{\omega^2}{\alpha_i^2}\phi_i = -\delta(x-y), \quad i = 1, 2, \tag{6.1.26}$$

$$\frac{\partial\phi_1}{\partial x_3} = \frac{\partial\phi_2}{\partial x_3}, \quad \rho_1\phi_1 = \rho_2\phi_2 \quad \text{at } x_3 = 0. \tag{6.1.27}$$

Conditions (6.1.27) mean that the normal displacement and the pressure are continuous across the interface. ϕ_1 (ϕ_2) in (6.1.26)–(6.1.27) denotes the value of the Green function in $\mathbb{R}^3_+$ $(\mathbb{R}^3_-)$.

In [60, pp. 95–96], the Green function is calculated:

$$\phi_2 = \frac{1}{2\pi}\int_0^\infty \frac{\exp(\nu_2 x_3 - \nu_1 y_3)J_0(k\rho)kdk}{\delta\nu_1 + \nu_2}, \quad x_3 < 0, \tag{6.1.28}$$

where $y_3 \geq 0$ (i.e., the source is in $\mathbb{R}^3_+$), $\nu_i = (k^2 - \omega^2\alpha_i^{-2})^{1/2}, \rho = |x^1 - y^1| = \{(x_1-y_1)^2 + (x_2-y_2)^2\}^{1/2}, \delta = \rho_2\rho_1^{-1}$. Take $y_3 = 0$ to obtain

$$\phi_2 = \frac{1}{2\pi}\int_0^\infty \frac{J_0(k\rho)k\exp(\nu_2 x_3)dk}{\nu_1\delta + \nu_2}, \quad x_3 < 0. \tag{6.1.29}$$

If $\omega \to 0$, then $\nu_i \to k$ and (6.1.29) takes the form

$$\begin{aligned}\phi_2 &= \frac{1}{2\pi(\delta+1)}\int_0^\infty J_0(k\rho)\exp(-k|x_3|)dk \\ &= \frac{2\rho_1}{\rho_2+\rho_1}\frac{1}{4\pi|x-y^1|}, \quad y^1 = (y_1, y_2, 0).\end{aligned} \tag{6.1.30}$$

Therefore, the basic integral equation (6.1.12) remains essentially the same:

$$\int \frac{v(z)dz}{|x^1 - z||y^1 - z|} = \left(\frac{\rho_2 + \rho_1}{2\rho_1}\right)^2 f(x^1, y^1)$$
$$= \frac{(1+\delta)^2}{4} f(x^1, y^1), \tag{6.1.31}$$

where $\delta = \rho_2/\rho_1$, and $f(x^1, y^1)$ is given in (6.1.10) with ϕ_2 in the place of g.

6.1.6. Inversion of the data given on a sphere.

Consider equation (6.1.6) in the Born approximation at an arbitrary positive frequency ω. This equation can be written as

$$\int g(x, z, \omega)v(z)g(z, y, \omega)dz = f(x, y), \quad x, y \in S_{R_1}, \tag{6.1.32}$$

where the right-hand side is the datum

$$f(x, y) = (u - g)\omega^{-2}, \tag{6.1.33}$$

and $R_1 > R$, so that the data are collected on the sphere S_{R_1}, $\int = \int_{B_R}$.

Let us use the well-known formulas (see [75])

$$g(x, z, \omega) = \sum_{l=0}^{\infty} C_l \psi_l^{\mathrm{re}}(z)\psi_l(x)$$
$$= i\omega \sum_{l=0}^{\infty} \sum_{m=-l}^{l} j_l(\omega|z|)Y_{lm}(z^0)h_l^{(1)}(\omega|x|)\overline{Y_{lm}(x^0)},$$
$$|x| > |z|, \tag{6.1.34}$$

$$\exp(i\omega n \cdot z) = \sum_{l=0}^{\infty} a_l \psi_l^{\mathrm{re}}(z)\overline{Y_l(n)}$$
$$= 4\pi \sum_{l=0}^{\infty} \sum_{m=-l}^{l} i^l j_l(\omega|z|)Y_{lm}(z^0)\overline{Y_{lm}(n)}, \tag{6.1.35}$$

where Y_l are the normalized spherical harmonics, ψ_l are the outgoing waves, and ψ_l^{re} are the regular waves, $j_l(r) = (\pi/2r)^{1/2} J_{l+1/2}(r)$, $h_l^{(1)}(r) = (\pi/2r)^{1/2} H_{l+1/2}^{(1)}(r)$, $\psi_l(x) = h_l^{(1)}(|x|)\overline{Y_{lm}(x^0)}$, $\psi_l^{\text{re}}(x) = j_l(|x|) Y_{lm}(x^0)$, $Y_{lm}(\theta,\phi) = \{(2l+1)/4\pi[(l-m)!]/[(l+m)!]\}^{1/2} e^{im\phi} P_{l,m}(\cos\theta)$, $x^0 = (\theta, \phi)$.

Substitute (6.1.34) into (6.1.32) to obtain

$$\sum_{l,l'=0}^{\infty} C_l C_{l'} \int v(z)\psi_l^{\text{re}}(z)\psi_{l'}^{\text{re}}(z)dz\psi_l(x)\psi_{l'}(y) = \sum_{l,l'=0}^{\infty} f_{ll'}\overline{Y_l(x^0)Y_{l'}(y^0)}, \tag{6.1.36}$$

where

$$f(x,y) = \sum_{l,l'=0}^{\infty} f_{ll'}\overline{Y_l(x^0)Y_{l'}(y^0)}, \tag{6.1.36$'$}$$

$$x^0 = xR_1^{-1}, \quad y^0 = yR_1^{-1}.$$

Thus, the numbers $v_{ll'} = f_{ll'}\{C_l C_{l'} h_l^{(1)}(\omega R_1) h_{l'}^{(1)}(\omega R_1)\}^{-1}$ are known, and

$$\int v(z)\psi_l^{\text{re}}(z)\psi_{l'}^{\text{re}}(z)dz = v_{ll'}. \tag{6.1.37}$$

Multiply (6.1.37) by $a_l a_{l'}\overline{Y_l(n')Y_{l'}(-n)}$, where n and n' are arbitrary unit vectors, and use (6.1.35) to obtain

$$\int v(z)\exp\{-i\omega(n-n')\cdot z\}dz = \sum_{l,l'=0}^{\infty} a_l a_{l'} v_{ll'}\overline{Y_l(n')Y_{l'}(-n)}. \tag{6.1.38}$$

Denote the right-hand side of (6.1.38) by $F(\omega(n-n'))$ and set $\omega(n-n') = p$. Then

$$\int_{B_R} v(z)\exp(-ip\cdot z)dz = F(p), \quad |p| \le 2\omega. \tag{6.1.39}$$

This equation is solved in Section 6.3 (see equation (6.3.9)).

Let us summarize the result.

Theorem 6.1.3. *Equation* (6.1.32) *has at most one solution* $v \in L^2(B_R)$. *If it is solvable, then the solution can be computed as follows:* (1) *compute numbers* $f_{ll'}$ *by formula* (6.1.36′), (2) *compute the right-hand side of* (6.1.38), (3) *solve equation* (6.1.39) *for* $v(z)$ *by the method of Section* 6.3.

In a similar way, one can solve the inverse problem in the case when x *and* y *run through a cylindrical surface,* $x_1^2+x_2^2 = R_1^2, R_1 > R$.

If $\omega = 0$, *then the above scheme should be modified because the right-hand side of* (6.1.39) *is known only at a point* $p = 0$ *if* $\omega = 0$.

If $\omega = 0$, *then the numbers*

$$\int v(z)|z|^{l+l'} Y_l(z^0) Y_{l'}(z^0) dz = v_{ll'}^0 \tag{6.1.40}$$

are uniquely determined by the data $f(x, y)$ *known for* $x, y \in S_{R_1}$. *These numbers define* $v(z)$ *uniquely.*

Lemma 6.1.4. *Equation* (6.1.32) *for* $\omega = 0$ *has at most one solution.*

Proof. Suppose $f(x, y) = 0, x, y \in S_{R_1}$. Then the numbers $v_{ll'}^0 = 0, \forall l, l' \geq 0$. We want to prove that in this case $v(z) = 0$. Let

$$v(z) = \sum_{l'=0}^{\infty} v_{l'}(r)\overline{Y_{l'}(z^0)}, \quad 0 \leq r = |z| \leq R. \tag{6.1.41}$$

Clearly, it is sufficient to prove that $v_l(r) = 0, \forall l \geq 0$. It follows from (6.1.40) and (6.1.41) that if $v_{ll'} = 0$ then

$$\sum_{l'=0}^{\infty} \int_0^R drr^{2+l+l'} v_{l'}(r) b(l, l', l'') = 0, \quad \forall l, l' \geq 0, \tag{6.1.42}$$

where

$$b(l, l', l'') = \int_{S^2} Y_l(z^0) Y_{l'}(z^0) \bar{Y}_{l'}(z^0) dz^0. \tag{6.1.43}$$

Let us consider for simplicity the two-dimensional analogue of (6.1.42)–(6.1.43):

$$\sum_{l=-\infty}^{\infty} \int_0^R drr^{1+|l|+|l'|} v_{l'}(r) \tilde{b}(l, l', l'') = 0 \quad \forall l, l', \tag{6.1.44}$$

where $l, l', l'' = 0, \pm 1, \pm 2, \ldots$, and

$$\tilde{b}(l, l', l'') = \frac{1}{(2\pi)^{3/2}} \int_0^{2\pi} \exp\{i(l + l' - l'')\phi\} d\phi$$
$$= \frac{1}{(2\pi)^{1/2}} \delta_{0, l+l'-l''}. \tag{6.1.45}$$

It follows from (6.1.44) and (6.1.45) that

$$\int_0^R drr^{1+|l|+|l'|} v_{l'}(r) = 0, \quad l'' = l + l'. \tag{6.1.46}$$

Let l'' be fixed. Then $1 + |l| + |l'|$ runs through all integers starting from $1+|l''|$ when l and l' run through the set $0, \pm 1, \pm 2$, and $l+l' = l''$ is fixed. Therefore,

$$\int_0^R drr^q v_{l''}(r) = 0, \quad q \geq 1 + |l''|. \tag{6.1.47}$$

From Müntz's theorem [2], it follows that $v_{l''}(r) = 0$ for all l''. Thus, $v(z) = 0$.

In the three-dimensional case, one uses the formula [132]

$$Y_{l_1 m_1}(\omega) Y_{l_2 m_2}(\omega)$$
$$= \sum_{l'=|l_1-l_2|}^{l_1+l_2} \rho(l_1, l_2, l')(l_1 l_2 m_1 m_2 | l' m') Y_{l'm'}(\omega), \quad m' = m_1 + m_2,$$

where $(l_1 l_2 m_1 m_2 | lm)$ are the Clebsch–Gordan coefficients, and $\rho(l_1, l_2, l) = \{(2l_1 + 1)(2l_2 + 1)/(2l + 1)\}^{1/2}(l_1 l_2 00 | l0)$. Multiply the above equation by $\bar{Y}_{l_3 m_3}(\omega)$ and integrate over S^2 to obtain:

$$\int_{S^2} Y_{l_1 m_1}(\omega) Y_{l_2 m_2}(\omega) \bar{Y}_{l_3 m_3}(\omega) d\omega$$
$$= \rho(l1, l_2, l_3)(l_1 l_2 m_1 m_2 | l_3 m_3), \tag{6.1.43'}$$

where the formula $\int_{S^2} Y_{l'm'} \bar{Y}_{lm} d\omega = \delta_{ll'} \delta_{mm'}$ was used. It follows now from (6.1.43′) and (6.1.42) that

$$0 = \sum_{l_3=0}^{\infty} \sum_{m_3=-l_3}^{l_3} \int_0^R drr^{2+l_1+l_2} v_{l_3 m_3}(r) \rho(l_1, l_2, l_3) \times (l_1 l_2 m_1 m_2 | l_3 m_3),$$

where we wrote l_1, l_2 and l_3 for l, l', and l''. Multiply this equation by $(l_1 l_2 m_1, m_2 | LM)$, sum over m_1 and m_2, and use the orthogonality condition [132]:

$$\sum_{m_1, m_2, m_1+m_2=m} (l_1 l_2 m_1 m_2 | lm)(l_1 l_2 m_1 m_2 | l'm') = \delta_{ll'} \delta_{mm'}$$

to obtain

$$0 = \rho(l_1, l_2, L) \int_0^R drr^{2+l_1+l_2} v_{LM}(r). \tag{6.1.48}$$

The coefficient $\rho(l_1, l_2, L)$ vanishes unless $l_1 + l_2 + L$ is even and $|l_1 - l_2| \leq L \leq l_1 + l_2$. Take $l_2 = l_1 + L$. Then the above conditions are met, and therefore $\rho(l_1, l_1 + L, L) \neq 0$. Thus, (6.1.48) yields

$$0 = \int_0^R drr^{2+2l_1+L} v_{LM}(r), \quad l_1 = 0, 1, 2, \ldots.$$

As above, Müntz's theorem yields $v_{LM}(r) = 0$. Since $LM, L = 0, 1, 2, \ldots, -L \leq M \leq L$, are arbitrary, $v(z) = 0$. □

Remark 6.1.3. The moment problem which consists in finding a function $f(r)$ from the knowledge of its moments f_m:

$$\int_0^R f(r) r^m dr = f_m, \quad m = m_0, \quad m_0 + 1, \quad m_0 + 2, \ldots$$

was studied extensively [3].

6.1.7. Computational aspects of the inversion scheme based on equation (6.1.21).

(a) If one inverts the Laplace transform using the Mellin contour $\mathrm{Re}\, q = c$, then the function

$$\psi(p_1, p_2, z_3) = (2\pi i)^{-1} \int_{c-i\infty}^{c+i\infty} \phi(p_1, p_2, q) \exp(-qz_3) dq \qquad (6.1.49)$$

is computed for some values p_1, p_2 and $z_3 < 0$. If q is a real number then, according to (6.1.17), $|p_1| \le p_3 + p_4 \le q$, $|p_2| \le q$. In particular, $|p_1| \le c$, and $|p_2| \le c$. Therefore, for a fixed $c > 0$, one obtains the values of the Fourier transform of $v(z_1, z_2, z_3)$ in a disk on the real two-dimensional space of the variables p_1 and p_2:

$$\begin{aligned} &\int_{-\infty}^{\infty} \int \exp(ip \cdot z^1) v(z_1, z_2, z_3) dz_1 dz_2 \\ &\quad = \pi^2 \psi(p_1, p_2, z_3), \quad |p_1|^2 + |p_2|^2 \le c^2. \end{aligned} \qquad (6.1.50)$$

Of course, the function $\psi(p_1, p_2, p_3)$ can be computed for any desired real point (p_1, p_2) by the Mellin integral provided that the parameter c in this integral is chosen so that $|p_1| \le c$ and $|p_2| \le c$. Therefore, the function $\psi(p_1, p_2, p_3)$ can be computed for all real p_1 and p_2.

However, it is of interest practically to invert the Fourier transform of a function v with a known compact support (say $v = 0$ for $|z| > R$) if the Fourier transform is known on a compact sets only, for example, on a disk $|p_1|^2 + |p_2|^2 \le c_2$. This problem is solved in Section 6.3.

(b) Let us briefly review some methods for inverting the Laplace transform (see [18, 44, 109]).

(1) Let

$$F(p) = Lf = \int_0^\infty \exp(-pt) f(t) dt, \quad p = \sigma + i\lambda, \qquad (6.1.51)$$

where $|f| \le c(1+t)^{-a}, a > 1, c = \text{const} > 0$. Then $F(p)$ is analytic in the half-plane $\sigma > 0$ and continuous up to the line $\sigma = 0$.

If $F(p), \sigma > \sigma_0 > 0$, is given, where σ_0 is an arbitrary fixed number, then

$$f(t) = \lim_{n\to\infty} \left[\frac{(-1)^n}{n!} \left(\frac{n}{t}\right)^{n+1} F^{(n)}\left(\frac{n}{t}\right) \right]. \tag{6.1.52}$$

This is a well-known formula [44]. Although there are higher order derivatives appeared in (6.1.52), this formula can be used in practice, e.g., in the case when $F(p)$ is a rational function.

Example.

$$F(p) = \frac{1}{p+1}, \quad f(t) = \lim_{n\to\infty} \left[\frac{(-1)^n}{n!} \frac{n^{n+1}}{t^{n+1}} \cdot \frac{(-1)^n n!}{(n/t+1)^{n+1}} \right]$$
$$= \lim_{n\to\infty} \frac{n^{n+1}}{(n+t)^{n+1}} = e^{-t}.$$

If we can approximate $F(p)$ by a rational function $P_\epsilon(p)Q_\epsilon^{-1}(p)$ so that

$$|F(p) - P_\epsilon(p)Q_\epsilon^{-1}(p)| \le \epsilon(1+|p|)^{-a}, \quad a > 1,\, \sigma > \sigma_0, \tag{6.1.53}$$

then

$$\max_{t\ge 0} |f(t) - f_\epsilon(t)| \le \frac{\epsilon}{(a-1)\pi} \tag{6.1.54}$$

where $f_\epsilon(t) = L^{-1}\{P_\epsilon(p)Q_\epsilon^{-1}(p)\}$. The difficulty is that the simple stability estimate (6.1.54) holds if (6.1.53) holds in the complex half-plane. It would be interesting to obtain an estimate for $|f(t) - f_\epsilon(t)|$ if an estimate of the type (6.1.53) is known on the real axis $\sigma > \sigma_0$ only (see [275, 391]).

(2) If $F(p)$ is known in a complex half-plane $\sigma > \sigma_0 > 0$, then the Mellin inversion formula can be used:

$$f(t) = \frac{1}{2\pi i} \int_{\sigma_0 - i\infty}^{\sigma_0 + i\infty} \exp(pt)F(p)dp. \tag{6.1.55}$$

Integral (6.1.55) can be easily reduced to the Fourier transform:

$$f(t) = \frac{1}{2\pi} \int_{-\infty}^{\infty} \exp(i\lambda t)F(\sigma_0 + i\lambda)d\lambda \cdot \exp(\sigma_0 t) \tag{6.1.56}$$

which is convenient for numerical computation. There are a number of quadrature formulas and the tables for numerical computation of (6.1.55) and the FFT routine for numerical computation of (6.1.56).

(3) Inversion methods which use the values of $F(p)$ on the real axis only, are of particular interest. Formula (6.1.52) is very sensitive to the errors in $F(p)$ because it requires high-order derivatives of $F(p)$. Therefore, the following methods, which are more stable towards the errors in $F(p)$, are of interest (see [109]).

Suppose that f is expanded in the Fourier–Laguerre series

$$f(t) = \sum_{n=0}^{\infty} A_n \frac{\exp(-t/2) L_n(t)}{n!}, \tag{6.1.57}$$

where $L_n(t)$ are the Laguerre polynomials

$$L_n(t) = (-1)^n \left[t^n - \frac{n^2}{1} t^{n-1} + \frac{n^2(n-1)^2}{2!} t^{n-2} + \cdots + (-1)^n n! \right]. \tag{6.1.58}$$

Then

$$A_n = \frac{1}{n!} \int_0^\infty \exp\left(-\frac{t}{2}\right) L_n(t) f(t) dt. \tag{6.1.59}$$

Let

$$a_m = \int_0^\infty \exp\left(-\frac{t}{2}\right) t^m f dt. \tag{6.1.60}$$

Then

$$a_m = F^{(m)}\left(\frac{1}{2}\right)(-1)^m \tag{6.1.61}$$

so that

$$A_n = \sum_{m=0}^{n} \binom{n}{m} \frac{(-1)^{n-m}}{(n-m)!} \alpha_{n-m} \tag{6.1.62}$$

and, by (6.1.57), (6.1.61), and (6.1.62),

$$f(t) = \exp\left(-\frac{t}{2}\right) \sum_{n=0}^{\infty} \sum_{m=0}^{n} \binom{n}{m} \frac{1}{(n-m)!} F^{(n-m)}\left(\frac{1}{2}\right) \frac{L_n(t)}{n!}. \tag{6.1.63}$$

The derivatives of F are present in this formula, but if the series (6.1.57) converges rapidly, so that only a few first terms are needed for approximating $f(t)$, then (6.1.63) might work better as an inversion algorithm than (6.1.52).

(4) One can use other expressions of $f(t)$ for inverting Laplace transform.

If $\exp(-t) = x$, then (6.1.51) becomes

$$\int_0^1 x^{p-1}\phi(x)dx = F(p), \quad \phi(x) = f(-lnx). \tag{6.1.64}$$

Let $p = 1, 2, 3, \dots$. Then

$$\int_0^1 x^m \phi(x)dx = F_m, \quad m = 0, 1, 2, \dots. \tag{6.1.65}$$

This is a well-known moments problem. Let $P_n(x)$ be the Legendre polynomials, $\tilde{P}_n(x) = P_n(2x-1)$ be the polynomials defined on the interval $[0, 1]$.

Let

$$f(x) = \sum_{n=0}^{\infty} (2n+1) C_n \tilde{P}_n(x), \tag{6.1.66}$$

where

$$C_n = \int_0^1 f(x)\tilde{P}_n(x)dx. \tag{6.1.67}$$

Let

$$\tilde{P}_n(y) = \sum_{m=0}^{n} P_{nm} y^m. \tag{6.1.68}$$

It follows from (6.1.68) and (6.1.65) that

$$C_n = \sum_{m=0}^{n} P_{nm} F_m. \tag{6.1.69}$$

The inversion formula is (6.1.66) with C_n given by (6.1.69).

(5) One can use expansion of ϕ in the series in Chebyshev's polynomials or Fourier series (see [109]). In the first case, one sets $\exp(-\tau) = (1 + \cos\theta)/2$ and $f(t) \to \phi(\theta), 0 \le \theta \le \pi$. In the second case, one sets $\exp(-\sigma t) = \cos\theta, \sigma > 0, f(t) = f(-\sigma^{-1} \ln \cos\theta) = \phi(\theta), 0 \le \theta \le \pi/2$.

(6) Finally, one can consider equation (6.1.51) with $p = \sigma > 0$ as an equation of the first kind and solve this equation for $f(t)$ numerically. This is an ill-posed problem and therefore one should use some regularization techniques (iterative methods, variational methods, singular value decomposition, or method of quasisolutions ([88, 226, 235, 369, 404, 458]).

6.1.8. Consider the two-dimensional analogue of the problem (6.1.1)–(6.1.3). The data now are the values of the field $u(x, y, k)$ on the line $L = \{x : x_2 = 0\}$ for all positions of x and y on L. The assumptions about $v(x)$ are the same as in Section 6.1.1. The governing equation is equation (6.1.6) with

$$g(x, y, \omega) = \frac{i}{4} H_0^{(1)}(\omega r_{xy}), \quad r_{xy} = |x - y|, \tag{6.1.70}$$

where $H_0^{(1)}$ is the Hankel function. The basic difference between the three-dimensional and two-dimensional problems is that the function (6.1.70) does not have a finite limit as $\omega \to 0$. One has

$$g(x, y) = \alpha(\omega) + g_0(x, y) + O\left(\omega^2 r^2 \log \frac{1}{\omega r}\right) \quad \text{as } \omega \to 0, \tag{6.1.71}$$

where

$$\alpha(\omega) = -(2\pi)(-1)\ln(\omega/2)\left(-\frac{\gamma}{2\pi}\right) + i/4, \quad g_0 = \frac{1}{2\pi}\ln\frac{1}{r_{xy}}, \tag{6.1.72}$$

and $\gamma = 0.5572\ldots$ is Euler's constant.

Let $u - g = \psi$. Equation (6.1.6) can be written as

$$\psi(x, y) = \omega^2 \int g(x, z)v(z)\psi(z, y)dz + \omega^2\psi_0(x, y), \tag{6.1.73}$$

where $\int = \int_{\mathcal{D}}, \mathcal{D} = \{x : v(x) \neq 0\}$,

$$\psi_0(x, y) = \int g(x, z)v(z)g(z, y)dz. \tag{6.1.74}$$

The field ψ is the scattered field. Equation (6.1.73) is uniquely solvable in $C(\mathbb{R}^2)$ by iterations if ω is sufficiently small. It follows from (6.1.71)–(6.1.74) that

$$\begin{aligned}\psi(x, y) = \omega^2 \bigg\{\alpha^2 \int v dz + \alpha \bigg(\int g_0(x, z)v(z)dz \\ + \int g_0(y, z)v(z)dz\bigg) + \int g_0(x, z)v(z)g_0(z, y)\bigg\} \\ + o(\omega^2), \quad \omega \to 0.\end{aligned} \tag{6.1.75}$$

Therefore,

$$f_0 = \lim_{\omega\to 0}(\omega\alpha)^{-2}\psi(x, y, \omega) = \int v(z)dz = v_0, \tag{6.1.76}$$

$$\begin{aligned}f_1 = \lim_{\omega\to 0}\frac{\psi - \alpha^2\omega^2 v_0}{\omega^2\alpha} = \int g_0(x, z)v(z)dz \\ + \int g_0(y, z)v(z)dz\end{aligned} \tag{6.1.77}$$

and

$$f_2 = \lim_{\omega\to 0}\frac{\psi - \alpha^2\omega^2 v_0 - \alpha\omega^2 f_1}{\omega^2} = \int g_0(x, z)v(z)g_0(z, y)dz. \tag{6.1.78}$$

The quantities $f_j, j = 0, 1, 2$ can be measured, and they are our data. In particular, f_0 gives the number v_0. If $x = y$, then equation (6.1.77) becomes

$$\int_{\mathcal{D}} g_0(x,z)v(z)dz = \frac{1}{2}f_1(x,x), \quad x \in L, \quad \mathcal{D} = \{x : v(x) \neq 0\}. \tag{6.1.79}$$

This equation is not uniquely solvable in general. Indeed, the right-hand side of (6.1.79) is the potential of the mass distribution $v(z)$. This potential is given on the line L. These data and v_0 determine the potential uniquely in the half-plane $x_2 > 0$. By the unique continuation property, the potential is uniquely determined in $\Omega = \mathbb{R}^2 \setminus \mathcal{D}$. However, there are many distributions v which produce the zero potential in Ω. For example, take the origin in $\mathcal{D}$ and let the disk $\mathcal{D}_{r_1}$ belong to $\mathcal{D}$. Choose r_2 and r_3 such that $0 < r_3 < r_2 < r_1$. Define $v = \{\pi(r_2^2 - r_3^2)\}^{-1}$ if $r_3 \leq r < r_2, v = -\{\pi(r_1^2 - r_2^2)\}^{-1}$ if $r_2 \leq r \leq r_1, v = 0$ if $r > r_1$ or $r < r_3, r = |x|$. Then the left-hand side of (6.1.79) vanishes for $r > r_1$, but $v \neq 0$. Under some special assumptions on $v(z)$, equation (6.1.79) can be uniquely solvable. For example, if $v = v(x_2)$ for $a \leq x_1 \leq b$ and $-R < x_2 < 0$ and is zero otherwise, then taking the Fourier transform of (6.1.79) yields

$$\begin{aligned} \int_{-\infty}^{0} & v(z_2)\exp(|\lambda|z_2)dz_2 \\ &= \frac{2i\pi\lambda|\lambda|F_1(\lambda)}{\exp(i\lambda b) - \exp(i\lambda a)}, \quad -\infty < \lambda < \infty, \end{aligned} \tag{6.1.80}$$

where

$$F_1(\lambda) = (2\pi)^{-1}\int_{-\infty}^{\infty} f_1(x,x)\exp(i\lambda x)dx. \tag{6.1.81}$$

Inverting the Laplace transform (6.1.80) yields $v(z_2)$. Thus, under the above assumption about $v(z) = v(z_2)$, an explicit analytical inversion procedure is given in terms of the back-scattered data measured on the line $x_2 = 0$.

If no special assumptions on v are made, then one can use equation (6.1.78) to recover v:

$$\int \log|x-z| \log|y-z| v(z) dz$$
$$= f(x,y), \quad x, y \in L, \quad f(x,y) = 4\pi^2 f_2(x,y). \quad (6.1.82)$$

This equation is similar to (6.1.9). Taking the Fourier transform in $x = (x, 0)$ and $y = (y, 0)$ yields

$$\int v(z) \exp\{i(\lambda+\mu)z_1 - (|\lambda| + |\mu|)|z_2|\} dz = 4|\lambda||\mu| F(\lambda, \mu), \quad (6.1.83)$$

where

$$F(\lambda, \mu) = (2\pi)^{-2} \int_{-\infty}^{\infty} \int f(x,y) \exp\{i(\lambda x + \mu y)\} dx dy, \quad (6.1.84)$$

and

$$\frac{1}{2\pi} \int_{-\infty}^{\infty} \exp(i\lambda x) \ln \sqrt{(x-z_1)^2 + z_2^2} dx$$
$$= -\frac{\exp(i\lambda z_1 - |\lambda||z_2|)}{2|\lambda|}, \quad \lambda \neq 0. \quad (6.1.85a)$$

Integral (6.1.85a) can be calculated as follows. Note that

$$(2\pi)^{-1} \int_{-\infty}^{\infty} dx \exp(i\lambda x) \ln \sqrt{(x-z_1)^2 + z_2^2}$$
$$= \exp(i\lambda z_1)(2\pi)^{-1} \int_{-\infty}^{\infty} \exp(i\lambda y) \ln \sqrt{y^2 + z_2^2} dy \quad (6.1.85b)$$

and

$$\widehat{\ln y} = (2\pi)^{-1} \int_{-\infty}^{\infty} \ln|y| \exp(i\lambda y) dy = -(2|\lambda|)^{-1}. \quad (6.1.85c)$$

Indeed,

$$-i\lambda\widehat{\ln|y|} = (2\pi)^{-1}\int_{-\infty}^{\infty} y^{-1}\exp(i\lambda y)dy$$
$$= (2\pi)^{-1}\cdot i\pi\,\mathrm{sgn}\,\lambda = (i\,\mathrm{sgn}\,\lambda)/2,$$

and (6.1.85c) follows. Formula (6.1.85a) follows from (6.1.85b) and (6.1.85c) if one takes into account that integral (6.1.85a) is a harmonic function of (z_1, z_2). Another way to derive (6.1.85a) is to differentiate the integral in the right-hand side of (6.1.85a) in z_2, calculate the resulting standard integral and integrate in z_2. Integral (6.1.85c) serves as an initial condition at $z_2 = 0$.

Let

$$p = \lambda + \mu \quad \text{and} \quad q = |\lambda| + |\mu|. \tag{6.1.86}$$

As in Section 6.1.3, one can prove that equation (6.1.83) has at most one solution $v \in L^2$. The Jacobian of the mapping (6.1.86), $\{\lambda, \mu\} \to \{p, q\}$ is $K = \mathrm{sgn}\,\mu - \mathrm{sgn}\,\lambda \neq 0$ if $\lambda\mu < 0$. Take $\mu > 0$ and $\lambda < 0$. Then (6.1.83) becomes

$$\int v(z)\exp(ipz_1 + qz_2)dz = |p^2 - q^2|\mathcal{F}(p, q), \quad z_2 < 0, \tag{6.1.87}$$

where

$$\mathcal{F}(p, q) = F\left(\frac{p-q}{2}, \frac{p+q}{2}\right). \tag{6.1.88}$$

If $\lambda < 0$ and $\mu > 0$, then the corresponding point (p, q) lies inside the wedge $0 < q < \infty, -q < p < q$.

Equation (6.1.87) can be written as

$$\int_0^{\infty} \exp(-q\xi)\left(\int_{-\infty}^{\infty} v(z_1, -\xi)\exp(ipz_1)dz_1\right)d\xi = \phi(p, q), \tag{6.1.89}$$

where

$$\phi(p, q) = |p^2 - q^2|\mathcal{F}(p, q). \tag{6.1.90}$$

Equation (6.1.89) is analogous to (6.1.21), and one can repeat the analysis given in Sections 6.1.3 and 6.1.6.

6.1.9. Consider the one-dimensional case.

The problem now is to find $v(x), x \in \mathbb{R}^1$, from the measurements of $u(x, y, \omega)$ for $x \geq 0$ and $y \geq 0$. We assume that $v(x) \in L^2, v(x) = 0$ if $x < -R$ or $x > -\epsilon$ where ϵ and R are positive constants, $\epsilon < R$. The Green function is given by

$$g = \frac{i}{2\omega} \exp(i\omega|x-y|) = \frac{i}{2\omega} + g_1$$
$$= \frac{i}{2\omega} + g_0 + O(\omega r_{xy}^2) \quad \text{as } \omega \to 0, \tag{6.1.91}$$

where $g_0 = -|x-y|/2, r_{xy} = |x-y|$. The integral equation (6.1.6) becomes

$$u = g(x, y) + \frac{i\omega}{2} \int \exp(i\omega|x-z|)v(z)u(z, y)dz. \tag{6.1.92}$$

Let $u - g = \psi$. Then

$$\psi(x, y) = \frac{i\omega}{2} \int \exp(i\omega|x-z|)v(z)\psi(z, y)dz + \psi_0(x, y), \tag{6.1.93}$$

where

$$\psi_0(x, y) = -\frac{1}{4} \int \exp(i\omega|x-z| + |y-z|)v(z)dz. \tag{6.1.94}$$

Equation (6.1.93) is uniquely solvable in $C(\mathbb{R}^1)$ by iterations if $(\omega/2) \cdot \int |v|dz < 1$, in particular, for sufficiently small ω. Thus, ψ is analytic in ω in a neighborhood of the point $\omega = 0$. The function ψ is the scattered field, it is the datum.

It follows from (6.1.93) that

$$\lim_{\omega \to 0} \psi = -\frac{1}{4} \int v dz = -\frac{v_0}{4}, \quad \text{where} \quad v_n = \int z^n v(z)dz. \tag{6.1.95}$$

Let

$$\psi = \psi(x, y, \omega) = \sum_{j=0}^{\infty} C_j(x, y)(i\omega)^j, \tag{6.1.96}$$

and

$$\psi_0 = -\frac{1}{4}\sum_{j=0}^{\infty} \psi_{0j}(i\omega)^j,$$
$$\psi_{0j} = \frac{1}{j!}\int (|x-z|+|y-z|)^j v(z)dz. \tag{6.1.97}$$

Substitute (6.1.96) and (6.1.97) into (6.1.93) and compare the coefficients in front of $(i\omega)^j$ to obtain the recurrence formula

$$C_j(x,y) = \frac{1}{2}\sum_{m+n+1=j} \frac{1}{m!}\int |x-z|^m v(z)C_n(z,y)dz$$
$$-\frac{1}{4}\psi_{0j}, \quad j=0,1,2,\dots. \tag{6.1.98}$$

If $j=0$, then (6.1.98) gives

$$C_0(x,y) = -\frac{\psi_{00}}{4} = -\frac{v_0}{4}. \tag{6.1.99}$$

If $j=1$, then

$$C_1(x,y) = -\frac{v_0^2}{8} + \frac{v_1}{2} - \frac{x+y}{4}v_0. \tag{6.1.100}$$

If $j=2$, then

$$C_2(x,y) = -\frac{v_0^3}{16} + \frac{3v_0v_1}{8} - \frac{v_0^2(x+y)}{8} - \frac{v_0^2 x}{4}$$
$$-\frac{v_0(x+y)}{8} + \frac{v_1(x+y)}{2} - \frac{v_2}{2}. \tag{6.1.101}$$

The coefficients C_j can be considered as data. Equations (6.1.99)–(6.1.100) determine recurrently v_0, v_1, and v_2. In principle, all the moments v_n (see (6.1.95)) can be determined. It is a classical problem, Hamburger's moment problem [3], to find $v(z)$ from the knowledge of its moments v_n:

$$\int_{-R}^{0} v(z)z^n dz = v_n, \quad n=0,1,2,\dots. \tag{6.1.102}$$

Similar problem we had in Section 6.1.5, equation (6.1.48). Since the system $\{z^n\}, n=0,1,2,\dots$ is closed in $L^2(-R,0)$, the moment problem (6.1.102) has at most one solution. Necessary and sufficient conditions are known for the solvability of problem (6.1.102) [2]. A numerical solution similar to the T-matrix approach studied in [248] can be suggested. Namely, take a basis $\{\phi_j(z)\}$ of $L^2(-R,0)$, look for an approximate solution in the form

$$v_m(z) = \sum_{j=0}^{m} a_j^{(m)} \phi_j(z), \tag{6.1.103}$$

substitute (6.1.103) into (6.1.102) and use the first m equations (6.1.102) to obtain a linear system for the coefficients $a_j^{(m)}$:

$$\sum_{j=0}^{m} b_{nj} a_j^{(m)} = v_n, \quad b_{nj} = \int_{-R}^{0} z^n \phi_j(z) dz. \tag{6.1.104}$$

We will not go into further detail because it is probably not so easy to measure higher order coefficients $C_j(x,y), j > 2$, and therefore find the higher order moments.

It is worth mentioning that an inversion procedure within the Born approximation can be constructed easily. In the Born approximation, one has $\psi \approx \psi_0$, so that

$$\int \exp\{i\omega(|x-z| + |y-z|)\} v(z) dz = f(x,y), \quad x,y > 0, \tag{6.1.105}$$

where

$$f(x,y,\omega) = -4\psi(x,y) \tag{6.1.106}$$

is the data. Set $x = y$ in (6.1.105) to obtain

$$\int_{-R}^{0} \exp(-2i\omega z) v(z) dz = \exp(-2i\omega x) f(x,x,\omega) = f(x,\omega). \tag{6.1.107}$$

Let us fix arbitrary $x \geq 0$ and consider the data $f(x,\omega) = \phi(\omega)$ as a function of ω. Then one can Fourier-invert the data $\phi(\omega)$ and compute $v(z)$. Since $v(z)$ is compactly supported, the Fourier inversion can be done also in the case when the data $\phi(\omega)$ are given on a finite interval of frequencies, $\omega_0 \le \omega \le \omega_1$. This is done in Section 6.3.

In conclusion, it is worth mentioning that if the incident field is a plane wave and the measured quantity is the reflection coefficient which is known at all frequencies, then the one-dimensional problem can be reduced by the Liouville transform to the problem of inverse scattering by a potential. This problem was studied in the literature [356].

6.1.10. Inversion of the data given on a cylinder.

Consider equation (6.1.32) with $x, y \in C_{R_1}$, where $C_{R_1} = \{x : \rho = R_1\}, R_1 > R, \rho = (x_1^2 + x_2^2)^{1/2}$. We have

$$g(x,z) = \sum_{n=-\infty}^{\infty} (2\pi)^{-2} \exp\{in(\phi_z - \phi_x)\} \int_{-\infty}^{\infty} \exp\{i\lambda(z_3 - x_3)\} \times \frac{i\pi}{2} J_n(\alpha\rho_z) H_n(\alpha\rho_x) d\lambda, \quad \rho_z < \rho_x, \tag{6.1.108}$$

where $\alpha = (k^2 - \lambda^2)^{1/2}$, and $H_n(r) = H_n^{(1)}(r)$ is the Hankel function, $\rho_x = \rho_y = R_1, \rho_x = (x_1^2 + x_2^2)^{1/2}$,

$$\exp\{i(\lambda x_3 + \alpha\rho\cos(\phi - \phi'))\} = \exp(i\lambda x_3) \sum_{n=-\infty}^{\infty} i^n \exp\{in(\phi - \phi')\} J_n(\alpha\rho). \tag{6.1.109}$$

Equation (6.1.32) can be written as

$$-\frac{1}{64\pi^2} \sum_{n=-\infty}^{\infty} \sum_{n'=\infty}^{\infty} \iint_{-\infty}^{\infty} \exp\{-i(n\phi_x + n'\phi_y) - (\lambda x_3 + \lambda' y_3)\} H_n(\alpha R_1) H_{n'}(\alpha' R_1) \cdot \int dz v(z) \exp\{i(n + n')\phi_z + i(\lambda + \lambda')z_3\} J_n(\alpha\rho_z) J_{n'}(\alpha'\rho_z) d\lambda d\lambda'$$

$$= \sum_{n=-\infty}^{\infty} \sum_{n'=-\infty}^{\infty} \iint_{-\infty}^{\infty} d\lambda d\lambda' f_{\lambda n \lambda' n'}$$

$$\cdot \exp\{-i(n\phi_x + n'\phi_y) - i(\lambda x_3 + \lambda' y_3)\}, \tag{6.1.110}$$

where $\alpha' = (k^2 - \lambda'^2)^{1/2}$, and

$$f_{\lambda n \lambda' n'} = \frac{1}{(2\pi)^4} \int_0^{2\pi} \int_0^{2\pi} \int_{-\infty}^{\infty} \int_{-\infty}^{\infty} \exp\{i(n\phi_x + n'\phi_y)$$
$$+ i(\lambda x_3 + \lambda' y_3)\} f(\phi_x, x_3, \phi_y, y_3) d\phi_x d\phi_y dx_3 dy_3.$$

Here, $f(\phi_x, x_3, \phi_y, y_3)$ are the data, the right-hand side of equation (6.1.32). From (6.1.110), it follows that

$$\int dz v(z) \exp\{i(n+n')\phi_z\} J_n(\alpha\rho_z) J_{n'}(\alpha'\rho_z)$$

$$= -\frac{64\pi^2 f_{\lambda n \lambda' n'}}{H_n(\alpha R_1) H_{n'}(\alpha' R_1)}. \tag{6.1.111}$$

Multiply (6.1.111) by $i^{n+n'} \exp\{-i(n\phi + n'\phi')\}$, sum over n, n', and use (6.1.109), to get:

$$\int_{|z|\le R} dz v(z) \exp\{ik(m+m')\cdot z\} = F(k(m+m')), \quad k > 0, \tag{6.1.112}$$

where

$$F(k(m+m')) = -64\pi^2 \sum_{n,n'=-\infty}^{\infty} \frac{f_{\lambda n \lambda' n'} i^{n+n'} \exp\{-i(n\phi + n'\phi')\}}{H_n(\alpha R_1) H_n(\alpha' R_1)},$$

$$m = (\sin\theta\cos\phi, \sin\theta\sin\phi, \cos\theta), \quad \lambda = k\cos\theta, \quad \alpha = k\sin\theta,$$
$$m' = (\sin\theta'\cos\theta', \sin\theta'\sin\phi', \cos\theta'), \quad \lambda' = k\cos\theta', \quad \alpha' = k\sin\theta'.$$

Equation (6.1.112) is of the type (6.3.9) and can be solved as in Section 6.3.

Equation (6.1.9) with $x, y \in C_{R_1}$ is equation (6.1.32) with $\omega = 0$. Let us prove that the homogeneous equation (6.1.9) with $x, y \in C_{R_1}$ has only the trivial solution. If

$$w(x,y) = \int_{|z|\le R} dzv(z)|x-z|^{-1}|y-z|^{-1} = 0, \quad x, y \in C_{R_1},$$

then for any fixed $y, |y| > R$, and for $|x| > R$, the function $w(x,y)$ solves the problem

$$\begin{aligned} \nabla_x^2 w(x,y) = 0,\ \text{if } |x| \ge R \text{ and } |y| \ge R; \quad w(x,y) = 0, \\ \text{if } y \in C_{R_1} \text{ and } x \in C_{R_1}; \quad w(\infty, y) = 0. \end{aligned} \tag{6.1.113}$$

Thus, $w(x,y) = 0$ if $|x| \ge R$ and $y \in C_{R_1}$, because (6.1.113) has only the trivial solution by the maximum principle.

Fix $x, |x| \ge R$, and note that $w(x,y)$ as a function of y solves the problem

$$\begin{aligned} \nabla_y^2 w(x,y) = 0,\ \text{if } |x| \ge R \text{ and } |y| \ge R; \quad w(x,y) = 0 \\ \text{if } y \in C_{R_1} \text{ and } x \ge R; \quad w(x,\infty) = 0. \end{aligned}$$

This problem has only the trivial solution $w(x,y) = 0, y_1^2 + y_2^2 \ge R_1^2, |x| \ge R$. Therefore, by the unique continuation property for solutions to Laplace's equation $w(x,y) = 0$ if $|x| \ge R$ and $|y| \ge R$. In particular, $w(x,y) = 0$ if $x, y \in P$, where P is a plane which does not intersect the support of v. This implies that $v = 0$ as is proved in Theorem 6.1.2.

6.1.11. Inversion of the backscattering data.

Consider equations (6.1.32)–(6.1.33) with $x = y \in P$, where $P = \{x : x_3 = 0\}$. The data $f(x^1, x^1, \omega), x^1 \in P, 0 < \omega < \infty$, are the backscattering data. In this section, we assume that the data are given for all $\omega > 0$. The problem is to find $v(z)$ from equation (6.1.32) with the right side $f(x^1, x^1, \omega)$. Let us differentiate (6.1.32) in ω, denote $2\omega = k$, and $-2\pi i \partial f/\partial\omega|_{\omega=k/2} = F(x^1, k)$, to obtain

$$\int g(x^1, z, k)v(z)dz = F(x^1, k), \quad x^1 \in P, \quad 0 < k < \infty. \tag{6.1.114}$$

We solve equation (6.1.114) for $v(z)$ analytically. Take the Fourier transform of (6.1.114) in x^1 to obtain

$$\frac{i}{2k_3(2\pi)^2}\int dzv(z)\exp\{i(\hat{k}\cdot z^1+|z_3|k_3)\}=\tilde{F}(\hat{k},k), \tag{6.1.115}$$

where $k_3=(k^2-|\hat{k}|^2)^{1/2}\geq 0$ if $k\geq|\hat{k}|$, and k_3 is defined on the complex k-plane with the cut $(-|\hat{k}|,|\hat{k}|)$ by analytic continuation. Here, we use the formula

$$\begin{aligned}\tilde{g}&=\frac{1}{(2\pi)^2}\iint_{-\infty}^{\infty}\exp(i\hat{k}\cdot x^1)g(x^1,z,k)dx^1\\&=\frac{i\exp\{i(\hat{k}\cdot z^1+|z_3|k_3)\}}{2k_3(2\pi)^2},\end{aligned} \tag{6.1.116}$$

which follows from formula (6.2.18) in the next section.

Let us write (6.1.115) as

$$(2\pi)^{-3}\int dzv(z)\exp(i\alpha\cdot z)=A(\alpha),\quad \alpha=(\hat{k},-k_3), \tag{6.1.117}$$

where we took into account that $|z_3|=-z_3$ if $z_3<0$, i.e., if $v(z)\neq 0$, and define

$$\begin{aligned}A(\alpha)&=-\frac{i}{\pi}k_3\tilde{F}(\hat{k},k)=\frac{i}{\pi}\alpha_3\tilde{F}(\hat{\alpha},(|\hat{\alpha}|^2+\alpha_3^2)^{1/2}),\\\hat{\alpha}&=(\alpha_1,\alpha_2)=\hat{k}.\end{aligned}$$

Note that $A(\alpha)$ is an entire function of α of exponential type since $v(z)$ is compactly supported. Thus, a necessary and sufficient condition for $F(x',k)$ to be the exact data, i.e., for the solvability of equation (6.1.114) in the class of compactly supported $v(z)$, is the requirements that $\alpha_3F(\hat{\alpha},(|\hat{\alpha}|^2+\alpha_3^2)^{1/2})$ is an entire function of α of exponential type. If this condition is satisfied, then

$$v(z)=\int A(\alpha)\exp(-i\alpha\cdot z)d\alpha. \tag{6.1.118}$$

Let us give an example of the inversion procedure.

Example. Let $v(z) = \delta(z)$. Then

$$F(x^1, k) = \frac{\exp(ik|x^1|)}{4\pi|x^1|}, \quad \tilde{F}(\hat{k}, k) = \frac{i}{8\pi^2 k_3}, \text{ and } A(\alpha) = \frac{1}{8\pi^3}.$$

Thus, by formula (6.1.118), one recovers $v(z) = \delta(z)$. This example shows that the inversion procedure is valid for distributions as well as for integrable functions.

Let us summarize the result.

Proposition 6.1.5. *Equation* (6.1.32) *with the right side* $f(x^1, x^1, \omega)$, $x^1 \in P$, $0 < \omega < \infty$, *has at most one solution in the class of compactly suported distributions. This equation has a compactly supported solution* $v(z)$ *if the function* $\alpha_3 \tilde{F}(\hat{\alpha}, (|\hat{\alpha}|^2 + \alpha_3^2)^{1/2})$ *is an entire function of* α *of exponential type. If a compactly supported solution exists, it can be found by formula* (6.1.118).

Remark 6.1.4. Equation (6.1.114) is equivalent to the problem of finding a compactly supported function $v(z)$ from the knowledge of the integrals $\int_{S(x^1,t)} v(z)ds$ of $v(z)$ over the spheres centered at x^1 and of radius $t > 0$, where x^1 runs through the plane P and t runs through the positive semi-axis.

Indeed, define $F(x^1, -k) = F^*(x^1, k), k \geq 0$, where the star denotes complex conjugate. Multiply (6.1.114) by $1/2\pi \exp(-ikt)$ and integrate in k over $(-\infty, \infty)$ to obtain

$$\int dz v(z) \frac{\delta(|x^1 - z| - t)}{4\pi|x^1 - z|} = h(x^1, t), \tag{6.1.119}$$

where $\delta(|x^1 - z| - t)$ is the delta function and

$$h = (2\pi)^{-1} \int_{-\infty}^{\infty} F(x^1, k) \exp(-ikt)dk.$$

Note that $h = 0$ if $t < 0$. Equation (6.1.119) can be written as

$$\int_{S(x^1,t)} v(z)ds = 4\pi t h(x^1, t). \tag{6.1.120}$$

Thus, if one can solve (6.1.114), one can solve (6.1.120). Conversely, taking the Fourier transform of (6.1.120) in t yields (6.1.114). Let us summarize the result.

Proposition 6.1.6. *Let*

$$\int_{S(x^1,t)} v(z)ds = H(x^1,t), \quad x^1 \in P, \quad t \geq 0. \tag{6.1.121}$$

Define

$$h(x^1,t) = (4\pi t)^{-1} H(x^1,t),$$
$$F(x^1,k) = \int_{-\infty}^{\infty} h(x^1,t)\exp(ikt)dt. \tag{6.1.122}$$

Then solve equation (6.1.114) *for* $v(z)$ *by the procedure given in Proposition* 6.1.5. *The function* $v(z)$ *solves equation* (6.1.121).

We leave the discussion of the conditions on $H(x^1,t)$, which are sufficient for $v(z)$ to have a compact support and to be a bounded function, to the reader.

In [43, 91], the problem of finding $v(z)$ from the knowledge of its integrals over the spheres $S(x,1), x \in \mathbb{R}^3$, was considered. In [118], the problem of finding $v(z)$ from the knowledge of its integrals over the spheres $S(x^1,t)$ was considered in $\mathbb{R}^2$, i.e., x^1 run through the line $-\infty < x^1 < \infty$, and $S(x^1,t)$ was the circle. The solution given in [118] requires differentiation of the data infinitely many times, and therefore, it is not suitable for practice.

The problems mentioned are the problems of integral geometry. The best known problem of this type is the Radon problem of finding a function from the knowledge of its integrals over planes ([91, 330]).

Let us prove that the data $f(x,x,\omega_j), |x| = R_1 > R, 0 < \omega_j, 1 \leq j \leq n$, do not determine $v(z)$ uniquely.

Let $v = v(r), r = |z|$. One has

$$\int_{|z|\leq 1} \frac{\exp(2i\omega|x-z|)}{|x-z|^2} v(r)dz$$

$$
\begin{aligned}
&= \int_0^1 drr^2 v(r) \\
&\quad \cdot \int_0^{2\pi} d\phi \int_0^{\pi} d\theta \sin\theta \frac{\exp[2i\omega(|x|^2 + r^2 - 2r|x|\cos\theta)^{1/2}]}{|x|^2 + r^2 - 2r|x|\cos\theta} \\
&= \int_0^1 drr^2 v(r) h(r, R_1, \omega), \quad R_1 = |x|,
\end{aligned}
$$

where h is the integral in the angular variables. Therefore, any function $v(r)$ which is orthogonal in $L^2([0,1])$ to $r^2 h(r, R_1, \omega)$ solves the equation

$$
\int_{|z|\le 1} \frac{\exp(2i\omega|x-z|)}{|x-z|^2} v(z)dz = 0, \quad |x| = R_1. \tag{6.1.123}
$$

Clearly, there are infinitely many nontrivial solutions of equation (6.1.123) for any finite number of $\omega = \omega_j, 1 \le j \le n$ and $R_j, 1 \le j \le m$. On the other hand, the reader can prove that if the data $f(x, x, \omega_j), |x| = R_1 > R$ are given on a set $\{\omega_j\}, 1 \le j < \infty$, and the set $\{\omega_j\}$ has a finite limit point ω_0, then $v(z)$ is uniquely determined. The same is true if the data $f(x, x, \omega)$ are given at a fixed $\omega > 0$ and for $|x| = R_j$, where the set $\{R_j\}$ has a limit point (not necessarily finite).

Exercise. The data $f(x^1, x^1, \omega_0), x^1 \in P, \omega_0 > 0$ is fixed, do not determine $v(z)$ uniquely.

6.2. Two-Parameter Inversion

6.2.1. In Section 6.1, one unknown function $n(x)$ was found from the seismic data. In this section, we use the data at two frequencies to find the density of the inhomogeneity and the velocity. Consider the equation

$$
\frac{\omega^2 G}{k(x)} + \nabla \cdot \left(\frac{1}{\rho(x)} \nabla G \right) = -\delta(x-y), \tag{6.2.1}
$$

where $x, y \in \mathbb{R}^3$, $\rho(x)$ is the density, and $k(x)$ is the bulk modulus. Let us assume that

$$\frac{1}{k(x)} = \frac{1}{k_r} - \frac{a_1(x)}{\rho_r}, \quad \frac{1}{\rho} = \frac{1}{\rho_r} - \frac{a_2(x)}{\rho_r}, \tag{6.2.2}$$

where k_r and ρ_r are positive constants, and $a_j(x), j = 1, 2$, are functions with compact support, i.e., $a_j(x) = 0$ if $|x| > R$, where $R > 0$ is an arbitrarily large fixed number. We assume that $a_1 \in L^2(B_R), B_R = \{x : |x| \leq R\}$, and $a_2 \in H^1(B_R)$.

Equation (6.2.1) can be written as

$$\left(\nabla^2 + \frac{\omega^2}{c^2}\right) G - \omega^2 a_1 G - \nabla \cdot (a_2 \nabla G) = -\rho_r \delta(x - y), \tag{6.2.3}$$

where $c^2 = k_r \rho_r^{-1}$. Let us choose units in which

$$c = 1 \quad \text{and} \quad \rho_r = 1. \tag{6.2.4}$$

This is possible: take $x = ax', y = ay'$, and $\omega = \beta\omega'$ where α and β are constants. Then (6.2.3) takes the form

$$\left(\alpha^{-2}\nabla'^2 + \beta^2 \frac{\omega'^2}{c^2}\right) G - \beta^2 \omega^2 a_1 G - \alpha^{-2} \nabla' \cdot (a_2 \nabla' G)$$
$$= -\rho_r^{-3} \alpha \delta(x' - y'). \tag{6.2.5}$$

Let $\alpha = \rho_r, \beta = c\rho_r^{-1}$ and define $a_1' = c^2 a_1$ and $a_2' = a_2$. Then (6.2.5) takes the form

$$(\nabla'^2 + \omega'^2)G - \omega^2 a_1'(x)G - \nabla' \cdot (a_2' \nabla' G) = -\delta(x' - y'). \tag{6.2.6}$$

Therefore, one can study the equation

$$(\nabla^2 + \omega^2)G - \omega^2 a_1(x)G - \nabla \cdot (a_2 \nabla G) = -\delta(x - y). \tag{6.2.7}$$

This equation can be written as the integral equation (which is analogous to equation (6.1.6)):

$$G = g - \omega^2 \int g a_1 G dz - \int g \nabla_z \cdot (a_2(z) \nabla_z G) dz, \tag{6.2.8}$$

where $\int = \int_{\mathbb{R}^3}$, $dz = dz_1 dz_2 dz_3$, and

$$g = \frac{\exp(i\omega|x-y|)}{4\pi|x-y|}. \tag{6.2.9}$$

In operator form, equation (6.2.8) can be written as

$$G = g - \omega^2 g a_1 G - g\nabla \cdot (a_2 \nabla G) = g - gVG, \tag{6.2.10}$$

where the potential V is given by

$$V = \omega^2 a_1 + \nabla \cdot a_2 \nabla. \tag{6.2.11}$$

Note that $G = g - GVg$ so that $GVg = gVG$. Let us define the T-matrix

$$T = V - VGV. \tag{6.2.12}$$

The T-matrix here is not to be confused with the T-matrix in Chapter 3.5. It is easy to check that

$$T = V - VgT \tag{6.2.13}$$

and

$$Tg = VG. \tag{6.2.14}$$

The operator V is an integral operator in impulse space with kernel

$$(2\pi)^{-3}\tilde{V}(k,p) = (2\pi)^{-3} \iint dx dy V(x,y) \exp(ik \cdot x - p \cdot y),$$

where $V(x,y)$ is the kernel of V. This kernel may be a distribution. For V defined in (6.2.11), we have $V(x,y) = \omega^2 a_1(x)\delta(x-y) + \nabla \cdot a_2(y)\nabla\delta(x-y)$, so that

$$\tilde{V}(k,p) = \omega^2 \tilde{a}_1(k-p) - k \cdot p \tilde{a}_2(k-p), \tag{6.2.15}$$

where $\tilde{a}(k) = \int \exp(ik \cdot x) a(x) dx$. The scattered field $G_s = G - g$ according to equations (6.2.10) and (6.2.14) can be written as

$$G_s = -gVG = -gTg. \tag{6.2.16}$$

Let us keep the x and y variables (the positions of the receiver and the source) on the plane $x_3 = 0$ and Fourier transform (6.2.16) in

$\hat{x} = (x_1, x_2)$ and $\hat{y} = (y_1, y_2)$. Let $\hat{k} \cdot \hat{x} = k_1 x_1 + k_2 x_2, d\hat{x} = dx_1 dx_2$, etc. Then

$$
\begin{aligned}
\tilde{G}_s(\hat{k}, \hat{p}, \omega) &= \int \exp(i\hat{k} \cdot \hat{x} - i\hat{p} \cdot \hat{y}) G_s(\hat{x}, \hat{y}, \omega) d\hat{x} d\hat{y} \\
&= -\int \exp(i\hat{k} \cdot \hat{x} - i\hat{p} \cdot \hat{y}) gTg d\hat{x}\, d\hat{y} \\
&= -\iint dz'\, dz'' T(z', z'') \\
&\quad \times \int \frac{\exp(i\omega|z' - \hat{x}| + i\hat{k}\hat{x})}{4\pi|z' - \hat{x}|} d\hat{x} \\
&\quad \cdot \int \frac{\exp(i\omega|z' - \hat{y}| - i\hat{p} \cdot \hat{y})}{4\pi|z'' - \hat{y}|} d\hat{y} \\
&= -\iint dz'\, dz'' T(z', z'') \exp(i\hat{k}\hat{z}' - i\hat{p} \cdot \hat{z}'') \\
&\quad \cdot h(|z_3'|, |\hat{k}|) \cdot h(|z_3''|, |\hat{p}|),
\end{aligned} \tag{6.2.17}
$$

where $T(z', z'')$ is the kernel of T,

$$
h(|z_3'|, |\hat{k}|) = \begin{cases} \dfrac{i}{2} \dfrac{\exp\{i|z_3'|(\omega^2 - |\hat{k}|^2)^{1/2}\}}{(\omega^2 - |\hat{k}|^2)^{1/2}}, & \omega > |\hat{k}|, \\ \dfrac{1}{2} \dfrac{\exp\{-|z_3'|(|\hat{k}|^2 - \omega^2)^{1/2}\}}{(|\hat{k}|^2 - \omega^2)^{1/2}}, & \omega < |\hat{k}|, \\ \infty, & \omega = |\hat{k}|. \end{cases} \tag{6.2.18}
$$

Let us give the details of the computation of integral (6.2.18). One has, for example,

$$
\begin{aligned}
&\int \frac{\exp(i\hat{k} \cdot \hat{x} + i\omega|z - \hat{x}|)}{4\pi|\hat{x} - z|} d\hat{x} \bigg|_{\hat{x} = \hat{u} + \hat{z}} \\
&\quad = \exp(i\hat{k} \cdot \hat{z}) \cdot \int \frac{\exp\{i|\hat{k}||\hat{u}| \cos\phi + i\omega(|\hat{u}|^2 + z_3^2)^{1/2}\}}{4\pi(|\hat{u}|^2 + z_3^2)^{1/2}} d\hat{u}
\end{aligned}
$$

$$= \frac{\exp(i\hat{k}\cdot\hat{z})}{4\pi}\int_0^\infty \frac{d\rho\rho\exp\{i\omega(|\hat{u}|^2+z_3^2)^{1/2}\}}{(\rho^2+z_3^2)^{1/2}}$$

$$\times\int_0^{2\pi} d\phi\exp(i|\hat{k}|\rho\cos\phi) = \frac{\exp(i\hat{k}\cdot\hat{z})}{4\pi}\frac{2\pi}{|\hat{k}|^{1/2}}$$

$$\times\int_0^\infty d\rho\rho^{1/2}\exp\{i\omega(\rho^2+z_3^2)^{1/2}\}\sqrt{\rho|\hat{k}|}J_0(\rho|\hat{k}|)$$

$$= \frac{\exp(i\hat{k}\cdot\hat{z})}{2|\hat{k}|^{1/2}}\chi = \frac{i\exp\{i\hat{k}\cdot\hat{z}+i|z_3|(\omega^2-|\hat{k}|^2)^{1/2}\}}{2(\omega^2-|\hat{k}|^2)^{1/2}},$$

$$\omega > |\hat{k}|,$$

where (see [58, formula (8.6.21)])

$$\chi = \begin{cases} i\left(\frac{\pi|z_3|}{2}\right)^{1/2}|\hat{k}|^{1/2}\frac{H^{(1)}_{-1/2}(|z_3|\sqrt{\omega^2-|\hat{k}|^2})}{(\omega^2-|\hat{k}|^2)^{1/4}}, & \omega > |\hat{k}|, \\ \left(\frac{2|z_3|}{\pi}\right)^{1/2}|\hat{k}|^{1/2}\frac{K_{1/2}(|z_3|\sqrt{|\hat{k}|^2-\omega^2})}{(|\hat{k}|^2-\omega^2)^{1/4}}, & \omega < |\hat{k}|, \end{cases}$$

and

$$H^{(1)}_{-1/2}(z) = \left(\frac{2}{\pi z}\right)^{1/2}\exp(iz), \quad K_{1/2}(z) = \left(\frac{\pi}{2z}\right)^{1/2}\exp(-z).$$

Formula (6.2.18) follows from the last three formulas. Let us define

$$k_3 = \begin{cases} (\omega^2-|\hat{k}|^2)^{1/2}, & |\hat{k}| < \omega, \\ i(|\hat{k}|^2-\omega^2)^{1/2}, & |\hat{k}| > \omega, \end{cases} \tag{6.2.19}$$

and p_3 in the same way. Then

$$|\hat{k}|^2 + k_3^2 = \omega^2, \quad p_3^2 + |\hat{p}|^2 = \omega^2. \tag{6.2.20}$$

Vectors $\hat{k}$ and $\hat{p}$ are real valued but k_3 and p_3 may be complex numbers. If V has compact support (and this was our assumption), then T has compact support as follows from (6.2.12). Therefore, the Fourier transform of T is meaningful for complex p_3 and k_3. From

(6.2.17) and (6.2.19), one obtains

$$\tilde{G}_s = \frac{1}{4k_3p_3}\iint dz'dz''T(z',z'')\exp(i\hat{k}\cdot\hat{z}' \\ +ik_3|z_3'| - i\hat{p}\cdot\hat{z}'' + ip_3|z_3''|). \tag{6.2.21}$$

Let us assume that the support of V belongs to the half-space $z_3 < 0$. Then $|z_3| = -z_3$ and (6.2.21) takes the form

$$\tilde{G}_s = \frac{1}{4k_3p_3}\tilde{T}(k',p), \quad \text{where } |k| = |p| = \omega, \\ k' = (\hat{k}, -k_3), \quad |k'| = |k|, \tag{6.2.22}$$

and

$$\tilde{T}(k,p) = \iint dz'dz''T(z',z'')\exp(ik\cdot z' - ip\cdot z''). \tag{6.2.23}$$

The basic problem can be formulated as follows. *Given the scattering data $\tilde{G}(\hat{x}, x_3 = 0, \hat{y}, y_3 = 0, \omega)$ for all $\hat{x}$ and $\hat{y}$ and two distinct frequencies $\omega_j, j = 1,2$, or small ω, find $a_j(x), j = 1,2$.*

Formulas (6.2.12), (6.2.15), and (6.2.22) are basic for our first inversion scheme with the data given at two distinct frequencies. Our second inversion scheme is based on the method given in Section 6.1.

The inversion schemes can be generalized to include the dissipative terms (corresponding to first derivative in time with a coefficient depending on x but not on t in the time-dependent problem). Our method is based on the Born approximation. Since the potential (6.2.11) contains second derivatives, it does not become small even for $\omega \to 0$.

6.2.2. Let us describe the first inversion scheme. Formula (6.2.15) shows that under the assumption

$$|k| = |p| = \omega, \tag{6.2.24}$$

which means that the T-matrix in (6.2.22) is known on shell, $\tilde{V}$ can be written as

$$\tilde{V}(k,p) = \omega^2 W(k,p), \quad W = \tilde{a}_1(k-p) - k^0\cdot p^0\tilde{a}_2(k-p), \tag{6.2.25}$$

where $k^0 = k|k|^{-1}$ and $p^0 = p|p|^{-1}$ are unit vectors which do not depend on ω. Since the functions a_1 and a_2 have compact support, their Fourier transforms $\tilde{a}_1$ and $\tilde{a}_2$ are entire functions of the three complex variables k_1, k_2 and k_3, which decay at infinity in the real space in $\mathbb{R}^3$.

In the Born approximation, $T = V$ because one neglects the quadratic terms in V in (6.2.14), and (6.2.22) becomes in this approximation

$$\tilde{G}_s = \frac{1}{4k_3p_3}\tilde{V}(k', p) = \frac{\omega^2}{4k_3p_3}W(k', p). \tag{6.2.26}$$

Thus,

$$4\tilde{G}_s k_3 p_3 \omega^{-2} = W(k', p). \tag{6.2.27}$$

Let

$$k' - p = q, \quad k' + p = s, \tag{6.2.28}$$

$$k' = (s + q)/2, \quad p = (s - q)/2. \tag{6.2.29}$$

Then

$$k' \cdot p = (|s|^2 - |q|^2)/4, \quad s \cdot q = 0, \tag{6.2.30}$$

$$k'^0 \cdot p^0 = (|s|^2 - |q|^2)(4\omega^2)^{-1}, \tag{6.2.31}$$

$$\begin{aligned} &|q|^2 = 2\omega^2(1 - k'^0 \cdot p^0), \\ &|s|^2 = |k'|^2 + |p|^2 + 2k' \cdot p = 2\omega^2(1 + k'^0 \cdot p^0), \quad \omega > 0, \end{aligned} \tag{6.2.32}$$

$$|s|^2 - |q|^2 = 4\omega^2 - 2|q|^2. \tag{6.2.33}$$

Therefore,

$$\begin{aligned} W = W(k', p) &= \tilde{a}_1(q) - \tilde{a}_2(q)\frac{|s|^2 - |q|^2}{4\omega^2} \\ &= \tilde{a}_1(q) - \tilde{a}_2(q)\left(1 - \frac{|q|^2}{2\omega^2}\right), \quad \omega > 0. \end{aligned} \tag{6.2.34}$$

Let us take two arbitrary frequencies ω_1 and $\omega_2 \neq \omega_1$ and solve the two equations

$$W_1 = \tilde{a}_1(q) - \tilde{a}_2(q)\left(1 - \frac{|q|^2}{2\omega_1^2}\right), \tag{6.2.35}$$

$$W_2 = \tilde{a}_1(q) - \tilde{a}_2(q)\left(1 - \frac{|q|^2}{2\omega_2^2}\right) \tag{6.2.36}$$

for $\tilde{a}_1$ and $\tilde{a}_2(q)$. The functions $W_j, j = 1, 2$, are the data according to (6.2.27). One obtains

$$\tilde{a}_2 = \frac{2(W_1 - W_2)}{|q|^2}\frac{\omega_1^2\omega_2^2}{\omega_2^2 - \omega_1^2}, \tag{6.2.37}$$

$$\tilde{a}_1 = W_1 + (W_1 - W_2)\frac{2\omega_1^2\omega_2^2}{(\omega_2^2 - \omega_1^2)|q|^2}\left(1 - \frac{|q|^2}{2\omega_1^2}\right). \tag{6.2.38}$$

Taking the inverse Fourier transform of the functions (6.2.37) and (6.2.38), one obtains $a_2(x)$ and $a_1(x)$. It is clear from (6.2.34) that the following conclusion holds.

Conclusion: One cannot find both functions a_1 and a_2 if the data are given at only one frequency.

Indeed, it follows from (6.2.34) that for a fixed frequency the data depends on q only, and there is no parameter to vary in order to find both $\tilde{a}_1$ and $\tilde{a}_2$. The conclusion that one cannot recover both functions $a_j, j = 1, 2$, from the data given at one frequency is not at all obvious: at first glance one can think that the two conditions $|k| = |p| = \omega$ leave 4 degrees of freedom in the six-dimensional space $\mathbb{R}^3_k \times \mathbb{R}^3_p$ which should be enough to determine two functions $\tilde{a}_j, j = 1, 2$, of 3 variables. This argument, however, is not valid as one saw above. The reason is that the function $W(k', p)$ has a very special structure as a function of two vectors k' and p. Our reasoning applies to the theory in the Born approximation. It does not imply that the conclusion holds in exact theory. The exact theory is currently out of reach.

Another important point concerns the numerical problem of inverting the Fourier transforms of $\tilde{a}_j, j = 1, 2$. Let $\omega_1 < \omega_2$. It follows from (6.2.32) that the real valued vectors q run through the ball $B_{2\omega} = \{q : |q| \leq 2\omega\}$ when k'^0 and p^0 run through the unit sphere S^2, so that the number $k'^0 \cdot p^0$ runs through the interval $[-1, 1]$. Therefore, for real-valued vectors q, formulas (6.2.37) and (6.2.38) determine the Fourier transforms $a_j, j = 1, 2$, in the ball $B_{2\omega_1}$ only. Since $a_j(x)$ are compactly supported, the functions $\tilde{a}_j(q)$ are entire analytic functions of q. Therefore, if one knows these functions in a ball $B_{2\omega_1}$, one can uniquely determine these functions everywhere in $\mathbb{R}^3$ by analytic continuation. However, the basic numerical problem is as follows.

Problem. *Given the Fourier transform $\tilde{a}(q)$ of a function $a(x), a(x) = 0$ if $|x| \geq R$, in the ball $B_{2\omega} = \{q, |q| < 2\omega\}$ compute $a(x)$ in the ball $B_R = \{x : |x| \leq R\}$ with a prescribed accuracy.*

This problem will be treated in Section 6.3.

6.2.3. Let us describe the second inversion scheme. If $\omega \to 0$, the limit of equation (6.2.8) is

$$G(x, y) = g_0(x, y) - \int g_0 \nabla \cdot (a_2 \nabla G) dz, \qquad g_0 = \frac{1}{4\pi|x - y|}. \tag{6.2.39}$$

The scattered field $G_s = G - g_0$ in the Born approximation for x and y on the plane $x_3 = 0$ is of the form:

$$G_s(\hat{x}, \hat{y}) = -\frac{1}{16\pi^2} \int \frac{1}{|\hat{x} - z|} \nabla \cdot \left(a_2(z) \nabla \frac{1}{|z - \hat{y}|} \right) dz. \tag{6.2.40}$$

This equation can be solved analytically by the method given in Section 6.1. One takes the Fourier transform in $\hat{x}$ and $\hat{y}$ of equation (6.2.40), uses formula (6.1.13), and obtains

$$f(\lambda, \mu) = \frac{1}{(2\pi)^4} \iint d\hat{x} d\hat{y} \exp(i\lambda \cdot \hat{x} + i\mu \cdot \hat{y})(-16\pi^2 G_s(\hat{x}, \hat{y}))$$

$$= \frac{1}{(2\pi)^2}\int dz \frac{\exp(i\lambda\cdot\hat{z} - |\lambda||z_3|)}{|\lambda|}\nabla_z$$

$$\cdot\left(a_2(z)\nabla_z \frac{\exp(i\mu\cdot\hat{z} - |\mu||z_3|)}{|\mu|}\right)$$

$$= \frac{-1}{(2\pi)^2|\lambda||\mu|}\int dz \exp\{i(\lambda+\mu)\cdot\hat{z}$$

$$+ (|\lambda| + |\mu|)z_3\}a_2(z)(i\lambda + |\lambda|e_3)\cdot(i\mu + |\mu|e_3)$$

$$= \frac{(\lambda\cdot\mu - |\lambda||\mu|)}{(2\pi)^2|\lambda||\mu|}\int dz a_2(z)\exp(ip\cdot\hat{z} + q\cdot z_3). \tag{6.2.41}$$

Here, we integrated once by parts, used the equality $|z_3| = -z_3$ and set $p = \lambda + \mu, q = |\lambda| + |\mu|, (p_1, p_2), p_3 = |\lambda|, p_4 = |\mu|$. These p and q should not be confused with p and q in Section 6.2.2. They are the same as in Section 6.1.

Define

$$F(p,q) = \frac{(2\pi)^2|\lambda||\mu|f(\lambda,\mu)}{\lambda\cdot\mu - |\lambda||\mu|} = \int dz a_2(z)\exp(ip\cdot\hat{z} + q\cdot z_3)$$

$$= \int_{-\infty}^{\infty}\int dp \exp(ip\cdot\hat{z})\int_0^{\infty} d\xi a_2(\hat{z}, -\xi)\exp(-q\xi) \tag{6.2.42}$$

as in Section 6.1. Here, $\xi = -z_3$. The function $F(p,q)$ is the datum. Taking the inverse Fourier and Laplace transforms of $F(p,q)$, one finds $a_2(z) = a_2(\hat{z}, -\xi)$:

$$a_2(z) = \frac{1}{(2\pi)^2}\int_{-\infty}^{\infty}\int dp \exp(-ip\cdot z)$$

$$\times \frac{1}{2\pi i}\int_{c-i\infty}^{c+i\infty} dq F(p,q)\exp(-qz_3), \quad c > 0. \tag{6.2.43}$$

If $a_2(z)$ is found, then equation (6.2.8) in the Born approximation, and for $x = (\hat{x}, 0), y = (\hat{y}, 0)$, can be written as

$$-16\pi^2 \lim_{\omega\to 0}\frac{G_s - Q}{\omega^2} = \int \frac{a_1(z)dz}{|\hat{x} - z||\hat{y} - z|}, \tag{6.2.44}$$

where

$$Q = -\int g(\hat{x}, z)\nabla_z \cdot (a_2(z)\nabla_z g(z, \hat{y}))dz \tag{6.2.45}$$

and $a_2(z)$ is given by (6.2.43).

Equation (6.2.44) was solved in Section 6.1. Let us summarize the second inversion scheme. First, one computes a_2 by formula (6.2.43), then a_1 by solving equation (6.2.44).

6.2.4. In this section, we give sufficient conditions for convergence of the Born series. These conditions can be considered as conditions under which the Born approximation is valid. Error estimates will be obtained.

Consider the equation

$$u = f + \omega^2 T_1 u + T_2 u, \tag{6.2.46}$$

where

$$\begin{aligned} T_1 u &= \int g a_1 u dz, \\ T_2 u &= \int g \nabla \cdot (a_2(z)\nabla u) dz. \end{aligned} \tag{6.2.47}$$

The assumptions about $a_j(x)$ are:

$$\begin{aligned} &a_j(x) = 0 \quad \text{if } |x| \geq R, \\ &|a_1| \leq c_1, \\ &|a_2| + |\nabla a_2| \leq c_2. \end{aligned} \tag{6.2.48}$$

The Born series for equation (6.2.47) is of the form $u = \sum_{n=0}^{\infty}(\omega^2 T_1 + T_2)^n f$. It converges in some space H if $\|\omega^2 T_1 + T_2\| < 1$ where $\|T\|$ is the norm of a linear operator T acting in this space. Let $H = H^2(B_R)$ be the Sobolev space, $B_R = \{x : |x| \leq R\}$. Since the functions $a_j(x)$ vanish outside of B_R, the values of the functions $T_j f$ are determined entirely by the values of f on B_R. The following estimate

$$\|gf\|_{2,R} \leq c_R \|f\|_{0,R} \tag{6.2.49}$$

is known (see, e.g., [96]). Here $\|u\|_{m,R} = \|u\|_{H^m(B_R)}$. If $\omega \to 0$, the inequality $\|\omega^2 T_1 + T_2\| < 1$ holds if $\|T_1\| \leq c$ and $\|T_2\| < 1$. Here and below we denote by c various constants. One has, using (6.2.49),

$$\|T_1 f\|_{2,R} \leq c_R \|a_1 f\|_{0,R} \leq c_R c_1 \|f\|_{0,R} \leq c\|f\|_{2,R}. \tag{6.2.50}$$

Thus, $\|T_1\| \leq c$ if $|a_1| \leq c_1$. Actually, $\|T_1\| \leq c$ if $\|a_1\|_{0,R} \leq c_1$. Furthermore,

$$\|T_2 f\|_{2,R} \leq c_R \|\nabla \cdot (a_2 \nabla f)\|_{0,R} \leq c_R \cdot c_2 \|f\|_{2,R}. \tag{6.2.51}$$

Therefore, $\|T_2\| < 1$ if $c_R c_2 < 1$. Let us summarize the result.

Proposition 6.2.1. *If* (6.2.48) *holds and* $q = \omega^2 c_1 c_R + c_2 c_R < 1$, *then the Born series* $\sum_{n=0}^{\infty} (\omega^2 T_1 + T_2)^n f$ *converges in* H_R^2 *and is majorized by the series* $\sum_{n=0}^{\infty} q^n$.

6.2.5. Consider the inverse scattering problem for equation (6.2.7) with the scattering amplitude as the data. If the incident field is the plane wave $u_0 = \exp(i\omega s_0 \cdot x)$, then the scattering amplitude is given by

$$f(s_0, s, \omega) = -\frac{1}{4\pi} \int \exp(-i\omega s \cdot y)[a_1(y)u(y)\omega^2 + \nabla \cdot (a_2(y)\nabla u)]dy. \tag{6.2.52}$$

In the Born approximation $u = u_0$, and, integrating by parts the second term in (6.2.52), one obtains for the scattering amplitude

$$\begin{aligned} f &= f_B(s_0, s, \omega) \\ &= -\frac{\omega^2}{4\pi} \int \exp[i\omega(s_0 - s) \cdot y]\{a_1(y) - s_0 \cdot s a_2(y)\}dy. \end{aligned} \tag{6.2.53}$$

Consider $D = -4\pi\omega^{-2} f_B$ as the data. Assume that the data are known at $\omega = \omega_j, j = 1, 2, \omega_1 < \omega_2$. Then the Fourier transforms $\tilde{a}_j$ of $a_j(y), j = 1, 2$, can be found from (6.2.53). Indeed, set $\omega(s_0 - s) = \omega p$, where $|p|^2 = 2 - 2\gamma$ and $\gamma = s \cdot s_0 = 1 - |p|^2/2$. Then

$$D_1 = \tilde{a}_1(\omega_1 p) - \left(1 - \frac{|p|^2}{2}\right)\tilde{a}_2(\omega_1 p), \tag{6.2.54}$$

$$D_2 = \tilde{a}_1(\omega_2 p) - \left(1 - \frac{|p|^2}{2}\right)\tilde{a}_2(\omega_2 p). \tag{6.2.55}$$

Set $q = \omega_1 p$. Then (6.2.54) and (6.2.55) can be written as

$$D_1(q) = \tilde{a}_1(q) - \left(1 - \frac{|q|^2}{2\omega_1^2}\right)\tilde{a}_2(q), \tag{6.2.56}$$

$$D_2\left(\frac{q}{\omega_2}\right) = \tilde{a}_1(q) - \left(1 - \frac{|q|^2}{2\omega_2^2}\right)\tilde{a}_2(q). \tag{6.2.57}$$

From (6.2.56) and (6.2.57), one finds $\tilde{a}_j(q)$ and, by applying the Fourier inversion, $a_j(y), j = 1, 2$. Again the problem formulated at the end of subsection 6.2.2 arises.

6.3. An Inversion Formula in Scattering Theory

6.3.1. In this section, we solve a three-dimensional inverse scattering problem in the Born approximation, and the problem in Section 6.2.2.

Let

$$(\nabla^2 + k^2 - q(x))u = 0 \quad \text{in } \mathbb{R}^3, \tag{6.3.1}$$

$$u = \exp(iks \cdot x) + v, \tag{6.3.2}$$

$$v \sim \frac{\exp(ikr)}{r} f(s, s', k) \text{ as } r = |x| \to \infty,\ xr^{-1} = s'. \tag{6.3.3}$$

The integral equation for u is

$$u = \exp(iks \cdot x) - \int gqu dy, \quad g = \frac{\exp(ik|x-y|)}{4\pi|x-y|}. \tag{6.3.4}$$

Therefore,

$$f(s, s', k) = -\frac{1}{4\pi}\int \exp(-iks' \cdot y) qu dy. \tag{6.3.5}$$

In the Born approximation ($u = \exp(iks \cdot x)$), one has

$$f = f_B = -\frac{1}{4\pi} \int \exp\{ik(s - s') \cdot y\} q(y) dy. \tag{6.3.6}$$

Let us assume that

$$q(x) \in L^2(B_R), \quad q(x) = 0 \text{ if } |x| > R. \tag{6.3.7}$$

Then one obtains for $q(y)$ the integral equation

$$\int_{|y| \leq R} \exp\{-ik(s' - s) \cdot y\} q(y) dy = F(k(s' - s)), \tag{6.3.8}$$

where $F(k(s' - s)) = -4\pi f_B$. In (6.3.8), $k > 0$ is fixed and s and s' run through the unit sphere S^2. Let $p = k(s' - s)$. When s and s' run through S^2, vector p runs through the ball $B_{2k} = \{p : |p| \leq 2k\}$. Equation (6.3.8) can be written as

$$\int_{|y| \leq R} \exp(-ip \cdot y) q(y) dy = F(p), \quad |p| \leq 2k. \tag{6.3.9}$$

This equation is solvable in the class of functions q satisfying conditions (6.3.7) if and only if $F(p) \in W_R$, where W_R is the set of entire functions of exponential type $\leq R$ square integrable over $\mathbb{R}^3$, that is, of entire functions such that

$$|F(p)| \leq c \exp(R|p|), \quad \int_{\mathbb{R}^3} |F(p)|^2 dp < \infty. \tag{6.3.10}$$

This is a variant of the Paley–Wiener theorem. If

$$\tilde{q}(p) = \int \exp(-ip \cdot y) q dy = 0 \quad \text{for } |p| > 2k_0, \tag{6.3.11}$$

then one can find $q(y)$, the solution of equation (6.3.9), by taking the inverse Fourier transform of the function $F(p)$. But (6.3.11) is equivalent to the assumption that $q(y)$ is an entire function of class W_{2k_0} which is not the case in applications where assumptions (6.3.7) are more realistic. The idea of what follows is to approximate $q(x)$ satisfying conditions (6.3.7) by an entire function $q_N(x) \in W_{2k_0}$ with arbitrary accuracy on $B_R : \|q_N - q\|_R \leq \epsilon(N) \to 0, N \to \infty$. Here $\|\cdot\|_R$ is the norm in $L^2(B_R)$ or $C(B_R)$ depending on whether

$q \in L^2(B_R)$ or $C(B_R)$. The function $q_N(x)$ will be constructed explicitly analytically from the data.

6.3.2. The data in (6.3.8) are given on $S^2 \times S^2$ and it is useful to have a formula which transforms the three-dimensional Fourier integral into an integral over $S^2 \times S^2$. Such a formula in $\mathbb{R}^d, d \geq 2$, is known (see [45] for $d = 2$ and [22] for $d \geq 2$):

$$\begin{aligned}(2\pi)^{-d} &\int_{|p|\leq 2k} F(p)\exp(ip\cdot x)dp \\ &= \frac{k^d}{8\pi^d \omega_{d-1}} \int_{S^{d-1}} \int_{S^{d-1}} \\ &\quad \times \frac{|s-s'| \exp\{ik(s-s')\cdot x\} F(k(s-s'))dsds'}{(4-|s-s'|^2)^{(d-3)/2}}, \end{aligned} \tag{6.3.12}$$

where ds is the element of the area of the unit sphere S^{d-1}. The identity (6.3.12) holds for any $F \in L^2(B_{2k})$, S^{d-1} is the unit sphere in $\mathbb{R}^d$, ω_d is the area of S^{d-1}, and $\omega_d = 2\pi^{d/2}/\Gamma(d/2)$, where $\Gamma(x)$ is the Gamma function. For $d = 3$, one has

$$\begin{aligned}(2\pi)^{-3} &\int_{|p|\leq 2k} F(p)\exp(ip\cdot x)dp \\ &= \frac{k^3}{32\pi^4} \int_{S^2} \int_{S^2} |s-s'| \exp\{ik(s-s')\cdot x\} \\ &\quad \times F(k(s-s'))dsds'. \end{aligned} \tag{6.3.13}$$

For $d = 2$, one has

$$\begin{aligned}(2\pi)^{-2} &\int_{|p|\leq 2k} F(p)\exp(ip\cdot x)dp \\ &= \frac{k^2}{16\pi^3} \int_{S^1} \int_{S^1} |s-s'|(4-|s-s'|^2)^{1/2} \\ &\quad \times \exp\{ik(s-s')\cdot x\} F(k(s-s'))dsds'. \end{aligned} \tag{6.3.14}$$

If (6.3.11) holds, then (6.3.8) is solvable by the formula

$$\begin{aligned} q(x) &= (2\pi)^{-3} \int \exp(ip \cdot x) F(p) dp \\ &= \frac{k^3}{32\pi^4} \int_{S^2} \int_{S^2} |s - s'| \exp\{ik(s - s') \cdot x\} \\ &\quad \times F(k(s - s')) ds ds', \end{aligned} \tag{6.3.15}$$

where formula (6.3.13) was used.

Before we pass to the formula for q_N, let us prove the following two lemmas.

Lemma 6.3.1. *Equation* (6.3.8) *has no more than one solution.*

Lemma 6.3.2. *Formula* (6.3.12) *holds.*

Proof of Lemma 6.3.1. Equation (6.3.8) is equivalent to (6.3.9). If $F(p) = 0, |p| \le 2k$, then

$$\int_{|y| \le R} \exp(-ip \cdot y) q(y) dy = 0, \quad p \in \mathbb{R}^3. \tag{6.3.16}$$

Indeed, the left-hand side in (6.3.9) is an entire function of p. If this function vanishes in the ball B_{2k}, it vanishes everywhere. From (6.3.16), one concludes that $q = 0$. □

Proof of Lemma 6.3.2. In this proof, we use the identity for the spherical means (see [91, p. 81]). Let us define the spherical mean for a continuous function F at the point x:

$$I(x, r) = \frac{1}{\omega_d} \int_{S^{d-1}} F(x + rs) ds. \tag{6.3.17}$$

Clearly, if F is continuous, then

$$I(x, 0) = F(x). \tag{6.3.18}$$

Let us define the iterated spherical mean

$$M(x, \lambda, \mu) = \frac{1}{\omega_d^2} \int_{S^{d-1}} \int_{S^{d-1}} F(x + \lambda s + \mu s') ds ds'. \tag{6.3.19}$$

One has $M(x,\lambda,\mu)=M(x,\mu,\lambda)$, $M(x,\lambda,0)=M(x,0,\lambda)=I(x,\lambda)$, $M(x,0,0)=F(x)$. The identity which we need is of the form $(\lambda,\mu>0)$:

$$M(x,\lambda,\mu)=\frac{2\omega_{d-1}}{\omega_d(2\lambda\mu)^{d-2}}\cdot\int_{\mu-\lambda}^{\mu+\lambda}[(r+\mu-\lambda)(r+\mu+\lambda)\cdot(r-\mu+\lambda)(\lambda-r+\mu)]^{(d-3)/2}rI(x,r)dr. \quad (6.3.20)$$

Set $\lambda=\mu=k$ in (6.3.20) to obtain

$$M(x,k,k)=\frac{\omega_{d-1}}{\omega_d k^{2d-4}2^{d-3}}\cdot\int_0^{2k}(4k^2-r^2)^{(d-3)/2}r^{d-2}I(x,r)dr. \quad (6.3.21)$$

Let $h_y(p)=|p|(4k^2-|p|^2)^{-(d-3)/2}\hat{F}(p)\exp(ip\cdot y)$, where $\hat{F}(p)$ is a continuous function vanishing outside B_{2k}. Taking the spherical mean of $h_y(p)$ yields:

$$I_y(0,r)=\frac{1}{\omega_d}\frac{r}{(4k^2-r^2)^{(d-3)/2}}\cdot\int_{S^{d-1}}\hat{F}(rs)\exp(irs\cdot y)ds. \quad (6.3.22)$$

In formula (6.3.21), the iterated spherical mean of $h_y(p)$ is

$$M(0,k,k)=\frac{\omega_{d-1}}{\omega_d^2 k^{2d-4}2^{d-3}}\int_0^{2k}drr^{d-1}\cdot\int_{S^{d-1}}\hat{F}(rs)\exp(irs\cdot y)ds$$
$$=\frac{\omega_{d-1}(2\pi)^d}{\omega_d^2 k^{2d-4}2^{d-3}}\frac{1}{(2\pi)^d}\int_{B_{2k}}dp\exp(ip\cdot y)\hat{F}(p). \quad (6.3.23)$$

Let

$$F(y) = \frac{1}{(2\pi)^d} \int_{B_{2k}} dp \exp(ip \cdot y) \hat{F}(p) dp,$$
$$\hat{F}(p) = \int_{\mathbb{R}^3} dy \exp(-ip \cdot y) F(y). \tag{6.3.24}$$

Let us write $M(0, k, k)$ for function $h_y(p)$ using definition (6.3.19) and then equate the expression to (6.3.23). This will lead to (6.3.15). Take $F(p) = h_y(p)$ in (6.3.19) to obtain

$$M(0, k, k) = \frac{1}{\omega_d^2} \int_{S^{d-1}} \int_{S^{d-1}} |k(s + s')| \cdot (4k^2 - k^2|s + s'|^2)^{-(d-3)/2} \hat{F}(k(s + s')) \cdot \exp\{ik(s + s') \cdot y\} ds ds'. \tag{6.3.25}$$

Compare (6.3.25) and (6.3.23) to obtain

$$\frac{1}{(2\pi)^d} \int_{B_{2k}} dp \exp(ip \cdot y) \hat{F}(p) = \frac{k^d}{8\pi^d \omega_{d-1}} \int_{S^{d-1}} \int_{S^{d-1}} (4 - |s + s'|^2)^{-(d-3)/2} \cdot |s + s'| \hat{F}(k(s + s')) \exp\{ik(s + s') \cdot y\} ds ds'. \tag{6.3.26}$$

If one changes s' for $-s'$ in (6.3.26), then one obtains (6.3.12). □

6.3.3. Let us pass to the construction of $q_N(x)$ which approximates the solution $q(y)$ of equation (6.3.8). We assume that the data $F(k(s' - s))$ are exact, so that equation (6.3.8) is solvable. The calculations are given for the general case of d-dimensional space.

Let us multiply (6.3.8) by $\exp\{ik(s' - s) \cdot x\} h_N(k(s' - s))$ and integrate over $S^{d-1} \times S^{d-1}$ to obtain

$$\int_{|y| \le R} dy q(y) \int_{S^{d-1}} \int_{S^{d-1}} \exp\{ik(s' - s) \cdot (x - y)\} h_N(k(s' - s)) ds ds'$$

$$= \int_{S^{d-1}} \int_{S^{d-1}} F(k(s'-s))h_N(k(s'-s)) \cdot \exp\{ik(s'-s)\cdot x\}dsds'. \tag{6.3.27}$$

Here, $h_N(p)$ is some function which will be specified later.

Our problem will be solved if the sequence defined by the formula

$$\delta_N(x) = \int_{S^{d-1}} \int_{S^{d-1}} \exp\{ik(s'-s)\cdot x\}h_N(k(s'-s))dsds' \tag{6.3.28}$$

is a delta-sequence in $L^2(B_R)$ or $C(B_R)$ in the sense that

$$\|q(x) - q_N(x)\|_R = \epsilon(N) \to 0 \quad \text{as} \quad N \to \infty. \tag{6.3.29}$$

Here,

$$q_N(x) = \int_{|y|\le R} \delta_N(x-y)q(y)dy \tag{6.3.30}$$

and the norm $\|\cdot\|_R$ in (6.3.29) is the $L^2(B_R)(C(B_R))$ norm if $q(x) \in L^2(B_R)$(or $q(x) \in C(B_R)$). One can estimate $\epsilon(N)$ if some additional information about $q(x)$ is known, e.g., bounds $|q| \le a$, $|\nabla q| \le b$.

Let us choose h_N such that the sequence (6.3.28) becomes a delta-sequence in the sense defined above. It follows from (6.3.12) that

$$\int_{S^{d-1}} \int_{S^{d-1}} \exp\{ik(s'-s)\cdot x\}h_N(k(s'-s))dsds' = \frac{8\pi^d \omega_{d-1}}{k^{2d-4}(2\pi)^d} \int_{B_{2k}} dp \exp(ip\cdot x) \cdot (4k^2 - |p|^2)^{(d-3)/2}|p|^{-1}h_n(p). \tag{6.3.31}$$

Let us denote

$$h_N(p)|p|^{-1}(4k^2 - |p|^2)^{(d-3)/2}\frac{8\omega_{d-1}\pi^d}{k^{2d-4}} = A_N(p). \tag{6.3.32}$$

Then, comparing (6.3.28), (6.3.31), and (6.3.32) yields

$$\delta_N(x) = \frac{1}{(2\pi)^d}\int_{B_{2k}} dp \exp(ip\cdot x)A_N(p). \tag{6.3.33}$$

Therefore, we need to construct a delta-sequence in $L^2(B_R)$ of functions $\delta_N \in W_{2k}$. Such a sequence was constructed in [226, 269]:

$$\delta_N(x) = \left\{ \frac{1}{|B_{2k}|} \int_{|p|\le 2k} dp \exp\left(\frac{ip\cdot x}{2N+d}\right) \right\}^{2N+d} \times \left(1-\frac{|x|^2}{4R^2}\right)^N \left(\frac{N}{4\pi R^2}\right)^{d/2}. \qquad (6.3.34)$$

Here, $|B_k|$ is the volume of the ball of radius k, $|B_k| = k^d\pi^{d/2}/\Gamma((d+2)/2)$, and

$$\int_{|p|\le 2k} dp \exp\left(\frac{ip\cdot x}{2N+d}\right) = \frac{(2k)^{d/2}(2N+d)^{d/2}(2\pi)^{d/2}}{|x|^{d/2}} J_{d/2}\left(\frac{2k|x|}{2N+d}\right), \qquad (6.3.35)$$

where $J_m(x)$ is the Bessel function. Thus,

$$\delta_N(x) = \left(1-\frac{|x|^2}{4R^2}\right)^N \left(\frac{N}{4\pi R^2}\right)^{d/2} \times \left\{ 2^{d/2}\Gamma\left(\frac{d}{2}+1\right) \frac{J_{d/2}\left(\frac{2k|x|}{2N+d}\right)}{\left(\frac{2k|x|}{2N+d}\right)^{d/2}} \right\}^{2N+d}, \qquad (6.3.36)$$

where $\Gamma(x)$ is the Gamma function.

In particular, for $d=3$, one has

$$\delta_N(x) = \left(1-\frac{|x|^2}{4R^2}\right)^N \left(\frac{N}{4\pi R^2}\right)^{3/2} \left(\frac{\sin\lambda - \lambda\cos\lambda}{\lambda^3/3}\right)^{2N+3}, \quad \lambda = \frac{2k|x|}{2N+3}. \qquad (6.3.37)$$

Note that the sequence (6.3.34) is a delta sequence on $L^2(B_R)$ but not on $L^2(\mathbb{R}^d)$. One cannot construct a delta-sequence $\delta_N \in W_{2k}$ on $L^2(\mathbb{R}^d)$ because the Fourier transform $\hat{\delta}_N(p)$ of a function $\delta_N \in W_{2k}$ should have its support in the ball B_{2k} while the support of the Fourier transform of a delta sequence on $L^2(\mathbb{R}^d)$ cannot stay as $N \to \infty$ within a fixed compact. It follows from (6.3.32) and (6.3.33) that

$$A_N(p) = \int_{\mathbb{R}^d} \delta_N(x) \exp(-ip \cdot x) dx \tag{6.3.38}$$

and

$$h_N(p) = |p|(4k^2 - |p|^2)^{-(d-3)/2} A_N(p) k^{2d-4} (8\pi^d \omega_{d-1})^{-1}. \tag{6.3.39}$$

It follows from (6.3.27), (6.3.28), and (6.3.30) that

$$\begin{aligned} q_N(x) = \int_{S^{d-1}} \int_{S^{d-1}} & F(k(s'-s)) h_N(k(s'-s)) \\ & \cdot \exp\{ik(s'-s) \cdot x\} ds ds' \end{aligned} \tag{6.3.40}$$

Let us formulate the result.

Theorem 6.3.3. *If equation* (6.3.8) *is solvable, then formula* (6.3.40) *gives an approximate analytical solution to equation* (6.3.8) *in the sense that equation* (6.3.29) *holds. The sequence* $h_N(x)$ *in* (6.3.40) *is given by formulas* (6.3.36), (6.3.38), *and* (6.3.39).

Remark 6.3.1. If

$$|q| \leq a \quad \text{and} \quad |\nabla q| \leq b \tag{6.3.41}$$

then, in the norm of $C(B_R)$, the estimate for $\epsilon(N)$ in (6.3.29) is of the form:

$$\epsilon(N) \leq \Gamma\left(\frac{d+1}{2}\right) \left\{\Gamma\left(\frac{d}{2}\right)\right\}^{-1} bRN^{-1/2} + O(N^{-1}), \quad N \to \infty. \tag{6.3.42}$$

This is proved in [226, p. 211].

6.3.4. Let $d = 3$. Suppose that the data are given with some error η, so that $F_\eta(k|s' - s|)$ is known such that $|F - F_\eta| < \eta$.

If one applies the inversion formula (6.3.40) to the noisy data F_η, then one obtains a function $q_{N,\eta}$ such that

$$\|q_{N,\eta} - q\|_R \leq \|q_{N,\eta} - q_N\|_R + \|q_N - q\|_R \leq \eta b_N + \epsilon(N), \tag{6.3.43}$$

where the norm is taken in $C(B_R), \epsilon(N)$, given by (6.3.29), is the error of the inversion formula (6.3.40) in the absence of noise, and

$$b_N = \int_{S^2}\int_{S^2} |h_N(k(s' - s))|dsds', \tag{6.3.44}$$

where, for $d = 3$, from (6.3.39) one has $h_N(p) = |p|k^2 A_N(p)\,(16\pi^4)^{-1}$, $A_N(p)$ is given by (6.3.38) and δ_N is given by (6.3.37). For a fixed $\eta > 0$, one can find $N = N(\eta)$ for which

$$\alpha(N, \eta) = \eta b_N + \epsilon(N) = \min. \tag{6.3.45}$$

Such $N(\eta)$ does exist because $\epsilon(N) \to 0$ and $b_N \to \infty$ as $N \to \infty$. For this $N(\eta)$, the error $\alpha(\eta) = \alpha(N(\eta), \eta) \to 0$ as $\eta \to 0$. Therefore,

$$\|q_{N(\eta),\eta}(x) - q(x)\|_R \to 0 \quad \text{as } \eta \to 0 \tag{6.3.46}$$

so that $q_{N(\eta),\eta}(x)$ is a stable approximation of $q(x)$ on B_R.

For computing b_N, the following lemma is useful.

Lemma 6.3.4.

$$\int_{S^{d-1}}\int_{S^{d-1}} h(k(s' - s))dsds' = \frac{\omega_{d-1}}{2^{d-3}k^{2d-4}} \int_{|p|\leq 2k} dph(p)|p|^{-1}(4k^2 - |p|^2)^{(d-3)/2}. \tag{6.3.47}$$

Proof. The proof is similar to the proof of Lemma 6.3.2 and therefore is left to the reader. □

From (6.3.44) and (6.3.47), it follows that

$$b_N = \frac{2\pi}{k^2} \int_{|p|\le 2k} dp |h_N(p)||p|^{-1}$$
$$= \frac{1}{8\pi^3} \int_{|p|\le 2k} |A_N(p)| dp. \tag{6.3.48}$$

A rough estimate for b_N, which follows from (6.3.48), is

$$b_N \le \frac{|B_{2k}|}{8\pi^3} \cdot \max_{|p|\le 2k} |A_N(p)| \le \frac{4k^3}{3\pi^2} \int_{\mathbb{R}^3} |\delta_N(x)| dx. \tag{6.3.49}$$

Another estimate for b_N follows directly from (6.3.44):

$$b_N \le (4\pi)^2 \max_{|p|\le 2k} |h_N(p)|. \tag{6.3.50}$$

Estimate (6.3.50) can also be obtained from the rough estimate of the first integral in (6.3.48).

6.3.5. In this section, we solve the following problems which are of interest in many applications.

Problem 6.3.5. *Given the Fourier transform $F(\omega)$ of a function $f(x) \in L^2, f(x) = 0$ if $|x| > R$, on a finite interval $-\omega_0 \le \omega \le \omega_0$, find $f(x)$.*

We have

$$(2\pi)^{-1} \int_{-R}^{R} f(t) \exp(-i\omega t) dt = F(\omega), \quad -\omega_0 \le \omega \le \omega_0. \tag{6.3.51}$$

The data are

$$F(\omega), \quad -\omega_0 \le \omega \le \omega_0, \quad \text{and} \quad R > 0. \tag{6.3.52}$$

Problem 6.3.6. *Let*

$$\int_0^R f(t) \exp(-pt) dt = F(p), \quad 0 \le p \le b. \tag{6.3.53}$$

Given

$$R \text{ and } F(p), \quad 0 \le p \le b, \tag{6.3.54}$$

find $f(t)$.

One can treat equations (6.3.51) and (6.3.53) as integral equations of the first kind and solve them numerically using a regularization procedure.

We give some explicit analytical formulas for calculating $f_N(t)$ such that

$$\|f_N(t) - f(t)\| \le \epsilon(N) \to 0 \quad \text{as } N \to \infty, \tag{6.3.55}$$

where $\|\cdot\|$ is the norm in $L^2(-R, R)$.

Solution to Problem 6.3.5:

$$f_N(t) = \int_{-\omega_0}^{\omega_0} F(\omega) h_N(\omega) \exp(i\omega t) d\omega, \tag{6.3.56}$$

where

$$h_N(\omega) = \int_{-\infty}^{\infty} \delta_N(t) \exp(-i\omega t) dt \tag{6.3.57}$$

and

$$\delta_N(t) = \left[\frac{N}{4\pi R^2}\right]^{1/2} \left[1 - \frac{t^2}{4R^2}\right]^N \times \left\{\frac{\sin[\omega_0 t(2N+q)^{-1}]}{\omega_0 t(2N+q)^{-1}}\right\}^{2N+q}, \quad q \ge 1. \tag{6.3.58}$$

Formula

$$F_N(\omega) = (2\pi)^{-1} \int_{-R}^{R} f_N(t) \exp(-i\omega t) dt \tag{6.3.59}$$

solves the problem of spectral extrapolation, of extrapolation of $F(\omega)$ in the region $|\omega| > \omega_0$.

Let us prove that the function (6.3.56) satisfies (6.3.55) and thus solves Problem 6.3.5. Multiply (6.3.51) by $\exp(i\omega t)h_N(\omega)$ and integrate over $(-\omega_0, \omega_0)$ to obtain

$$\int_{-R}^{R} dt' f(t') \delta_N(t - t') = f_N(t), \tag{6.3.60}$$

where

$$\delta_N(t) = (2\pi)^{-1} \int_{-\omega_0}^{\omega_0} h_N(\omega) \exp(i\omega t) d\omega \tag{6.3.61}$$

so that formula (6.3.57) holds. Formula (6.3.55) is valid if $\delta_N(t)$ is a delta sequence in $L^2(-R, R)$. One can easily check that formula (6.3.58) defines a delta sequence in $L^2(-R, R)$ which is a sequence of entire functions of t of exponential type $\leq \omega_0$, so that by Paley–Wiener theorem, the function $h_N(\omega)$ defined by (6.3.57) vanishes if $|\omega| > \omega_0$, $\operatorname{Im}\omega = 0$. Details can be found in [226, pp. 211–215], where the estimate $\epsilon(N) \leq AN^{-1/2} + O(N^{-1})$ is obtained under the assumption $|f| + |f'| \leq c$, where $c = \text{const}$ and A is an explicitly written constant proportional to c (see formula (6.3.42)). □

Solution to Problem 6.3.6:

$$f_N(t) = \int_0^b F(p) H_N\left[p - \frac{b}{2}\right] \exp(pt) dp \tag{6.3.62}$$

where

$$H_N(p) = (2\pi)^{-1} \int_{-\infty}^{\infty} \Delta_N(iy) \exp(-iyp) dy, \tag{6.3.63}$$

$$H_N(p) = 0 \quad \text{if } |p| > \frac{b}{2}, \tag{6.3.64}$$

and

$$\Delta_N(iy) = \left[\frac{N}{4\pi R^2}\right]^{1/2} \left[1 + \frac{y^2}{4R^2}\right]^N \times \left[\frac{\sin \frac{yb}{2(2N+q)}}{\frac{yb}{2(2N+q)}}\right]^{2N+q}, \quad q \geq 1. \tag{6.3.65}$$

Let us prove that function (6.3.62) satisfies (6.3.55) and thus solves Problem 6.3.6. Multiply (6.3.53) by

$$H_N\left[p - \frac{b}{2}\right] \exp(pt)$$

and integrate in p over $(0, b)$ to get

$$\int_0^R dt' f(t') d_N(t - t') = f_N(t), \tag{6.3.66}$$

where

$$\begin{aligned} d_N(t) &= \int_0^b H_N\left[p - \frac{b}{2}\right] \exp(pt) dp \\ &= \int_{-b/2}^{b/2} H_N(p) \exp(pt) dp \exp\left[\frac{b}{2} t\right]. \end{aligned} \tag{6.3.67}$$

Let us show that the sequence

$$a_N(t) = \int_{-b/2}^{b/2} H_N(p) \exp(pt) dp \tag{6.3.68}$$

is a delta sequence in $L^2(-R, R)$. Then

$$d_N(t) = a_N(t) \exp\left[\frac{b}{2} t\right]$$

is also a delta sequence in $L^2(-R, R)$. It follows from (6.3.63) and (6.3.68) that

$$\begin{aligned} &a_N(t) = \Delta_N(t), \quad t = iy, \\ &\Delta_N(t) = \left(\frac{N}{4\pi R^2}\right)^{1/2} \left(1 - \frac{t^2}{4R^2}\right)^N \left\{\frac{\sinh[tb/(4N + 2q)]}{tb/(4N + 2q)}\right\}^{2N+q}, \\ &\qquad q \geq 1. \end{aligned} \tag{6.3.69}$$

The sequence $\Delta_N(t)$, defined by formula (6.3.69), is a delta-sequence in $L^2(-R, R)$. This is checked in the same way as for the sequence (6.3.58). Therefore, $a_N(t)$ and $d_N(t)$ are delta sequences in $L^2(-R, R)$, and the proof is complete. □

Formulas (6.3.56) and (6.3.62) give an analytical solution to Problems 6.3.5 and 6.3.6, respectively, if the data are exact. If the data are noisy, one can use formulas (6.3.56) and (6.3.62) in order to obtain a stable approximation of $f(t)$. For example, suppose $F_\eta(\omega)$ is given such that $|F_\eta(\omega) - F(\omega)| \leq \eta, -\omega_0 \leq \omega \leq \omega_0$, then one can find $N(\eta)$ such that formula (6.3.56) with $F_\eta(\omega)$ in place of $F(\omega)$ and $N = N(\eta)$ will give $f_\eta(t)$ such that

$$\|f_\eta(t) - f(t)\| \to 0 \quad \text{as } \eta \to 0. \tag{6.3.70}$$

Indeed

$$\begin{aligned} \|f_\eta(t) - f(t)\| &\leq \|f_\eta - f_{N(\eta)}\| + \|f_{N(\eta)} - f\| \\ &\leq \left\| \int_{-\omega_0}^{\omega_0} (F_\eta - F) h_{N(\eta)}(\omega) \exp(i\omega t) d\omega \right\| + \epsilon(N(\eta)) \\ &\leq \eta c(N(\eta)) + \epsilon(N(\eta)), \quad c(N) = \int_{-\omega_0}^{\omega_0} |h_N(\omega)| d\omega. \end{aligned} \tag{6.3.71}$$

For a fixed η, one can find $N = N(\eta)$ for which

$$\eta c(N) + \epsilon(N) = \min = \alpha(\eta). \tag{6.3.72}$$

Since $\epsilon(N) \to 0$ as $N \to \infty$, and $c(N) \to \infty$, one concludes that if $N(\eta)$ solves problem (6.3.72) then $N(\eta) \to \infty$ as $\eta \to 0$, and $\alpha(\eta) \to 0$ as $\eta \to 0$. This argument is similar to the one given in Section 6.3.4.

Let us summarize the result:

If the data are given with an error η, $|F_\eta - F| \leq \eta$, then the function

$$f_\eta(t) = \int_{-\omega_0}^{\omega_0} F_\eta(\omega) h_{N(\eta)}(\omega) \exp(i\omega t) d\omega$$

satisfies (6.3.70). *Here, $N(\eta)$ solves problem* (6.3.72).

Practically, Problems 6.3.5 and 6.3.6 are ill-posed and the given above stable approximation to $f(t)$, the function $f_\eta(t)$, may be of limited value because the function $c(N)$ grows fast and in order for $\alpha(\eta)$ to become sufficiently small, one may have to choose η very small. From the practical point of view, this is not always possible since the precision of the measurements may be limited.

6.4. A Model Inverse Problem of Induction Logging

6.4.1. Consider a medium properties of which we want to find from the knowledge of the field on the ray $x_3 < 0$. One can think about a borehold, a cylindrical hole in the medium; along the x_3-axis the source and receiver can be moved. They can be kept either at a fixed distance d, or their positions can vary arbitrarily. The problem is to recover the properties of the medium from the measurements recorded by the receiver.

In induction logging, the medium is characterized by the dielectric and magnetic parameters $\epsilon = 1, \mu = \text{const}$, and by the conductivity $\sigma = \sigma_0 + v(X)$, where $\sigma_0 = \text{const}$ and $v(X)$ is the inhomogeneous part of σ. The displacement currents $\partial\mathcal{D}/\partial t$ are negligible. The vector potential A satisfies the equation $(\nabla^2 + k^2)A = -j, k^2 = i\omega\mu\sigma(X)$, and j is the source current. Let $X = (\rho, \phi, \xi)$ be the cylindrical coordinates of the point x. In induction logging, it is always assumed that j is a current in a circular loop of wire. The loop is perpendicular to the x_3-axis, it has a small radius a, and its center is at a point $(0, 0, \xi)$. The vector potential A under these assumptions and under the additional assumption $v = v(\rho, z)$, has only one nonzero component, A_ϕ, in the cylindrical coordinates, $\nabla \cdot A = 0, E = i\omega\mu A, H = \nabla \times A$.

We consider a model scalar acoustical problem in which the field u satisfies the equation

$$\nabla^2 u + k^2(X)u = -\delta(X - Y), \quad X, Y \in \mathbb{R}^3,$$

$k^2(X) = i\omega\sigma(x), \sigma(X) = \sigma_0 + v(X), \sigma_0 = \text{const} > 0, \omega > 0, v(X) = 0$ if $x_3 > 0$ or $|X| > R$, where $R > 0$ is a fixed (arbitrary large) number, $v \in L^2$, $\operatorname{Im} k(X) > 0, k_0 = (i\omega\sigma_0)^{1/2}$. The integral equation for u is

$$u(X, Y, \omega) = g(X, Y) + i\omega \int g(X, Z)v(Z)u(Z, Y)dZ, \qquad (6.4.1)$$

where

$$g = \frac{\exp(ik_0|X - Y|)}{4\pi|X - Y|}.$$

If ω is sufficiently small then the integral equation is uniquely solvable by iterations and u is analytic in the variable $p = \omega^{1/2}$ in a neighborhood of the point $p = 0$.

Therefore, the following limit exists

$$f(X,Y) = \lim_{\omega \to 0} 16\pi^2 \frac{u-g}{i\omega} = \int \frac{v(Z)dZ}{|X-Z||Y-Z|}. \tag{6.4.2}$$

The function $f(X,Y)$ is the datum. Let $X = (0,0,x)$ and $Y = (0,0,y)$. Under various assumptions on $v(Z)$, we give inversion algorithms for recovering $v(Z)$ from the data $f(x,y)$ measured on the x_3-axis.

6.4.2. Inversion in the one-dimensional case.

Assume that $v(Z) = v(z_1, z_2, z_3) = v(z_3), |z_3| \leq R, r \leq R$, and $v = 0$ otherwise. Here, $r = (z_1^2 + z_2^2)^{1/2}$. Let $X = (0,0,x), Y = (0,0,y), z_3 = \xi, y = x - d, d = \text{const}$. Then (6.4.2) takes the form

$$\int_{-\infty}^{\infty} d\xi v(\xi) h(x-\xi) = \psi(x), \quad -\infty < x < \infty, \tag{6.4.3}$$

where

$$\psi(x) = (2\pi)^{-1} f(x, x-d),$$

$$h(x-\xi) = \int_0^R \frac{dr\, r}{[(\xi-x)^2 + r^2]^{1/2}[(\xi - x + d)^2 + r^2]^{1/2}}. \tag{6.4.4}$$

Equation (6.4.3) is of convolution type and can be solved by Fourier's transform. If the data is noisy, i.e., ψ_δ is known instead of ψ, $|\psi_\delta - \psi| \leq \delta$, then a regularizing algorithm should be used for numerical solution of (6.4.3). For example, one can solve (6.4.3) with ψ_δ in the right-hand side by the formula

$$v_\delta(\xi) = (2\pi)^{-1} \int_{-\infty}^{\infty} d\lambda \exp(i\lambda\xi) \tilde{\psi}_\delta(\lambda) \tilde{h}^{-1}(\lambda) \exp(-\alpha(\delta)|\lambda|),$$

where $\alpha(\delta) \to 0$ and $\delta\alpha^{-1}(\delta) \to 0$ (see [226, p. 207], for details and for an optimal choice of $\alpha(\delta)$). One can solve also the Wiener–Hopf type equation

$$\int_0^\infty d\xi v(\xi)h(x-\xi) = \psi(x), \quad 0 \le x < \infty, \tag{6.4.3$'$}$$

which models the case when measurements are taken strictly inside the borehole.

The same method is valid for inversion of the data at a fixed frequency $\omega > 0$ in the Born approximation. The function analogous to $h(x-\xi)$ in (6.4.4) will be $h_\omega(x-\xi) = \int_0^R dr\, r \exp(i\omega\beta)\alpha^{-1}$, where $\alpha = [(\xi-x)^2 + r^2]^{1/2} \cdot [(\xi - x + d)^2 + r^2]^{1/2} := \alpha_1\alpha_2, \beta = \alpha_1 + \alpha_2$.

6.4.3. The two-dimensional case.

(1) The results of this section show that the data in the Born approximation

$$\begin{aligned} F(X,Y) &= \frac{16\pi^2(u-g)}{i\omega\mu} \\ &= \int \frac{\exp\{ik_0(|X-Z| + |Z-Y|)\}v(Z)}{|X-Z||Z-Y|} dZ \end{aligned} \tag{6.4.5}$$

measured for all X, Y belonging to the ξ-axis and at a fixed frequency $\omega > 0$ does determine $v(r,z)$ in some cases. Assume that $v = v(r,z) = 0$ if $r \ge R$ or $|z| \ge R$.

Let us use the spheroidal coordinates as in [192]: $z_3 = lst + (x_3+y_3)/2, z_2 = l(s^2-1)^{1/2}(1-t^2)^{1/2}\cos\theta + (x_2+y_2)/2, z_1 = l(s^2-1)^{1/2}(1-t^2)^{1/2}\sin\theta + (x_1+y_1)/2$, where $l = |X-Y|/2, |X-Z| + |Z-Y| = 2ls, 1 \le s < \infty, |X-Z| - |Z-Y| = 2lt, -1 \le t \le 1$, θ is the angle between the plane spanned by $X-Z$ and $Y-Z$ and a fixed plane containing X and Y. The Jacobian of the transformation $Z \to (s,t,\theta)$ is $J = l^3(s^2-t^2)$. In our case, X and Y belong to the ξ-axis. As the fixed plane, we take the plane $z_1 \cdot z_3, z_3 = \xi$, so that θ is the angular variable in the cylindrical coordinate system. In the new variables, (6.4.5) takes the form:

$$2\pi l \int_1^\infty ds \exp(2ik_0 ls) \int_{-1}^1 dtv \left(l(s^2-1)^{1/2}(1-t^2)^{1/2}, lst + \frac{x+y}{2} \right) \\ = F(x,y), \qquad (6.4.6)$$

where x and y denote the ξ components of the vectors X and Y, $(x-y)/2 = l$ if $x \geq y$. From (6.4.6), it is clear that under our assumption $v = v(r,z)$ the data depends on $m = (x+y)/2$ and $l = |x-y|/2$.

(2) Let us assume that F is given for all frequencies ω, i.e., for all $k_0, k_0 = [(1+i)/\sqrt{2}]\sigma_0^{1/2}\omega^{1/2}$. Then the function $p(s,l,m) = \int_{-1}^1 dtv(l(s^2-1)^{1/2}(1-t^2)^{1/2}, lst+m)$ can be found from (6.4.6) by taking the inverse Fourier transform in k_0. Therefore, one obtains the equation for $v(r,\xi)$

$$\int_{-1}^1 v(l(s^2-1)^{1/2}(1-t^2)^{1/2}, lst+m)dt \\ = p(s,l,m), \quad s \geq 1, \quad l \geq 0, \qquad (6.4.7)$$

where $p(s,l,m)$ is known.

Let $\xi = l(s^2-1)^{1/2}(1-t^2)^{1/2}$ and $\eta = lst+m$. Consider the ellipse $E(s,l)$ on the plane (ξ,η) given by the equation $\xi^2(s^2-1)^{-1} + (\eta - m)^2 s^{-2} = l^2$. Equation (6.4.7) can be written as

$$\int_{E(s,l)=\{\xi,\eta:\xi^2(s^2-1)^{-1}+(\eta-m)^2s^{-2}=l^2\}} w(\xi,\eta)dS \\ = p(s,l,m), \quad s \geq 1, \quad l \geq 0. \qquad (6.4.8)$$

Here, dS is the arc length of the ellipse,

$$dS = l((s^2-t^2)/(1-t^2))^{1/2}dt, w(\xi,\eta) = v(\xi,\eta)((1-t^2)/(s^2-t^2))^{1/2}1/l,$$

so $dS = v(\xi,\eta)a(\xi,\eta)$, and $a(\xi,\eta)$ is the function $((1-t^2)/(s^2-t^2))^{1/2}1/l$ expressed in the variables ξ,η,

$$s^2+t^2 = \xi^2 l^{-2} + (\eta-m)^2 l^{-2} + 1, \quad s^2t^2 = (\eta-m)^2 l^{-2}.$$

Problem (6.4.8) is an integral geometry problem. If w is found from (6.4.8), then v is found since $a(\xi,\eta)$ is known. It seems that equation (6.4.8) was not studied in the literature. A similar equation in which the ellipse had one focus at the origin and the other running through one of the coordinate axes was studied in [118, 119], but their results are not applicable to equation (6.4.8) because the families of ellipses are different.

(3) Consider the inversion scheme for the case when $v(r,z) = v(z)$. If $v = v(z)$ does not depend on r, then (6.4.7) takes the form

$$\int_{-1}^{1} dt v(lst+m) = p(s,l,m), \quad s \geq 1,\ l \geq 0. \tag{6.4.9}$$

Take $s = m/l$ and $m(t+1) = \tau$. Then (6.4.9) can be written as

$$\begin{aligned}\int_{0}^{2m} v(\tau)d\tau &= mp\left(\frac{m}{l}, l, m\right) \\ &= h(m,l), \quad l \geq 0,\ m \geq l.\end{aligned} \tag{6.4.10}$$

From (6.4.10), one finds by differentiation that

$$v(2m) = \frac{1}{2}\frac{dh(m,l)}{dm}, \quad v(z) = \frac{1}{2}\left(\frac{dh}{dm}\right)\bigg|_{m=z/2}, \quad m \geq l. \tag{6.4.11}$$

Thus, the inversion scheme in this subsection can be summarized as follows. First, take the inverse Fourier transform of the data $F(x,y,k_0)$ in k_0 variable (see equation (6.4.6)) to obtain (6.4.7). Then, take $s = ml^{-1}$ in (6.4.7) and find $v(z)$ by formula (6.4.11). In this scheme, the data is assumed to be three-dimensional (a function of l, m and k_0), while the conductivity is assumed to be a function of z only, i.e., a function of one variable. The differentiation in (6.4.11) makes the inversion an ill-posed problem. In [197, 356], an algorithm for stable differentiation of noisy data is given. There is considerable freedom in the choice of the method for solving (6.4.9). For example, one could set $m = l$ and $s = 1$ and solve (6.4.9) as in Section 6.6.

(4) Consider the two-dimensional inhomogeneity $v(r, z)$ and assume that the dependence on r is linear near the borehole, that is, $v(r, z) = v(0, z)+(\partial v(0, z))/\partial r\, r = v(z)+w(z)r$. Then (6.4.7) can be written as

$$\int_{-1}^{1} v(lst+m)dt + l\int_{-1}^{1} w(lst+m)(1-t^2)^{1/2}dt(s^2-1)^{1/2}$$
$$= p(s, l, m), \quad s \geq 1,\ l \geq 0. \tag{6.4.12}$$

Take $s = 1$ in (6.4.12) and solve (6.4.12) for v (see Section 6.6). If $v(z)$ is known, then (6.4.12) can be written as an equation for w:

$$\int_{-1}^{1} w(lst+m)(1-t^2)^{1/2}dt = p_1(s, l, m),$$
$$p_1 = \left[p(s, l, m) - \int_{-1}^{1} v(lst+m)dt\right](s^2-1)^{-1/2}l^{-1}. \tag{6.4.13}$$

Equation (6.4.13) can be solved for w. Note that the function $p_1(s, l, m)$ is continuous as $s \to 1$ because of the choice of v. One has, if $m = l$, $p_1(1, l) = p_1(1, l, l)$,

$$\int_{0}^{2} w(l\tau)(1-(\tau-1)^2)^{1/2}d\tau = p_1(1, l), \quad l \geq 0. \tag{6.4.14}$$

Take the Mellin transform of (6.4.14) to obtain

$$\hat{w}(u)a(u) = b(u), \quad \hat{w}(u) = b(u)a^{-1}(u), \tag{6.4.15}$$

where $\hat{w}(u) = \int_0^\infty w(x)x^{u-1}dx$, $b(u) = \hat{p}_1(1, l)$, $a(u) = \int_0^2 \tau^{-u}(1-(\tau-1)^2)^{1/2}d\tau$, and the formula $\hat{w}(x\tau) = \hat{w}(u)\tau^{-u}$ was used. To find $w(z)$, take the inverse Mellin transform of $a^{-1}(u)b(u)$.

(5) A general idea for inversion. Let $A = A_0 + i\omega\mu\int gvAdZ$ be equation (6.4.1) corresponding to the source which produces potential A_0 in the absense of the inhomogeneity v. $A - A_0 = A'$ is the scattered field. In the Born approximation, one has $A' = i\omega\mu\int gvA_0dZ$ (*). One can try to find such A_0 that (*) can be solved for v.

(6) A way to solve the equation

$$\int_{-1}^{1} v(lt+m)dt = p(1,l,m), \quad l \geq 0,$$

is to take $m = l$, set $l(t+1) = \tau$, write the equation as

$$\int_{0}^{2l} v(\tau)d\tau = lp(1,l,l) = h(l),$$

and find

$$v(2l) = \frac{1}{2}\frac{dh(l)}{dl}, \quad v(z) = \frac{1}{2}\frac{dh(l)}{dl}\bigg|_{l=z/2}.$$

(7) Assume $v = v(r)$. Then (6.4.7) becomes

$$\int_{-1}^{1} v(l[(s^2-1)(1-t^2)]^{1/2})dt = p(s,l), \quad s \geq 1, \quad l \geq 0 \tag{6.4.16}$$

(p does not depend on m). This equation can be solved analytically even in the case when s (or l) is fixed. Indeed, let $l(s^2-1)^{1/2} = z$. Then (6.4.16) becomes

$$\int_{0}^{1} v(z(1-t^2)^{1/2})dt = f(z), \quad f(z) = 0.5p(s, z(s^2-1)^{-1/2},$$
$$s > 1 \text{ is fixed.} \tag{6.4.17}$$

Let $(1-t^2)^{1/2} = y, zy = x$. Then (6.4.17) becomes

$$\int_{0}^{z} v(x)x(z^2-x^2)^{-1/2}dx = zf(z), \quad z \geq 0. \tag{6.4.18}$$

This equation can be solved by the formula

$$v(z) = \frac{1}{z}\frac{2}{\pi}\frac{d}{dz}\int_{0}^{z}\frac{x^2 f(x)dx}{\sqrt{z^2-x^2}}. \tag{6.4.19}$$

To derive (6.4.19), multiply (6.4.18) by $2z(y^2-z^2)^{-1/2}$, integrate in z over $(0,y)$, and use the formula $\int_{x}^{y}(2z\,dz/[(y^2-z^2)(z^2-x^2)]^{1/2}) = \pi$. This is a variant of Abel's equation.

(8) Assume $v(r,z) = v_0(z) + v_1(z)h(r_0 - r)$ where $v_0(z), v_1(z)$ are to be found, $r_0 > 0$ is to be found, $h(x) = 1, x \geq 0, h(x) = 0, x < 0$. This model is used in geophysics. Then (6.4.7) becomes

$$\begin{aligned}\int_{-1}^{1} & v_0(lst + m)dt \\ & + \int_{-1}^{1} v_1(lst + km)h(r_0 - l(s^2-1)^{1/2}(1-t^2)^{1/2})dt \\ & = p(s,l,m). \end{aligned} \tag{6.4.20}$$

Set $s = 1$ and find $v_0 + v_1$ as in Section 6.6. Take $l(s^2-1)^{1/2}$ large, then the second integral in (6.4.20) is small and (6.4.20) reduces to (6.4.9) with $v = v_0$. Solve for v_0. Since $v_0 + v_1$ and v_0 are found, both functions v_0 and v_1 are found. Equation (6.4.20) can be written as

$$\begin{aligned}\int_{1 \geq |t| > \{1 - r_0^2 l^{-2}(s^2-1)^{-1}\}^{1/2}} & v_1(lst + m)dt \\ & = p_1(s,l,m) = p(s,l,m) - \int_{-1}^{1} v_0(lst + m)dt, \end{aligned} \tag{6.4.21}$$

where $p_1(s,l,m)$ is known. Let $l^{-2}(s^2-1)^{-1} = \beta$. Then (6.4.21) becomes

$$V(m + ls\sqrt{1 - r_0^2\beta}) - V(m - ls\sqrt{1 - r_0^2\beta}) = p_2(s,\beta,m), \tag{6.4.22}$$

where $p_2 = -[p_1(s,l,m)ls - V(m+ls) + V(m-ls)]$ and V is an antiderivative of $v_1, V' = v_1$. Choose $ls(1 - r_0^2\beta)^{1/2} \ll 1$. This is possible: $ls = l(1 + l^{-2}\beta^{-1})^{1/2}$ and one can take l small and $\beta \sim r_0^{-2}$. Then Taylor's formula and (6.4.22) yield

$$2ls\sqrt{1 - r_0^2\beta}v_1(m) = p_2(s,\beta,m), \quad l^{-2}(s^2-1)^{-1} = \beta. \tag{6.4.23}$$

Therefore,

$$r_0 = \beta^{-1/2}\left\{1 - \left(\frac{p_2(s,\beta,m)}{2lsv_1(m)}\right)^2\right\}^{1/2}, \quad \begin{array}{l} ls(1-r_0^2\beta)^{1/2} \ll 1, \\ l^{-2}(s^2-1)^{-1} = \beta. \end{array} \tag{6.4.24}$$

Example. Assume that $v(X) = 1$ if $r \leq 1, -1 < x_3 < 0, v(X) = 0$ otherwise, $\mu = \sigma_0 = 1$, and $d = 0$, so that the positions of the source and the receiver are the same. The inversion procedure from Section 6.4.2 is based on equation (6.4.3). Under our assumptions, this equation is

$$\begin{aligned} 2\pi \int_{-\infty}^{\infty} dz v(z) \int_0^1 \frac{r dr}{(z-x)^2 + r^2} \\ = \pi \int_{-\infty}^{\infty} dz v(z) \ln\left(1 + \frac{1}{(z-x)^2}\right) = \psi(x). \end{aligned} \tag{6.4.25}$$

Let $\tilde{\psi}(y) = \int_{-\infty}^{\infty} \psi(x)\exp(ixy)dx$. Taking the Fourier transform of (6.4.25), one obtains

$$\begin{aligned} \pi\tilde{v}(y) \int_{-\infty}^{\infty} \exp(iyx) \ln\left(1 + \frac{1}{x^2}\right) dx \\ = \pi\tilde{v}(y) 2\pi |y|^{-1}[1 - \exp(-|y|)] = \tilde{\psi}(y), \end{aligned} \tag{6.4.26}$$

where formula (1.5.10) from [58] was used. Therefore,

$$\begin{aligned} \tilde{v}(y) &= |y|\tilde{\psi}(y)\{2\pi^2[1 - \exp(-|y|)]\}^{-1} \\ &= \tilde{\psi}/\tilde{h}, \quad \tilde{h} = |y|^{-1}2\pi^2[1 - \exp(-|y|)]. \end{aligned} \tag{6.4.27}$$

In particular, under the above assumptions about v, one has

$$\psi(x) = \int_{-1}^{0} dz \ln\left(1 + \frac{1}{(z-x)^2}\right),$$

$$\tilde{\psi}(y) = \pi \frac{1 - \exp(-iy)}{iy} 2\pi |y|^{-1}[1 - \exp(-|y|)].$$

From (6.4.27) and the last formula, one finds

$$\tilde{v}(y) = \frac{1 - \exp(-iy)}{iy},$$

$$v(x) = (2\pi)^{-1} \int_{-\infty}^{\infty} \exp(-ixy)\tilde{v}(y)dy = \begin{cases} 1, & -1 < x < 0, \\ 0 & \text{otherwise.} \end{cases}$$

Therefore, the inhomogeneity is recovered from the exact data analytically. Suppose now that the data are noisy: ψ_δ is given instead of ψ, $\|\psi_\delta - \psi\| \le \delta$, where $\delta > 0$ is known, and $\|\psi\| = (\int_{-\infty}^{\infty} |\psi|^2 dx)^{1/2}$. Let

$$v_\delta(x) = (2\pi)^{-1} \int_{-\infty}^{\infty} \exp(-iyx)\tilde{\psi}_\delta(y) \frac{|y| \exp(-\alpha(\delta)|y|)}{2\pi^2(1 - \exp(-|y|))} dy. \tag{6.4.28}$$

We will show that under an appropriate choice of $\alpha(\delta)$, the function $v_\delta(x)$ gives a stable approximation to the solution $v(x)$ (corresponding to the exact data) in the sense that $\|v_\delta(x) - v(x)\| \to 0$ as $\delta \to 0$. The regularization parameter $\alpha(\delta)$ will be chosen optimally. Let us assume that the unknown solution $v(x)$ satisfies the inequality $\int_{-\infty}^{\infty} |\tilde{v}|^2 (1 + y^2)^b dy \le M^2$, where b and M are given positive constants. This assumption means that the solution lies in a certain compact set in L^2. Assume that $b > 0$. It will be clear soon how this assumption is used. Parseval's equality yields

$$\begin{aligned} \|v_\delta - v\| &= (2\pi)^{-1/2} \|\tilde{v}_\delta - \tilde{v}\| \\ &\le (2\pi)^{-1/2} \left(\|\tilde{\psi}_\delta - \tilde{\psi}\| \max_{y \ge 0} \frac{\exp(-\alpha|y|)|y|}{1 - \exp(-|y|)} \right. \\ &\quad \left. + \left\| \frac{\tilde{\psi}}{\tilde{h}} (1 + y^2)^{b/2} \right\| \max_{y \ge 0} \frac{1 - \exp(-\alpha|y|)}{(1 + y^2)^{b/2}} \right) \\ &\le c(\delta\alpha^{-1} + M\alpha^\gamma), \ c = \text{ const} > 0, \ \gamma = \min(1, b), \ b > 0. \end{aligned}$$

The estimate $(1 - \exp(-\alpha|y|))/(1 + y^2)^{b/2} \leq c\alpha^{\gamma}$ holds if $b > 0$. Let us find $\alpha = \alpha(\delta)$ such that $\delta\alpha^{-1} + M\alpha^{\gamma} = \min$. One has $\alpha(\delta) = c_1\delta^{1/(\gamma+1)}, c_1 = (\gamma M)^{-1/(\gamma+1)}$. For this optimal $\alpha(\delta)$, the error of the approximate solution is $\|v_\delta - v\| \leq \text{const}\ \delta^{\gamma/(\gamma+1)}$, as $\delta \to 0$.

In [36, 151, 173], one can find many references on the work done in induction logging theory.

Chapter 7

Scattering by Obstacles with Infinite Boundaries

7.1. Statement of the Problem

Let $D \subset \mathbb{R}^3$ be an infinite domain with infinite (non-compact) boundary Γ. Consider the problem

$$[\Delta + (k + i\epsilon)^2]u_\epsilon = -f \text{ in } D, \quad k > 0, \quad \epsilon > 0,\ f \in C_0^\infty(D), \tag{7.1.1}$$

$$u_\epsilon = 0 \quad \text{on } \Gamma, \tag{7.1.2}$$

$$u_\epsilon \in L^2(D). \tag{7.1.3}$$

We prove that the weak solution to problem (7.1.1)–(7.1.3) exists and is unique for an arbitrary domain D. The weak solution to problem (7.1.1)–(7.1.3) is a function $u \in L^2(D)$ such that

$$\int \nabla u \overline{\nabla \eta} dx - (k + i\epsilon)^2 \int u \bar{\eta} dx$$
$$= \int f \bar{\eta} dx,\ \forall \eta \in C_0^\infty(D), \quad \int = \int_D. \tag{7.1.4}$$

Under some assumptions about D, we prove that the limit

$$\lim u_\epsilon = u, \quad \epsilon \to +0 \tag{7.1.5}$$

exists in an appropriate sense, the function u solves the equation

$$(\nabla^2 + k^2)u = -f \quad \text{in } D \tag{7.1.6}$$

and satisfies equation (7.1.2). This statement is the limiting arsorption principle for the Dirichlet Laplacian in D.

For domains with boundaries which approach asymptotically the boundary of a cone, we prove the eigenfunction expansion theorem similar to the Theorem 1.3.2. It is assumed that the boundary of D is starshaped near infinity (i.e., in the region $|x| > R$, where $R > 0$ is an arbitrarily large fixed number) with respect to a point $0 \in \Omega = \mathbb{R}^3 \backslash D$. For domains with infinite boundaries, we develop a theory of compact and smooth perturbations. The problem is to show that if the Dirichlet Laplacian in D has absolutely continuous spectrum and the eigenfunction expansion theorem holds, then this is true for the domain D_1, which has the same boundary as D in the region $B'_R = \mathbb{R}^3 \backslash B_R, B_R = \{x : |x| \leq R\}$. We also study the spectral properties of the Dirichlet Laplacian and give criteria for its essential self-adjointness and for the discreteness and continuity of its spectrum.

7.2. Spectral Properties of the Laplacians

7.2.1. Let $L_0 = -\Delta$ be the Laplacian defined on $D(L_0) = C_0^\infty(D)$. Clearly L_0 is symmetric and nonnegative on $D(L_0)$. Therefore, there exists a self-adjoint nonnegative extension L of L_0 (Friedrichs extension).

The operator L_0 is not essentially self-adjoint if $D \neq \mathbb{R}^3$, i.e., its closure $\overline{L}_0$ is not self-adjoint. Indeed, the closure of $C_0^\infty(D)$ in the graph norm $\|L_0 u\| + \|u\|$ consists of the functions which are in $H^2_{\text{loc}}(D)$, while, e.g., in the case of a finite domain D the domain of definition of the Friedrichs extension of the Dirichlet Laplacian is $\mathring{H}^1(D) \cap H^2(D)$, The index of defect of $\overline{L}_0, D(\overline{L}_0) = \dot{H}^2(D)$, in the case of a finite domain D is infinite.

For an arbitrary infinite domain D with a smooth boundary $\Gamma, \Gamma \in C^2_{\text{loc}}$ we prove that $-\Delta$, defined on the set M_0 of functions which vanish in a neighborhood of infinity and belong to $\mathring{H}^1(D) \cap H^2_{\text{loc}}(\bar{D})$, is essentially self-adjoint.

First we prove that if $D = \mathbb{R}^3$, then the operator $L_0 = -\Delta, D(L_0) = C_0^\infty(\mathbb{R}^3)$, is essentially self-adjoint. Indeed, this will be the case if and only if the equation

$$(L_0^* + 1)u = 0 \quad \text{in } \mathbb{R}^3, \quad u \in D(L_0^*) \tag{7.2.1}$$

has only the trivial solution. Equation (7.2.1) means that

$$\int u\overline{(-\Delta h + h)}dx = 0, \quad \forall h \in D(L_0), \quad u \in L^2(\mathbb{R}^3), \quad \int = \int_{\mathbb{R}^3}. \tag{7.2.2}$$

Lemma 7.2.1. *Let* (7.2.2) *hold and* $D = \mathbb{R}^3$. *Then* $u = 0$ *in* $\mathbb{R}^3$.

Proof. Let $u_\epsilon = up(\epsilon x)$ where $p \in C_0^\infty(\mathbb{R}^3), 0 \le p \le 1$,

$$p(x) = p(|x|) = \begin{cases} 0 & \text{if } |x| > 2a, \\ 1 & \text{if } |x| \le a, \end{cases}$$

$p(r)$ decreases, monotonically on the interval $[a, 2a]$ and $B_{2a} = \{x : |x| \le 2a\}$. Clearly $u_\epsilon \in D(L_0)$ since any u which satisfies (7.2.2) is locally C^∞. One can assume that u and p_ϵ are real-valued functions. Then

$$\begin{aligned} \int (up_\epsilon)^2 dx &\le \int [-\Delta(up_\epsilon) + up_\epsilon]up_\epsilon dx \\ &= \int (-2\nabla u \nabla p_\epsilon u p_\epsilon - u^2 \Delta p_\epsilon \cdot p_\epsilon)dx = \int u^2(\nabla p_\epsilon)^2 dx, \end{aligned} \tag{7.2.3}$$

where we integrate by parts:

$$-2\int \nabla u \nabla p_\epsilon u p_\epsilon dx = \int u^2[p_\epsilon \Delta p_\epsilon + (\nabla p_\epsilon)^2]dx.$$

It follows from (7.2.3) that

$$\int_{|x|<a\epsilon^{-1}} u^2 dx < c^2\epsilon^2 \int u^2 dx \to 0, \quad \text{as } \epsilon \to 0. \tag{7.2.3$'$}$$

Here, $|\nabla p(x)| \le c$. It follows from (7.2.3$'$) that $u = 0$ in $\mathbb{R}^3$. □

Corollary 7.2.2. *The operator* $L_0 = -\Delta$ *is essentially self-adjoint on* $C_0^\infty(\mathbb{R}^3)$.

Now we turn to the proof of the general statement mentioned above. Assume that Γ is the smooth infinite boundary of an infinite domain $D \subset \mathbb{R}^3$, e.g., Γ belongs to $C^a_{\text{loc}}, a \geq 2$. Let M_0 denote that set of $H^2_{\text{loc}} \cap \mathring{H}^1$ functions vanishing outside of $D_R = D \cap B_R$, where $H^m = H^m(D)$ is the Sobolev space, and $R > 0$ is a number which depends on the functions. Let L_1 denote the symmetric operator defined in $H = L^2(D)$ as $L_1 u = (-\Delta + 1)u$ on $D(L_1) = M_0$.

Lemma 7.2.3. *The operator L_1 is essentially self-adjoint.*

Proof. It is sufficient to prove that the equation

$$L_1^* u = 0 \quad \text{or} \quad \int_D u(-\Delta + 1)\phi dx = 0, \quad \forall \phi \in M_0, \tag{7.2.1$'$}$$

has only the trivial solution, since in this case, the defect index of a positive definite operator L_1 is $(0, 0)$. Since $C_0^\infty(D) \subset M_0$, one concludes from (7.2.1$'$) that $u \in H^2_{\text{loc}}(\bar{D})$, and $(-\Delta + 1)u = 0$ in D (see [20, Theorem 4.6, p. 178]). Therefore, taking $\phi \in M_0$, and integrating by parts, one obtains

$$\begin{aligned} 0 &= \int_D u(-\Delta + 1)\phi dx = \int_D (-\Delta u + u)\phi dx - \int_\Gamma u \frac{\partial \phi}{\partial N} ds \\ &= -\int_\Gamma u \frac{\partial \phi}{\partial N} ds. \end{aligned} \tag{7.2.2$'$}$$

Here, we used the assumption $\phi \in \mathring{H}^1$ and the equation $-\Delta u + u = 0$. Actually the integral over Γ is taken over a compact part of Γ on which $\partial\phi/\partial N$ does not vanish. We claim that the set of $\partial\phi/\partial N$ is a dense set on $L^2(\Gamma)$ if ϕ runs through M_0. Indeed, $\partial\phi/\partial N$ can be an arbitrary compactly supported smooth function h on Γ. To see this, one solves the problem

$$L_1^2 \phi = 0, \quad \phi = 0 \text{ on } \Gamma_R \cup S_R, \quad \frac{\partial \phi}{\partial N} = h \text{ on } \Gamma_R,$$

$$\frac{\partial \phi}{\partial N} = 0 \quad \text{on } S_R,$$

where $S_R = \{x : |x| = R, x \in D\}$, and h is smooth on $\Gamma_R \cup S_R$. This Dirichlet boundary value problem is uniquely solvable. This proves the claim. It follows from (7.2.2′) that $u = 0$ on Γ. The rest of the proof is the same as the proof of Lemma 7.2.1. □

7.2.2. Let $L = -\Delta$ denote Friedrichs's extension of L_0, that is the Dirichlet Laplacian in D. The following is a simple sufficient condition on D for the spectrum of L to be discrete.

Lemma 7.2.4. *The operator L has a discrete spectrum if $\lambda_1(b) \to \infty$ as $b \to \infty$. Here, $\lambda_1(b)$ is the smallest eigenvalue of the Dirichlet Laplacian in the domain $d(b)$ which is the cross-section of D by the plane $x_3 = b$. We assume for simplicity that $d(b)$ is a finite domain for any b.*

Proof. By Lemma B.0.5, L has a discrete spectrum if and only if the embedding

$$i : \mathring{H}^1(D) \to H^0(D) \tag{7.2.4}$$

is compact. Since the embedding $i : \mathring{H}^1(D) \to H^0(D_R)$ is compact for any $R > 0, D_R = D \cup B_R$, it is sufficient to show that $\int_{D \setminus D_R} |u|^2 dx \to 0, R \to \infty$, uniformly with respect to u: $\|u\|_{\mathring{H}^1(D)} \leq 1$.

Since

$$\begin{aligned}
&\int_{x_3 \geq b} |\nabla u|^2 dx \\
&\quad = \int_b^\infty dx_3 \int_{d(x_3)} |\nabla u|^2 dx_1 dx_2 \\
&\quad \geq \int_b^\infty dx_3 \lambda_1(x_3) \int_{d(x_3)} |u|^2 dx_1 dx_2 \\
&\quad \geq \min_{x_3 \geq b} \int_{D \setminus D_b} |u|^2 dx, \text{ and } \min_{x_3 \geq b} \lambda_1(x_3) \to +\infty, \text{ as } b \to \infty,
\end{aligned}$$

the embedding (7.2.4) is compact. Lemma 7.2.4 is proved. □

Remark 7.2.1. A necessary and sufficient condition for L to have a discrete spectrum, i.e., for the compactness of the embedding $i : \mathring{H}^1(D) \to H^0(D)$ is known: the domain D should contain only a finite number of nonintersecting balls of radius r for any $r > 0$ [154]. In applications, the result in Lemma 7.2.4 is sufficient.

7.2.3. The following is a simple condition on D for the spectrum $\sigma(L)$ of L to be continuous.

Lemma 7.2.5. *If D contains balls of arbitrary large radii, then the spectrum of L is the semi-axis $[0, \infty)$.*

Proof. It is known (Lemma B.0.3) that λ is a point of continuous spectrum $\sigma_c(L)$ of L if there exists a sequence u_m, which is not compact and for which

$$\|Lu_m - \lambda u_m\| \to 0, \quad m \to 0, \quad \|u_m\| = 1. \tag{7.2.5}$$

As u_m, one can take $u_m = \exp[i\sqrt{\lambda}(x - x_m) \cdot n]\epsilon_m^{3/2} p(\epsilon_m(x - x_m))$, where n is a unit vector, x_m is the center of the ball $B_m = \{x : \epsilon_m|x - x_m| \le 1\}, \epsilon_m \to 0, B_m \subset D$, and $p(x) \in C_0^\infty$ is a fixed function, $0 \le p(x) \le 1, p(x) = 1$ if $|x| \le 1/2, p(x) = 0$ if $|x| \ge 1, \int |p|^2 dx = 1$. One has $\int |u_m|^2 dx = \epsilon_m^2 \int |p(\epsilon_m(x - x_m)|^2 dx = \int |p(z)|^2 dz = 1$.

Furthermore,

$$\begin{aligned}(\lambda + \Delta)u_m = {} & 2\nabla \exp[i\sqrt{\lambda}(x - x_m) \cdot n]\nabla p(\epsilon_m(x - x_m))\epsilon_m^{3/2} \\ & + \exp[i\sqrt{\lambda}(x - x_m) \cdot n]\Delta p(\epsilon_m(x - x_m)\epsilon_m^{3/2},\end{aligned}$$

so that (7.2.5) holds. □

The following is a simple condition on D for L to have no positive eigenvalues.

Lemma 7.2.6. *Let N be the normal to Γ pointing out of D, and e_3 be the unit vector of the x_3-axis. Assume that D lies in a half-space $x_3 > 0$ and*

$$N \cdot e_3 \le 0 \quad \text{on } \Gamma. \tag{7.2.6}$$

Then L does not have positive eigenvalues.

Proof. Let $u \not\equiv 0$,

$$\Delta u + \lambda u = 0 \quad \text{in } D, \quad u = 0 \quad \text{on } \Gamma, \quad u \in L^2(D), \quad \lambda > 0. \tag{7.2.7}$$

One can assume that u is real-valued. Integrate the identity

$$\begin{aligned} 0 &= (\Delta u + \lambda u)\frac{\partial u}{\partial x_3} \\ &= \frac{\lambda}{2}\frac{\partial u^2}{\partial x_3} - \frac{1}{2}\frac{\partial}{\partial x_3}(\nabla u)^2 + \sum_{j=1}^{3}\frac{\partial}{\partial x_j}\left(\frac{\partial u}{\partial x_j}\frac{\partial u}{\partial x_3}\right) \end{aligned}$$

over the part D_{0b} of D between $x_3 = 0$ and $x_3 = b$ and then use the Gauss formula to obtain

$$0 = \int_{\Gamma_b \cup P_b}\left(\frac{\partial u}{\partial x_3}\frac{\partial u}{\partial N} - \frac{1}{2}(\nabla u)^2 N\cdot e_3 + \frac{\lambda}{2}u^2 N\cdot e_3\right)ds, \tag{7.2.8}$$

where Γ_b is the part of Γ between $x_3 = 0$ and $x_3 = b$, and P_b is the part of the plane $x_3 = b$ which lies in D.

Since $u = 0$ on Γ, one has $\nabla u = (\partial u/\partial N)N$ on Γ. Therefore, (7.2.8) can be written as

$$0 = \int_{\Gamma_b}\left(\frac{\partial u}{\partial N}\right)^2 N\cdot e_3 ds + \int_{P_b}\left[\lambda u^2 + \left(\frac{\partial u}{\partial x_3}\right)^2 - (\hat{\nabla} u)^2\right]ds, \tag{7.2.9}$$

where $(\hat{\nabla} u)^2 = (\partial u/\partial x_1)^2 + (\partial u/\partial x_2)^2$.

It follows from (7.2.6) that

$$\begin{aligned} -\int_{\Gamma_b}\left(\frac{\partial u}{\partial N}\right)^2 N\cdot e_3 ds &\geq -\int_{\Gamma_{b_0}}\left(\frac{\partial u}{\partial N}\right)^2 N\cdot e_3 \\ &\geq d > 0 \quad \text{if } b \geq b_0. \end{aligned} \tag{7.2.10}$$

Therefore, (7.2.9) and (7.2.10) imply

$$\int_{P_b}\left[\lambda u^2 + \left(\frac{\partial u}{\partial x_3}\right)^2 - (\hat{\nabla} u)^2\right]ds \geq d > 0. \tag{7.2.11}$$

It follows from (7.2.11) that

$$\int_{P_b} [\lambda u^2 + (\nabla u)^2] ds \geq d > 0. \tag{7.2.12}$$

Integrate (7.2.12) in x_3 to obtain

$$\int_{D_{b_0 b}} [\lambda u^2 + (\nabla u)^2] dx \geq d(b - b_0), \quad b \geq b_0, \tag{7.2.13}$$

where $D_{b_0 b} = \{x : x \in D, b_0 \leq x_3 \leq b\}$. It follows from (7.2.13) that $u \notin L^2(D)$. To see that $u \notin L^2(D)$ if (7.2.13) holds, note that if $u \in L^2(D)$ and u solves (7.2.7) then $J(b) = \int_{P_b} u^2 ds$ cannot increase for all $b \geq b_0$, where $b_0 > 0$ is arbitrary.

Therefore, there exists a sequence $b_n \to \infty$, such that

$$J'(b_n) = 2 \int_{P_{b_n}} u \frac{\partial u}{\partial x_3} ds \leq 0, \quad b_n \to +\infty. \tag{7.2.14}$$

Since

$$\begin{aligned} 0 &= \int_{D_{0b_n}} (\Delta u + \lambda u) u dx \\ &= \int_{D_{0b_n}} [\lambda u^2 - (\nabla u)^2] dx + \int_{P_{b_n}} u \frac{\partial u}{\partial x_3} ds, \end{aligned} \tag{7.2.15}$$

it follows from (7.2.14) and (7.2.15) that

$$\int_{D_{0b_n}} (\nabla u)^2 dx \leq \lambda \int_{D_{0b_n}} u^2 dx < \infty. \tag{7.2.16}$$

Therefore,

$$\int_D [\lambda u^2 + |\nabla u|^2] dx < \infty. \tag{7.2.17}$$

Lemma 7.2.6 is proved. □

Remark 7.2.2. If Γ is a periodic surface with the equation $x_3 = f(x_1, x_2), f(x_1 + T_1, x_2 + T_2) = f(x_1, x_2), |f| \leq c$, then condition (7.2.6) holds and L does not have positive eigenvalues in the domain $D = \{x : x_3 \geq f(x_1, x_2)\}$.

Lemma 7.2.7. *Suppose that the problem*

$$\Delta u + k^2 u = 0 \text{ in } D,\ u = 0 \text{ on } \Gamma,\ k > 0, \tag{7.2.18}$$

$$\liminf_{R\to\infty} \int_{S_R(D)} (|u_r|^2 + k^2|u|^2)ds = 0, \tag{7.2.19}$$

where $S_R(D) = \{s : |s| = R,\ s \in D\}$, *has only the trivial solution. Let one of the following conditions*

$$\liminf_{R\to\infty} \int_{S_R(D)} |u_r - iku|^2 ds = 0, \tag{7.2.20}$$

or

$$\int_D |u_r - iku|^2 (1 + |x|)^{-1} dx < \infty, \tag{7.2.21}$$

hold. Then the problems (7.2.18), (7.2.20), *and* (7.2.18), (7.2.21) *have only the trivial solution.*

Proof. It follows from (7.2.20) that

$$\lim_{R\to\infty} \left\{ \int_{S_R(D)} (|u_r|^2 + k^2|u|^2)ds + ik \int_{S_R(D)} (u_r\bar{u} - \bar{u}_r u)ds \right\} = 0. \tag{7.2.22}$$

It follows from (7.2.18) that

$$\lim_{R\to\infty} \int_{S_R(D)} (u_r\bar{u} - \bar{u}_r u)ds = 0. \tag{7.2.23}$$

Equations (7.2.22) and (7.2.23) imply (7.2.19). Therefore, $u = 0$ in D. If (7.2.21) holds, then

$$\int_0^\infty ds(1 + r)^{-1} \int_{S_R(D)} |u_r - iku|^2 ds < \infty.$$

Thus,

$$\liminf_{R\to\infty} \int_{S_R(D)} |u_r - iku|^2 ds = 0. \tag{7.2.24}$$

Therefore, $u = 0$. □

7.3. Spectral Properties of the Dirichlet Laplacian in Semi-infinite Tubes

7.3.1. Let $D \subset \mathbb{R}^3$ be a cylinder if $x_3 \geq R$ (with the axis along the x_3-axis) and a bounded domain with a smooth boundary if $x_3 \leq R$. Let F_h denote the cross-section of D by the plane $x_3 = h, F_h = F$ if $h \geq R$. Let $D_h(D'_h)$ denote the part of D in the half-space $x_3 \leq h(x_3 \geq h), \Gamma_h = \partial D_h, \Gamma'_h = \partial D'_h$. We wish to prove two theorems.

Theorem 7.3.1. *The spectrum of the Dirichlet Laplacian $-\Delta_D$ in the above domain D, the semi-infinite tube, is continuous on the ray $[\alpha_1, \infty)$ and discrete on the segment $[0, \alpha_1)$. Here, α_1 is the lowest eigenvalue of the Dirichlet Laplacian in F.*

Let

$$d = \inf\left\{\int_D |\nabla u|^2 dx \left(\int_D |u|^2 dx\right)^{-1}\right\}, \tag{7.3.1}$$

where inf is taken over the set $\mathring{H}^1(D)$. Assume that

$$\alpha_1 \geq d + q, \quad q > 0. \tag{7.3.2}$$

Theorem 7.3.2. *If (7.3.2) holds, then $d>0$ is an eigenvalue of $-\Delta_D$.*

7.3.2. Proof of Theorem 7.3.1. The boundary ∂D is a compact perturbation of the boundary $\partial D'_R$. By separation of variables, one can check that $-\Delta_D$ in D'_R has a continuous spectrum $[\alpha_1, \infty)$. Compact perturbation of the boundary $\partial D'_R$ does not change the continuous spectrum but may introduce a discrete spectrum below α_1 and, possibly, on the ray $[\alpha_1, \infty)$, i.e., isolated eigenvalues of finite multiplicity. Let us prove these statements. First, let us prove that there are at most a finite number of eigenvalues of finite multiplicity of $-\Delta_D$ on any interval $(-\infty, \alpha)$. Suppose there are infinitely many eigenvalues λ_j on this interval, and let u_j be the corresponding normalized eigenfunctions. If $x_3 \geq R$, one can write

$$u_j = \sum_{m=1}^{\infty} c_{mj}(x_3) w_m(x_1, x_2), \quad x_3 \geq R, \tag{7.3.3}$$

where $\{w_j\}$ is the complete orthonormal in $L^2(F)$ system of eigenfunctions of the Dirichlet Laplacian $\hat{\Delta}_D$ in F, and $c_{mj}(x_3) = \int_F u_j \bar{w}_m dx_1 dx_2 = (u_j, w_m)_F$ are the Fourier coefficients. Substitute (7.3.3) into the equation

$$(\nabla^2 + \lambda_j)u_j = 0, \quad u_j \in L^2(D), \quad \|u_j\|_{L^2(D)} = 1, \tag{7.3.4}$$

to get

$$c''_{mj} + (\lambda_j - \alpha_m)c_{mj} = 0, \tag{7.3.5}$$

where α_m are the eigenvalues of $\hat{\Delta}_D$. The Parseval equality and the inclusion (7.3.4) yield

$$\sup_{1 \leq j < \infty} \int_R^{\infty} dx_3 \sum_{m=1}^{\infty} |c_{mj}|^2 \leq c < \infty. \tag{7.3.6}$$

It follows from (7.3.5) that

$$c_{mj}(x_3) = a_{mj} \exp\{(\alpha_m - \lambda_j)^{1/2} x_3\} + b_{mj} \exp\{-(\alpha_m - \lambda_j)^{1/2} x_3\}, \tag{7.3.7}$$

where a_{mj} and b_{mj} are constants. Thus, if, say, $\alpha = \alpha_n$, and $\lambda_j \geq \alpha_n$, then $c_{mj} = 0$ for $1 \leq m \leq n$, and if $\lambda_j < \alpha_n$, then $a_{mj} = 0, m \geq n$. Therefore, $\nabla u_j \in L^2(D)$, and

$$\sup_{1 \leq j < \infty} \sum_{m=1}^{\infty} |b_{mj}|^2 (1 + \alpha_m - \lambda_j) \leq c < \infty. \tag{7.3.8}$$

Inequality (7.3.8) and the fact that $\alpha_m \to +\infty$ imply that the set $\{b_j\}$ of the elements of l^2 is compact. Here, b_j is the sequence $b_{jm}, m = 1, 2, 3, \ldots$. Therefore, the set $\{u_j\}$ is compact in $L^2(D)$ which is a contradiction since $\{u_j\}$ is an orthonormal set in $L^2(D)$. This contradiction shows that on any interval $(-\infty, \alpha)$ there can be only a finite number of eigenvalues of $-\Delta_D$ in D. A similar argument and the result hold for $-\Delta_N$ in D. □

Proof of Theorem 7.3.2. Inequality (7.3.2) holds if D_R has sufficiently large volume. Indeed, in this case

$$d_R = \inf_{u \in \mathring{H}^1(D_R)} \left\{ \int |\nabla u|^2 dx \left(\int_{D_R} |u|^2 dx \right)^{-1} \right\} \geq d, \tag{7.3.9}$$

because any $u \in \mathring{H}^1(D_R)$ can be extended to D by setting $u = 0$ in D'_R and the extended $u \in \mathring{H}^1(D)$. If D_R has sufficiently large volume, $d_R > 0$ can be as small as we wish, so that $\alpha_1 \geq d_R + q \geq d + q, q > 0$. By Theorem 7.3.1, any point of the spectrum of $-\Delta_D$ which is smaller than α_1 is an eigenvalue of finite multiplicity. Therefore, we should prove that d is a point of the spectrum of $-\Delta_D$. But this follows from the following lemma. □

Lemma 7.3.3. *Let A be a self-adjoint operator in a Hilbert space H, and $d = \inf(Au, u), \|u\| = 1, d > -\infty$. Then $d \in \sigma(A)$, where $\sigma(A)$ is the spectrum of A.*

Proof. Without loss of generality, assume that $d = 0$. Otherwise, consider the operator $A - dI$ if $d > -\infty$. Assume that $0 \notin \sigma(A)$. Then there is a neighborhood $|\lambda| < \epsilon$, where $\epsilon > 0$ is sufficiently small, which does not contain any point of $\sigma(A)$. Therefore, $A \geq \epsilon I$ and $\inf(Au, u) \geq \epsilon > 0, \|u\| = 1$. This is a contradiction since $\inf(Au, u) = 0, \|u\| = 1$. □

Remark 7.3.1. It is not known if there can be eigenvalues of $-\Delta_D$ on the ray (α_1, ∞). If the boundary condition is the Robin condition at least on a part of the boundary, then eigenvalues on the continuous spectrum of the corresponding Laplacian can occur.

Example. Let $D = \{(x_1, x_2), 0 \leq x_1 < \infty, 0 \leq x_2 \leq \pi\}$, and the boundary condition is

$$u = 0 \quad \text{if } x_2 = 0 \text{ or } x_2 = \pi, \; -\frac{\partial u}{\partial x_1} = au \quad \text{if } x_1 = 0,\, a > 0. \tag{7.3.10}$$

Here, $\alpha_1 = 1$. Let $u = \exp(-ax_1)\sin nx_2$. Then

$$\Delta u + (n^2 - a^2)u = 0 \quad \text{in } D.$$

The boundary condition (7.3.10) is satisfied. Clearly $u \in L^2(D)$ if $a > 0$. Thus, $n^2 - a^2$ is an eigenvalue of the Laplacian in D defined by the boundary condition (7.3.10). Therefore, any real point λ can be an eigenvalue of this Laplacian with an appropriate a.

7.4. Absence of Positive Eigenvalues for the Dirichlet Laplacian under Local Assumptions at Infinity

7.4.1. The aim of this section is to establish that under certain assumptions on the boundary Γ'_R the Dirichlet Laplacian L in D has no positive eigenvalues. The number R can be arbitrarily large. Therefore, only local assumptions at infinity are imposed on Γ. Any compact part of Γ is an arbitrary sufficiently smooth (say Lyapunov type) surface. The local smoothness is not necessary but we will not go into details.

Let us first assume that

$$x \cdot N \leq 0 \quad \text{on } \Gamma'_R = \{x : x \in \Gamma, |x| \geq R\}, \tag{7.4.1}$$

Theorem 7.4.1. *If* (7.4.1) *holds, then L has no positive eigenvalues.*

Theorem 7.4.1 holds for general second-order elliptic operators (see Bibliographical notes). Geometrically condition (7.4.1) means that every ray which crosses Γ'_R remains in D. This means that Γ'_R is illuminated by a point source situated at the origin.

In order to prove Theorem 7.4.1, we need some lemmas.

Let

$$(\nabla^2 + k^2)u = 0 \quad \text{in } D, \quad k^2 > 0, \quad u = 0 \quad \text{on } \Gamma'_R. \tag{7.4.2}$$

In the spherical coordinates, equation (7.4.2) takes the form

$$u'' + \frac{2}{r}u' - \frac{Au}{r^2} + k^2 u = 0, \quad u' = \frac{\partial u}{\partial r}, \tag{7.4.3}$$

where $A = -\Delta^*$ is the angular part of the Laplacian,

$$A = -(\sin\theta)^{-1}\frac{\partial}{\partial\theta}\sin\theta\frac{\partial}{\partial\theta} - (\sin\theta)^{-2}\frac{\partial^2}{\partial\phi^2}.$$

Let $v = ru$. Then (7.4.3) becomes

$$v'' - r^{-2}Av + k^2 v = 0. \tag{7.4.4}$$

Let $v_m = r^m v$. Then

$$v_m'' - 2mr^{-1}v_m' + [m(m+1) - A]r^{-2}v_m + k^2 v_m = 0. \tag{7.4.5}$$

Let

$$\begin{aligned} F(m,t,r) = \|v_m'\|^2 + (k^2 - tr^{-1} + m(m+1)r^{-2})\|v_m\|^2 \\ - r^{-2}(Av_m, v_m), \end{aligned} \tag{7.4.6}$$

where m and t are positive parameters,

$$\|v\| = \|v\|_r = (v,v)_r^{1/2}, \quad (v,f)_r = (v,f) = \int_{S_r^2} v\bar{f}ds. \tag{7.4.7}$$

Here, S_r^2 is an open set on the unit sphere S^2 which is the central projection of $S_r(D) = S_r \cup D$ onto S^2. Condition (7.4.1) implies that

$$S_{r_2}^2 \supseteq S_{r_1}^2 \quad \text{if } r_2 \geq r_1 \geq R. \tag{7.4.8}$$

Therefore, the family of Hilbert spaces $L^2(S_r) = H_r$ depends on the parameter r and $H_{r_2} \supseteq H_{r_1}$ if $r_2 \geq r_1 \geq R$. The operator A is symmetric on the spaces $\mathring{H}^2(S_r^2)$.

Let $0 \leq s \leq 2$ be a real number. Let us compute the derivative

$$\begin{aligned} \frac{d}{dr}(r^s F) = sr^{s-1}F + r^s\{2\,\mathrm{Re}(v_m'', v_m') \\ + 2(k^2 - tr^{-1} + m(m+1)r^{-2})\,\mathrm{Re}(v_m', v_m) \\ + (tr^{-2} - 2m(m+1)r^{-3})\|v_m\|^2 + 2r^{-3}(Av_m, v_m) \\ - r^{-2}(Av_m', v_m) - r^{-2}(Av_m, v_m')\} + J, \end{aligned} \tag{7.4.9}$$

where

$$J = \lim_{\Delta r \to 0} \frac{1}{\Delta r} \int_{\Delta S_r^2} \{|v'_m|^2 + (k^2 - tr^{-1} + m(m+1)r^{-2})|v_m|^2 - r^{-2} A v_m \bar{v}_m\} ds$$
$$= \lim_{\Delta r \to 0} \frac{1}{\Delta r} \int_{\Delta S_r^2} |v'_m|^2 ds \geq 0, \quad r \geq R. \tag{7.4.10}$$

Here, $\Delta S_r^2 = S_{r+\Delta r}^2 \backslash S_r^2$, and the limit of the last two terms is zero because $v_m = 0$ on Γ. Substituting v''_m from (7.4.5) into (7.4.9) yields

$$\frac{d}{dr}(r^s F) = F_1 + F_2 + J, \tag{7.4.11}$$

where

$$F_1 = r^{s-1}\{(4m+s)\|v'_m\|^2 + (sk^2 + (1-s)r^{-1}t + (s-2)m(m+1)r^{-2})\|v_m\|^2 - 2t \operatorname{Re}(v'_m, v_m) + (2-s)r^{-2}(Av_m, v_m)\}, \tag{7.4.12}$$

and

$$F_2 = r^{s-2}[(v'_m, Av_m) - (Av'_m, v_m)]. \tag{7.4.13}$$

Integrating by parts yields

$$F_2 = -r^{s-2} \int_{\partial S_r^2} \frac{\partial v_m}{\partial n} v'_m dl$$
$$= -r^{s-2} \int_{\partial S_r^2} |\nabla v_m|^2 (n \cdot N)(N \cdot x^0) dl. \tag{7.4.14}$$

Here, $x^0 = x|x|^{-1}$, n is the normal to $\partial S_r(D)$ pointing out of $S_r(D)$, N is the normal to Γ pointing out of D, ∇v_m on Γ is directed either along N or in the opposite direction because $v_m = 0$ on Γ. Since N enters twice in the integrand in (7.4.14), it does not matter whether $\nabla v_m = N|\nabla v_m|$ on Γ or $\nabla v_m = -N|\nabla v_m|$ on Γ.

Any condition on Γ'_R which implies that

$$(N \cdot n)(N \cdot x^0) \leq 0 \tag{7.4.15}$$

is sufficient for the inequality

$$F_2 \geq 0 \tag{7.4.16}$$

to hold. In particular, if (7.4.1) holds, then $n \cdot N \geq 0$ and (7.4.16) holds. Let $m \geq m_0$, $r \geq r_0$, and $s = 2$, where m_0 and r_0 are sufficiently large. Then

$$F_1 \geq 0 \quad \text{for all } m \geq m_0, \quad r \geq r_0, \quad s = 2. \tag{7.4.17}$$

Indeed, $(Av_m, v_m)(2-s)r^{-2} = 0$ if $s = 2$, and the remaining part of F_1 is nonnegative, if

$$(4m + s) \cdot (sk^2 + (1-s)r^{-1}t + (s-2)m(m+1)r^{-2}) \geq t^2 \tag{7.4.18}$$

and

$$4m + s > 0, \quad sk^2 + (1-s)r^{-1}t + (s-2)m(m+1)r^{-2} > 0. \tag{7.4.19}$$

Since $s = 2, k > 0, m \geq m_0, r \geq r_0$, and m_0, r_0 are sufficiently large, inequalities (7.4.18) and (7.4.19) hold. Thus, (7.4.17) holds. We have proved the following lemma.

Lemma 7.4.2. *If* (7.4.1) *holds and* $t = t_0 > 0$ *is fixed, then*

$$\frac{d}{dr}(r^2 F) \geq 0 \quad \textit{for all } m \geq m_0, \quad r \geq r_0. \tag{7.4.20}$$

Let us prove the following lemma.

Lemma 7.4.3. *There exist* m_0, t_0, *and* r_0 *such that*

$$F(m_0, t_0, r) > 0 \quad \textit{for all } r \geq r_0. \tag{7.4.21}$$

Proof. We have

$$\begin{aligned} F(m,t,r) = r^{2m}\{&\|v' + mr^{-1}v\|^2 \\ &+ (k^2 - tr^{-1} + m(m+1)r^{-2})\|v\|^2 - r^{-2}(Av, v)\}. \end{aligned} \tag{7.4.22}$$

Since $v \neq 0$, there is a $r_1 > r_0$ such that $\|v(r_1)\| > 0$. Therefore, $F(m, t, r_1) > 0$ for all sufficiently large m. We can increase m_0, if necessary, and then $F(m,t,r_1) > 0$ for all $m \geq m_0$. By Lemma 7.4.2, this implies that $r^2F(m,t_0,r) \geq r_1^2F(m,t_0,r_1) > 0$ for all $r \geq r_1$. Lemma 7.4.3 is proved. □

Now we can prove the following lemma.

Lemma 7.4.4. *If $\|v(r)\|_r^2$ does not increase monotonically for $r \geq R$, where $R > 0$ is arbitrarily large fixed number, then there exists a sequence $r_m \to +\infty$ such that*

$$F(0,0,r_n) > 0. \tag{7.4.23}$$

Proof. Since $\|v(r)\|_r^2$ does not increase monotonically for $r \geq R$, there exists a sequence $r_n \to +\infty$ such that

$$\frac{d}{dr}\|v(r)\|_r^2 = 2\,\mathrm{Re}(v',v)_r \leq 0 \quad \text{at } r = r_n. \tag{7.4.24}$$

It follows from (7.4.24) that

$$\begin{aligned}\|v' + mr^{-1}v\|^2 &= \|v'\|^2 + m^2r^{-2}\|v\|^2 + 2mr^{-1}\,\mathrm{Re}(v',v)\\ &\leq \|v'\|^2 + m^2r^{-2}\|v\|^2 \quad \text{at } r = r_n.\end{aligned} \tag{7.4.25}$$

Thus,

$$\begin{aligned}F(m,t,r) \leq r^{2m}\{&\|v'\|^2 + (k^2 - tr^{-1} + m(2m+1)r^{-2})\|v\|^2\\ &- r^{-2}(Av,v)\}, \quad r = r_n.\end{aligned} \tag{7.4.26}$$

This and (7.4.21) imply

$$0 < r^{2m_0}\{\|v'\|^2 + k^2\|v\|^2 - r^{-2}(Av,v)\} = r^{2m_0}F(0,0,r),\ r = r_n \tag{7.4.27}$$

provided that r_n is sufficiently large so that $-tr_n^{-1} + m_0(2m_0 + 1)r_n^{-2} < 0$.

Inequality (7.4.23) follows immediately from (7.4.27). □

Lemma 7.4.5. *For $\epsilon > 0$, however small, one has*

$$\lim_{r\to\infty} r^{\epsilon}(\|v'\|^2 + k^2\|v\|^2) = \infty. \tag{7.4.28}$$

Moreover,

$$\liminf_{r\to\infty}(\|v'\|^2 + k^2\|v\|^2) > 0 \tag{7.4.29}$$

provided that $v \not\equiv 0$.

Proof. Since $(Av, v) \geq 0$, one has

$$F(0,0,r) \leq \|v'\|^2 + k^2\|v\|^2. \tag{7.4.30}$$

One can assume that $\|v\|^2 = \|v(r)\|_r^2$ does not increase monotonically for $r \geq R$ (since if it does then (7.4.28) and (7.4.29) are obvious). Therefore, by Lemma 7.4.4, there exists $r = r_1$, as large as we wish, such that $F(0,0,r_1) > 0$. Let $s > 0$ be arbitrarily small.

In equalities, (7.4.18) and (7.4.19) hold for $m = t = 0$ provided that r is sufficiently large. Therefore,

$$\frac{d}{dr} r^s F(0,0,r) \geq 0, \quad r \geq r_1. \tag{7.4.31}$$

Thus,

$$r^s F(0,0,r) \geq r_1^s F(0,0,r_1) > 0, \quad r \geq r_1. \tag{7.4.32}$$

Inequalities (7.4.30) and (7.4.32) imply (7.4.28). To prove (7.4.29), set $s = 0$ in (7.4.11). Then (7.4.10) and (7.4.16) are still valid, and $F_1 = 2r^{-2}(Av, v) \geq 0$. Thus,

$$\frac{d}{dr} F(0,0,r) \geq 0. \tag{7.4.33}$$

From (7.4.23), (7.4.30), and (7.4.33), the inequality (7.4.29) follows. □

There are at least two ways to proceed with the proof of Theorem 7.4.1. One way is to prove the following lemma, which is of independent interest.

Lemma 7.4.6. *If $u \in L^2(D)$ and conditions* (7.4.1) *and* (7.4.2) *hold, then*

$$\nabla u \in L^2(D). \tag{7.4.34}$$

The second way is based on the following lemma.

Lemma 7.4.7. *If $v \neq 0$ solves* (7.4.4) *and* (7.4.1) *holds, then*

$$\lim_{R\to\infty} \int_R^{R+b} \|v\|^2 dr > 0 \tag{7.4.35}$$

for any $b > 0$.

Let us first prove Theorem 7.4.1 assuming either Lemma 7.4.6 or Lemma 7.4.7.

Proof of Theorem 7.4.1. (a) If u solves (7.4.2), (7.4.1) holds, and $u \in L^2(D)$, then, by Lemma 7.4.6, $\nabla u \in L^2$. Thus,

$$\int^\infty drr^2(\|u\|^2 + \|u'\|^2) < \infty. \tag{7.4.36}$$

Therefore,

$$\int^\infty dr(\|v\|^2 + \|v'\|^2) < \infty. \tag{7.4.37}$$

Indeed, $v = ru, u' = r^{-1}v' - r^{-2}v, v' = ru' + r^{-1}v, \|v'\| \leq r\|u'\| + r^{-1}\|v\|$, so that (7.4.36) implies (7.4.37).

If (7.4.37) holds and $v \neq 0$, then

$$\liminf_{r\to\infty}(\|v\|^2 + \|v'\|^2) = 0. \tag{7.4.38}$$

This is impossible because of (7.4.29). Therefore, $u \in L^2(D)$ implies that $u = 0$. Theorem 7.4.1 is proved.

(b) If u solves (7.4.2), (7.4.1) holds, and $u \in L^2(D)$, then $\int^\infty dr\|v\|^2 < \infty$, and therefore

$$\liminf_{r\to\infty} \int_r^{r+b} \|v\|^2 dr = 0, \quad b > 0. \tag{7.4.39}$$

This is impossible because of (7.4.35). Theorem 7.4.1 is proved. □

We now prove Lemma 7.4.6.

Proof of Lemma 7.4.6. Multiply (7.4.2) by $\bar{u}$ and integrate over D_r to obtain

$$\int_{D_r} |\nabla u|^2 dx = k^2 \int_{D_r} |u|^2 dx + \int_{\Gamma_R} \bar{u}\frac{\partial u}{\partial N} ds + \int_{S_r(D)} \bar{u}\frac{\partial u}{\partial r} ds. \tag{7.4.40}$$

Since $u = 0$ on Γ'_R, one can take the second integral on the right side of (7.4.40) over Γ_R rather than over Γ_r. It follows from (7.4.40) that in order to prove (7.4.34), it is sufficient to prove that

$$\liminf_{r\to\infty} \mathrm{Re} \int_{S_r(D)} \bar{u}\frac{\partial u}{\partial r} ds < \infty. \tag{7.4.41}$$

Let

$$\begin{aligned} f(r) &= \mathrm{Re} \int_{S_r(D)} \bar{u}\frac{\partial u}{\partial r} ds \\ &= \frac{1}{2}\left(\frac{d}{dr}\int_{S_r(D)} |u|^2 ds - 2r\int_{S_r^2} |u|^2 d\omega\right). \end{aligned} \tag{7.4.42}$$

Here, $ds = r^2 d\omega$ and $d\omega$ is an element of the area of S_r^2. Thus,

$$\frac{1}{2}\frac{d}{dr}\int_{S_r(D)} |u|^2 ds = f(r) + r\int_{S_r^2} |u|^2 d\omega \geq f(r). \tag{7.4.43}$$

Therefore, (7.4.41) will be proved if we show that

$$\liminf_{r\to\infty} \frac{d}{dr}\int_{S_r(D)} |u|^2 ds < \infty. \tag{7.4.44}$$

Denote $h(r) = \int_{S_r(D)} |u|^2 ds$. The assumption $u \in L^2(D)$ implies that

$$\int^{\infty} dr h(r) < \infty. \tag{7.4.45}$$

If (7.4.44) does not hold, then

$$\frac{dh}{dr} = \frac{d}{dr}\int_{S_r(D)} |u|^2 ds \geq a > 0, \quad r \geq r_0, \tag{7.4.46}$$

where a is a constant and r_0 is a sufficiently large positive number. Inequality (7.4.46) cannot hold because of (7.4.45). Therefore, (7.4.44) holds. Thus, (7.4.41) is valid and Lemma 7.4.6 is proved.

□

Remark 7.4.1. Lemma 7.4.6 remains valid if the boundary condition $u = 0$ on Γ'_R is replaced by either

$$\frac{\partial u}{\partial N} = 0 \quad \text{on } \Gamma'_R, \tag{7.4.47}$$

or

$$\frac{\partial u}{\partial N} + \eta u = 0 \quad \text{on } \Gamma'_R, \quad \eta \geq 0. \tag{7.4.48}$$

The proof is essentially the same, but equation (7.4.43) should be modified:

$$\frac{dh}{dr} = 2f + 2r \int_{S_r^2} |u|^2 d\omega + \lim_{\Delta r \to \infty} \frac{1}{\Delta r} \int_{\Delta S_r(D)} |u|^2 ds, \tag{7.4.49}$$

where $\Delta S_r(D) = S_{r+\Delta r}(D) \backslash S_r(D)$. If (7.4.1) holds, then the last term in (7.4.49) is nonnegative, and the proof of Lemma 7.4.6 remains valid. If condition (7.4.48) holds, then the integral

$$\int_{\Gamma_r} \bar{u} \frac{\partial u}{\partial N} ds = \int_{\Gamma_R} \bar{u} \frac{\partial u}{\partial N} ds - \int_{\Gamma_r \backslash \Gamma_R} \eta |u|^2 ds \leq \int_{\Gamma_R} \bar{u} \frac{\partial u}{\partial N} ds, \tag{7.4.50}$$

so that the argument in the proof of Lemma 7.4.6 is valid for the boundary condition (7.4.48) as well.

Let us prove Lemma 7.4.7.

Proof of Lemma 7.4.7. One has

$$\begin{aligned} \frac{d^2 \|v\|^2}{dr^2} &= 2\,\mathrm{Re} \frac{d}{dr}(v', v) = 2\,\mathrm{Re}[(v'', v) + (v', v')] \\ &= 2\|v'\|^2 + 2r^{-2}(Av, v) - 2k^2 \|v\|^2. \end{aligned} \tag{7.4.51}$$

Therefore,

$$\frac{d^2\|v\|^2}{dr^2} + 4k^2\|v\|^2 = 2(\|v'\|^2 + k^2\|v\|^2 + r^{-2}(Av, v))$$
$$\geq c > 0. \qquad (7.4.52)$$

The last inequality follows from (7.4.29) and the inequality $(Av, v) \geq 0$. The conclusion of Lemma 7.4.7 follows immediately from (7.4.52) and the following lemma. □

Lemma 7.4.8. *Let $y(t) \geq 0$ and*

$$y'' + k^2 y \geq c > 0, \quad t > t_0, \qquad (7.4.53)$$

where $t_0 \geq 0$ is a number. Then

$$\liminf_{R\to\infty} \int_R^{R+b} y dt > 0 \qquad (7.4.54)$$

for any $b > 0$.

The main difficulty in the proof of Lemma 7.4.7 is the proof of Lemma 7.4.8. If one takes $y = \|v\|^2$, then (7.4.52) and Lemma 7.4.8 imply (7.4.35).

Proof of Lemma 7.4.8. Let $z(t) = y(t) - k^{-2}c \geq -k^{-2}c$. Then

$$z'' + k^2 z \geq 0. \qquad (7.4.55)$$

The graph of any solution to this inequality consists of an alternating succession of negative and nonnegative parts. The length of the interval corresponding to a negative (nonnegative) part is not larger (smaller) than πk^{-1}. The area under a negative part is not larger than the area under each of its two neighboring positive parts. These properties we will establish later. Given these properties, one sees that the area under any negative part of the graph of z is not larger than $\pi k^{-1} \cdot k^{-2}c = \pi c k^{-3}$, and

$$\int_R^{R+b} z(t)dt \geq -\pi c k^{-3} \quad \text{for any } b > 0. \qquad (7.4.56)$$

Therefore,

$$\int_R^{R+b} y(t)dt \geq k^{-2}cb - \pi ck^{-3} = k^{-2}c(b - \pi k^{-1}). \tag{7.4.57}$$

If $y \geq 0$ and (7.4.53) holds with some $k > 0$, then it certainly holds for any larger value of k, say $k_1 > k$, and the inequality (7.4.57) holds with k_1 in place of k.

Fix any $b > 0$ and choose $k_1 > \pi b^{-1}$. Then inequality (7.4.57) (with k_1 in place of k) yields

$$\int_R^{R+b} y(t)dt \geq k_1^{-2}c(b - \pi k_1^{-1}) > 0. \tag{7.4.58}$$

Since the middle term in (7.4.58) increases when b increases, one obtains (7.4.54) for any $b > 0$. To complete the proof, let us establish the properties of the graph of $z(t)$ used in the proof.

Assume that the graph of $z(t)$ has a negative part. Otherwise inequality (7.4.56) becomes $\int_R^{R+b} z(t)dt \geq 0$ and the argument after (7.4.56) becomes even simpler. The negative part must have at least one finite end, say t_0. Otherwise $z'' \geq -k^2 z \geq 0, -\infty < t < \infty$, so that z is convex and therefore $z > 0$ for sufficiently large t. Without loss of generality, one can assume that (7.4.55) holds for $-\infty < t < \infty$. Otherwise one could extend the graph so that (7.4.55) would hold for $-\infty < t < \infty$. One can also assume without loss of generality that the finite end of the negative part is its right end. Therefore, $z(t_0) = 0, z'(t_0) \geq 0$.

Let

$$x(t) = \frac{kz(t)}{\sin k(t - t_0)}, \quad -k^{-1}\pi \leq t - t_0 \leq k^{-1}\pi. \tag{7.4.59}$$

Then $x(t)$ is continuous on the interval (7.4.59), $x(t_0) = z'(t_0) \geq 0$, and

$$\begin{aligned} x'(t) &= k\frac{z'(t)\sin k(t - t_0) - kz(t)\cos k(t - t_0)}{\sin^2 k(t - t_0)} \\ &= k\frac{h(t)}{\sin^2 k(t - t_0)}. \end{aligned} \tag{7.4.60}$$

Clearly

$$h' = (z'' + k^2 z) \sin k(t - t_0).$$

Thus,

$$\begin{aligned} h'(t) &\geq 0 \quad \text{if } 0 \leq t - t_0 \leq \pi k^{-1}, \\ h'(t) &\leq 0 \quad \text{if } -\pi k^{-1} \leq t - t_0 \leq 0. \end{aligned} \tag{7.4.61}$$

One has $h(t_0) = 0$. Therefore, (7.4.61) implies that $h(t) \geq 0$ on the interval (7.4.59) and $x'(t) \geq 0$ on this interval. Thus,

$$\begin{aligned} x(t) &\geq x(t_0) \quad \text{if } 0 \leq t - t_0 \leq \pi k^{-1}, \\ x(t) &\leq x(t_0) \quad \text{if } -\pi k^{-1} \leq t - t_0 \leq 0. \end{aligned} \tag{7.4.62}$$

From (7.4.59) and (7.4.62), it follows that

$$\begin{aligned} z(t) &= k^{-1} x(t) \sin k(t - t_0) \geq k^{-1} x(t_0) \sin k(t - t_0) \\ &= k^{-1} z'(t_0) \sin k(t - t_0). \end{aligned} \tag{7.4.63}$$

Therefore, the graph of $z(t)$ lies above the graph $k^{-1} z'(t_0) \sin k(t-t_0)$ on the interval (7.4.59). This implies the properties of the graph $z(t)$ used in the proof. □

Remark 7.4.2. Absence of positive eigenvalues was established for general elliptic operators of the second order in $\mathbb{R}^N$ not necessarily self-adjoint

$$\begin{aligned} lu &= -\sum_{p,j=1}^{N} D_p(a_{pj}(x) D_j u) + \sum_{j=1}^{N} b_j(x) D_j u + Q(x) u, \\ D_j &= \frac{\partial}{\partial x_j}, \end{aligned} \tag{7.4.64}$$

under the following assumptions ([152, 153]):

(1) $a_{pj}(x)$ and $b_j(x)$ are real-valued functions, $a_{pj} \in C^2(\bar{D}), b_j \in C^1$ $(\bar{D})$, $a_{pj} = a_{jp}$, the matrix $a_{ij}(x)$ is positive definite uniformly in $\bar{D}$, $d_1|t|^2 \leq a_{pj} t_j \, \bar{t}_p \leq d_2 |t|^2$, where $d_2 \geq d_1 > 0, |t|^2 = |t_1|^2 + \cdots + |t_N|^2$, and one sumes over the repeated indices; $a_{pj} - \delta_{pj} = O(|x|^{-a})$,

$|\nabla a_{pj}| = O(|x|^{-1-a}), |(\partial^2 a_{pj})/(\partial x_m \partial x_n)| = O(|x|^{-2-a}), |x| \gg 1, a > 0; Q(x) \in L^2_{\text{loc}}(D), |Q(x)| = O(|x|^{-1-a}), |x| \gg 1$;

(2) there exists an $m, 2 \le m \le N+1$, such that

$$D \cap \left\{ x : \sum_{j=1}^{m-1} x_j^2 \ge R^2 \right\} \subseteq \left\{ x : \sum_{j=m}^{N} x_j^2 \le d \left(\sum_{j=1}^{m-1} x_j^2 \right)^b \right\}, \tag{7.4.65}$$

where $0 < b \le 1$.

Condition (7.4.65) allows the domain D to be a paraboloid, in which case condition (1) is not valid for sufficiently large x.

Remark 7.4.3. In the case when condition (7.4.1) is replaced by the local condition (7.2.6):

$$N \cdot e_3 \le 0 \quad \text{on } \Gamma'_R \tag{7.4.66}$$

and $u = 0$ on Γ, it is not known if there are eigenvalues of $-\Delta_D$ embedded in its continuous spectrum.

Remark 7.4.4. Zero can be an eigenvalue of the operator

$$-\nabla^2 + Q(x) \text{ in } L^2(\mathbb{R}^3) \quad \text{even if } Q \in C_0^\infty(\mathbb{R}^3).$$

Example 7.4.1. Let $u = r^{-2}\cos\theta$ if $r \ge 1$. Then $\nabla^2 u = 0$ in $\Omega_1 = \{x : |x| \ge 1\}, |x| = r$. Let us extend u into B_1 smoothly and so that $Q = \nabla^2 u/u$ is smooth. For example, set $u = f(r)\cos\theta, r \le 1$. Then $\nabla^2 u = (f'' + 2r^{-1}f'(r) - 2fr^{-2})\cos\theta, Q = f^{-1}(f'' + 2r^{-1}f'(r) + 2fr^{-2}), r \le 1, Q = 0, r \ge 1$. One can choose f such that $Q \in C^\infty$. Take $f = r(1 + h(r)), h^{(n)}(R) = O(r^{p-n}), r \to \infty, n \le p, p \ge 2$. Then $Q \in C^\infty$ in a neighborhood of the point $r = 0$. On the interval $(0, 1)$, one can choose $f > 0, f \in C^\infty$, and such that $Q \in C^\infty$ in a neighborhood of the point $r = 1$. Thus,

$$\nabla^2 u - Qu = 0 \text{ in } \mathbb{R}^3,\ u \in L^2(\mathbb{R}^3),\ Q \in C^\infty,\ Q = 0 \text{ if } r \ge 1. \tag{7.4.67}$$

A similar construction gives a solution u to equation (7.4.67) which decays as $O(r^{-n-1})$ with any given $n = 0, 1, 2, \ldots, u = r^{-(n+1)}Y_{nm}(\omega)$ if $r \geq 1$, where Y_{nm} are the spherical harmonics. If $n = 0$, then $u = r^{-1}$ in Ω_1, so that $u \notin L^2(\mathbb{R}^3)$.

7.4.2. Let us show that the spectrum of L is continuous if condition (7.4.1) holds.

Lemma 7.4.9. *If* (7.4.1) *holds, then the spectrum of* L *fills the ray* $[0, \infty)$ *and does not contain eigenvalues.*

Proof. If (7.4.1) holds, then D contains a cone. Therefore, the conclusion of Lemma 7.4.9 follows from Lemma 7.2.4 and Theorem 7.4.1. □

We are now interested in the following questions:

(1) *Is the spectrum of* L *absolutely continuous?*
(2) *Is the limiting absorption principle* (*LAP*) *for the operator* L *valid for all* $k^2 > 0$?

If LAP for the operator L is valid for all $k > 0$, then the positive spectrum of L is absolutely continuous. In the next section, we will prove that LAP for the operator L is valid for all $k^2 > 0$ provided that assumption (7.4.1) and some additional assumptions concerning Γ'_R are satisfied.

Remark 7.4.5. If D is a finite domain, then any solution $u \in L^2(\Omega)$ of the Helmholtz equation

$$(\nabla^2 + k^2)u = 0 \quad \text{in } \Omega, \quad k > 0, \quad u \in L^2(\Omega)$$

is $u = 0$ in Ω (Lemma 1.2.1). Here, no boundary condition u on Γ is imposed. If Γ is infinite, one needs, in general, a boundary condition on u in order to prove that L has no positive eigenvalues. However, if D contains a half-space, say $\mathbb{R}^3_+$, then again no boundary condition is necessary in order to prove that L has no positive eigenvalues.

7.5. LAP and Compact Perturbations of the Boundary

7.5.1. In this section, we always assume that condition (7.4.1) holds so that the Dirichlet Laplacian L in D has continuous spectrum $[0, \infty)$, does not have positive eigenvalues, and the radiation condition guarantees uniqueness of the solution to the Dirichlet problem in D.

By a canonical domain, we mean a domain in which the resolvent kernel of L can be studied (e.g., by separation of variables). For instance, a cone is a canonical domain.

Let $G(x, y, k)$ be the resolvent kernel of L, that is, the kernel of the operator $(L - k^2 - i0)^{-1}, k > 0$, in D. This kernel solves the problem

$$(-\nabla_x^2 - k^2)G(x, y, k) = \delta(x - y) \quad \text{in } D, \tag{7.5.1}$$

$$G(x, y, k) = 0, \quad x \in \Gamma, \tag{7.5.2}$$

$$G(x, y, \sqrt{k^2 + i\epsilon}) \in L^2(D) \quad \text{if } \epsilon > 0, \tag{7.5.3}$$

$$G(x, y, k) = \lim_{\epsilon \to +0} G(x, y, \sqrt{k^2 + i\epsilon}), \tag{7.5.4}$$

where the limit is attained uniformly on compact set $|x - y| \geq a > 0$, $k \geq b > 0$, a and b are arbitrarily small fixed numbers.

By Lemma 7.2.5, the radiation condition

$$\lim_{r \to \infty} \int_{S_r(D)} \left| \frac{\partial G}{\partial r} - ikG \right|^2 ds = 0 \tag{7.5.5}$$

guarantees uniqueness of the solution to problem (7.5.1) and (7.5.2). Let us assume that (7.5.5) holds and

$$|G(x, y, k)| \leq \frac{c}{|x - y|}, \quad |\nabla G| \leq \frac{c}{|x - y|^2}, \ |x - y| \leq 1 \tag{7.5.6}$$

$$|G| + |\nabla G| + |D^2 G| \leq \frac{c}{|x - y|}, \quad |x - y| \geq 1, \tag{7.5.7}$$

where $D^2 G$ stands for an arbitrary second derivative of G.

The local estimates for G are well known and easy to establish. The estimates for $|x-y| \geq 1$ are more difficult to prove. The estimates (7.5.6) and (7.5.7) are uniform with respect to $k, 0 < b \leq k \leq d$, where $d > b$ is any number.

We first prove the following general result.

Theorem 7.5.1. *Let $\tilde{D}$ denote any domain which is obtained from D by a compact perturbation, so that $\tilde{\Gamma}'_R = \Gamma'_R$ for some R, where $\tilde{\Gamma} = \partial\tilde{D}$. Assume that the resolvent kernel of L in D satisfies estimates* (7.5.5)–(7.5.7) *and the radiation condition guarantees uniqueness of the solution to the Dirichlet problem in $\tilde{D}$. Then the operator L in $\tilde{D}$ has absolutely continuous spectrum and its resolvent kernel $\tilde{G}$ satisfies estimates* (7.5.6) *and* (7.5.7).

Note that if $\partial\tilde{D}_R \subset D$ then one can follow the scheme of Chapter and prove the existence of $\tilde{G}$ by the integral equation method. If some part of $\partial\tilde{D}_R$ does not lie in D, then it is not clear if the integral equation method can be used. The reason is that the kernel $G(x, y, k)$ is defined only for $x, y \in D$. In some special cases, e.g., if $D = \mathbb{R}^3_+$, one can construct G which is defined for all $x, y, x \neq y$. In these cases, the integral equation method is convenient.

The method we wish to use in the general case is based on some *a priori* estimates. It is similar to the method used in the proof of Theorem 1.2.3. The proof of Theorem 7.5.1 consists of three steps.

Step 1. First, we note that it is sufficient to prove that the boundary value problem

$$(-\nabla^2 - k^2)u = f \quad \text{in } \tilde{D}, \quad f \in C_0^\infty(\tilde{D}), \tag{7.5.8}$$

$$u = 0 \quad \text{on } \tilde{\Gamma}, \tag{7.5.9}$$

$$\lim_{r\to\infty} \int_{S_r(\tilde{D})} \left| \frac{\partial u}{\partial r} - iku \right|^2 ds = 0 \tag{7.5.10}$$

is uniquely solvable, and its solution can be obtained by the LAP:

$$u(x, k) = \lim_{\epsilon\to+0} u(x, \sqrt{k^2 + i\epsilon}), \tag{7.5.11}$$

where the limit is attained uniformly with respect to $x \in \tilde{D}_r$ and $b \leq k \leq d$, i.e., uniformly on compact sets in $\tilde{D} \times \mathbb{R}_+, x \in \tilde{D}, k \in \mathbb{R}_+$. Indeed, since

$$u(x,k) = \int_{\tilde{D}} \tilde{G}(x,y,k) f(y) dy, \tag{7.5.12}$$

the knowledge of $u(x,k)$ for all $f \in C_0^\infty(\tilde{D})$ defines the kernel $\tilde{G}(x,y,k)$ uniquely as a distribution depending on a parameter x. The differential equation (7.5.8) and boundary condition (7.5.9) imply that

$$(-\nabla_x^2 - k^2)\tilde{G}(x,y) = \delta(x-y) \text{ in } \tilde{D}, \quad \tilde{G} = 0 \text{ on } \Gamma. \tag{7.5.13}$$

The standard regularity properties of the solution to equation (7.5.13) imply that $\tilde{G}(x,y,k) \in C^\infty(\tilde{D}\backslash y)$ and satisfies estimates (7.5.6). The radiation condition allows one to use the standard proof of the symmetry of $\tilde{G}$:

$$\tilde{G}(x,y,k) = \tilde{G}(y,x,k). \tag{7.5.14}$$

Namely, in the sense of distributions, one has

$$\int_{\tilde{D}} \tilde{G}(z,y)(\nabla_z^2 + k^2)\tilde{G}(z,x)dz = \int_{\tilde{D}} (\nabla_z^2 + k^2)\tilde{G}(z,y) \cdot \tilde{G}(z,x)dz. \tag{7.5.15}$$

Because of (7.5.13), equation (7.5.15) reduces to (7.5.14). The same proof in classical term (without using the theory of distributions) requires an application of Green's formula to $\tilde{G}(z,y)$ and $\tilde{G}(z,x)$ in the domain $\tilde{D}_{x,y} = \tilde{D}\backslash(B_\epsilon(x) \cup B_\epsilon(y))$, where $B_\epsilon(x) = \{z : |z-x| < \epsilon\}$, $\epsilon > 0$, and $B_\epsilon(x) \subset \tilde{D}$, $B_\epsilon(y) \subset \tilde{D}$.

The radiation condition for $\tilde{G}$ is easy to establish as follows. Write Green's formula in $D'_R = \tilde{D}'_R$ for $G(x,y,\sqrt{k^2+i\epsilon}), \epsilon > 0$, and then pass to the limit $\epsilon \to +0$. This yields

$$\tilde{G}(x,y) = \alpha(y)G(x,y) + \int_{S_R(D)} \left(G(s,x)\frac{\partial \tilde{G}(s,y)}{\partial N_s} - \tilde{G}(s,y)\frac{\partial G(s,x)}{\partial N_s} \right) ds, \tag{7.5.16}$$

where

$$\alpha(y) = 1 \text{ if } y \in D'_R, \quad \alpha(y) = 0 \quad \text{if } y \notin D'_R. \tag{7.5.17}$$

The right side of (7.5.16) satisfies the radiation condition as a function of x because the integral is taken over a compact set and $G(s, x)$ satisfies the radiation condition (7.5.5) by assumption.

Step 2. Secondly, we prove that problem (7.5.9)–(7.5.10) is uniquely solvable and the solution can be obtained by the LAP (7.5.11). Consider the problem

$$(-\nabla^2 - k^2 - i\epsilon)u_\epsilon = f \text{ in } \tilde{D},\ \epsilon > 0,\ u_\epsilon = 0 \text{ on } \tilde{\Gamma}, \tag{7.5.18}$$

$$u_\epsilon \in L^2(\tilde{D}), \tag{7.5.19}$$

This problem is uniquely solvable in $L^2(\tilde{D})$ since L is self-adjoint in $L^2(\tilde{D})$ (see Section 7.2.1). Suppose that the following *a priori* estimate holds

$$N^2(u_\epsilon) = \int_{\tilde{D}} (1 + |x|)^{-1-a} |u_\epsilon|^2 dx \le c, \tag{7.5.20}$$

where $c > 0$ is a constant which does not depend on ϵ, and $a > 0$. Then one can prove that

$$u_\epsilon \xrightarrow[H^2_{\text{loc}}(\tilde{D})]{} u \quad \text{as } \epsilon \to +0. \tag{7.5.21}$$

If (7.5.21) holds, then one can pass to the limit $\epsilon \to +0$ in (7.5.18) and see that the function u, defined in (7.5.21), satisfies equation (7.5.8) and (7.5.9). Applying Green's formula to $G_\epsilon = G(x, y, \sqrt{k^2 + i\epsilon})$) and u_ϵ in D'_R yields:

$$u_\epsilon(x, k) = \int_{D'_R} G_\epsilon(x, y) f dy + \int_{S_R(D)} \left(u_\epsilon \frac{\partial G_\epsilon}{\partial N} - G_\epsilon \frac{\partial u_\epsilon}{\partial N} \right) ds. \tag{7.5.22}$$

Passing to the limit $\epsilon \to +0$ in (7.5.22) yields

$$u(x, k) = \int_{D'_R} G f dy + \int_{S_R(D)} \left(u \frac{\partial G}{\partial N} - G \frac{\partial u}{\partial N} \right) ds. \tag{7.5.23}$$

We used the continuity of the imbedding $i : H^2(D_R) \to H^1(S_R(D))$ to pass to the limit in the last integral in (7.5.23). Since the integrals in (7.5.23) are taken over compact sets and $G(x,y,k)$ satisfies the radiation condition (7.5.5), the function $u(x,k)$ satisfies the radiation condition (7.5.10). To complete the argument in Step 2, we should prove (7.5.21) and (7.5.20). Equation (7.5.21) follows from the standard inequality

$$\|u\|_{H^2(D_1)} \leq c(\|\nabla^2 u\|_{L^2(D_2)} + \|u\|_{L^2(D_2)}), \quad D_1 \subset D_2, \tag{7.5.24}$$

where $c = c(D_1, D_2) > 0$, D_1 and D_2 are arbitrary compact domains, and $D_1 \subset D_2$ is a strictly inner subdomain of D_2. If $u = 0$ on ∂D_2, then $D_1 = D_2$ is allowed in (7.5.24). If $u = 0$ on a part of ∂D_2, then part of ∂D_1 is allowed to belong to this part of ∂D_2. If (7.5.20) holds, then

$$\|u_\epsilon\|_{L^2(\tilde{D}_r)} + \|\nabla^2 u_\epsilon\|_{L^2(\tilde{D}_r)} \leq c, \tag{7.5.25}$$

where c denotes different positive constants, $r > 0$ is an arbitrary number, and we used equation (7.5.18). From (7.5.25), it follows that

$$\|u_\epsilon\|_{L^2(\tilde{D}_r)} \leq c. \tag{7.5.26}$$

From (7.5.26) and the compactness of the imbedding $i : H^2(\tilde{D}_r) \to H^0(\tilde{D}_r)$, it follows that the set of functions $u_\epsilon, 0 < \epsilon \leq 1$ contains a subsequence which converges in $H^0(\tilde{D}_r) = L^2(\tilde{D}_r)$ as $\epsilon \to +0$. This subsequence we denote u_{ϵ_n}. Equation (7.5.18) shows that $\nabla^2 u_{\epsilon_n}$ converges in $L^2(\tilde{D}_r)$ as $\epsilon_n \to 0$. This and inequality (7.5.24) imply that u_{ϵ_n} converges in $H^2(\tilde{D}_r)$ as $\epsilon_n \to +0$. Since the limiting function $u(x,k)$ solves problem (7.5.8)–(7.5.10), as we have already proved, and since, by assumption, the solution to problem (7.5.8)–(7.5.10) is unique, one concludes that not only the subsequence u_{ϵ_n} but the function u_ϵ itself converges in H^2_{loc} as $\epsilon \to +0$.

Finally, estimate (7.5.20) should be proved.

Suppose (7.5.20) is not valid. Then there exists a sequence $\epsilon_n \to +0$ such that $N(u_n) \geq n$, $n = 1, 2, \ldots$, $u_n = u_{\epsilon_n}$.

Let $u_n/N(u_n) = v_n$. Then

$$N(v_n) = 1 \tag{7.5.27}$$

$$(-\nabla^2 - k^2 - i\epsilon_n)v_n = f_n, \quad f_n = \frac{f}{N(u_n)} \xrightarrow[n\to\infty]{} 0, \tag{7.5.28}$$

$$v_n = 0 \quad \text{on } \tilde{\Gamma}. \tag{7.5.29}$$

Our argument above shows that

$$v_n \xrightarrow[H^2_{\text{loc}}(\tilde{D})]{} v, \quad n \to \infty \tag{7.5.30}$$

and v solves the problem

$$(-\nabla^2 - k^2)v = 0 \text{ in } \tilde{D}, \quad v = 0 \text{ on } \tilde{\Gamma}, \tag{7.5.31}$$

$$\lim_{r\to\infty} \int_{S_R(\tilde{D})} \left| \frac{\partial v}{\partial r} - ikv \right|^2 ds = 0. \tag{7.5.32}$$

Therefore,

$$v = 0 \quad \text{in } \tilde{D}. \tag{7.5.33}$$

If we show that

$$\lim_{n\to\infty} N(v_n) = 0, \tag{7.5.34}$$

then (7.5.33) and (7.5.27) cannot hold simultaneously. This contradiction proves estimate (7.5.20). To prove (7.5.34), one writes Green's formula for v_n

$$v_n = \int_{D'_R} G_{\epsilon_n}(x,y) f_n dy + \int_{S_R(D)} \left(v_n \frac{\partial G_{\epsilon_n}}{\partial N} - G_{\epsilon_n} \frac{\partial v_n}{\partial N} \right) ds. \tag{7.5.35}$$

Each of the terms in the right-hand side of (7.5.35) can be estimated in the same way. The estimate is possible because the integrals in (7.5.35) are taken over compact sets and because $G_{\epsilon_n}(x,y)$ satisfies the inequality

$$|G_{\epsilon_n}(x,y)| \le \frac{c}{|x-y|}, \quad |x-y| \ge 1. \tag{7.5.36}$$

This inequality we need only for y running through compact sets. Let us estimate, for example, the first integral in (7.5.35).

We have

$$N\left(\int_{D'_R} G_{\epsilon_n}(x,y) f_n dy\right)$$
$$\le c\|f_n\|_{L^2(D'_R)} \left(\int_{\tilde{D}}\int_{D'_R} \frac{dxdy}{(1+|x|)^{1+a}|x-y|^2}\right)^{1/2}$$
$$\le c_1\|f_n\|_{L^2(D'_R)} \to 0, \quad n\to\infty. \tag{7.5.37}$$

The last relation follows from the second equation (7.5.28). It follows from (7.5.30) and (7.5.33) that

$$v_n \xrightarrow[H^1(S_r(D))]{} 0, \quad n \to \infty, \tag{7.5.38}$$

so that the other two integrals in (7.5.35) can be estimated similarly.

Step 3. To finish the proof of Theorem 7.5.1, we should prove that $\tilde{G}$ satisfies inequality (7.5.7). This inequality is assumed to be valid for G and therefore formula (7.5.16) implies inequality (7.5.7) for $\tilde{G}(x,y)$ when y runs through compact sets. In order to prove that

$$|\tilde{G}(x,y)| \le \frac{c}{|x-y|}, \quad |x-y|\ge 1, \quad x,y\in\tilde{D}, \tag{7.5.39}$$

we assume that (7.5.39) is not valid. In this case, there exist sequences $x_n, y_n \in \tilde{D}$, such that

$$|\tilde{G}(x_n,y_n)| \ge \frac{n}{|x_n-y_n|}, \quad |x_n-y_n|\ge 1. \tag{7.5.40}$$

Without loss of generality, one can assume that $|x_n - y_n| \to \infty$. By assumption,

$$|G(x_n,y_n)||x_n-y_n| \le c. \tag{7.5.41}$$

It follows from (7.5.16) and (7.5.7) that

$$|\tilde{G}(x,y)| \le \frac{c}{|x|}, \quad |x|\to\infty, \quad |y|\le c_1. \tag{7.5.42}$$

Therefore, one can assume that

$$|x_n|\to\infty \quad \text{and} \quad |y_n|\to\infty. \tag{7.5.43}$$

Multiply equation (7.5.16) by $|x_n - y_n|$, take $x = x_n, y = y_n$, use (7.5.40), and note that $\alpha(y_n) = 1$. This yields

$$\begin{aligned} n &\le |x_n - y_n|\,|\tilde{G}(x_n, y_n)| \\ &\le |x_n - y_n|\,|G(x_n, y_n)| + c|x_n - y_n| \int_{S_R(D)} \frac{ds}{|s - x_n|\,|s - y_n|}. \end{aligned} \tag{7.5.44}$$

The right side of (7.5.44) is bounded and we have a contradiction. This proves (7.5.39). The other estimates (7.5.7) can be proved similarly. Theorem 7.5.1 is proved. □

Remark 7.5.1. In the proof of Theorem 7.5.1, we used estimate (7.5.7) only for y running through compact sets.

7.5.2. Suppose now that the perturbation of Γ is not compact, i.e., there is no such $R > 0$ that $\Gamma'_R = \tilde{\Gamma}'_R$. Let us assume that

$$\tilde{\Gamma}'_R \subset D'_R, \text{ and } \rho(r) \le c(1 + r)^{-a}, \quad a > 2, \tag{7.5.45}$$

where $\rho(r)$ is the distance between $\tilde{\Gamma}'_r \cap S_r$ and $\Gamma'_r \cap S_r$. Let us also assume that condition (7.4.1) holds for Γ'_R and $\tilde{\Gamma}'_R$. Without loss of generality, we can assume that $\tilde{D} \subset D$. Otherwise, we can compactly perturb D so that the above condition is satisfied. Let us denote $D\backslash\tilde{D}$ by d. The spectrum of the Laplacian $-\Delta_d$ defined in d by the boundary conditions $u = 0$ on $\Gamma, \partial u/\partial N = 0$ on $\tilde{\Gamma}$, where N is the normal on $\tilde{\Gamma}$ pointing into $\tilde{D}$, is discrete because of (7.5.45), and without loss of generality, we assume that k^2 is not an eigenvalue of $-\Delta_d$. Otherwise we can compactly perturb D so that k^2 will not be an eigenvalue of $-\Delta_d$. Green's formula for G and $\tilde{G}$ in $\tilde{D}$ yields

$$\tilde{G}(x, y) = G(x, y) - \int_{\tilde{\Gamma}} G(x, s)\mu(s, y)ds, \; \mu = \frac{\partial \tilde{G}(s, y)}{\partial N_s}. \tag{7.5.46}$$

Take the normal derivative and use the formula for the normal derivative of the single-layer potential to obtain

$$\mu + A\mu = 2\frac{\partial G}{\partial N}, \quad A\mu = \int_{\tilde{\Gamma}} \frac{\partial G(s, t)}{\partial N_s}\mu(t)dt. \tag{7.5.47}$$

Under the assumption (7.5.45), one can check that A is compact in $C(\tilde{\Gamma})$ and $L^2(\tilde{\Gamma})$. The homogeneous equation

$$\mu + A\mu = 0 \tag{7.5.48}$$

has only the trivial solution $\mu = 0$. Otherwise, k^2 is an eigenvalue of $-\Delta_d$. Therefore, by Fredholm's alternative, equation (7.5.47) is uniquely solvable and the kernel $\tilde{G}$ of the Dirichlet Laplacian in $\tilde{D}$ is given by formula (7.5.46). To justify this argument completely, one should establish the following lemma.

Lemma 7.5.2. *The operator* (7.5.47) *is compact in* $C(\tilde{\Gamma})$ *(and in* $L^2(\tilde{\Gamma})$*) provided that* (7.5.45) *holds.*

The proof of Lemma 7.5.2 is relatively long. The reader can find it in [182]. In this proof, the following estimate is essential:

$$\frac{\partial G(s,t)}{\partial N_s} \leq \begin{cases} c|s-t|^{2-\lambda}, & \lambda > 0,\ |s-t| \leq 1, \\ c|s-t|^{-1}\rho(t,\Gamma), & |s-t| \geq 1, \end{cases} \tag{7.5.49}$$

where $\rho(t,\Gamma)$ is the distance from the point $t \in \tilde{\Gamma}$ to $\Gamma, \rho(t,\Gamma) = \inf_{x\in\Gamma} |t-x|$,

$$\rho(t,\Gamma) \leq c(1+|t|)^{-a}, \quad a > 2 \tag{7.5.50}$$

(see (7.5.45)). The constant λ in (7.5.49) depends on the smoothness of $\tilde{\Gamma}$: if the equation of $\tilde{\Gamma}$ in the local coordinates is $s_3 = f(s_1, s_2)$ and $f \in C^{1,\lambda}$, then this λ is used in (7.5.49). We assume that λ can be chosen uniformly on $\tilde{\Gamma}$. The inequality (7.5.49) for $|s-t| \leq 1$ is standard in potential theory, and for $|s-t| \geq 1$ can be obtained as follows. Let $\rho(t',t) = \rho(t,\Gamma), t' \in \Gamma$. By the mean value (Lagrange) formula, one has

$$\frac{\partial G(s,t)}{\partial N_s} = \frac{\partial G(s,t')}{\partial N_s} + \nabla_t \frac{\partial G(s,z)}{\partial N_s} \cdot (t-t'), \tag{7.5.51}$$

where z is the "mean point" in the mean value formula. Since $t' \in \Gamma$, the first term in the right-hand side of (7.5.51) is zero. Estimating the second term and using (7.5.7), one obtains inequality (7.5.49) for $|s-t| \geq 1$.

7.6. Eigenfunction Expansion in Canonical Domains

In this section, we sketch the eigenfuntion expansion theory similar to the one presented in Section 1.3. Let us assume that D is a cone K, which plays the role of a canonical domain. Then $G(x,y,k)$ can be constructed explicitly by separation of variables and its asymptotics is

$$G(x,y,k) = \gamma(|y|)u_0(x,n,k)(1+o(1)), \quad |y| \to \infty, \frac{y}{|y|} = n, \tag{7.6.1}$$

where $\gamma(r) = (4\pi r)^{-1}\exp(ikr)$ and n is a unit vector in the cone K. This formula is analogous to (1.3.7). From (7.6.1) and (7.5.46), it follows that

$$\tilde{G}(x,y,k) = \gamma(|y|)u(x,n,k)(1+o(1)), \quad |y| \to \infty, \quad \frac{y}{|y|} = n, \tag{7.6.2}$$

where

$$\begin{aligned} u(x,n,k) &= u_0(x,n,k) - \int_{\tilde{\Gamma}} u_0(s,n,k)\mu(s,x,k)ds, \\ \mu &= \frac{\partial \tilde{G}(s,x,k)}{\partial N_s}. \end{aligned} \tag{7.6.3}$$

As in Section 1.3, the function $u(x,n,k)$ is the solution to the scattering problem in $\tilde{D}$:

$$(\nabla^2 + k^2)u = 0 \text{ in } \tilde{D}, \quad u = 0 \text{ on } \tilde{\Gamma}, \tag{7.6.4}$$

$$u = u_0 + v, \tag{7.6.5}$$

where v satisfies the radiation condition

$$\lim_{r\to\infty} \int_{S_r(\tilde{D})} \left| \frac{\partial v}{\partial r} - ikv \right|^2 ds = 0. \tag{7.6.6}$$

As in Section 1.3, one can prove the eigenfunction expansion theorem. This theorem is analogous to Theorem 1.3.2.

Theorem 7.6.1. *Let $f \in L^2(\tilde{D})$. Define*

$$Ff = \tilde{f}(\kappa) = \frac{1}{(2\pi)^{3/2}} \int_{\tilde{D}} f(x)\overline{u(x,\kappa)}dx, \quad \kappa = kn, \tag{7.6.7}$$

$u(x,\kappa) = u(x,k,n)$. Then

$$F^{-1}f = f(x) = \frac{1}{(2\pi)^{3/2}} \int_K \tilde{f}(\kappa)u(x,\kappa)d\kappa. \tag{7.6.8}$$

The operator $F : L^2(\tilde{D}) \to L^2(K)$ is an isometric isomorphism, $F^{-1} = F^*$, and

$$F(-\Delta_D(\tilde{D}))F^{-1} = k^2, \tag{7.6.9}$$

where $-\Delta_D(\tilde{D})$ is the Dirichlet Laplacian in $\tilde{D}$.

The proof of Theorem 7.6.1 is similar to that of Theorem 1.3.2 and is omitted.

Remark 7.6.1. One can construct the wave operators and the S-matrix as in Section 1.5, formulas (1.5.5); (1.5.6). The role of F_0 in (1.5.5) is played by the Fourier transform in D, that is, the operator

$$F_0f = \frac{1}{(2\pi)^{3/2}} \int_K f(x)\overline{u_0(x,\kappa)}dx. \tag{7.6.10}$$

Part II

Scattering by Potentials

Chapter 8

Scattering by Potentials

8.1. Introduction: Existence and Uniqueness of the Scattering Solution — Uniqueness of the Solution to Inverse Scattering Problem

In this part, we are interested mostly in the inverse scattering problem. By this reason, it is natural and even necessary to assume the potential compactly supported. The direct scattering problem consists of solving the equation

$$(\nabla^2 + k^2 - q(x))u = 0 \quad \text{in } \mathbb{R}^3, \tag{8.1.1}$$

$$u = e^{ik\alpha\cdot x} + v, \quad v_r - ikv = o\left(\frac{1}{r}\right), \quad r = |x| \to \infty, \tag{8.1.2}$$

where $k = \text{const} > 0$, $\alpha \in S^2, S^2$ is the unit sphere in $\mathbb{R}^3, u_r = \frac{\partial u}{\partial |x|}, q(x)$ is a real-valued function, $q \in Q_a$, that is,

$$Q_a := \{q : q \in L^2(\mathbb{R}^3),\ q(x) = 0 \text{ for } |x| > a,\ q = \bar{q}\}. \tag{8.1.3}$$

Theorem 8.1.1. *Problem* (8.1.1)–(8.1.2) *has a unique solution.*

Proof. If and only if u solves problem (8.1.1)–(8.1.2), u solves the equation

$$u = e^{i\alpha\cdot x} - \int g(x,y)q(y)u(y)dy, \quad \int := \int_{|y|\le a}, \quad g(x,y) = \frac{e^{ik|x-y|}}{4\pi|x-y|}. \tag{8.1.4}$$

Equation (8.1.4) is of Fredholm type since the operator

$$Tu := \int g(x,y)q(y)u(y)dy$$

is compact in $C(B_a)$, where $B_a = \{x : |x| \leq a\}$ and $C(B_a)$ is the space of continuous functions in B_a with the norm $\|u\| = \max_{x\in B_a} |u(x)|$.

By the Fredholm alternative, equation (8.1.4) is uniquely solvable if the homogeneous equation (8.1.4) has only the trivial (zero) solution. If v is the solution to equation $v = -Tv$, then v solves equation (8.1.1) and satisfies the radiation condition $v_r - ikv = o(\frac{1}{r}), r \to \infty$.

Multiply equation (8.1.1) for v by $\bar{v}$, the bar stands for complex conjugate, integrate over B_R and then take $R \to \infty$ and use the radiation condition to get

$$\int_{B_R} (k^2|u|^2 - |\nabla u|^2 - q(x)|u|^2)dx + \int_{|s|=R} \bar{u}u_r ds = 0. \tag{8.1.5}$$

The radiation condition yields

$$\int_{|s|=R} \bar{u}u_r ds = ik \int_{|s|=R} |\bar{u}|^2 ds + o(1), \quad R \to \infty. \tag{8.1.6}$$

Taking imaginary part of (8.1.5) and using (8.1.6), one gets

$$\lim_{R\to\infty} \int_{|s|=R} |v|^2 ds = 0. \tag{8.1.7}$$

Since $q(x) = 0$ for $|x| > a$, v solves equation

$$(\nabla^2 + k^2)v = 0, \quad \text{if } 5|x| > a. \tag{8.1.8}$$

From (8.1.7) and (8.1.8), it follows by Lemma 1.2.1 that $v = 0$ in $|x| > a$. By the unique continuation property for solutions to equation (8.1.1), it follows that $v = 0$ in $\mathbb{R}^3$. Theorem 8.1.1 is proved. □

It follows from equation (8.1.4) that

$$v = \frac{e^{ik|x|}}{|x|}A(\beta,\alpha,k) + O\left(\frac{1}{|x|^2}\right), \quad |x| \to \infty, \quad \frac{x}{|x|} = \beta, \qquad (8.1.9)$$

and

$$-4\pi A(\beta,\alpha,k) = \int e^{-ik\beta\cdot y} q(y)u(y,\alpha,k)dy, \qquad (8.1.10)$$

where $A(\beta,\alpha,k)$ is the scattering amplitude.

The inverse scattering problem consists of finding q given the scattering data. These data are the values of $A(\beta,\alpha,k)$ on some subsets of $S^2 \times S^2 \times \mathbb{R}_+, \mathbb{R}_+ = (0,\infty)$. It is easy to prove that the values of $A(\beta,\alpha,k)$ known for all $\beta \in S^2, \alpha \in S^2$ and $k \in \mathbb{R}_+$ determine q uniquely. In 1987, the author announced [286] and in 1988 published a detailed proof of the following result [295].

Theorem 8.1.2. *If $q = \bar{q} \in L^2(B_a), q = 0$ if $|x| > a$, then the data $A(\beta,\alpha,k_0)$, known for all α on an open subset of S^2, all β on an open subset of S^2, and a fixed $k = k_0 > 0$, determine q uniquely.*

Much work was done by the author [317, 345, 356] for obtaining the reconstruction algorithm and its error estimate. These results will be presented in this chapter in detail and in a sharpened form.

The data in Theorem 8.1.2 is a function on a four-dimensional set $S^2 \times S^2$, while the unknown q is a function on a three-dimensional set $\{x : |x| \leq a, x \in \mathbb{R}^3\}$. Such an inverse scattering problem is called the problem with overdetermined data. The uniqueness theorems for three-dimensional inverse scattering with non-overdetermined data were not known for many decades. Such theorems were proved for the first time by the author for inverse potential scattering [395, 396, 401] in 2010–2011 and for inverse obstacle scattering [420] in 2016. These results for inverse potential scattering are presented in Chapter 8.

Let us now explain why it is not only natural but also necessary to assume that q is compactly supported when one deals with inverse scattering problems. The reason is: the data are always noisy, and if $q(x)$ decays rapidly at infinity then its tail contributes to the scattering data less than the noise, so this contribution cannot be distinguished from the noise. To make this conclusion concrete, let us assume that $|q(x)| \leq \frac{c}{(1+|x|)^p}, p > 3$. Then the contribution A_t of the tail of q to the scattering amplitude is

$$A_t := -\frac{1}{4\pi} \int_{|y|>R} e^{-ik\beta\cdot y} q(y) u(y, \alpha, k) dy,$$

$$|A_t| \leq c \int_R^\infty \frac{r^2 dr}{(1+r)^p} \leq \frac{c}{R^{p-3}}, \tag{8.1.11}$$

so $A_t \to 0$ as $R \to \infty$. Here, we have used the estimate

$$\max |u(y, \alpha, k)| \leq c. \tag{8.1.12}$$

By c *various* estimation constants are denoted. It is clear from estimate (8.1.11) that for sufficiently large R, the contribution of the tail of the potential becomes less than the noise level. Therefore, it is necessary to assume q compactly supported when one deals with the inverse scattering problem.

The uniqueness of the solution to the three-dimensional (3D) inverse scattering problem with q decaying as $O(\frac{1}{(1+|x|)^p})$, $p > 3$ and fixed-energy scattering data, is not known. For the inverse scattering problem with exact scattering data and q decaying faster than exponential, a uniqueness theorem is known [320].

In the next sections, Theorem 8.1.2 is proved and stability estimates for the solution of inverse scattering problem are obtained. Furthermore, the inverse scattering problems with non-overdetermined data are considered and uniqueness theorems for their solutions are derived.

8.2. Global Perturbation Formula: A New Definition of the Scattering Solution — High-Frequency Approximation

Theorem 8.2.1 (Global Perturbation Formula). *If $A_m(\beta, \alpha, k), m = 1, 2,$ are the scattering amplitudes corresponding to the* *potentials* $q_m \in Q_a := \{q : q = \bar{q}, q \in L^2(B_a), q = 0 \text{ for } |x| > a\}$, *then*

$$\begin{aligned} &-4\pi(A_1(\beta, \alpha, k) - A_2(\beta, \alpha, k)) \\ &\quad = \int_{B_a} (q_1(y) - q_2(y)) u_1(x, \alpha, k) u_2(x, -\beta, k) dx. \end{aligned} \tag{8.2.1}$$

We prove Theorem 8.2.1 later.

Let $G(x, y, k)$ be the resolvent kernel of the self-adjoint operator $L = -\nabla^2 + q(x)$ in $L^2(\mathbb{R}^3)$. This kernel can be defined as the (unique) solution to the problem

$$G(x, y, k) = g(x, y, k) - \int_{\mathbb{R}^3} g(x, z, k) q(z) G(z, y, k) dz, \tag{8.2.2}$$

where

$$g(x, y, k) = \frac{e^{ik|x-y|}}{4\pi|x - y|}, \quad \text{Im } k \geq 0. \tag{8.2.3}$$

The kernel G can also be defined as the solution to the problem

$$(\nabla^2 + k^2 - q(x)) G(x, y, k) = -\delta(x - y) \quad \text{in } \mathbb{R}^3, \tag{8.2.4}$$

$$\frac{\partial G}{\partial |x|} - ikG = o\left(\frac{1}{|x|}\right), \quad |x| \to \infty. \tag{8.2.5}$$

Theorem 8.2.2. *One has*

$$G(x, y, k) = g(x, 0, k) u(y, \alpha, k) + O\left(\frac{1}{|x|^2}\right), \; |x| \to \infty, \; \frac{x}{|x|} = -\alpha, \tag{8.2.6}$$

where $u(y, \alpha, k)$ *is the scattering solution, that is, the unique solution to the problem*

$$(\nabla^2 + k^2 - q(x)) u = 0 \quad in \ \mathbb{R}^3, \tag{8.2.7}$$

$$u = e^{ik\alpha \cdot x} + v, \tag{8.2.8}$$

$$\begin{aligned} &v(x, \alpha, k) = A(\beta, \alpha, k) \frac{e^{ik|x|}}{|x|} + O\left(\frac{1}{|x|^2}\right), \\ &\quad |x| \to \infty, \quad \frac{x}{|x|} = \beta. \end{aligned} \tag{8.2.9}$$

Let us prove Theorem 8.2.2 first.

Proof of Theorem 8.2.2. Let $|x| \to \infty$, $\frac{x}{|x|} = -\alpha$ in formula (8.2.2). The right-hand side of this formula admits asymptotic

$$\frac{e^{ik|x|}}{4\pi|x|}\left(e^{ik\alpha\cdot y} - \int_{\mathbb{R}^3} e^{ik\alpha\cdot z}q(z)G(z,y,k)dz\right) + O\left(\frac{1}{|x|^2}\right). \tag{8.2.10}$$

Denote by $u(y,\alpha,k)$, the expression in the parentheses in (8.2.10):

$$u(y,\alpha,k) = e^{ik\alpha\cdot y} - \int_{\mathbb{R}^3} e^{ik\alpha\cdot z}q(z)G(z,y,k)dz. \tag{8.2.11}$$

Let us check that this $u(y,\alpha,k)$ is the scattering solution. Condition (8.2.8) is clearly fulfilled because $G(z,y,k)$ satisfies the radiation condition (8.2.5) also with respect to y variable. Equation (8.2.4) is also satisfied: apply $\nabla_y^2 + k^2$ to equation (8.2.11) and get

$$\begin{aligned}
&(\nabla_y^2 + k^2)u(y,\alpha,k) \\
&\quad = -\int_{\mathbb{R}^3} e^{ik\alpha\cdot z}q(z)[q(y)G(z,y,k) - \delta(z-y)]dz \\
&\quad = e^{ik\alpha\cdot y}q(y) - q(y)\int_{\mathbb{R}^3} e^{ik\alpha\cdot z}q(z)G(z,y,k)dz = q(y)u(y,\alpha,k).
\end{aligned}$$

Thus, $u(y,\alpha,k)$ solves problem (8.2.7)–(8.2.9). Since this problem has at most one solution, $u(y,\alpha,k)$ is the scattering solution.

Theorem 8.2.2 is proved. □

Proof of Theorem 8.2.1. Consider Green's formula

$$\begin{aligned}
&G_1(x,y,k) - G(x,y,k) \\
&\quad = \int_{\mathbb{R}^3} (q_2(z) - q_1(z))G_1(x,z,k)G_2(z,y,k)dz,
\end{aligned} \tag{8.2.12}$$

where the boundary term over the sphere $|z| = R$ tends to zero as $R \to \infty$ due to the radiation condition for G_1 and G_2. Take

$|y| \to \infty$, $\frac{y}{|y|} = \beta$ in (8.2.12) and use (8.2.6) to get

$$\begin{aligned} &u_1(x, -\beta, k) - u_2(x, -\beta, k) \\ &= \int_{\mathbb{R}^3} p(z) G_1(x, z, k) u_2(z, -\beta, k) dz, \ p(z) := q_2 - q_1. \end{aligned} \tag{8.2.13}$$

Take $|x| \to \infty$, $\frac{x}{|x|} = -\alpha$ in (8.2.13) and use (8.2.6) and (8.2.10) to get

$$4\pi[A_1(-\alpha, -\beta) - A_2(-\alpha, -\beta)] = \int_{\mathbb{R}^3} p(z) u_1(z, \alpha, k) u_2(z, -\beta, k) dz. \tag{8.2.14}$$

Formula (8.2.14) is equivalent to formula (8.2.1) because

$$A(-\alpha, -\beta, k) = A(\beta, \alpha, k) \tag{8.2.15}$$

as was proved in Section 1.1.4.

Theorem 8.1.1 is proved. □

Relation (8.2.15) is well-known, it is called reciprocity relation. If $q = \bar{q}$, then

$$\overline{A(\beta, \alpha, k)} = A(\beta, \alpha, -k), \quad k > 0 \text{ (reality relation)}. \tag{8.2.16}$$

The unitarity relation is

$$\text{Im } A(\beta, \alpha, k) = \frac{k}{4\pi} \int_{S^2} A(\beta, \nu, k) \overline{A(\alpha, \nu, k)} d\nu, \quad k > 0. \tag{8.2.17}$$

A particular case of (8.2.17) is the optical theorem

$$\text{Im } A(\alpha, \alpha, k) = \frac{k}{4\pi} \int_{S^2} |A(\alpha, \nu, k)|^2 d\nu, \quad k > 0. \tag{8.2.18}$$

If $q \in Q_a$, then $A(\beta, \alpha, k)$ is an analytic function of β and α on the variety

$$\mathcal{M} := \{\alpha \in \mathbb{C}^3, \alpha \cdot \alpha = 1\}, \quad \alpha \cdot \alpha := \sum_{j=1}^{3} \alpha_j^2. \tag{8.2.19}$$

The variety $\mathcal{M}$ is noncompact. By $\mathcal{M}_1 \subset \mathcal{M}$, we define the subset of $\mathcal{M}$ which consists of the numbers

$$\{\sin\vartheta\cos\phi, \sin\vartheta\sin\phi, \cos\vartheta\}, \quad \vartheta, \phi \in \mathbb{C}. \tag{8.2.20}$$

Since

$$\sin^2\vartheta\cos^2\phi + \sin^2\vartheta\sin^2\phi + \cos^2\vartheta = 1, \tag{8.2.21}$$

it follows that $\mathcal{M}_1 \subset \mathcal{M}$.

Lemma 8.2.3. *For any $\zeta \in \mathbb{R}^3$, one can find (nonuniquely) vectors $\theta, \theta' \in \mathcal{M}$ such that*

$$\theta' - \theta = \zeta, \quad \theta', \theta \in \mathcal{M}, \quad |\theta| \to \infty. \tag{8.2.22}$$

Proof. To check this, choose the coordinate system in which $\zeta = te_3, t > 0$, e_j are orthonormal basis vectors of $\mathbb{R}^3$. Let $\theta = A+iB$, A, B are vectors in $\mathbb{R}^3$. For $\theta \in \mathcal{M}$, it is necessary and sufficient that $A^2 - B^2 = 1$ and $A \cdot B = 0$. Therefore, $\theta' - \theta = te_3$ holds if $\theta_3' - \theta_3 = t$. Let $\theta = A_2e_2 + iB_1e_1 - \frac{t}{2}e_3, \theta' = A_2'e_2 + iB_1'e_1 + \frac{t}{2}e_3, A_2 = A_2' = A$, and $B_1 = B_1' = B$. This is just one of infinitely many possible choices of $\theta, \theta' \in \mathcal{M}$, such that $\theta' - \theta = te_3, |\theta| \to \infty$. To satisfy the relations $\theta \cdot \theta = \theta' \cdot \theta' = 1$, one has to require $A_2^2 - B_1^2 + \frac{t^2}{4} = 1$. To satisfy the requirement $|\theta| \to \infty$, it is necessary and sufficient to require $A_2 \to \infty$. Lemma 8.2.3 is proved. □

The function $G(x, y, k)$ is meromorphic on the complex plane k, analytic function of k in the upper half-plane Im $k > 0$ except, possibly, a finite number of simple poles at the points ik_j, where $k_j > 0, -k_j^2$ are eigenvalues of the self-adjoint operator $L = \nabla^2 + q(x)$ in $H = L^2(\mathbb{R}^3)$.

In the half-plane $\operatorname{Im} k < 0$, there are countably many poles of $G(x, y, k)$. The semi-axis $k \geq 0$ consists of points of continuous spectrum of L. For any $k > 0$, there exists the limit

$$\lim_{\epsilon \downarrow 0} G(x, y, k + i\epsilon) = G(x, y, k), \quad x \neq y, \quad k > 0. \tag{8.2.23}$$

One has

$$\sup_{x\in\mathbb{R}^3,\alpha\in S^2} |u(x,\alpha,k)| \le c, \tag{8.2.24}$$

where the constant c can be chosen uniformly on the interval $(k_0,\infty), k_0 > 0$ is an arbitrary fixed number.

The eigenfunction expansion formula is known:

If $f \in L^2(\mathbb{R}^3)$ is arbitrary, then

$$f(x) = \int_{\mathbb{R}^3} \tilde{f}(\xi)u(x,\xi)d\xi + \sum_{j=1}^{J}(f,u_j)u_j(x), \tag{8.2.25}$$

where $\xi = k\alpha, \alpha \in S^2, k = |\xi|, k \in [0,\infty)$,

$$\tilde{f}(\xi) = \int_{\mathbb{R}^3} f(x)\overline{u(x,\xi)}dx, \quad (f,u_j) = \int_{\mathbb{R}^3} f(x)\overline{u_j(x)}dx. \tag{8.2.26}$$

A proof of this theorem one can find, for example, in [365, Chapter 6], see also this book, Chapter 1, Section 1.3.

If E_λ is the resolution of the identity of the operator L, then the kernel $E_\lambda(x,y)$ satisfies the relation:

$$\begin{aligned}\frac{dE_\lambda(x,y)}{d\lambda} &= \frac{1}{\pi}\operatorname{Im} G(x,y,\sqrt{\lambda}) = \frac{\sqrt{\lambda}}{16\pi^3}\\ &\quad\times \int_{S^2} u(x,\alpha,\sqrt{\lambda})\overline{u(y,\alpha,\sqrt{\lambda})}d\alpha, \ \lambda > 0.\end{aligned} \tag{8.2.27}$$

The scattering solution $u(x,\alpha,k)$ satisfies the equation

$$u(x,\alpha,k) = e^{ik\alpha\cdot x} - \int_{\mathbb{R}^3} g(x,y,k)q(y)u(y,\alpha,k)dy, \ g = \frac{e^{ik|x-y|}}{4\pi|x-y|}. \tag{8.2.28}$$

The function $\phi(x,\alpha,k) := e^{-ik\alpha\cdot x}u(x,\alpha,k)$ solves the equation:

$$\phi(x,\alpha,k) = 1 - \int_{\mathbb{R}^3} \frac{e^{ik[|x-y|-\alpha\cdot(x-y)]}}{4\pi|x-y|}q(y)\phi(y,\alpha,k)dy. \tag{8.2.29}$$

Theorem 8.2.4. *Assume that $q \in Q_a^1$, that is $q \in Q_a$ and $\|\nabla q\|_{C(B_a)} \le c$. Then*

$$\phi(x, \alpha, k) = 1 + \frac{1}{2ik}\int_0^\infty q(x - r\alpha)dr + o\left(\frac{1}{k}\right), \quad k \to \infty. \tag{8.2.30}$$

First, let us prove a lemma.

Lemma 8.2.5. *Denote*

$$T(k)f := \int_{\mathbb{R}^3} \frac{e^{ik[|x-y|-\alpha\cdot(x-y)]}}{4\pi|x-y|} q(y)f(y)dy. \tag{8.2.31}$$

If $q \in Q_1$, then

$$\sup_{\alpha\in S^2} \|T^2(k)\| \to 0 \quad \textit{as } k \to \infty. \tag{8.2.32}$$

Proof. The kernel of $T^2(k)$ is

$$B(x, y, k) := \int_{\mathbb{R}^3} \frac{e^{ik[|x-z|+|z-y|]}}{16\pi^2|x-z||z-y|} q(z)dzq(y)e^{ik\alpha\cdot(x-y)}. \tag{8.2.33}$$

Use the coordinates

$$z_1 = lst + \frac{x_1 + y_1}{2}, \qquad z_2 = l\sqrt{(s^2-1)(1-t^2)}\cos\psi + \frac{x_2 + y_2}{2},$$
$$z_3 = l\sqrt{(s^2-1)(1-t^2)}\sin\psi + \frac{x_3 + y_3}{2}, \tag{8.2.34}$$

where

$$l = \frac{|x-y|}{2}, \ |x-z| + |z-y| = 2ls,$$
$$|x-z| - |z-y| = 2lt, \ J = l^3(s^2 - t^2), \tag{8.2.35}$$

and J is the Jacobian of the transform

$$(z_1, z_2, z_3) \to (s, t, \psi), \quad 1 \le s < \infty, \ -1 \le t \le 1, \ 0 \le \psi \le 2\pi. \tag{8.2.36}$$

In these coordinates, one gets

$$|B(x,y,k)| \leq \frac{1}{16\pi^2}|q(y)|l\left|\int_1^\infty e^{2ikls}p(s)ds\right|, \tag{8.2.37}$$

where

$$p(s) := \int_0^{2\pi} d\psi \int_{-1}^{1} dt q_1(s,t,\psi), \tag{8.2.38}$$

and $q_1(s,t,\psi) = q(z_1,z_2,z_3)$. Since $p \in L^2(1,\infty)$, the Riemann–Lebesgue's lemma yields the desired conclusion (8.2.32). Since we assumed $q(z)$ compactly supported and $|\nabla q| \leq c$, the function $p(s)$ is compactly supported and $|B(x,y,k)| = O\left(\frac{1}{k}\right)$ as $k \to \infty$.

Lemma 8.2.5 is proved. □

Proof of Theorem 8.2.4. By Lemma 8.2.5, equation (8.2.29) can be uniquely solved by iterations in the space $C(\mathbb{R}^3)$ if k is sufficiently large, and

$$\phi(x,\alpha,k) = 1 - \int_{\mathbb{R}^3} \frac{e^{ik[|x-y|-\alpha\cdot(x-y)]}}{4\pi|x-y|}q(y)dy + O(\|T^2(k)\|). \tag{8.2.39}$$

Let $y = x + r\beta, \beta \in S^2$. Then

$$\int_{\mathbb{R}^3} \frac{e^{ik[|x-y|-\alpha\cdot(x-y)]}}{4\pi|x-y|}q(y)dy = \int_0^\infty dr r^2 \frac{e^{ikr}}{4\pi r}\int_{S^2} e^{ikr\beta\cdot\alpha}q(x+r\beta)d\beta. \tag{8.2.40}$$

Let us use now the formula

$$\int_{S^2} e^{ikr\beta\cdot\alpha}f(\beta)d\beta = 2\pi i\left[\frac{e^{-ikr}}{kr}f(-\alpha) - \frac{e^{ikr}}{kr}f(\alpha)\right] + o\left(\frac{1}{k}\right),\ k \to \infty, \tag{8.2.41}$$

where $f \in C(S^2)$. From (8.2.39) and (8.2.41), one obtains, as $k \to \infty$, formula (8.2.30).

Theorem 8.2.4 is proved. □

Remark 8.2.1. It follows from (8.2.30) that

$$q(x) = \alpha \cdot \nabla_x \lim_{k\to\infty} \{2ik[\phi(x, \alpha, k) - 1]\} \tag{8.2.42}$$

and

$$u(x, \alpha, k) = e^{ik\alpha\cdot x}\left[1 + \frac{1}{2ik}\int_0^\infty q(x - r\alpha)dr\right] + o\left(\frac{1}{k}\right), \quad k \to \infty. \tag{8.2.43}$$

Another formula is:

$$A(\beta, \alpha, k) = -\frac{1}{4\pi}\int_{B_a} e^{ik(\alpha-\beta)\cdot x} q(x)dx + O\left(\frac{1}{k}\right), \quad k \to \infty. \tag{8.2.44}$$

Taking $\alpha = \beta$ in (8.2.44), one gets

$$A(\alpha, \alpha, k) = -\frac{1}{4\pi}\int_{B_a} q(x)dx + O\left(\frac{1}{k}\right), \quad k \to \infty. \tag{8.2.45}$$

8.3. Completeness Properties of the Scattering Solutions

Theorem 8.3.1. *Let $h \in L^2(S^2)$ be arbitrary. If $k > 0$ is fixed and*

$$\int_{S^2} h(\alpha)u(x, \alpha, k)d\alpha = 0, \quad \forall x \in B'_R := \{x : |x| > R\}, \tag{8.3.1}$$

then $h = 0$.

Proof. Since equation (8.2.7) holds in $\mathbb{R}^3$, the unique continuation principle for solutions to equation (8.2.7) implies that (8.3.1) holds everywhere, that is, $\forall x \in \mathbb{R}^3$. Since $u = (I + T_1)e^{ik\alpha\cdot x}$, where $I + T_1 := (I + T)^{-1}$, see equation (8.2.28) where $Tu := \int_{\mathbb{R}^3} g(x, y, k)q(y)u(y)dy$ and $I + T$ is boundedly invertible in $C(\mathbb{R}^3)$, one may apply the operator $I + T$ to (8.3.1) and get an equivalent equation

$$\int_{S^2} h(\alpha)e^{ik\alpha\cdot x}d\alpha = 0, \quad \forall x \in \mathbb{R}^3. \tag{8.3.2}$$

This equation shows that the Fourier transform of the distribution $\frac{h(\alpha)\delta(\lambda-k)}{\lambda^2}$ equals to zero. Thus, $\frac{h(\alpha)\delta(\lambda-k)}{\lambda^2} = 0$. Consequently, $h(\alpha)=0$.

Theorem 8.3.1 is proved. □

Let $L = -\nabla^2 + q(x) - k^2, D \subset \mathbb{R}^3$ is a bounded domain, $k > 0$ is a fixed number,

$$N_D(L) := \{w : Lw = 0 \text{ in } D, w \in H^2(D)\}. \tag{8.3.3}$$

Theorem 8.3.2. *The set $\{u(x,\alpha,k)\}_{\forall\alpha\in S^2}$ is total in $N_D(L)$. In other words, for any $\epsilon > 0$, however small, and for any $w \in N_D(L)$, there exists a linear combination of $u(x,\alpha,k)$ such that*

$$\left\| w - \int_{S^2} u(x,\alpha,k)\nu(\alpha)d\alpha \right\|_{L^2(D)} < \epsilon, \tag{8.3.4}$$

where $\nu(\alpha) \in L^2(S^2)$ plays the role of the coefficients of the linear combination.

Proof. Assume that (8.3.4) is false. Then there is an element $w \in N_D(L), w \neq 0$, orthogonal to every $u(x,\alpha,k)$:

$$\int_D w(x)u(x,\alpha,k)dx = 0, \quad \forall\alpha \in S^2,\ w \in N_D(L). \tag{8.3.5}$$

Using the lifting Lemma 2.1.9, one concludes that relation (8.3.5) implies

$$\eta(y) := \int_D w(x)G(x,y,k)dx = 0, \quad \forall y : |y| > R, \tag{8.3.6}$$

where R is a large number so that $D \subset B_R$.

Applying the operator L to (8.3.6), one gets

$$L\eta = w, \quad \eta = 0 \quad \text{in } D'. \tag{8.3.7}$$

Multiply (8.3.7) by $\bar{w}$ and integrate over D and then by parts to obtain

$$\int_D |w|^2 dx = \int_D L\eta\bar{w}dx = \int_D \eta\overline{Lw}dx - \int_{\partial D}(\eta_N\bar{w} - \eta\bar{w}_N)ds = 0, \tag{8.3.8}$$

where we have used the assumption $Lw = 0$ in D and the conditions $\eta = \eta_N = 0$ on ∂D, which follow from the relations $\eta \in H^2_{\text{loc}}(\mathbb{R}^3)$ and $\eta = 0$ in $D' = \mathbb{R}^3 \setminus D$. From (8.3.8), it follows that $w = 0$. This contradicts our assumption $w \neq 0$.

Theorem 8.3.2 is proved. □

8.4. Property C as the Tool for Proving Uniqueness Theorems in Inverse Scattering

If the scattering data, corresponding to two potentials $q_j \in Q$, $j = 1, 2$, are the same, then the global perturbation formula (8.2.1) yields

$$\int_{B_a} [q_1(x) - q_2(x)] u_1(x, \alpha, k) u_2(x, -\beta, k) dx = 0 \tag{8.4.1}$$

for all α, β, k for which the scattering data are the same. For example, If these data are the same for all $\alpha, \beta \in S^2$ and a fixed $k > 0$, then (8.4.1) holds for all $\alpha, \beta \in S^2$ and a fixed $k = k_0 > 0$.

If the set $\{u_1(x, \alpha, k_0) u_2(x, \beta, k_0)\}_{\forall \alpha, \beta \in S^2}$ is total in $L^2(B_a)$, then (8.4.1) implies $q_1 = q_2$, that is, a uniqueness theorem for inverse scattering problem with fixed-energy data.

Let us introduce *Property C* for a pair of linear differential operators $\{L_1, L_2\}$, where

$$L_j = \sum_{|m|=0}^{M_j} a_{mj}(x) \partial^m, \quad \partial^m := \frac{\partial^{|m|}}{\partial x_1^{m_1} \dots \partial x_3^{m_3}}, \ |m| = m_1 + m_2 + m_3. \tag{8.4.2}$$

Define the null-space of L_j in a bounded domain D:

$$N_j := N_D(L_j) = \{u : L_j u = 0 \text{ in } D\}. \tag{8.4.3}$$

Definition 8.4.1. The pair $\{L_1, L_2\}$ has Property C if and only if the set $\{u_1 u_2\}_{\forall u_j \in N_j}$ is total in $L^2(D)$.

If $a_{mj}(x) = a_{mj} = \text{const}$, then the author found a necessary and sufficient condition for a pair $\{L_1, L_2\}$ to have Property C.

Property C stands for completeness of the set of products of the elements of the null-spaces of the operators $L_j, j = 1, 2$.

To formulate the result, let us introduce the algebraic varieties:

$$V_j := \left\{ z : z \in \mathbb{C}^3, \sum_{|m|=0}^{M_j} a_{mj} z^m = 0 \right\}. \tag{8.4.4}$$

Note that $e^{x \cdot z} \in N_j$ if and only if $z \in V_j, j = 1, 2$.

Definition 8.4.2. We say that V_1 is transversal to V_2 and write $V_1 \nparallel V_2$ if and only if there exist at least one point $z^{(j)} \in V_j, j = 1, 2$, such that the tangent spaces $T_j, j = 1, 2$, at the points $z^{(j)}$ to the varieties V_j are transversal, that is, the hyperplanes T_j are not parallel.

Theorem 8.4.1. *The pair of partial differential operators with constant coefficients $\{L_1, L_2\}$ has Property C if and only if $V_1 \nparallel V_2$.*

Proof. *Sufficiency.* Assume that

$$\int_D f(x) u_1(x) u_2(x) dx = 0 \quad \forall u_j \in N_j, \quad f \in L^2(D). \tag{8.4.5}$$

We want to derive from (8.4.5) that $f = 0$ provided that $V_1 \nparallel V_2$. Take $u_j = e^{x \cdot z_j}, z_j \in V_j$. If $V_1 \nparallel V_2$, then the set $\{z_1 + z_2\}_{\forall z_j \in V_j}$ covers an open subset of $\mathbb{C}^3$ and (8.4.5) in this case says that the Fourier transform $\tilde{f}$ of a compactly supported f vanishes on an open set in $\mathbb{R}^3$. Since $\tilde{f}$ is an entire function, it must vanish identically. Therefore, $f = 0$ by the injectivity of the Fourier transform.

Necessity. Assume that (8.4.5) implies $f = 0$ for any $f \in L^2(D)$. Let us prove that $V_1 \nparallel V_2$. Suppose the contrary: $V_1 \parallel V_2$. Then the set $\{z_1 + z_2\}_{\forall z_j \in V_j}$ is a plane in $\mathbb{C}^3$. This plane is a two-dimensional subspace in $\mathbb{C}^3$. Therefore, one can find $f \neq 0$ which is orthogonal to every vector in the above plane. This contradicts our assumption that (8.4.5) implies $f = 0$.

Theorem 8.4.1 is proved. □

Definition 8.4.3. *We say that a linear partial differential operator L has Property C if and only if the pair $\{L, L\}$ has this property.*

Example 8.4.1. (a) If $L = \nabla^2$, then $V = \{z : z \in \mathbb{C}^3, z \cdot z = 0\}$. Let $z^{(1)} = (1, 0, i), z^{(2)} = (i, 0, 1)$. Then $T^{(1)}$ is $z_1 - 1 + i(z_3 - i) = 0$, $T^{(2)}$ is $i(z_1 - i) + (z_3 - 1) = 0$. Clearly, $T^{(1)}$ is not parallel to $T^{(2)}$. Therefore, $L = \nabla^2$ has Property C.

(b) If $L = \nabla^2 + k^2$, then $V = \{z : z \in \mathbb{C}^3, z \cdot z + k^2 = 0\}$. One checks, as before, that L has Property C.

(c) If $L = \partial_t - \nabla^2$, then $V = \{z : z \in \mathbb{C}^4, z_4 - z_1^2 - z_2^2 - z_3^2 = 0\}$. This L has Property C.

(d) If $L = \partial_{tt}^2 - \nabla^2$, then $V = \{z : z \in \mathbb{C}^4, z_4^2 - z_1^2 - z_2^2 - z_3^2 = 0\}$. This L has Property C.

(e) If $L = i\partial_t - \nabla^2$, then $V = \{z : z \in \mathbb{C}^4, iz_4 - z_1^2 - z_2^2 - z_3^2 = 0\}$. This L has Property C.

Therefore, Property C is a generic property; classical partial differential operators (Laplace, Helmholtz, heat, wave, Schrödinger) have this property, see also [274, 286, 298, 304, 306, 310, 317, 356].

The above ideas and results can be used for proving that the operator $L = \nabla^2 + k^2 - q(x), q \in Q_a$, does have Property C. The proof is based on the following result.

Theorem 8.4.2. *There exists a (special) solution ψ to the equation*

$$L\psi := (\nabla^2 + k^2 - q(x))\psi = 0 \quad in\ \mathbb{R}^3, \tag{8.4.6}$$

such that

$$\psi := \psi(x, \theta) = e^{ik\theta \cdot x}(1 + R(x, \theta)), \quad \theta \in \mathcal{M}, \tag{8.4.7}$$

where

$$\sup_{x \in B_b} |R(x, \theta)| \leq \epsilon(\theta), \quad \epsilon(\theta) \leq c\sqrt{\frac{\ln|\theta|}{|\theta|}}, \quad \theta \to \infty, \tag{8.4.8}$$

$b > 0$ is an arbitrary fixed number, $B_b := \{x : |x| \leq b\}$.

A proof of this theorem is given in the next section. Assuming that the conclusions (8.4.7)–(8.4.8) of this theorem hold, it is easy to prove that a pair $\{L_1, L_2\}$ has Property C.

Theorem 8.4.3. *If $q \in Q_a$, then the pair $\{L_1, L_2\}$ has Property C.*

Proof. One has $\psi_j \in N_j, \psi_1\psi_2 = e^{ik(\theta_1+\theta_2)\cdot x}(1+R_1+R_2+R_1R_2)$. It follows from Lemma 8.2.3 that for any $\xi \in \mathbb{R}^3$ there exist $\theta_1, \theta_2 \in \mathcal{M}$ such that $\theta_1+\theta_2 = \xi$ and $|\theta_1| \to \infty$. Therefore, using (8.4.8), one gets

$$\psi_1\psi_2 = e^{i\xi\cdot x}(1+\epsilon(\theta)), \qquad \lim_{|\theta|\to\infty} \epsilon(\theta) = 0. \tag{8.4.9}$$

Since the set $\{e^{i\xi\cdot x}\}_{\forall\xi\in\mathbb{R}^3}$ is total in $L^2(B_b)$, so is the set $\{\psi_1\psi_2\}_{\forall\psi_j\in N_j}$.

Theorem 8.4.3 is proved. □

As we have explained already below formula (8.4.9), Property C for the pair $\{L_1, L_2\}$ and the global perturbation formula (8.2.1) yield the uniqueness theorem for inverse scattering problem.

Theorem 8.4.4. *Assume that $q_j \in Q_a$ and that $A_1(\beta, \alpha, k_0) = A_2(\beta, \alpha, k_0)$, $\forall\alpha, \beta \in S^2$. Then $q_1 = q_2$.*

Remark 8.4.1. Since $A_j(\beta, \alpha, k_0)$ is analytic function of β and α on the variety $\mathcal{M}$, a knowledge of $A_j(\beta, \alpha, k_0)$ on an open subset $S_1^2 \times S_2^2$ of $S^2 \times S^2$ allows one to uniquely recover $A_j(\beta, \alpha, k_0)$ on $\mathcal{M} \times \mathcal{M}$. Therefore, Theorem (8.4.8) remains valid if $A_1(\beta, \alpha, k_0) = A_2(\beta, \alpha, k_0)$ on $S_1^2 \times S_2^2$.

8.5. Existence of Special Solutions

In this section, we prove Theorem 8.4.2. If $L\psi := (\nabla^2 + k^2 - q(x))\psi = 0$ and $\psi = e^{i\theta\cdot x}(1+R)$, then

$$[\nabla^2 + 2ik\theta\cdot\nabla - q(x)]R = q(x). \tag{8.5.1}$$

Let $LR := \nabla^2 + 2ik\theta\cdot\nabla$. Then

$$LR - q(x)R = q(x). \tag{8.5.2}$$

For simplicity of writing, let us take (without loss of generality) $k = 1$ from now on. Then the equation

$$Lw := (\nabla^2 + 2i\theta\cdot\nabla)w = f \tag{8.5.3}$$

after taking the Fourier transform

$$w(x) = \frac{1}{(2\pi)^3}\int_{\mathbb{R}^3} e^{i\xi\cdot x}\tilde{w}(\xi)d\xi, \quad \tilde{w}(\xi) := \int_{\mathbb{R}^3} e^{-i\xi\cdot x}w(x)dx, \tag{8.5.4}$$

takes the form

$$-(\xi^2 + 2\theta\cdot\xi)\tilde{w} = \tilde{f}; \quad w(x) = L^{-1}f := -\frac{1}{(2\pi)^3}\int_{\mathbb{R}^3}\frac{e^{i\xi\cdot x}\tilde{f}(\xi)}{\xi^2 + 2\theta\cdot\xi}d\xi. \tag{8.5.5}$$

From (8.5.2), one obtains

$$R = L^{-1}qR + L^{-1}q. \tag{8.5.6}$$

We prove the following theorem:

Theorem 8.5.1. *One has*

$$\|L^{-1}f\|_{L^\infty(D_1)} \le c\sqrt{\frac{\ln|\theta|}{|\theta|}}\|f\|_{L^2(D)}, \quad \theta\in\mathcal{M},\ |\theta|\to\infty, \tag{8.5.7}$$

where D_1 and D are bounded domains and $c > 0$ depends on D_1 and D but does not depend on $\theta, |\theta| \gg 1$.

This theorem is proved later. Assuming its validity, let us prove the existence of R and, therefore, the existence of the special solution.

Theorem 8.5.2. *If $\theta\in\mathcal{M}$ and $|\theta|$ is sufficiently large, then there exists a unique solution R to equation (8.5.2) in the space $L^\infty(D_1)$, where D_1 is an arbitrary fixed domain, $D_1 \supset B_1$, and $B_a = \mathrm{supp}\, q$.*

Proof. It is sufficient to prove existence and uniqueness in the space $L^\infty(D_1)$ of the solution to equation (8.5.6), because a solution to this equation solves equation (8.5.2): just apply the operator L to the solution of (8.5.6). Let us prove, using estimate (8.5.7), that the operator $L^{-1}q$ has small norm in $L^\infty(D_1)$ as $|\theta|\to\infty$. Indeed,

$$\|L^{-1}qR\|_{L^\infty(D_1)} \le c\epsilon(|\theta|)\|qR\|_{L^2(B_a)} \le c\epsilon(|\theta|)\|q\|_{L^2(B_1)}\|R\|_{L^\infty(D_1)}, \tag{8.5.8a}$$

where

$$\epsilon := \epsilon(|\theta|) = \sqrt{\frac{\ln|\theta|}{|\theta|}}. \tag{8.5.8b}$$

If $c\epsilon(|\theta|)\|q\|_{L^2(B_a)} < 1$, then $\|L^{-1}q\|$ in $L^\infty(D_1)$ is less than 1. Therefore, equation (8.5.6) is uniquely solvable in $L^\infty(D_1)$. Consequently, the special solution does exist and is unique in $L^\infty(D_1)$ in any bounded domain D_1 if $\theta \in \mathcal{M}$ and $|\theta|$ is sufficiently large. One has

$$\|R\|_{L^\infty(D_1)} \leq c\epsilon(|\theta|). \tag{8.5.8$'$}$$

Theorem 8.5.2 is proved. □

Let us now prove Theorem 8.5.1. This proof is relatively long.

Proof of Theorem 8.5.1. If $\theta \in \mathcal{M}$, then $\theta = A + iB, A, B \in \mathbb{R}^3, A \cdot B = 0, A^2 - B^2 = 1$. Choose the coordinate system such that $A = \tau e_2, B = te_1, \tau = (1+t^2)^{1/2}, e_j, 1 \leq j \leq 3$, are the orthonormal basis vectors. Then

$$\begin{aligned} \xi^2 + 2\theta \cdot \xi &= \xi_1^2 + \xi_2^2 + \xi_3^2 + 2\tau\xi_2 + 2it\xi_1 \\ &= \xi_1^2 + (\xi_2 + \tau)^2 + \xi_3^2 - \tau^2 + 2it\xi_1. \end{aligned} \tag{8.5.9}$$

This function vanishes if and only if

$$\xi_1 = 0, \quad (\xi_2 + \tau)^2 + \xi_3^2 = \tau^2. \tag{8.5.10}$$

Equation (8.5.10) defines a circle C_τ of radius τ in the plane $\xi_1 = 0$ centered at $(0, -\tau, 0)$. Let T_δ be a toroidal neighborhood of C_τ, where the section of the torus by a plane orthogonal to C_τ is a square with size 2δ and the center at C_τ.

Denote $u(x) := L^{-1}f$, where $L^{-1}f$ is defined in (8.5.5). One has

$$\begin{aligned} |u(x)| \leq{}& \frac{1}{(2\pi)^3}\left|\int_{T_\delta} \frac{\tilde{f}(\xi)e^{i\xi\cdot x}d\xi}{\xi^2 + 2\xi\cdot\theta}\right| \\ &+ \frac{1}{(2\pi)^3}\left|\int_{\mathbb{R}^3\setminus T_\delta} \frac{\tilde{f}(\xi)e^{i\xi\cdot x}d\xi}{\xi^2 + 2\xi\cdot\theta}\right| := I_1 + I_2, \end{aligned} \tag{8.5.11}$$

$$\begin{aligned}
I_1 &\le c\|\tilde{f}\|_{L^\infty(\mathbb{R}^3)} \int_{T_\delta} \frac{d\xi}{|\xi^2 + 2\xi\cdot\theta|} \\
&\le c\|f\|_{L^1(\mathbb{R}^3)} \int_{-\delta}^{\delta} d\xi_1 \int_0^{2\pi} d\varphi \int_{\tau-\delta}^{\tau+\delta} \frac{\rho d\rho}{\sqrt{4t^2\xi_1^2 + (\xi_1^2+\rho^2-\tau^2)^2}} \\
&= c\|f\|_{L^1(\mathbb{R}^3)} \int_0^{\delta} d\xi_1 \int_{\xi_1^2-2\tau\delta+\delta^2}^{\xi_1^2+2\tau\delta+\delta^2} \frac{d\mu}{\sqrt{4t^2\xi_1^2+\mu^2}} \\
&\le c(D)\|f\|_{L^2(D)} \int_0^{\delta} d\xi_1 \int_0^{3\tau\delta} \frac{d\mu}{\sqrt{4t^2\xi_1^2+\mu^2}}, \quad 0 < \delta < \frac{\tau}{2},
\end{aligned} \tag{8.5.12}$$

where

$$\rho^2 = (\xi_2+\tau)^2 + \xi_3^2, \quad \mu^2 = (\xi_1^2+\rho^2-\tau^2)^2,$$

we have used the Cauchy inequality

$$\|f\|_{L^1(\mathbb{R}^3)} = \|f\|_{L^1(D)} \le c(D)\|f\|_{L^2(D)}$$

and an elementary inequality $\xi_1^2 + 2\tau\delta + \delta^2 \le 3\tau\delta$, which holds if $\xi_1^2 \le \delta^2$ and $\tau > 2\delta$.

Let $\beta := 2t\xi_1$. Then

$$\begin{aligned}
&\frac{1}{2t}\int_0^{2t\delta} d\beta \int_0^{3\tau\delta} \frac{d\mu}{\sqrt{\beta^2+\mu^2}} \\
&\quad \le \frac{1}{2t}\int_0^{3(t+\tau)\delta} d\rho\rho \int_0^{\pi/2} d\varphi \frac{1}{\rho} = \frac{\pi}{4t} 3(t+\tau)\rho \le c\delta,
\end{aligned} \tag{8.5.13}$$

where we have used the relations $\frac{\tau}{t} \to 1$ as $t \to \infty$ and took into account that $t \to \infty$ if $|\theta| \to \infty$. From (8.5.12) and (8.5.13), one gets

$$I_1 \le c\|f\|_{L^2(D)}\delta. \tag{8.5.14}$$

By $c > 0$, we denote various constants independent of δ and t.

Let us estimate I_2:

$$I_2^2 \le c\|\tilde{f}\|^2_{L^2(\mathbb{R}^3)} \int_{\mathbb{R}^3\setminus T_\delta} \frac{d\xi}{|\xi^2 + 2\theta\cdot\xi|} = c\|f\|^2_{L^2(D)}\mathcal{J}, \tag{8.5.15}$$

where the Parseval equality was used and by $\mathcal{J}$ the integral in (8.5.15) is denoted. One has

$$\mathcal{J} \leq \int_{|\xi_1|>\delta} \frac{d\xi}{|\xi^2 + 2\theta \cdot \xi|^2} + \int_{|\xi_1<\delta, |\rho-\tau|\geq\delta} \frac{d\xi}{|\xi^2 + 2\theta \cdot \xi|^2} := j_1 + j_2. \tag{8.5.16}$$

Let us estimate j_1:

$$\begin{aligned} j_1 &\leq c \int_\delta^\infty d\xi_1 \int_0^\infty \frac{\rho d\rho}{4\xi_1^2 t^2 + (\xi_1^2 + \rho^2 - \tau^2)^2} \\ &\leq c \int_\delta^\infty d\xi_1 \int_{\xi_1^2-\tau^2}^\infty \frac{d\mu}{4\xi_1^2 t^2 + \mu^2} \\ &\leq c \int_\delta^\infty \frac{d\xi_1}{\xi_1 t} \left(\frac{\pi}{2} - \arctan \frac{\xi_1^2 - \tau^2}{2\xi_1 t} \right). \end{aligned} \tag{8.5.17}$$

Let $\frac{\xi_1}{2t} = x$. Then the integral on the right-hand side of (8.5.17) can be written as:

$$j_1 \leq \frac{c}{t} \int_{\frac{\delta}{2t}}^\infty \frac{dx}{x} \left[\frac{\pi}{2} - \arctan \left(x - \frac{\tau^2}{4t^2} \frac{1}{x} \right) \right]. \tag{8.5.18}$$

If $t \to \infty$, then $\frac{\tau^2}{4t^2} \to \frac{1}{4}$. Let us use the elementary inequalities:

$$\frac{\pi}{2} - x \leq \arctan \frac{1}{x}, \quad 0 < x \leq \frac{\pi}{2}; \tag{8.5.19}$$

$$\arctan \frac{1}{x} \leq \frac{\pi}{2} - \frac{x}{2}, \quad x \to +0. \tag{8.5.20}$$

Then

$$\frac{\pi}{2} - \frac{1}{y} \leq \arctan y \leq \frac{\pi}{2} - \frac{1}{2y}, \quad y \to +\infty. \tag{8.5.21}$$

Thus, with $A' := \frac{\tau^2}{4t^2}$, one gets

$$\frac{1}{x} \left[\frac{\pi}{2} - \arctan \left(x - \frac{A'}{x} \right) \right] \leq \frac{1}{x} \left(x - \frac{A'}{x} \right)^{-1} \leq \frac{c}{x^2}, \quad x \to +\infty, \tag{8.5.22}$$

and

$$\frac{1}{x}\left[\frac{\pi}{2} - \arctan\left(x - \frac{A'}{x}\right)\right] \le \frac{c}{x}, \quad x \to +0. \tag{8.5.23}$$

From (8.5.19)–(8.5.23), and (8.5.18) one gets

$$\begin{aligned} j_1 &\le \frac{c}{t}\left(\int_{\frac{\delta}{2t}}^{1} \frac{dx}{x} + c\right) \le \frac{c}{t}\left|\ln\frac{\delta}{2t}\right|, \quad t \to +\infty, \\ &\le c\frac{|\ln\frac{\delta}{2t}|}{t}, \quad t \to +\infty. \end{aligned} \tag{8.5.24}$$

Let us estimate j_2:

$$\begin{aligned} j_2 &= \int_{-\delta}^{\delta} d\xi_1 \int_{\rho>\tau+\delta,\, 0<\rho\le\tau-\delta} d\rho\rho \int_0^{2\pi} d\varphi \frac{1}{4\xi_1^2 t^2 + (\xi_1^2+\rho^2-\tau^2)^2} \\ &\le c\int_0^{\delta} d\xi_1 \left\{\int_{\xi_1^2+(\tau+\delta)^2-\tau^2}^{\infty} \frac{d\mu}{4\xi_1^2t^2+\mu^2} + \int_{2\tau\delta-\delta^2-\xi_1^2}^{\tau^2-\xi_1^2} \frac{d\nu}{4\xi_1^2t^2+\nu^2}\right\}, \end{aligned} \tag{8.5.25}$$

where $\mu = \xi_1^2 + \rho^2 - \tau^2$ and $\nu = \tau^2 - \rho^2 - \xi_1^2$.
One has:

$$\begin{aligned} j_2 &\le \frac{c}{t}\int_0^{\delta} \frac{d\xi_1}{\xi_1}\left(\frac{\pi}{2} - \arctan\frac{\xi_1^2+2\delta\tau+\delta^2}{2\xi_1 t} + \arctan\frac{\tau^2-\xi_1^2}{2\xi_1 t}\right. \\ &\qquad \left. - \arctan\frac{2\tau\delta-\delta^2-\xi_1^2}{2\xi_1 t}\right) \\ &\le \frac{c}{t}\int_0^{\delta} \frac{d\xi_1}{\xi_1}\left(\frac{\pi}{2} - \arctan\frac{\delta}{\xi_1} + \arctan\frac{t}{\xi_1} - \arctan\frac{\delta}{2\xi_1}\right), \end{aligned} \tag{8.5.26}$$

where we have used the monotonicity of $\arctan x$, for example,

$$\arctan\frac{\xi_1^2+2\delta\tau+\delta^2}{2\xi_1 t} \ge \arctan\frac{\delta}{\xi_1}, \tag{8.5.27}$$

etc., and the relation, $\frac{\tau}{t} \to 1$ as $t \to +\infty, \tau > t$.

By (8.5.19)–(8.5.21), one gets:

$$\frac{\pi}{2} - \arctan\frac{\delta}{\xi_1} \leq \frac{\xi_1}{\delta}, \quad \arctan\frac{t}{\xi_1} \leq \frac{\pi}{2} - \frac{\xi_1}{2t},$$

$$\arctan\frac{\delta}{2\xi_1} \geq \frac{\pi}{2} - \frac{2\xi_1}{\delta}. \tag{8.5.28}$$

From (8.5.26) and (8.5.28), one derives:

$$\begin{aligned} j_2 &\leq \frac{c_1}{t}\int_0^{\delta}\frac{d\xi_1}{\xi_1}\left(\frac{\xi_1}{\delta} - \frac{\xi_1}{2t} + \frac{2\xi_1}{\delta}\right) \\ &= \frac{c_1}{t}\left(1 - \frac{\delta}{2t} + \frac{2\delta}{\delta}\right) = \frac{c_1}{t}\left(3 - \frac{\delta}{2t}\right) \leq \frac{c_1}{t}. \end{aligned} \tag{8.5.29}$$

Remember that $c > 0$ stands for various constants independent of δ and θ. Thus,

$$j_2 \leq \frac{c}{t}, \quad t \to +\infty. \tag{8.5.30}$$

From (8.5.11), (8.5.14), (8.5.24), and (8.5.30), one gets:

$$|u(x)| \leq c\|f\|_{L^2(D)}\left[\delta + \left(\frac{|\ln\frac{\delta}{2t}|}{t}\right)^{1/2} + \frac{1}{t^{1/2}}\right]. \tag{8.5.31}$$

Choose $\delta = \frac{1}{t}$. Then (8.5.31) yields

$$|u(x)| \leq c\|f\|_{L^2(D)}\left(\frac{\ln|\theta|}{|\theta|}\right)^{1/2}, \quad |\theta| \to \infty, \quad \theta \in \mathcal{M}. \tag{8.5.32}$$

Estimate (8.5.7) is proved.

Theorem 8.5.1 is proved. □

8.6. Reconstruction Formula for the Potential

We have proved (Theorem 8.1.2) uniqueness of the solution to inverse scattering problem with fixed energy ($k = k_0$) data. Now a reconstruction formula for the potential will be derived.

First, let us explain the idea. We start with the formula for the scattering amplitude

$$-4\pi A(\beta,\alpha,k)=\int_{B_a}e^{-ik\beta\cdot x}q(x)u(x,\alpha,k)dx, \tag{8.6.1}$$

where $k=k_0>0$ is fixed. The integral in (8.6.1) depends on β analytically on the variety

$$\mathcal{M}:=\{\theta:\theta\in\mathbb{C}^3,\theta\cdot\theta=1\},\quad \text{where } \theta\cdot\theta:=\sum_{j=1}^{3}\theta_j^2.$$

Therefore, one can write

$$-4\pi A(\theta',\alpha,k)=\int_{B_a}e^{-ik\theta'\cdot x}q(x)u(x,\alpha,k)dx,\quad \theta'\in\mathcal{M}. \tag{8.6.2}$$

Multiply (8.6.2) by a $\nu(\alpha)\in L^2(S^2)$ and integrate over S^2:

$$-4\pi\int_{S^2}A(\theta',\alpha,k)\nu(\alpha)d\alpha=\int_{B_a}e^{-ik\theta'\cdot x}\int_{S^2}u(x,\alpha,k)\nu(\alpha)d\alpha q(x)dx. \tag{8.6.3}$$

Let us drop in what follows the k-dependence since k is fixed and introduce

$$\rho:=\rho(x,\theta):=e^{-ik\theta\cdot x}\int_{S^2}u(x,\alpha,k)\nu(\alpha)d\alpha-1. \tag{8.6.4}$$

Then (8.6.3) can be written as

$$-4\pi\int_{S^2}A(\theta',\alpha)\nu(\alpha)d\alpha=\int_{B_a}e^{-i\xi\cdot x}(\rho+1)q(x)dx, \tag{8.6.5}$$

where

$$\xi=\theta'-\theta,\quad \theta',\theta\in\mathcal{M},\quad k=1. \tag{8.6.6}$$

We have already proved (Lemma 8.2.3) that for any $\xi\in\mathbb{R}^3$ there exist $\theta',\theta\in\mathcal{M}$ such that (8.6.6) holds and $|\theta|\to\infty$. It follows from (8.6.4) that

$$\int_{S^2}u(x,\alpha)\nu(x)d\alpha=e^{i\theta\cdot x}(1+\rho). \tag{8.6.7}$$

Since $u(x, \alpha)$ for $k = 1$ satisfies the equation

$$Lu := (\nabla^2 + 1 - q(x))u = 0 \quad \text{in } \mathbb{R}^3, \quad k = 1, \tag{8.6.8}$$

a simple calculation yields

$$(L - q)\rho := (\nabla^2 + 2i\theta \cdot \nabla)\rho - q\rho = q, \quad \theta \in \mathcal{M}, \tag{8.6.9}$$

so, for sufficiently large $|\theta|$, one has

$$\rho = L^{-1}q\rho + L^{-1}\rho, \tag{8.6.10}$$

and by Theorem 8.5.2 one concludes that $\rho \in L^\infty(D_1)$ is uniquely defined and satisfies the estimate

$$\|\rho\|_{L^\infty(D_1)} \leq c\epsilon(|\theta|), \quad \epsilon(|\theta|) = \sqrt{\frac{\ln|\theta|}{|\theta|}}, \quad |\theta| \gg 1. \tag{8.6.11}$$

If $w := \rho - R$, where R is defined in Theorem 8.4.2, then

$$Lw - qw = 0. \tag{8.6.12}$$

Therefore, if $|\theta|$ is sufficiently large so that $\|L^{-1}q\|_{L^\infty(D_1)\to L^\infty(D_1)} < 1$, then $w = 0$ and $\rho = R$. It follows from equation (8.6.5) that

$$\tilde{q}(\xi) = -4\pi \int_{B_a} A(\theta', \alpha)\nu(\alpha, \theta)d\alpha + O\left(\|q\|_{L^2(B_a)}\|\rho\|_{L^2(B_a)}\right). \tag{8.6.13}$$

From (8.6.13) and (8.6.11), it follows that

$$\tilde{q}(\xi) = \lim_{\theta'-\theta=\xi,\, \theta',\theta\in\mathcal{M},\, |\theta|\to\infty} -4\pi \int_{B_a} A(\theta', \alpha)\nu(\alpha)d\alpha, \tag{8.6.14}$$

where $\nu(\alpha) = \nu(\alpha, \theta)$.

Let us formulate the result.

Theorem 8.6.1. *There exists $\nu(\alpha) = \nu(\alpha, \theta)$ such that formula (8.6.14) holds.*

The important question is as follows:

How can one calculate $\nu(\alpha) = \nu(\alpha, \theta)$?

To answer this question, we need some properties of the Fourier coefficients of the scattering amplitude

$$A(\beta, \alpha) = \sum_{l=0}^{\infty} A_l(\alpha) Y_l(\beta), \quad A_l(\alpha) := \int_{S^2} A(\beta, \alpha) \overline{Y_l(\beta)} d\beta. \tag{8.6.15}$$

Here and below, $\sum_{l=0}^{\infty} := \sum_{l=0}^{\infty} \sum_{m=-l}^{l}$,

$$Y_l(\alpha) := Y_{lm}(\alpha) := \frac{(-1)^m i^l}{\sqrt{4\pi}} \left[\frac{(2l+1)(l-m)!}{(l+m)!} \right]^{1/2} e^{im\varphi} P_{l,m}(\cos\vartheta), \tag{8.6.16}$$

$$\overline{Y_{l,m}(\alpha)} = (-1)^{l+m} Y_{l,-m}(\alpha), \quad Y_{l,m}(-\alpha) = (-1)^l Y_{l,m}(\alpha). \tag{8.6.17}$$

Here, $P_{l,m}(\cos\vartheta) = (\sin\vartheta)^m \frac{d^m P_l(\cos\vartheta)}{(d\cos\vartheta)^m}, 0 \leq m \leq l$, $P_l(x)$ is the Legendre polynomial, $0 \leq \vartheta \leq \pi, 0 \leq \varphi \leq 2\pi$, are spherical coordinates on S^2, $P_{l,-m}(\cos\vartheta) = (-1)^m \frac{(l-m)!}{(l+m)!} P_{l,m}(\cos\vartheta), 0 \leq m \leq l$.

The set $\{Y_l(\alpha)\}_{l=0}^{\infty}$ forms an orthonormal basis of $L^2(S^2)$. The phase in the definition (8.6.16) is chosen so that formulas (8.6.17) hold.

We use the spherical Bessel functions

$$j_l(r) := \sqrt{\frac{\pi}{2r}} J_{l+\frac{1}{2}}(r) = \frac{1}{2\sqrt{2}} \frac{1}{l} \left(\frac{er}{2l} \right)^l [1 + o(1)], \quad l \to \infty, \tag{8.6.18}$$

where $J_l(r)$ is the Bessel function regular at $r = 0$,

$$h_l(r) := e^{i\frac{\pi}{2}(l+1)} \sqrt{\frac{\pi}{2r}} H^{(1)}_{l+\frac{1}{2}}(r) = \frac{e^{ir}}{r} [1 + o(1)], \quad r \to \infty, \tag{8.6.19}$$

where $H_l(r)$ is the Hankel function, and one has:

$$h_l(r) = -\frac{i^l \sqrt{2}}{r} \left(\frac{2l}{er} \right)^2 [1 + o(1)], \quad l \to \infty. \tag{8.6.20}$$

The function $|h_l(r)|^2$ is a monotonically increasing function of l and monotonically decreasing function of r if $l > 0$.

Lemma 8.6.2. *If $q \in Q_a$, then*

$$\max_{\alpha \in S^2} |A_l(\alpha)| \le c\left(\frac{ae}{2l}\right)^l, \quad l \to \infty. \tag{8.6.21}$$

Proof. We start with formula (8.6.1) with $k = 1$, multiply this formula by $\overline{Y_l(\beta)}$, integrate over S^2 and use the known formula

$$e^{i\beta\cdot x} = \sum_{l=0}^{\infty} 4\pi i^l Y_l(\beta)\overline{Y_l(x^0)}j_l(r), \quad r = |x|, \quad x^0 := \frac{x}{r}. \tag{8.6.22}$$

The result is

$$|A_l(\alpha)| \le \left|\int_{B_a} \overline{Y_l(x^0)}j_l(r)q(x)u(x,\alpha)dx\right|. \tag{8.6.23}$$

Using the Cauchy inequality, (8.6.14) and (8.6.18), one gets

$$\left|\int_{B_a} Y_l(x^0)j_l(r)q(x)u(x,\alpha)dx\right| \le c\frac{1}{l}\left(\frac{ea}{2l}\right)^l \le c\left(\frac{ea}{2l}\right)^l, \quad l \to \infty. \tag{8.6.24}$$

Lemma 8.6.2 is proved. □

Lemma 8.6.3. *One has*

$$|Y_l(\theta)| \le \frac{e^{\kappa r}}{|j_l(r)|}, \quad \forall r > 0, \quad \kappa = |\operatorname{Im}\theta|, \quad \theta \in \mathcal{M}. \tag{8.6.25}$$

Proof. Take $\beta = \theta \in \mathcal{M}$ in formula (8.6.22). This formula remains valid on $\mathcal{M}$. Multiply this formula by $\overline{Y_l(x^0)}$ and integrate over S^2 with respect to x^0. The result is

$$\int_{S^2} e^{i\theta\cdot x}\overline{Y_l(x^0)}dx^0 = 4\pi i^l j_l(r)\overline{Y_l(\theta)}. \tag{8.6.26}$$

It follows from (8.6.26) that

$$|Y_l(\theta)| \le \frac{1}{4\pi|j_l(r)|}\left(\int_{S^2} |e^{i\theta\cdot x}|^2 dx^0\right)^{1/2} \le \frac{e^{\kappa r}}{|j_l(r)|}. \tag{8.6.27}$$

Lemma 8.6.3 is proved. □

The next lemma gives a formula for the scattering solution outside the support of the potential.

Lemma 8.6.4. *In the region* $r := |x| > a$, *one has*

$$u(x,\alpha) = e^{i\alpha\cdot x} + \sum_{l=0}^{\infty} A_l(\alpha) Y_l(x^0) h_l(r), \quad r > a. \tag{8.6.28}$$

Proof. In the region $r > a$, the scattering solution satisfies the homogeneous Helmholtz equation with $k = 1$ and is of the form $e^{i\alpha\cdot x} + v$, where v satisfies the radiation condition,

$$v = \frac{e^{ir}}{r} A(x^0, \alpha) + o\left(\frac{1}{r}\right), \quad r \to \infty. \tag{8.6.29}$$

By formula (8.6.19), one conclude s that v is the sum in formula (8.6.28).

Lemma 8.6.4 is proved. $\square$

Consider the problem

$$F(\nu) := \|\rho\|^2_{L^2(B_b \setminus B_a)} = \min := d, \tag{8.6.30}$$

where $b > a$ is an arbitrary fixed number and minimization is taken over $\nu(\alpha) \in L^2(S^2)$.

Let

$$F(\nu_n) = \|\rho_n\|^2_{L^2(B_b\setminus B_a)} \xrightarrow[n\to\infty]{} d. \tag{8.6.31}$$

Since

$$L\rho_n - q\rho_n = q, \tag{8.6.32}$$

one applies elliptic inequality

$$\|\rho_n\|_{H^2(B_a)} \le c\left(\|q\|_{L^2(B_b)} + \|\rho_n\|_{L^2(B_b)}\right) \le c, \tag{8.6.33}$$

where $a < a_1 < b$, a_1 is arbitrary and $c > 0$ denotes various constants. By the embedding theorem,

$$\rho_n \xrightarrow[H^m(B_{a_1})]{} \rho_0, \quad m < 2, \quad a_1 < b. \tag{8.6.34}$$

Actually, ρ_n converges to ρ_0 in $H^2(B_{a_1})$. To check this, apply inequality (8.6.33) to $\rho_{nm} := \rho_n - \rho_m, L\rho_{nm} - q\rho_{nm} = 0$. This inequality yields

$$\|\rho_{nm}\|_{H^2(B_{a_1})} \le c\|\rho_{nm}\|_{L^2(B_b)}. \tag{8.6.35}$$

Since a_1 and b are arbitrary, it follows that $\rho_n \to \rho_0$ in $L^2(D_1)$ in an arbitrary bounded domain D_1 and inequality (8.6.35) shows that $\rho_n \to \rho_0$ in $H^2(D), D \subset D_1$.

Recall that if L is an elliptic second-order partial differential operator, then the elliptic inequality (8.6.33) that we used can be formulated as follows:

$$\|u\|_{H^2(D)} \le c(\|Lu\|_{L^1(D_1)} + \|u\|_{L^2(D_1)}), \quad D \subset D_1, \tag{8.6.36}$$

where $c = c(D, D_1)$ does not depend on u, and D is a strictly inner subdomain of a bounded domain D_1.

If $\rho_n \to \rho_0$ in $H^2(B_b)$, then one can pass to the limit $n \to \infty$ in equation (8.6.32) and get

$$L\rho_0 - q\rho_0 = q \quad \text{in } B_b, \quad \|\rho_0\|_{C(B_b)} \le c. \tag{8.6.37}$$

From (8.6.37) for $\theta \in \mathcal{M}$ sufficiently large, so that $\|L^{-1}q\|_{L^\infty(B_b)\to L^\infty(B_b)} < 1$, it follows that $\rho_0 = R$ and

$$\|\rho_0\| \le c\epsilon(|\theta|) = c\sqrt{\frac{\ln|\theta|}{|\theta|}}. \tag{8.6.38}$$

Let $\nu_n = \nu_n(\alpha, \theta)$ correspond to ρ_n. *Will ν_n converge in some sense?* This is a highly nontrivial question because ρ_n and ν_n are related by the equation

$$B\nu_n := \int_{S^2} u(x,\alpha)\nu_n(\alpha,\theta)d\alpha = e^{i\theta\cdot x}(1+\rho_n), \quad k=1, \tag{8.6.39}$$

and B is a compact operator from $L^2(S^2)$ into $L^2(B_b)$. Therefore, there exists a function ν with a very large norm $\|\nu\|_{L^2(S^2)}$ such that $\|B\nu\|_{L^2(B_b)}$ is very small.

Let us prove, nevertheless, that (8.6.39) and convergence of ρ_n in $H^2(B_b)$ imply convergence of $\nu_{nl} := \int_{S^2} \nu_n(\alpha,\theta)\overline{Y_l(\alpha)}d\alpha$ as $n \to \infty$. We start the proof with the equation

$$(I+T)u = e^{i\alpha\cdot x}, \quad Tu := \int_{B_a} \frac{e^{i|x-y|}}{4\pi|x-y|} q(y)u(y)dy, \quad k=1, \tag{8.6.40}$$

where the operator $I+T$ has a bounded inverse in $L^2(B_b), b > a$. Apply the operator $I+T$ to equation (8.6.39) and get

$$\int_{S^2} e^{i\alpha\cdot x}\nu_n(\alpha,\theta)d\alpha = (I+T)(e^{i\theta\cdot x}(1+\rho_n)). \tag{8.6.41}$$

Substitute formula (8.6.22) with $\beta = \alpha$ into (8.6.41), denote

$$\nu_{nl} := \int_{S^2} \nu_n(\alpha,\theta)\overline{Y_l(\alpha)}d\alpha, \tag{8.6.42}$$

multiply the resulting formula by $\overline{Y_l(x^0)}$, and integrate over S^2 with respect to x^0. The result is

$$4\pi i^l j_l(r)\nu_{nl} = \int_{S^2} (I+T)(e^{i\theta\cdot x}(1+\rho_n))\overline{Y_l(x^0)}dx^0. \tag{8.6.43}$$

The integral in formula (8.6.43) converges as $n \to \infty$ if ρ_n converges in $H^2(B_b)$. Indeed, the function $(I+T)(e^{i\theta\cdot x}\rho_n)$ converges in $H^2(B_b)$. By the embedding theorem, its trace converges in $H_{3/2}(S^2)$. Thus, the integral in (8.6.43) converges, as $n \to \infty$, for any $r \in [0,b)$.

Let us return to problem (8.6.30). We have proved that the sequence ρ_n from (8.6.31) contains a convergent in $H^2(B_b)$ subsequence denoted ρ_n again, where $b > a$ is an arbitrary fixed number. If $\theta \in \mathcal{M}$ and $|\theta|$ is sufficiently large, then estimate (8.6.38) holds.

Lemma 8.6.5. *If $\theta \in \mathcal{M}$ and $|\theta|$ is sufficiently large, then $\rho_0 = R$, where R is defined in Theorem* 8.4.2.

Proof. Subtract equation (8.6.37) from equation (8.5.2). Then, for $\eta := R - \rho$ one gets the equation

$$L\eta - q\eta = 0. \tag{8.6.44}$$

If $|\theta|$ is sufficiently large, then $\|L^{-1}q\|$ is small (see the proof of Theorem 8.5.2). Thus, $\eta = 0$.

Lemma 8.6.5 is proved. □

Remark 8.6.1. The reason we have chosen the domain $B_b \setminus B_a$ in (8.6.30) is simple: in this domain, the scattering data $A(\beta, \alpha)$ allows one to write ρ explicitly using formula (8.6.28) for $u(x, \alpha)$. Therefore, the functional $F(\nu)$ in (8.6.30) can be written explicitly as a quadratic functional of ν. For sufficiently large $|\theta|$, a minimizing sequence ρ_n for problem (8.2.29) converges to R and formulas (8.6.13) and (8.6.14) are the inversion formulas for the exact scattering data.

8.7. Stability Estimate for the Inversion Formulas

Let us assume that $q_j \in Q_a$ and

$$\sup_{\beta,\alpha \in S^2} |A_1(\beta, \alpha) - A_2(\beta, \alpha)| < \delta. \tag{8.7.1}$$

We want to estimate $p(x) := q_1(x) - q_2(x)$.

Theorem 8.7.1. *If* $q_j \in Q_a, j = 1, 2,$ *and* (8.7.1) *holds, then*

$$\sup_{\xi \in \mathbb{R}^3} |\tilde{q}_1(\xi) - \tilde{q}_2(\xi)| \le c\epsilon\left(\frac{1}{c}\ln\frac{1}{\delta}\right), \quad \delta \to 0. \tag{8.7.2}$$

Here, the function ϵ *is defined in formula* (8.5.8b).

Proof. Multiply formula (8.2.1) by $\nu_n^{(1)}(\alpha, \theta_1)\nu_m^{(2)}(-\beta, \theta_2)$ and integrate over S^2 with respect to α and β. Choose $\theta_j \in \mathcal{M}, \theta_1 + \theta_2 = \xi, |\theta_1| \to \infty$. The result is

$$-4\pi \int_{S^2}\int_{S^2} [A_1(\beta, \alpha) - A_2(\beta, \alpha)]\nu_n^{(1)}(\alpha, \theta_1)\nu_m^{(2)}(-\beta, \theta_2)d\alpha d\beta$$

$$= \int_{B_a} p(x) \left(\int_{S^2} u_1(x,\alpha)\nu_{n_1}^{(1)}(\alpha,\theta_1)d\alpha \right.$$

$$\left. \times \int_{S^2} u_2(x,-\beta)\nu_{n_2}^{(2)}(-\beta,\theta_2)d\beta \right) dx. \tag{8.7.3}$$

Let $n_j \to \infty$. Then (see formula (8.6.7)):

$$\int_{S^2} u_j(x,\alpha)\nu_{n_j}^{(j)}(\alpha,\theta_j)d\alpha = e^{i\theta_j\cdot x}(1+\rho^{(j)}), \quad j=1,2. \tag{8.7.4}$$

Therefore, (8.7.4) yields

$$\left| \int_{B_a} p(x)e^{i\xi\cdot x}dx \right| \le O(|\rho^{(1)}| + |\rho^{(2)}| + |\rho^{(1)}\rho^{(2)}|)$$

$$+ c\delta \max_{j=1,2} \|\nu^{(j)}\|^2_{L^2(S^2)}. \tag{8.7.5}$$

Let us use the following lemma.

Lemma 8.7.2. *There exists $\nu_n(\alpha,\theta)$ such that*

$$\|\nu_n(\alpha,\theta)\| \le ce^{c|\theta|}. \tag{8.7.6}$$

A proof of Lemma 8.7.2 is given later.

From (8.7.5), (8.7.6), and estimate (8.6.38), one gets

$$\sup_{\xi\in\mathbb{R}^3} |\tilde{p}(\xi)| \le c(\epsilon(|\theta|) + \delta e^{c|\theta|}). \tag{8.7.7}$$

Minimizing the right-hand side of (8.7.7) with respect to $s := |\theta|$, $s \gg 1$, one gets

$$\min_{s>e}(\epsilon(s) + \delta e^{cs}) \le c\epsilon\left(\frac{1}{c}\ln\left(\frac{1}{\delta}\right)\right), \quad \delta \to 0. \tag{8.7.8}$$

From (8.7.7) and (8.7.8), the desired estimate (8.7.2) follows.

Theorem 8.7.1 is proved. □

Proof of Lemma 8.7.2. Denote

$$\inf \|\nu\| := m(\epsilon,\theta), \tag{8.7.9}$$

where the infimum is taken over $\nu \in L^2(S^2)$ such that

$$\|e^{i\theta\cdot x}(1+\rho) - \int_{S^2} u(x,\alpha)\nu(\alpha)d\alpha\|_{L^2(B_b)} \le \epsilon, \tag{8.7.10}$$

where $u(x,\alpha)$ is the scattering solution, $\theta \in \mathcal{M}$. Since $(I+T)u = e^{i\alpha\cdot x}$ and $(I+T)^{-1}$ is a bounded operator in $L^2(B_b)$, inequality (8.7.10) is equivalent to the inequality

$$\left\|(I+T)(e^{i\theta\cdot x}(1+\rho)) - \int_{S^2} e^{i\alpha\cdot x}\nu(\alpha)d\alpha\right\|_{L^2(B_b)} \le c\epsilon. \tag{8.7.11}$$

The function $e^{i\theta\cdot x}(1+\rho) := w$ solves equation

$$(\nabla^2 + 1 - q(x))w = 0 \tag{8.7.12}$$

because ρ solves equation (8.6.32). Therefore,

$$\phi := (I+T)w \tag{8.7.13}$$

solves the equation

$$(\nabla^2 + 1)\phi = 0. \tag{8.7.14}$$

Apply the operator $\nabla^2 + 1$ to equation (8.7.1) and get

$$(\nabla^2+1)\phi = (\nabla^2+1)w + (\nabla^2+1)Tw = (\nabla^2+1)w - qw = 0 \text{ in } \mathbb{R}^3. \tag{8.7.15}$$

Therefore, ϕ admits a representation similar to (8.6.22):

$$\phi(x) = \sum_{l=0}^{\infty} 4\pi i^l \phi_l Y_l(x^0) j_l(r), \quad r = |x|,\, x^0 = \frac{x}{r}. \tag{8.7.16}$$

Inequality (8.7.11) can be written as

$$16\pi^2 \sum_{l=0}^{\infty} |\phi_l - \nu_l|^2 \int_0^b |j_l(r)|^2 r^2 dr \le c^2\epsilon^2. \tag{8.7.17}$$

Choose

$$\nu_l = \phi_l \quad \text{for} \quad l < n, \quad \nu_l = 0 \text{ for } l \ge n. \tag{8.7.18}$$

Then (8.7.17) yields

$$16\pi^2 \sum_{l=n}^{\infty} |\phi_l|^2 \int_0^b |j_l(r)|^2 r^2 dr \le c^2 \epsilon^2. \tag{8.7.19}$$

If $|\theta| \to \infty$, then the main part of ϕ, according to formula (8.7.13), is the function $e^{i\theta\cdot x}$ in a neighborhood of $|x| = b$. The function $e^{i\theta\cdot x}$ admits a representation similar to (8.6.22):

$$e^{i\theta\cdot x} = \sum_{l=0}^{\infty} 4\pi i^l \overline{Y_l(\theta)} Y_l(x^0) j_l(r). \tag{8.7.20}$$

Therefore, the coefficients ϕ_l in (8.7.16) are of the order of $|Y_l(\theta)|$ for large $|\theta|, \theta \in \mathcal{M}$:

$$|\phi_l| \le c|Y_l(\theta)|. \tag{8.7.21}$$

Using estimates (8.6.18) and (8.6.25), one obtains

$$\int_0^b |j_l(r)|^2 r^2 dr \le c\left(\frac{eb}{2l}\right)^{2l}, \quad |Y_l(\theta)| \le c\frac{e^{\kappa r}}{|j_l(r)|}, \tag{8.7.22}$$

where $\kappa = |\operatorname{Im}\theta|$ and $r > b$. Therefore, inequality (8.7.19) holds if

$$e^{\kappa r}\left(\frac{b}{r}\right)^n \le c\epsilon. \tag{8.7.23}$$

Minimize the left-hand side of (8.7.23) with respect to r and get $r = \frac{n}{\kappa}$ as the minimizer, and

$$\min_{r\ge 0} e^{\kappa r}\left(\frac{b}{r}\right)^n = e^n\left(\frac{b\kappa}{n}\right)^n. \tag{8.7.24}$$

What is the maximal $n = n(\epsilon)$ for which inequality (8.7.24) holds? To find $n(\epsilon)$, let us solve asymptotically, as $\epsilon \to 0$, the equation

$$e^n\left(\frac{b\kappa}{n}\right)^n = c\epsilon. \tag{8.7.25}$$

Write this equation as $\frac{1}{\epsilon} = ce^{n\ln n[1+o(1)]}$ and take log to get

$$\ln\frac{1}{\epsilon} = n\ln n[1+o(1)], \quad \ln\ln\frac{1}{\epsilon} = \ln n[1+o(1)]. \tag{8.7.26}$$

Thus,

$$n(\epsilon) = \frac{\ln \frac{1}{\epsilon}}{\ln \ln \frac{1}{\epsilon}}[1 + o(1)], \quad \epsilon \to 0. \tag{8.7.27}$$

By Parseval's equality one has, using (8.7.18), (8.7.21), and (8.6.27),

$$\begin{aligned} \|\nu\|^2_{L^2(S^2)} &= \sum_{l=0}^{n-1} |\nu_l|^2 \leq c \sum_{l=0}^{n-1} |Y_l(\theta)|^2 \\ &\leq ce^{2\kappa r} \left(\frac{2n}{er}\right)^{2n} n^2, \quad n = n(\epsilon). \end{aligned} \tag{8.7.28}$$

Here, the relation $\sum_{l=0}^{n-1} \sum_{m=-l}^{l} 1 = n^2$ was used. Therefore,

$$\begin{aligned} \inf \|\nu\| := m(\epsilon, \theta) &\leq c \min_r e^{\kappa r} \left(\frac{2n(\epsilon)}{er}\right)^{n(\epsilon)} n(\epsilon) \\ &\leq ce^{n(\epsilon)} \left(\frac{2n(\epsilon)\kappa}{en(\epsilon)}\right)^{n(\epsilon)} n(\epsilon) \\ &= c(2\kappa)^{n(\epsilon)} n(\epsilon). \end{aligned} \tag{8.7.29}$$

Choose in (8.7.10) ϵ as follows

$$\epsilon = e^{-\kappa b}, \quad b > a, \tag{8.7.30}$$

and find $n(\epsilon)$ as a function of κ:

$$n(\epsilon) = \frac{\ln \frac{1}{\epsilon}}{\ln \ln \frac{1}{\epsilon}}[1 + o(1)] = \frac{\kappa b}{\ln \kappa}[1 + o(1)], \quad \kappa \to \infty. \tag{8.7.31}$$

From (8.7.29) and (8.7.31), one gets

$$m(\epsilon(\theta), \theta) \leq ce^{n \ln \kappa [1+o(1)]} \leq ce^{\kappa b[1+o(1)]}, \quad \kappa \to \infty, \quad \kappa = |\mathrm{Im}\, \theta|. \tag{8.7.32}$$

Estimate (8.7.32) implies (8.7.6).

Lemma 8.7.2 is proved. □

8.8. Inversion of Noisy Scattering Data

If the data $A_\delta(\beta,\alpha)$ are noisy, so that the exact data $A(\beta,\alpha)$ are unknown but $\delta>0$ is known and

$$\sup_{\beta,\alpha\in S^2} |A(\beta,\alpha)-A_\delta(\beta,\alpha)|<\delta, \tag{8.8.1}$$

then the estimate (8.6.21) is not valid for A_δ and our inversion theory cannot be applied.

Our goal is to derive a theory for finding $\hat{q}_\delta$ such that

$$\sup_{\xi\in\mathbb{R}^3} |\hat{q}_\delta-\tilde{q}(\xi)|<\eta(\delta), \quad \lim_{\delta\to 0}\eta(\delta)=0, \tag{8.8.2}$$

and estimate the rate at which $\eta(\delta)$ tends to zero. We need several quantities. Define

$$N(\delta)=\left[\frac{|\ln\delta|}{\ln|\ln\delta|}\right], \tag{8.8.3}$$

where $[x]$ is the nearest to $x>0$ integer,

$$\hat{A}_\delta(\theta',\alpha):=\sum_{l=0}^{N(\delta)} A_{\delta l}(\alpha)Y_l(\theta'),$$

$$A_{\delta l}(\alpha)=\int_{S^2} A_\delta(\beta,\alpha)\overline{Y_l(\beta)}d\beta, \tag{8.8.4}$$

$$\theta'-\theta=\xi, \quad \theta',\theta\in\mathcal{M}, \quad \kappa:=|\mathrm{Im}\,\theta|, \tag{8.8.5}$$

$$u_\delta(x,\alpha):=e^{i\alpha\cdot x}+\sum_{l=0}^{N(\delta)} A_{\delta l}(\alpha)Y_l(x^0)h_l(r),$$

$$r=|x|,\ x^0=\frac{x}{r}, \tag{8.8.6}$$

$$\rho_\delta(x):=e^{-i\theta\cdot x}\int_{S^2} u_\delta(x,\alpha)\nu(\alpha)d\alpha-1, \tag{8.8.7}$$

$$\mu(\delta):=e^{-\gamma N(\delta)}, \quad \gamma:=\ln\frac{b}{a_1}>0, \quad a<a_1<b, \tag{8.8.8}$$

where $\xi\in\mathbb{R}^3$ is an arbitrary fixed vector.

Consider the problem

$$|\theta| = \sup := \nu(\delta) \tag{8.8.9}$$

under the constraints (8.8.5) and the following constraint:

$$\|\rho_\delta\| + \|\nu\|_{L^2(S^2)} e^{\kappa b}\mu(\delta) \leq c\epsilon(|\theta|), \tag{8.8.10}$$

where $\epsilon(|\theta|)$ is defined in (8.5.8a). The supremum in (8.8.9) is taken over $\theta \in \mathcal{M}$ and $\nu \in L^2(S^2)$ under the constraints (8.8.5) and (8.8.10), where $c > 0$ in (8.8.10) is a sufficiently large constant and $b > a$ is an arbitrary fixed number, $q \in Q_a$, $\|\rho\| = \|\rho\|_{L^2(B_b \setminus B_a)}$.

Suppose that a pair $\{\theta(\delta), \nu_\delta = \nu_\delta(\alpha)\}$ is an approximate solution to problem (8.8.9)–(8.8.10) in the sense

$$\theta(\delta) \geq \frac{\nu(\delta)}{2}, \quad \|\nu_\delta\| \leq ce^{c|\theta|}. \tag{8.8.11}$$

Calculate

$$\hat{q}_\delta := -4\pi \int_{S^2} \hat{A}_\delta(\theta', \alpha)\nu_\delta(\alpha)d\alpha. \tag{8.8.12}$$

Theorem 8.8.1. *If a pair $\{\theta(\delta), \nu_\delta\}$ is an approximate solution to problem (8.8.9)–(8.8.10), then (8.8.2) holds with $\hat{q}_\delta$ defined in (8.8.12).*

A proof of this theorem requires some preparations. Note that

$$\hat{q}_\delta - \tilde{q}(\xi) = -4\pi \int_{S^2} A(\theta', \alpha)\nu_\delta(\alpha)d\alpha - \tilde{q}(\xi)$$
$$+4\pi \int_{S^2} [A(\theta', \alpha) - \hat{A}_\delta(\theta', \alpha)]\nu_\delta(\alpha)d\alpha. \tag{8.8.13}$$

Lemma 8.8.2. *One has*

$$I := \left\| e^{-i\theta\cdot x} \int_{S^2} [u(x, \alpha) - u_\delta(x, \alpha)]\nu_\delta(\alpha)d\alpha \right\|_{L^2(B_b \setminus B_{a_1})}$$
$$\leq c\|\nu_\delta(\alpha)\|_{L^2(S^2)} e^{\kappa b}\mu(\delta), \quad a < a_1 < b. \tag{8.8.14}$$

Lemma 8.8.3. *One has*

$$\left| \int_{S^2} [A(\theta', \alpha) - \hat{A}_\delta(\theta', \alpha)] \nu_\delta(\alpha) d\alpha \right| \le c \|\nu_\delta(\alpha)\|_{L^2(S^2)} e^{\kappa b} \mu(\delta). \tag{8.8.15}$$

Lemma 8.8.4. *If*

$$\sup_{\beta, \alpha \in S^2} |A_1(\beta, \alpha) - A_2(\beta, \alpha)| < \delta \tag{8.8.16}$$

then

$$\sup_{\alpha \in S^2, x \in B'_a} |u_1(x, \alpha) - u_2(x, \alpha)| \le c\mu(\delta). \tag{8.8.17}$$

Proof of Lemma 8.8.2. Estimate $|e^{i\theta \cdot x}| \le e^{\kappa b}$, estimate the integral in (8.8.14) by the Cauchy inequality and get

$$I \le e^{\kappa b} \|\nu_\delta\|_{L^2(S^2)} \left\| \, \|u(x, \alpha) - u_\delta(x, \alpha)\|_{L^2(S^2)} \, \right\|_{L^2(B_b \setminus B_{a_1})}. \tag{8.8.18}$$

Let us estimate the last factor in (8.8.18). One has

$$\begin{aligned} J &:= \|u(x, \alpha) - u_\delta(x, \alpha)\|^2_{L^2(S^2)} \\ &= \sum_{l=0}^{n-1} \int_{S^2} |A_l(\alpha) - \hat{A}_{\delta l}(\alpha)|^2 d\alpha |Y_l(x^0)|^2 |h_l(r)|^2 \\ &\quad + \sum_{l=n}^{\infty} \int_{S^2} |A_l(\alpha)|^2 d\alpha |Y_l(x^0)|^2 |h_l(r)|^2. \end{aligned} \tag{8.8.19}$$

By estimate (8.8.1) and (8.6.21), one gets

$$\|J\|^2_{L^2(B_b \setminus B_{a_1})} \le \delta^2 \sum_{l=0}^{n-1} \int_{a_1}^b |h_l(r)|^2 r^2 dr + c \sum_{l=n}^{\infty} \left(\frac{ae}{2l} \right)^{2l} \int_{a_1}^b |h_l(r)|^2 r^2 dr. \tag{8.8.20}$$

By estimate (8.6.20), one has

$$\int_{a_1}^b |h_l(r)|^2 r^2 dr \le c \left(\frac{2l}{ea_1} \right)^{2l}. \tag{8.8.21}$$

From (8.8.20) and (8.8.21), it follows that

$$\|J\|^2_{L^2(B_b\setminus B_{a_1})} \leq \delta^2 \left(\frac{2n}{ea_1}\right)^{2n} n^2 + c\left(\frac{a}{a_1}\right)^{2n}. \tag{8.8.22}$$

If $A, B > 0$, then $\sqrt{A^2+B^2} \leq A + B$. Thus, (8.8.22) implies

$$\|J\|_{L^2(B_b\setminus B_{a_1})} \leq \delta \left(\frac{2n}{ea_1}\right)^{n} n + \sqrt{c}\left(\frac{a}{a_1}\right)^{n}. \tag{8.8.23}$$

Minimizing the right-hand side of (8.8.23) with respect to n, one gets minimal value of the right-hand side of (8.8.23) at $n = N(\delta)$, where $N(\delta)$ is defined in (8.8.3) and this minimal value is of the order of $\mu(\delta)$, where $\mu(\delta)$ is defined in (8.8.8).

Lemma (8.8.2) is proved. □

Proof of Lemma 8.8.3. Inequality (8.8.15) will be proved if one proves that

$$\|A(\theta',\alpha) - \hat{A}_\delta(\theta',\alpha)\|_{L^2(S^2)} \leq ce^{\kappa b}\mu(\delta). \tag{8.8.24}$$

To prove (8.8.24), one starts with the inequality:

$$\begin{aligned}\|A(\theta',\alpha) - \hat{A}_\delta(\theta',\alpha)\|_{L^2(S^2)} &\leq \left\|\sum_{l=0}^{N(\delta)-1}[A_l(\alpha) - \hat{A}_{\delta l}(\alpha)]Y_l(\theta')\right\|_{L^2(S^2)} \\ &\quad + \left\|\sum_{l=N(\delta)}^{\infty} A_l(\alpha)Y_l(\theta')\right\| := I_1 + I_2.\end{aligned} \tag{8.8.25}$$

One has, using estimates (8.6.27) and (8.6.21),

$$I_1 \leq \delta e^{\kappa b}\left(\frac{2N(\delta)}{eb}\right)^{N(\delta)}, \quad I_2 \leq ce^{\kappa b}\left(\frac{a}{b}\right)^{N(\delta)}. \tag{8.8.26}$$

Thus,

$$I_1 + I_2 \leq e^{\kappa b}\left(\delta N(\delta)\left(\frac{2N(\delta)}{eb}\right)^{N(\delta)} + ce^{-\gamma N(\delta)}\right), \quad \gamma = \ln\frac{b}{a} > 1. \tag{8.8.27}$$

The expression in the parentheses is of the order of $\mu(\delta)$. One could take in (8.8.25) and (8.8.26) n in place of $N(\delta)$ and minimize the expression in parentheses in (8.8.27) with respect to n. The minimizer is $n = N(\delta)$ and the minimum is of the order of $\mu(\delta)$.

Lemma 8.8.3 is proved. □

Proof of Lemma 8.8.4. This proof is similar to the proof given below formula (8.8.19). It leads to estimate (8.8.17).

Lemma 8.8.4 is proved. □

Let us estimate $\nu(\delta)$ defined in (8.8.9).

Theorem 8.8.5. *One has*

$$\nu(\delta) > cN(\delta), \tag{8.8.28}$$

where $N(\delta)$ is defined in (8.8.3) and $c > 0$ is a constant independent of δ.

Proof. Note that

$$\begin{aligned}\|\rho_\delta\| &\le \|\rho\| + \|\rho - \rho_\delta\| \\ &= \|\rho\| + \left\| e^{-i\theta\cdot x} \int_{S^2} [u(x,\alpha) - u_\delta(x,\alpha)]\nu(\alpha)d\alpha \right\|,\end{aligned} \tag{8.8.29}$$

where $\|\cdot\| = \|\cdot\|_{L^2(B_b \setminus B_{a_1})}, a < a_1 < b$. Also

$$\|\rho\| \le \|\rho_\delta\| + \|\rho - \rho_\delta\|. \tag{8.8.30}$$

By Lemma 8.8.2, one has

$$\|\rho - \rho_\delta\| \le c\|\nu\|_{L^2(S^2)} e^{\kappa b}\mu(\delta). \tag{8.8.31}$$

For sufficiently large $|\theta|, \theta \in \mathcal{M}$, an estimate (8.5.8′) holds for ρ:

$$\|\rho\| \le c\epsilon(|\theta|). \tag{8.8.32}$$

From (8.8.29), (8.8.31), and (8.8.32), one obtains

$$\|\rho_\delta\| \le c\epsilon(|\theta|) + c\|\nu\|_{L^2(S^2)} e^{\kappa b}\mu(\delta). \tag{8.8.33}$$

By estimate (8.7.6), one has

$$\|\nu\|_{L^2(S^2)} \le ce^{c|\theta|}. \tag{8.8.34}$$

For the term $\|\nu\|_{L^2(S^2)}e^{\kappa b-\gamma N(\delta)}$ in (8.8.33) to go to zero for maximal $|\theta|$, it is necessary that

$$c|\theta| + \kappa b - \gamma N(\delta) < 0, \tag{8.8.35}$$

that is

$$\max|\theta| \sim cN(\delta), \tag{8.8.36}$$

where $c > 0$ denotes various constants. Relation (8.8.36) yields (8.8.28).

Theorem 8.8.5 is proved. □

8.9. Estimates of ρ

It was proved in Theorem 8.5.2, estimate (8.5.8′), that

$$\|R\|_{L^\infty(D_1)} \le c\epsilon(|\theta|), \tag{8.9.1}$$

where $\epsilon(|\theta|)$ is defined in (8.5.8b).

Let us prove that $\|\rho\|_{L^\infty(D_1)}$ is of the same order of magnitude when $|\theta|$ is sufficiently large. Since $e^{i\theta\cdot x}(1+\rho)$ solves equation (8.1.1), for any $\epsilon > 0, \epsilon = e^{-\kappa b_1}, b_1 > b$, however small, there is a $\nu \in L^2(S^2)$ such that

$$\|e^{i\theta\cdot x}(1+R) - e^{i\theta\cdot x}(1+\rho)\|_{L^2(B_b)} < e^{-\kappa b_1}, \quad \kappa = |\mathrm{Im}\,\theta|. \tag{8.9.2}$$

Thus,

$$e^{-\kappa b}\|R-\rho\|_{L^2(B_b)} < e^{-\kappa b_1}. \tag{8.9.3}$$

So,

$$\|R-\rho\|_{L^2(B_b)} < e^{-\kappa(b_1-b)}, \quad b_1 > b. \tag{8.9.4}$$

Therefore,

$$\|\rho\|_{L^2(B_b)} \le \|R\|_{L^2(B_b)} + \|R-\rho\|_{L^2(B_b)} \le \epsilon(|\theta|) + e^{-\kappa(b_1-b)} \le c\epsilon(|\theta|), \tag{8.9.5}$$

where $\epsilon(|\theta|)$ is defined in (8.5.8b).

On the other hand,

$$\|\rho\|_{L^2(B_b)} \geq \|R\|_{L^2(B_b)} - \|R-\rho\|_{L^2(B_b)}$$
$$\geq \epsilon(|\theta|) - e^{-\kappa(b_1-b)} \geq c\epsilon(|\theta|), \tag{8.9.6}$$

where $c > 0$ stand for different constants independent of $\theta, \theta \in \mathcal{M}$. We have proved the following result:

Theorem 8.9.1. *There exists $\nu \in L^2(S^2)$ such that*

$$c_1\epsilon(|\theta|) \leq \|\rho\|_{L^2(B_b)} \leq c_2\epsilon(|\theta|), \quad c_1 > 0. \tag{8.9.7}$$

8.10. Construction of the Dirichlet-to-Neumann Map from the Scattering Data and Vice Versa

The main point of this section is to make clear that computation methods for solving inverse scattering problem, based on the Dirichlet-to-Neumann (D–N) map are far inferior to the method described above. The difficulties of the method, described in this section, are:

(1) Computing function (8.10.7): the integrand is not absolutely integrable and to calculate numerically this integral is quite difficult,
(2) Calculating the limit (8.10.13) where $t(\theta', \theta)$ is defined in (8.10.12) is quite difficult,
(3) Solving equation (8.10.10) for ψ is difficult since Λ is not given explicitly and $\theta \in \mathcal{M}$ is large,
(4) Solving equation (8.10.20) for σ is very difficult: this Fredholm integral equation of the first kind with a smooth kernel is extremely ill-posed.

Consider a ball $B_a \supset D = \operatorname{supp} q(x)$ and assume that the problem

$$[\nabla^2 + 1 - q(x)]w = 0 \text{ in } B_a, \quad w = f \text{ on } S_a := \partial B_a, \tag{8.10.1}$$

is uniquely solvable for any $f \in H^{\frac{3}{2}}(S_a)$, where $H^l(S_a)$ is the Sobolev space.

Then the D–N map is defined as

$$\Lambda : f \to w_N, \tag{8.10.2}$$

where w_N is the normal derivative of w on S_a, N is the normal to S_a pointing into $B'_a := \mathbb{R}^3 \setminus B_a$.

If Λ is known, then $q(x)$ can be found as follows.

The special solution $\psi(x)$ satisfies the equation:

$$\psi(x) = e^{i\theta\cdot x} - \int_{B_a} G(x-y)q(y)\psi(y)dy, \tag{8.10.3}$$

where

$$G(x) := e^{i\theta\cdot x}G_0(x) \tag{8.10.4}$$

and

$$\nabla^2 G(x) + G(x) = -\delta(x), \text{ in } \mathbb{R}^3. \tag{8.10.5}$$

Thus,

$$\nabla^2 G_0 + 2i\theta \cdot \nabla G_0 = -\delta(x), \tag{8.10.6}$$

so that $G_0(x-y)$ is the Green's function of the operator L, that is,

$$G_0(x) = \frac{1}{(2\pi)^3}\int_{\mathbb{R}^3} \frac{e^{i\xi\cdot x}}{\xi^2 + 2\xi\cdot\theta}. \tag{8.10.7}$$

The function $G(x)$ can be considered known, see (8.10.4).

Since $q\psi = (\nabla^2 + 1)\psi$, one can write, for $x \in B'_a$,

$$\begin{aligned}
&\int_{B_a} G(x-y)q(y)\psi(y)dy \\
&\quad = \int_{B_a} G(\nabla^2 + 1)\psi dy \\
&\quad = \int_{S_a} [G(x-s)\psi_N(s) - G_N(x-s)\psi(s)]ds
\end{aligned}$$

$$= \int_{S_a} G(x-s)(\Lambda - \Lambda_0)\psi(s)ds$$

$$+ \int_{S_a} [G(x-s)\Lambda_0\psi - G_N(x-s)\psi]ds$$

$$= \int_{S_a} G(x-s)(\Lambda - \Lambda_0)\psi(s)ds. \tag{8.10.8}$$

Here, Λ_0 is Λ for $q(x) = 0$, we have used Green's formula and took into account that

$$\int_{S_a} [G(x-s)\Lambda_0\psi - G_N(x-s)\psi]ds$$

$$= \int_{B_a} [G(\Delta+1)\varphi - \varphi(\Delta+1)G]dy = 0, \tag{8.10.9}$$

where φ solves problem (8.10.1) with $q(x) = 0$ and $\varphi = f$ on S_a.

From (8.10.3) and (8.10.8) taking $x \to s \in S_a$, one gets a linear Fredholm-type equation for $\psi|_{S_a}$:

$$\psi(s) = e^{i\theta\cdot s} - \int_{S_a} G(s-s')(\Lambda - \Lambda_0)\psi(s')ds'. \tag{8.10.10}$$

If Λ is known, one can find from (8.10.10) $\psi|_{S_a}$ and then find $q(x)$ using the following calculation:
Define

$$t(\theta', \theta) = \int_{B_a} e^{i\theta'\cdot y} q(y)\psi(y,\theta)dy. \tag{8.10.11}$$

By Green's formula, as in (8.10.8), one gets

$$t(\theta', \theta) = \int_{S_a} e^{-i\theta'\cdot s}(\Lambda - \Lambda_0)\psi(s,\theta)ds. \tag{8.10.12}$$

From (8.10.8), one gets

$$\lim_{\substack{|\theta|\to\infty \\ \theta'-\theta=\xi \\ \theta\in\mathcal{M}}} t(\theta', \theta) = \int_{B_a} e^{-i\xi\cdot x} q(x)dx := \tilde{q}(\xi). \tag{8.10.13}$$

Therefore, the knowledge of Λ allows one to recover $\tilde{q}(\xi)$ by formula (8.10.10), but first one has to solve equation (8.10.10). We leave to the reader to check that the homogeneous equation (8.10.10) has only the trivial solution so that Fredholm-type equation (8.10.10) is uniquely solvable in $L^2(S_a)$.

Practically, however, there are *essential difficulties*:

(a) the function $G(x,y)$ is not known, analytically and it is difficult to solve equation (8.10.10) by this reason,
(b) the $D-N$ map is not given analytically as well.

Let us show how to construct Λ from the scattering amplitude $A(\alpha',\alpha)$ and vice versa.

If Λ is given, then we have shown how to find $q(x)$ and if $q(x)$ is found then $A(\alpha',\alpha)$, the scattering amplitude, can be found.

Conversely, suppose $A(\alpha',\alpha)$ is known and show how Λ can be found.

If $A(\alpha',\alpha)$ is given, then the scattering solution can be calculated in B_a' by formula (8.6.28).

Let $f \in H^{\frac{3}{2}}(S_a)$ be given, $\mathcal{G}(x,y)$ be the Green's function of the operator $-\nabla^2+q(x)-1$ in $\mathbb{R}^3$ which satisfies *the radiation condition* and define

$$w(x) = \int_{S_a} \mathbb{G}(x,s)\sigma(s)ds, \tag{8.10.14}$$

such that

$$w = f \quad \text{on } S_a. \tag{8.10.15}$$

Since $(\nabla^2+1)w = 0$ in B_a', $w = f$ on S_a and w satisfies the radiation condition, one can find w in B_a' explicitly:

$$w(x) = \sum_{l=0}^{\infty} \frac{f_l}{h_l(a)} Y_l(\alpha')h_l(r), \quad r \geq a, \quad z = |x|, \quad \alpha' = \frac{x}{r}, \tag{8.10.16}$$

where f_l are the Fourier coefficients of f:

$$f(s) = \sum_{l=0}^{\infty} f_l Y_l(\alpha'), \quad s \in S_a. \tag{8.10.17}$$

Therefore, the function

$$w_N^- = \lim_{|x|\to a, x\in B_a'} \frac{\partial w(x)}{\partial r} \tag{8.10.18}$$

is known. By the jump formula for single-layer potentials, one has

$$w_N^+ = w_N^- + \sigma. \tag{8.10.19}$$

The map $\Lambda : f \to w_N^+$ is constructed as soon as we find $\sigma(s)$, because w_N^- is already found.

To find σ, consider the asymptotics of $w(x)$ as $|x| \to \infty$, $\frac{x}{|x|} = \beta$. Using (8.2.6) and (8.10.14), one gets

$$\frac{1}{4\pi} \int_{S_a} u(s, -\beta)\sigma(s)ds = \eta(\beta) := \sum_{l=0}^{\infty} \frac{f_l Y_l(\beta)}{h_l(a)}, \tag{8.10.20}$$

where we have used (8.10.16) and the asymptotics $h_l(r) \sim \frac{e^{ir}}{r}$ as $r \to +\infty$. As we have already mentioned, the function $u(s, \alpha')$ is known explicitly, and equation (8.10.20) is uniquely solvable for $\sigma(s)$. Analytical solution of equation (8.10.20) for $\sigma(s)$ can be obtained as a series

$$\sigma(s) = \sum_{l=0}^{\infty} \sigma_l Y_l(\alpha'), \quad \alpha' = \frac{s}{|s|}. \tag{8.10.21}$$

Substitute (8.7.20) with $\theta = -\beta$ into

$$u = e^{i\alpha\cdot x} + \sum_{l=0}^{\infty} A_l(\alpha) Y_l(\beta) h_l(r), \tag{8.10.22}$$

take $r = a$ in (8.10.22) and $\beta = \frac{s}{a}$, and substitute (8.10.22) into (8.10.20). By our choice of the spherical harmonics, both systems

$\{Y_l\}_{l=0,1,2,...}$ and $\{\bar{Y}_l\}_{l=0,1,2,...}$ form orthonormal bases of $L^2(S^2)$. Therefore, one gets

$$\frac{1}{4\pi}\sum_{l=0}^{\infty} 4\pi i^l \bar{Y}_l(-\beta) j_l(a) a^2 \int_{S^2} Y_l(\alpha')\sigma(a\alpha')d\alpha'$$
$$+\frac{1}{4\pi}\sum_{l=0}^{\infty} A_l(-\beta) h_l(a) a^2 \int_{S^2} Y_l(\alpha')\sigma(a\alpha')d\alpha'$$
$$=\sum_{l=0}^{\infty} \frac{f_l Y_l(\beta)}{h_l(a)}. \tag{8.10.23}$$

Denote

$$\int_{S^2} \sigma(a\alpha')Y_{l,m}(\alpha')d\alpha' := \sigma_{lm}. \tag{8.10.24}$$

Using (8.6.17) one gets

$$\begin{aligned} Y_{l,m}(-\beta) &= (-1)^l Y_{l,m}(\beta), \quad \overline{Y_{l,m}(-\beta)} = (-1)^l \overline{Y_{l,m}(\beta)} \\ &= (-1)^{l+l+m} Y_{l,-m}(\beta) = (-1)^m Y_{l,-m}(\beta). \end{aligned} \tag{8.10.25}$$

Define $A_{lm,l'm'}$ by the formula

$$A_{l,m}(-\beta) = \sum_{l',m'} A_{lm,l'm'} Y_{l',-m'}(\beta). \tag{8.10.26}$$

The above definition is used for convenience in this section.

Equating the coefficients in front of $Y_{l,-m}(\beta)$ in (8.10.23) one gets

$$i^l(-1)^m j_l(a) a^2 \sigma_{lm} + \frac{a^2}{4\pi}\sum_{l'=0}^{\infty}\sum_{m'=-l'}^{l'} A_{l'm',lm} h_{l'}(a)\sigma_{l'm'} = \frac{f_{l,-m}}{h_l(a)}, \tag{8.10.27}$$

or

$$\sigma_{lm} + \frac{(-1)^m(-i)^l}{4\pi j_l(a)}\sum_{l'=0}^{\infty}\sum_{m'=-l'}^{l'} A_{l'm',lm} h_{l'}(a)\sigma_{l'm'} = \frac{f_{l,-m}(-1)^m(-i)^l}{a^2 j_l(a) h_l(a)}. \tag{8.10.28}$$

The matrix of the linear system (8.10.28) is ill-conditioned.

8.11. Various Remarks

In the preceding sections, we only use the estimate (8.5.7). One could use an estimate of L^2-norm, which was done in [317, 345, 356], but we have chosen to make the presentation completely self-consistent and by this reason, we did not use the mentioned estimate.

The remaining part of our inverse scattering theory deals with inverse scattering when *the data are non-overdetermined.* This means that the scattering data depend on three variables and the unknown potential also depends on three variables.

This theory was developed by the author in [395, 396, 401, 420].

We present essentially the uniqueness theorems from [396, 401] and a method for numerical solution of the inverse scattering problem with non-overdetermined data.

The scattering data are the values $A(\beta, \alpha_0, k)$ for all β on an open subset $\tilde{S}^2$ of S^2 and all $k \in (a, b), 0 \leq a < b$, in [401], and the backscattering data $A(-\beta, \beta, k)$ for all $\beta \in \tilde{S}^2$ and all $k \in (a, b)$, $0 \leq a < b$.

8.12. Uniqueness of the Solution to Inverse Scattering Problem with Backscattering Data

In this section, we follow very closely paper [396]. Our goal is to prove a uniqueness theorem for inverse scattering problem with non-overdetermined data.

8.12.1. *Introduction*

The scattering solution $u(x, \alpha, k)$ solves the scattering problem:

$$[\nabla^2 + k^2 - q(x)]u = 0 \quad \text{in } \mathbb{R}^3, \tag{8.12.1}$$

$$u = e^{ik\alpha\cdot x} + A(\beta, \alpha, k)\frac{e^{ikr}}{r} + o\left(\frac{1}{r}\right), \quad r := |x| \to \infty, \ \beta := \frac{x}{r}. \tag{8.12.2}$$

Here, $\alpha, \beta \in S^2$ are the unit vectors, S^2 is the unit sphere, the coefficient $A(\beta, \alpha, k)$ is called the scattering amplitude, $q(x)$ is a real-valued compactly supported sufficiently smooth function. The inverse scattering problem of interest is to determine $q(x)$ given the backscattering data $A(-\beta, \beta, k)$ $\forall \beta \in S^2$, $\forall k > 0$. This problem is called *the inverse scattering problem with backscattering data.*

The function $A(-\beta, \beta, k)$ depends on one unit vector β and on the scalar k, i.e., on three variables. The potential $q(x)$ depends also on three variables $x \in \mathbb{R}^3$. This inverse problem is, therefore, not overdetermined in the sense that the data and the unknown $q(x)$ are functions of the same number of variables.

Assumption (A): *We assume that q is compactly supported, i.e., $q(x) = 0$ for $|x| > a$, where $a > 0$ is an arbitrary large fixed number; $q(x)$ is real-valued, i.e., $q = \overline{q}$; and $q(x) \in H_0^\ell(B_a)$, $\ell > 3$.*

Here, B_a is the ball centered at the origin and of radius a, and $H_0^\ell(B_a)$ is the closure of $C_0^\infty(B_a)$ in the norm of the Sobolev space $H^\ell(B_a)$ of functions whose derivatives up to the order ℓ belong to $L^2(B_a)$.

It was proved in [199] (see also Theorem 1.4.3) that if $q = \overline{q}$ and $q \in L^2(B_a)$ is compactly supported, then the resolvent kernel $G(x, y, k)$ of the Schrödinger operator $-\nabla^2 + q(x) - k^2$ is a meromorphic function of k on the whole complex plane k, analytic in Im $k \geq 0$, except, possibly, of a finitely many simple poles at the points ik_j, $k_j > 0$, $1 \leq j \leq n$, where $-k_j^2$ are negative eigenvalues of the self-adjoint operator $-\nabla^2 + q(x)$ in $L^2(\mathbb{R}^3)$. Consequently, the scattering amplitude $A(\beta, \alpha, k)$, corresponding to the above q, is a restriction to the positive semi-axis $k \in [0, \infty)$ of a meromorphic on the whole complex k-plane function.

It was proved by the author [295] (see also Theorem 8.4.4), that the *fixed-energy scattering data* $A(\beta, \alpha) := A(\beta, \alpha, k_0)$, $k_0 = \text{const} > 0$, $\forall \beta \in S_1^2$, $\forall \alpha \in S_2^2$, determine real-valued compactly supported $q \in L^2(B_a)$ uniquely. Here, S_j^2, $j = 1, 2$, are arbitrary small open subsets of S^2 (solid angles).

In [345] (see also monograph [356, Chapter 5], and [317] and Theorems 8.6.1, 8.7.1), an analytical formula is derived for the reconstruction of the potential q from exact fixed-energy scattering data, and from noisy fixed-energy scattering data, and stability estimates and error estimates for the reconstruction method are obtained. To the author's knowledge, these are the only known until now theoretical error estimates for the recovery of the potential from noisy fixed-energy scattering data in the three-dimensional inverse scattering problem.

In [324], stability results are obtained for the inverse scattering problem for obstacles.

The scattering data $A(\beta, \alpha)$ depend on four variables (two unit vectors), while the unknown $q(x)$ depends on three variables. In this sense, the inverse scattering problem, which consists of finding q from the fixed-energy scattering data $A(\beta, \alpha)$, is overdetermined.

In the beginning of the forties of the last century, physicists raised the following question: is it possible to recover the Hamiltonian of a quantum-mechanical system from the observed quantities, such as S-matrix? In the non-relativistic quantum mechanics, the simplest Hamiltonian $\mathbf{H} = -\nabla^2 + q(x)$ can be uniquely determined if one knows the potential $q(x)$. The S-matrix in this case is in one-to-one correspondence with the scattering amplitude A: $S = I - \frac{k}{2\pi i} A$, where I is the identity operator in $L^2(S^2)$, A is an integral operator in $L^2(S^2)$ with the kernel $A(\beta, \alpha, k)$, and $k^2 > 0$ is energy. Therefore, the question, raised by the physicists, is reduced to an inverse scattering problem:

Can one determine the potential $q(x)$ from the knowledge of the scattering amplitude?

We have briefly discussed this problem above.

Since the above question was raised, there were no uniqueness theorems for three-dimensional inverse scattering problems with non-overdetermined data. The goal of this section is to prove such a theorem.

Theorem 8.12.1. *If Assumption (A) holds, then the data $A(-\beta, \beta, k) \forall \beta \in S^2$, $\forall k > 0$, determine q uniquely.*

Remark 8.12.1. The conclusion of Theorem 8.12.1 remains valid if the data $A(-\beta,\beta,k)$ are known $\forall\beta\in S_1^2$ and $k\in(k_0,k_1)$, where $(k_0,k_1)\subset[0,\infty)$ is an arbitrary small interval, $k_1>k_0$, and S_1^2 is an arbitrary small open subset of S^2.

In Section 8.12.2, we formulate some known auxiliary results.

In Section 8.12.3, proof of Theorem 8.12.1 is given.

In Section 8.12.4, a technical estimate is proved.

A brief announcement of the result is given in [395]. We follow the outline of the ideas from [401], but our presentation is essentially self-contained and contains new arguments.

8.12.2. *Auxiliary results*

Let

$$F(g):=\tilde{g}(\xi)=\int_{\mathbb{R}^3}g(x)e^{i\xi\cdot x}dx,\quad g(x)=\frac{1}{(2\pi)^3}\int_{\mathbb{R}^3}e^{-i\xi\cdot x}\tilde{g}(\xi)d\xi. \tag{8.12.3}$$

If $f*g:=\int_{\mathbb{R}^3}f(x-y)g(y)dy$, then

$$F(f*g)=\tilde{f}(\xi)\tilde{g}(\xi),\quad F(f(x)g(x))=\frac{1}{(2\pi)^3}\tilde{f}*\tilde{g}. \tag{8.12.4}$$

If

$$G(x-y,k):=\frac{e^{ik[|x-y|-\beta\cdot(x-y)]}}{4\pi|x-y|}, \tag{8.12.5}$$

then

$$F(G(x,k))=\frac{1}{\xi^2-2k\beta\cdot\xi},\qquad \xi^2:=\xi\cdot\xi. \tag{8.12.6}$$

The scattering solution $u=u(x,\alpha,k)$ solves (uniquely) the integral equation

$$u(x,\alpha,k)=e^{ik\alpha\cdot x}-\int_{B_a}g(x,y,k)q(y)u(y,\alpha,k)dy, \tag{8.12.7}$$

where

$$g(x,y,k):=\frac{e^{ik|x-y|}}{4\pi|x-y|}. \tag{8.12.8}$$

If

$$v = e^{-ik\alpha\cdot x}u(x, \alpha, k), \tag{8.12.9}$$

then

$$v = 1 - \int_{B_a} G(x-y,k)q(y)v(y,\alpha,k)dy, \tag{8.12.10}$$

where G is defined in (8.12.5).

Define ϵ by the formula

$$v = 1 + \epsilon. \tag{8.12.11}$$

Then (8.12.10) can be rewritten as

$$\epsilon(x,\alpha,k) = -\int_{\mathbb{R}^3} G(x-y,k)q(y)dy - T\epsilon, \tag{8.12.12}$$

where

$$T\epsilon := \int_{B_a} G(x-y,k)q(y)\epsilon(y,\alpha,k)dy.$$

Fourier transform of (8.12.12) yields (see (8.12.4) and (8.12.6)):

$$\tilde{\epsilon}(\xi,\alpha,k) = -\frac{\tilde{q}(\xi)}{\xi^2 - 2k\alpha\cdot\xi} - \frac{1}{(2\pi)^3}\frac{1}{\xi^2 - 2k\alpha\cdot\xi}\tilde{q} * \tilde{\epsilon}. \tag{8.12.13}$$

An essential ingredient of our proof in Section 8.12.3 is lemma proved by the author in [356, p. 262], and in [345]. A proof of this lemma is given in Section 8.12.4.

Consider an algebraic variety $\mathcal{M}$ in $\mathbb{C}^3$ defined by the equation

$$\mathcal{M} := \{\theta\cdot\theta = 1, \quad \theta\cdot\theta := \theta_1^2 + \theta_2^2 + \theta_3^2, \quad \theta_j \in \mathbb{C},\ 1 \le j \le 3\}. \tag{8.12.14}$$

This is a noncompact variety, intersecting $\mathbb{R}^3$ over the unit sphere S^2.

Let $R_+ = [0,\infty)$. The following result is proved in [269, p. 62], see also Theorem 1.4.1 of this book.

Lemma 8.12.2. *If Assumption (A) holds, then the scattering amplitude $A(\beta,\alpha,k)$ is a restriction to $S^2 \times S^2 \times R_+$ of a function $A(\theta',\theta,k)$*

on $\mathcal{M} \times \mathcal{M} \times \mathbb{C}$, *analytic on* $\mathcal{M} \times \mathcal{M}$ *and meromorphic on* $\mathbb{C}$, θ', $\theta \in \mathcal{M}$, $k \in \mathbb{C}$.

The scattering solution $u(x, \alpha, k)$ is a meromorphic function of k in $\mathbb{C}$, analytic in $\text{Im } k \geq 0$, except, possibly, at the points $k = ik_j$, $1 \leq j \leq n$, $k_j > 0$, where $-k_j^2$ are negative eigenvalues of the self-adjoint Schrödinger operator, defined by the potential q in $L^2(\mathbb{R}^3)$. These eigenvalues can be absent, for example, if $q \geq 0$.

We need the notion of the Radon transform:

$$\hat{f}(\beta, \lambda) := \int_{\beta \cdot x = \lambda} f(x) d\sigma, \tag{8.12.15}$$

where $d\sigma$ is the element of the area of the plane $\beta \cdot x = \lambda$, $\beta \in S^2$, λ is a real number. The following properties of the Radon transform will be used:

$$\int_{B_a} f(x) dx = \int_{-a}^{a} \hat{f}(\beta, \lambda) d\lambda, \tag{8.12.16}$$

$$\int_{B_a} e^{ik\beta \cdot x} f(x) dx = \int_{-a}^{a} e^{ik\lambda} \hat{f}(\beta, \lambda) d\lambda, \tag{8.12.17}$$

$$\hat{f}(\beta, \lambda) = \hat{f}(-\beta, -\lambda). \tag{8.12.18}$$

These properties are proved in [330, pp. 12 and 15]. We also need the following Phragmen–Lindelöf lemma, which is proved in [127, p. 69].

Lemma 8.12.3. *Let* $f(z)$ *be holomorphic inside an angle* $\mathcal{A}$ *of opening* $< \pi$; $|f(z)| \leq c_1 e^{c_2 |z|}$, $z \in \mathcal{A}$, $c_1, c_2 > 0$ *are constants*; $|f(z)| \leq M$ *on the boundary of* $\mathcal{A}$; *and* f *is continuous up to the boundary of* $\mathcal{A}$. *Then* $|f(z)| \leq M$, $\forall z \in \mathcal{A}$.

8.12.3. *Proof of Theorem* 8.12.1

The scattering data in Remark 8.12.1 determine uniquely the scattering data in Theorem 8.12.1 by Lemma 8.12.2.

Let us outline the ideas of the proof of Theorem 8.12.1.

Assume that potentials q_j, $j = 1, 2$, generate the same scattering data:

$$A_1(-\beta, \beta, k) = A_2(-\beta, \beta, k), \qquad \forall \beta \in S^2, \ \forall k > 0,$$

and let

$$p(x) := q_1(x) - q_2(x).$$

Then by Theorem 8.4.1, one gets

$$0 = \int_{B_a} p(x) u_1(x, \beta, k) u_2(x, \beta, k) dx, \qquad \forall \beta \in S^2, \ \forall k > 0. \tag{8.12.19}$$

By (8.12.9) and (8.12.11), one can rewrite (8.12.19) as

$$\int_{B_a} e^{2ik\beta\cdot x}[1 + \epsilon(x, k)] p(x) dx = 0, \quad \forall \beta \in S^2, \ \forall k > 0, \tag{8.12.20}$$

where

$$\epsilon(x, k) := \epsilon := \epsilon_1(x, k) + \epsilon_2(x, k) + \epsilon_1(x, k)\epsilon_2(x, k).$$

The relations (8.12.19) and (8.12.20) hold for complex k,

$$k = \frac{\kappa + i\eta}{2}, \qquad \kappa + i\eta \neq 2ik_j, \quad \eta \geq 0, \tag{8.12.21}$$

where $-k_j^2$, $1 \leq j \leq J$, are the negative eigenvalues of the Schrödinger operator (8.12.1). Using formulas (8.12.3)–(8.12.4), one derives from (8.12.20) the relation

$$\tilde{p}((\kappa + i\eta)\beta) + \frac{1}{(2\pi)^3}(\tilde{\epsilon} * \tilde{p})((\kappa + i\eta)\beta) = 0, \qquad \forall \beta \in S^2, \ \forall \kappa \in \mathbb{R}, \tag{8.12.22}$$

where the notation $(f * g)(z)$ means that the convolution $f * g$ is calculated at the argument $z = (\kappa + i\eta)\beta$.

One has

$$\sup_{\beta \in S^2} |\tilde{\epsilon} * \tilde{p}| := \sup_{\beta \in S^2} \left| \int_{\mathbb{R}^3} \tilde{\epsilon}((\kappa + i\eta)\beta - s)\tilde{p}(s) ds \right| \leq \nu(\kappa, \eta) \sup_{s \in \mathbb{R}^3} |\tilde{p}(s)|, \tag{8.12.23}$$

where

$$\nu(\kappa, \eta) := \sup_{\beta \in S^2} \int_{\mathbb{R}^3} |\tilde{\epsilon}((\kappa + i\eta)\beta - s)| ds.$$

We prove that if $\eta = \eta(\kappa) = O(\ln \kappa)$ is suitably chosen, namely as in (8.12.28) below, then the following inequality holds:

$$0 < \nu(\kappa, \eta(\kappa)) < 1, \quad \kappa \to \infty. \tag{8.12.24}$$

We also prove that

$$\sup_{\beta \in S^2} |\tilde{p}((\kappa + i\eta(\kappa))\beta)| \geq \sup_{s \in \mathbb{R}^3} |\tilde{p}(s)|, \quad \kappa \to \infty, \tag{8.12.25}$$

and then it follows from (8.12.22)–(8.12.25) that $\tilde{p}(s) = 0$, so $p(x) = 0$, and Theorem 8.12.1 is proved. Indeed, it follows from (8.12.22) and (8.12.25) that, for sufficiently large κ and a suitable $\eta(k) = O(\ln k)$, one has

$$\sup_{s \in \mathbb{R}^3} |\tilde{p}(s)| \leq \frac{1}{(2\pi)^3} \nu(\kappa, \eta(\kappa)) \sup_{s \in \mathbb{R}^3} |\tilde{p}(s)|.$$

If (8.12.24) holds, then the above equation implies that $\tilde{p} = 0$. This and the injectivity of the Fourier transform imply that $p = 0$.

This completes the outline of the proof of Theorem 8.12.1.

Let us now give a detailed proof of estimates (8.12.24) and (8.12.25) that completes the proof of Theorem 8.12.1.

We assume that $p(x) \not\equiv 0$, because otherwise there is nothing to prove. Let

$$\max_{s \in \mathbb{R}^3} |\tilde{p}(s)| := \mathcal{P} \neq 0.$$

Lemma 8.12.4. *If Assumption (A) holds and $\mathcal{P} \neq 0$, then*

$$\limsup_{\eta \to \infty} \max_{\beta \in S^2} |\tilde{p}((\kappa + i\eta)\beta)| = \infty, \tag{8.12.26}$$

where $\kappa > 0$ is arbitrary but fixed. For any $\kappa > 0$, there is an $\eta = \eta(\kappa)$, such that

$$\max_{\beta \in S^2} |\tilde{p}((\kappa + i\eta(\kappa))\beta)| = \mathcal{P}, \tag{8.12.27}$$

where the number $\mathcal{P} := \max_{s\in\mathbb{R}^3} |\tilde{p}(s)|$, *and*

$$\eta(\kappa) = a^{-1}\ln\kappa + O(1) \quad as \quad \kappa \to +\infty. \tag{8.12.28}$$

Proof. By formula (8.12.17), one gets

$$\tilde{p}((\kappa + i\eta)\beta) = \int_{B_a} p(x)e^{i(\kappa+i\eta)\beta\cdot x}dx = \int_{-a}^{a} e^{i\kappa\lambda-\eta\lambda}\hat{p}(\beta,\lambda)d\lambda. \tag{8.12.29}$$

The function $\hat{p}(\beta,\lambda)$ is compactly supported, real-valued, and satisfies relation (8.12.18). Therefore,

$$\max_{\beta\in S^2} |\tilde{p}((\kappa + i\eta(\kappa))\beta)| = \max_{\beta\in S^2} |\tilde{p}((\kappa - i\eta(\kappa))\beta)|. \tag{8.12.30}$$

Indeed, using the properties (8.12.15)–(8.12.18) of the Radon transform, one gets:

$$\begin{aligned}
\max_{\beta\in S^2} |\tilde{p}((\kappa + i\eta(\kappa))\beta)| &= \max_{\beta\in S^2} \left| \int_{-a}^{a} e^{i\kappa\lambda-\eta\lambda}\hat{p}(\beta,\lambda)d\lambda \right| \\
&= \max_{\beta\in S^2} \left| \int_{-a}^{a} e^{-i\kappa\mu+\eta\mu}\hat{p}(\beta,-\mu)d\mu \right| \\
&= \max_{\beta'\in S^2} \left| \int_{-a}^{a} e^{-i\kappa\mu+\eta\mu}\hat{p}(-\beta',-\mu)d\mu \right| \\
&= \max_{\beta'\in S^2} \left| \int_{-a}^{a} e^{-i\kappa\mu+\eta\mu}\hat{p}(\beta',\mu)d\mu \right| \\
&= \max_{\beta\in S^2} |\tilde{p}((\kappa - i\eta)\beta)|. \qquad (8.12.31)
\end{aligned}$$

At the last step, we took into account that $\hat{p}(\beta,\lambda)$ is a real-valued function, so, by taking complex conjugate, one gets:

$$\begin{aligned}
\max_{\beta\in S^2} \left| \int_{-a}^{a} e^{-i\kappa\mu+\eta\mu}\hat{p}(\beta,\mu)d\mu \right| &= \max_{\beta\in S^2} \left| \int_{-a}^{a} e^{i\kappa\mu+\eta\mu}\hat{p}(\beta,\mu)d\mu \right| \\
&= \max_{\beta\in S^2} |\tilde{p}((\kappa - i\eta)\beta)|. \qquad (8.12.32)
\end{aligned}$$

If $p(x) \not\equiv 0$, then (8.12.29) and (8.12.30) imply (8.12.26), as follows from Lemma 8.12.4. Let us give a detailed proof of this statement.

Consider the function h of the complex variable $z := \kappa + i\eta$:

$$h := h(z, \beta) := \int_{-a}^{a} e^{iz\lambda} \hat{p}(\beta, \lambda) d\lambda. \tag{8.12.33}$$

If (8.12.26) is false, then

$$|h(z, \beta)| \leq c \quad \forall z = \kappa + i\eta, \quad \eta \geq 0, \quad \forall \beta \in S^2, \tag{8.12.34}$$

where $\kappa \geq 0$ is an arbitrary fixed number and the constant $c > 0$ does not depend on β and η.

Thus, $|h|$ is bounded on the ray $\{\kappa = 0, \eta \geq 0\}$, which is part of the boundary of the right angle $\mathcal{A}$, and the other part of its boundary is the ray $\{\kappa \geq 0, \eta = 0\}$. Let us check that $|h|$ is bounded on this ray also.

One has

$$|h(\kappa, \beta)| = \left| \int_{-a}^{a} e^{i\kappa\lambda} \hat{p}(\beta, \lambda) d\lambda \right| \leq \int_{-a}^{a} |\hat{p}(\beta, \lambda)| d\lambda \leq c, \tag{8.12.35}$$

where c stands for *various* constants. From (8.12.34)–(8.12.35), it follows that on the boundary of the right angle $\mathcal{A}$, namely, on the two rays $\{\kappa \geq 0, \eta = 0\}$ and $\{\kappa = 0, \eta \geq 0\}$ the entire function $h(z, \beta)$ of the complex variable z is bounded, $|h(z, \beta)| \leq c$, and inside $\mathcal{A}$ this function satisfies the estimate

$$|h(z, \beta)| \leq e^{a|\eta|} \int_{-a}^{a} |\hat{p}(\beta, \lambda)| d\lambda \leq c e^{a|\eta|}, \tag{8.12.36}$$

where c does not depend on β. Therefore, by Lemma 8.12.4, $|h(z, \beta)| \leq c$ in the whole angle $\mathcal{A}$.

By (8.12.30), the same argument is applicable to the remaining three right angles, the union of which is the whole complex z-plane $\mathbb{C}$. Therefore,

$$\sup_{z \in \mathbb{C}, \beta \in S^2} |h(z, \beta)| \leq c. \tag{8.12.37}$$

This implies, by the Liouville theorem, that $h(z, \beta) = c$, $\forall z \in \mathbb{C}$.

Since $\hat{p}(\beta, \lambda) \in L^1(-a, a)$, the relation

$$\int_{-a}^{a} e^{iz\lambda}\hat{p}(\beta, \lambda)d\lambda = c, \quad \forall z \in \mathbb{C}, \tag{8.12.38}$$

and the Riemann–Lebesgue lemma imply that $c = 0$, so $\hat{p}(\beta, \lambda) = 0$ $\forall \beta \in S^2$ and $\forall \lambda \in \mathbb{R}$. Therefore, $p(x) = 0$, contrary to our assumption. Consequently, relation (8.12.26) is proved.

Relation (8.12.27) follows from (8.12.26) because for large η the left-hand side of (8.12.27) is larger than $\mathcal{P}$ due to (8.12.26), while for $\eta = 0$ the left-hand side of (8.12.27) is not larger than $\mathcal{P}$ by the definition of the Fourier transform.

Let us derive estimate (8.12.28).

From the assumption $p(x) \in H_0^\ell(B_a)$, it follows that

$$|\tilde{p}((\kappa + i\eta)\beta)| \le c\frac{e^{a|\eta|}}{(1 + \kappa^2 + \eta^2)^{\ell/2}}. \tag{8.12.39}$$

This inequality is proved in Lemma 8.12.5 below.

The right-hand side of this inequality is of the order $O(1)$ as $\kappa \to \infty$ if $|\eta| = a^{-1}\ln\kappa + O(1)$ as $\kappa \to \infty$. This proves relation (8.12.28) and we specify $O(\ln\kappa)$ as in this relation.

Let us now prove inequality (8.12.39).

Lemma 8.12.5. *If* $p \in H_0^\ell(B_a)$, *then estimate* (8.12.39) *holds.*

Proof. Let $\partial_j p := \frac{\partial p}{\partial x_j}$. One has

$$\begin{aligned} &\left|\int_{B_a} e^{i(\kappa+i\eta)\beta\cdot x}\partial_j p\, dx\right| \\ &\quad = \left|-i(\kappa + i\eta)\beta_j \int_{B_a} p(x)e^{i(\kappa+i\eta)\beta\cdot x}dx\right| \\ &\quad = |\beta_j|(\kappa^2 + \eta^2)^{1/2}|\tilde{p}((\kappa + i\eta)\beta)|, \end{aligned} \tag{8.12.40}$$

and $\sum_{j=1}^{3} |\beta_j|^2 = 1$.

The left-hand side of the above formula admits the following estimate

$$\left| \int_{B_a} e^{i(\kappa+i\eta)\beta\cdot x} \partial_j p \, dx \right| \leq c e^{a|\eta|}, \tag{8.12.41}$$

where the constant $c > 0$ is proportional to $\|\partial_j p\|_{L^2(B_a)}$.

Square (8.12.40), sum up with respect to j, use the relation $\sum_{j=1}^{3} |\beta_j|^2 = 1$ and inequality (8.12.41), to obtain inequality

$$|\tilde{p}((\kappa + i\eta)\beta)| \leq c[1 + (\kappa^2 + \eta^2)]^{-1/2} e^{a|\eta|}. \tag{8.12.42}$$

Repeating this argument, one gets estimate (8.12.39). Lemma 8.12.4 is proved. □

Estimate (8.12.42) implies that if relation (8.12.28) holds and $\kappa \to \infty$, then the quantity $\sup_{\beta\in S^2} |\tilde{p}((\kappa + i\eta)\beta)|$ remains bounded as $\kappa \to \infty$.

If η is fixed and $\kappa \to \infty$, then $\sup_{\beta\in S^2} |\tilde{p}((\kappa + i\eta)\beta)| \to 0$ by the Riemann–Lebesgue lemma. This, the continuity of $|\tilde{p}((\kappa+i\eta)\beta)|$ with respect to η, and relation (8.12.26), imply the existence of $\eta = \eta(\kappa)$, such that equality (8.12.27) holds, and, consequently, inequality (8.12.25) holds. This $\eta(\kappa)$ satisfies (8.12.28) because $\mathcal{P}$ is bounded.

Lemma 8.12.2 is proved. □

To complete the proof of Theorem 8.12.1, one has to establish estimate (8.12.24). This estimate will be established if one proves the following relation:

$$\lim_{\kappa\to\infty} \nu(\kappa) := \lim_{\kappa\to\infty} \nu(\kappa, \eta(\kappa)) = 0, \tag{8.12.43}$$

where $\eta(\kappa)$ satisfies (8.12.28) and

$$\nu(\kappa, \eta) = \sup_{\beta\in S^2} \int_{\mathbb{R}^3} |\tilde{\epsilon}((\kappa + i\eta)\beta - s)| ds. \tag{8.12.44}$$

Our argument is valid for ϵ_1, ϵ_2 and $\epsilon_1\epsilon_2$, so we will use the letter ϵ and equation (8.12.13) for $\tilde{\epsilon}$.

Below we denote $2k := \kappa + i\eta$, and choose

$$\eta = \eta(\kappa) = a^{-1} \ln \kappa + O(1) \quad \text{as } \kappa \to \infty.$$

We prove that equation (8.12.12) can be solved by iterations if Im $\eta \geq 0$ and $|k + i\eta|$ is sufficiently large, because for such $k + i\eta$ the operator T^2 has small norm in $C(B_a)$, the space of functions, continuous in the ball B_a, with the sup-norm. Since equation (8.12.12) can be solved by iterations and the norm of T^2 is small, the main term in the series, representing its solution, as $|\kappa + i\eta| \to \infty$, $\eta \geq 0$, is the free term of equation (8.12.12). The same is true for the Fourier transform of equation (8.12.12), i.e., for equation (8.12.13). Therefore, the main term of the solution $\tilde{\epsilon}$ to equation (8.12.13) as $|\kappa + i\eta| \to \infty$, $\eta \geq 0$, is obtained by using the estimate of the free term of this equation. Thus, it is sufficient to check estimate (8.12.43) for the function $\nu(\kappa, \eta(\kappa))$ using, in place of $\tilde{\epsilon}$, the function $\tilde{q}(\xi)(\xi^2 - 2k\beta \cdot \xi)^{-1}$, with $2k$ replaced by $\kappa + i\eta$ and $\eta = a^{-1} \ln \kappa + O(1)$ as $\kappa \to \infty$.

For the proof of the above claim that equation (8.12.12) has the operator

$$T\epsilon = \int_{B_a} G(x - y, k) q(y) \epsilon(y, \beta, k) dy,$$

with the norm $\|T^2\|$ in the space $C(B_a)$, which tends to zero as $|\kappa + i\eta| \to \infty$, $\eta \geq 0$, see Section 8.12.4.

Thus, let us estimate the modulus of the factor $\nu(\kappa, \eta)$ in (8.12.23) with $\eta = \eta(\kappa)$ as in (8.12.28). Using inequality (8.12.39), and denoting $\xi = (\kappa + i\eta)\beta$, where $\beta \in S^2$ plays the role of α in (8.12.13), one obtains:

$$\begin{aligned} I &:= \sup_{\beta \in S^2} \int_{\mathbb{R}^3} \frac{|\tilde{q}((\kappa + i\eta)\beta - s)| ds}{|[(\kappa + i\eta)\beta - s)^2 - (\kappa + i\eta)\beta \cdot ((\kappa + i\eta)\beta - s)]|} \\ &\leq c e^{a|\eta|} \sup_{\beta \in S^2} \int_{\mathbb{R}^3} \frac{ds}{|s^2 - (\kappa + i\eta)\beta \cdot s|[1 + (\kappa\beta - s)^2 + \eta^2]^{\ell/2}} \\ &:= c e^{a|\eta|} J. \end{aligned} \tag{8.12.45}$$

Let us prove that

$$J = o\left(\frac{1}{\kappa}\right), \quad \kappa \to \infty.$$

If this estimate is proved and $\eta = a^{-1} \ln \kappa + O(1)$, then $I = o(1)$ as $\kappa \to \infty$, therefore relation (8.12.43) follows, and Theorem 1.1 is proved.

Let us write the integral J in the spherical coordinates with x_3-axis directed along vector β. We have

$$|s| = r, \qquad \beta \cdot s = r\cos\theta := rt, \quad -1 \le t \le 1.$$

Denote

$$\gamma := \kappa^2 + \eta^2.$$

Then

$$\begin{aligned} J &\le 2\pi \int_0^\infty drr \int_{-1}^1 \frac{dt}{[(r-\kappa t)^2 + \eta^2 t^2]^{1/2}(1+\gamma+r^2-2r\kappa t)^{\ell/2}} \\ &:= 2\pi \int_0^\infty drrB(r), \end{aligned} \tag{8.12.46}$$

where

$$\begin{aligned} B &:= B(r) = B(r,\kappa,\eta) \\ &:= \int_{-1}^1 \frac{dt}{[(r-\kappa t)^2 + \eta^2 t^2]^{1/2}(1+\gamma+r^2-2r\kappa t)^{\ell/2}}. \end{aligned}$$

Estimate of J we start with the observation

$$\tau := \min_{t\in[-1,1]} [(r-\kappa t)^2 + \eta^2 t^2] = \min\{r^2\eta^2/\gamma,\ (r-\kappa)^2 + \eta^2\}.$$

Let $\tau = r^2\eta^2/\gamma$, which is always the case if r is sufficiently small. In the case when $\tau = (r-\kappa)^2 + \eta^2$, the proof is considerably simpler and is left for the reader. If $\tau = r^2\eta^2/\gamma$, then

$$J \le 2\pi\gamma^{1/2}\eta^{-1} \int_0^\infty dr \int_{-1}^1 dt[1+\gamma+r^2-2\kappa rt]^{-\ell/2}.$$

Integrating over t yields

$$J \le 2\pi\gamma^{1/2}\eta^{-1}[(\ell-2)\kappa]^{-1}\mathcal{J},$$

where

$$\mathcal{J} := \int_0^\infty drr^{-1}[(1+\gamma+r^2-2\kappa r)^{-b} - (1+\gamma+r^2+2\kappa r)^{-b}],$$

and $b := \ell/2 - 1$.

Since $\eta = O(\ln\kappa)$, one has $\frac{\eta}{\kappa} = o(1)$ as $\kappa \to \infty$. Therefore,

$$\gamma^{1/2}\eta^{-1}\kappa^{-1} = O(\eta^{-1}) \qquad \text{as } \kappa \to \infty.$$

Since $\ell > 3$, one has $b > \frac{1}{2}$, and, as we prove below,

$$\mathcal{J} = o\left(\frac{1}{\kappa}\right) \quad \text{as } \kappa \to \infty. \tag{8.12.47}$$

This relation implies the desired inequality:

$$J \leq o\left(\frac{1}{\kappa}\right) \quad \text{as } \kappa \to \infty. \tag{8.12.48}$$

Let us derive relation (8.12.47). One has

$$\mathcal{J} = \int_0^1 + \int_1^\infty := J_1 + J_2,$$

$$J_1 \leq \int_0^1 drr^{-1}\frac{(w^2+2r\kappa+r^2)^b - (w^2-2r\kappa+r^2)^b}{(w^2+2r\kappa+r^2)^b(w^2-2r\kappa+r^2)^b},$$

where

$$w^2 := 1+\gamma = 1+\eta^2+\kappa^2.$$

Furthermore,

$$(w^2+2r\kappa+r^2)^b - (w^2-2r\kappa+r^2)^b \leq \frac{4br\kappa}{(w^2-2r\kappa+r^2)^{1-b}}.$$

Thus,

$$J_1 \leq 4b\kappa\int_0^1 dr\frac{1}{(w^2+2r\kappa+r^2)^b(w^2-2r\kappa+r^2)}.$$

This implies the following estimate:

$$J_1 \leq O(\kappa/w^{2+2b}) \leq O(\kappa^{-(1+2b)}),$$

because $w = \kappa[1 + o(1)]$ as $\kappa \to \infty$. Furthermore,

$$J_2 \le \int_1^\infty drr^{-1}[(1+\eta^2+(r-\kappa)^2)^{-b}-(1+\eta^2+(r+\kappa)^2)^{-b}] := J_{21}-J_{22}.$$

One has $J_{22} \le J_{21}$.

Let us estimate J_{21}. One obtains

$$J_{21} = \int_1^{\kappa/2} + \int_{\kappa/2}^\infty := j_1 + j_2,$$

and

$$j_1 \le \frac{1}{[W^2 + \frac{\kappa^2}{4}]^b} \ln \kappa = o\left(\frac{1}{\kappa}\right), \qquad W^2 := 1 + \eta^2, \qquad b > \frac{1}{2}.$$

Furthermore,

$$j_2 \le \frac{2}{\kappa} \int_{\kappa/2}^\infty \frac{dr}{[W^2 + (r-\kappa)^2]^b} \le \frac{2}{\kappa} \int_{-\infty}^\infty \frac{dy}{[W^2 + y^2]^b} = o\left(\frac{1}{\kappa}\right).$$

Thus, if $b > \frac{1}{2}$, then $J_2 = o(\frac{1}{\kappa})$ and $\mathcal{J} = J_1 + J_2 = o(\frac{1}{\kappa})$. Thus, relation (8.12.47) is proved.

Relation (8.12.47) yields the desired estimate

$$J = o\left(\frac{1}{\kappa}\right).$$

Thus, both estimates (8.12.47) and (8.12.48) are proved.

Note that the desired relation $\mathcal{J} = o(\frac{1}{\kappa})$ could have been obtained even by replacing W^2 by the smaller quantity 1 in the above argument.

Estimate (8.12.45) implies

$$I \le ce^{a|\eta|} o\left(\frac{1}{\sqrt{\kappa^2 + \eta^2}}\right), \quad \kappa \to \infty, \quad \eta = a^{-1} \ln \kappa + O(1). \tag{8.12.49}$$

The quantity $\eta = \eta(k) = a^{-1} \ln \kappa + O(1)$ was chosen so that if $\kappa \to \infty$, then the quantity $\frac{e^{|\eta|a}}{\sqrt{\kappa^2+\eta^2}}$ remains bounded as $\kappa \to \infty$. Therefore, estimate (8.12.49) implies

$$\lim_{\kappa\to\infty, \eta=a^{-1}\ln\kappa+O(1)} I = 0. \tag{8.12.50}$$

Consequently, estimate (8.12.43) holds.

Theorem 8.12.1 is proved. □

8.12.4. *Estimate of the norm of the operator T^2*

Let

$$Tf := \int_{B_a} G(x-y, \kappa+i\eta)q(y)f(y)dy. \tag{8.12.51}$$

Assume $q \in H_0^\ell(B_a)$, $\ell > 2$, $f \in C(B_a)$. Our goal is to prove that equation (8.12.12) can be solved by iterations for all sufficiently large κ.

Consider T as an operator in $C(B_a)$. One has:

$$\begin{aligned} T^2 f &= \int_{B_a} dz G(x-z, \kappa+i\eta)q(z)\int_{B_a} G(z-y, \kappa+i\eta)q(y)f(y)dy \\ &= \int_{B_a} dy f(y)q(y)\int_{B_a} dz q(z)G(x-z, \kappa+i\eta)G(z-y, \kappa+i\eta). \end{aligned} \tag{8.12.52}$$

Let us estimate the integral

$$\begin{aligned} I(x,y) &:= \int_{B_a} G(x-z, \kappa+i\eta)G(z-y, \kappa+i\eta)q(z)dz \\ &= \int_{B_a} \frac{e^{i(\kappa+i\eta)[|x-z|-\beta\cdot(x-z)+|z-y|-\beta\cdot(z-y)]}}{16\pi^2|x-z||z-y|}q(z)dz \\ &= \frac{1}{16\pi^2}\int_{B_a} \frac{e^{i(\kappa+i\eta)[|x-z|+|z-y|-\beta\cdot(x-y)]}}{|x-z||z-y|}q(z)dz \\ &:= \frac{e^{-i(\kappa+i\eta)\beta\cdot(x-y)}}{16\pi^2}I_1(x,y). \end{aligned} \tag{8.12.53}$$

Let us use the following coordinates (see [269, p. 391]):

$$z_1 = \ell st + \frac{x_1 + y_1}{2}, \quad z_2 = \ell\sqrt{(s^2-1)(1-t^2)}\cos\psi + \frac{x_2+y_2}{2}, \tag{8.12.54}$$

$$z_3 = \ell\sqrt{(s^2-1)(1-t^2)}\sin\psi + \frac{x_3+y_3}{2}. \tag{8.12.55}$$

The Jacobian J of the ransformation $(z_1, z_2, z_3) \to (\ell, t, \psi)$ is

$$J = \ell^3(s^2 - t^2), \tag{8.12.56}$$

where

$$\ell = \frac{|x-y|}{2}, \quad |x-z| + |z-y| = 2\ell s, \quad |x-z| - |z-y| = 2\ell t, \tag{8.12.57}$$

$$|x-z||z-y| = 4\ell^2(s^2-t^2), \quad 0 \le \psi < 2\pi,$$
$$t \in [-1,1], \ s \in [1,\infty). \tag{8.12.58}$$

One has

$$I_1 = \ell \int_a^\infty e^{2i(\kappa+i\eta)\ell s} Q(s)ds, \tag{8.12.59}$$

where

$$Q(s) := Q\left(s, \ell, \frac{x+y}{2}\right) = \int_0^{2\pi} d\psi \int_{-1}^1 dtq\left(z\left(s,t,\psi;\ell,\frac{x+y}{2}\right)\right), \tag{8.12.60}$$

and the function $Q(s) \in H_0^2(\mathbb{R}^3)$ for any fixed x, y. Therefore, an integration by parts in (8.12.59) yields the following estimate:

$$|I_1| = O\left(\frac{1}{|\kappa+i\eta|}\right), \quad |\kappa+i\eta| \to \infty. \tag{8.12.61}$$

From (8.12.52), (8.12.53), and (8.12.61) one gets:

$$\|T^2\| = O\left(\frac{1}{\sqrt{\gamma}}\right), \qquad \gamma := \kappa^2 + \eta^2 \to \infty. \tag{8.12.62}$$

Therefore, integral equation (8.12.12), with k replaced by $\frac{\kappa+i\eta}{2}$, can be solved by iterations if γ is sufficiently large and $\eta \ge 0$.

Consequently, integral equation (8.12.13) can be solved by iterations. Thus, estimate (8.12.43) holds if such an estimate holds for the free term in equation (8.12.13), that is, for the function $\frac{\tilde{q}}{\xi^2-(\kappa+i\eta)\beta\cdot\xi}$, namely, if estimate (8.12.50) holds.

8.13. Uniqueness of the Solution to Inverse Scattering Problem with Scattering Data at a Fixed Direction of the Incident Wave

8.13.1. *Introduction*

This section follows [401].[a] Let $q(x)$ be real-valued compactly supported sufficiently smooth function. It is proved that the scattering data $A(\beta, \alpha_0, k)$ $\forall\beta \in S^2$, $\forall k > 0$, determine q uniquely. Here, $\alpha_0 \in S^2$ is a fixed direction of the incident plane wave.

The scattering solution $u(x, \alpha, k)$ solves the scattering problem:

$$[\nabla^2 + k^2 - q(x)]u = 0 \quad in \quad \mathbb{R}^3, \tag{8.13.1}$$

$$u = e^{ik\alpha\cdot x} + A(\beta, \alpha, k)\frac{e^{ikr}}{r} + o\left(\frac{1}{r}\right), \quad r := |x| \to \infty, \; \beta := \frac{x}{r}. \tag{8.13.2}$$

Here, $\alpha, \beta \in S^2$ are the unit vectors, S^2 is the unit sphere, the coefficient $A(\beta, \alpha, k)$ is called the scattering amplitude, $q(x)$ is a real-valued compactly supported sufficiently smooth function. The inverse scattering problem of interest is to determine $q(x)$ given the scattering data $A(\beta, \alpha_0, k)$ $\forall\beta \in S^2$, $\forall k > 0$. This problem is called *the inverse scattering problem with fixed direction of the incident plane wave data.*

The function $A(\beta, \alpha_0, k)$ depends on one unit vector β and on the scalar k, i.e., on three variables. The potential $q(x)$ depends also on three variables $x \in \mathbb{R}^3$. This inverse problem is, therefore, not

---

[a]Reproduced from *J. Math. Phys.* **52**, 123506 (2011), with the permission of AIP Publishing.

overdetermined in the sense that the data and the unknown $q(x)$ are functions of the same number of variables.

The inverse scattering problem with fixed direction α_0 of the incident plane wave scattering data $A(\beta, \alpha_0, k)$, known for all $\beta \in S^2$ and all $k > 0$, has been open from the forties of the last century. In this chapter, we prove uniqueness of the solution to this inverse problem under the *Assumption* (A) formulated below. Although there is a large literature on inverse scattering (see [97, 356]), the above problem has been open from mid-forties of the last century, and the references we give are only to the papers directly related to our presentation.

Let B_a be the ball centered at the origin and of radius a, and $H_0^\ell(B_a)$ be the closure of $C_0^\infty(B_a)$ in the norm of the Sobolev space $H^\ell(B_a)$ of functions whose derivatives up to the order ℓ belong to $L^2(B_a)$.

Assumption (A): *We assume that q is compactly supported, i.e., $q(x) = 0$ for $|x| > a$, where $a > 0$ is an arbitrary large fixed number; $q(x)$ is real-valued, i.e., $q = \overline{q}$; and $q(x) \in H_0^\ell(B_a)$, $\ell > 3$.*

It was proved in [184, 199] (see also [365, Chapter 6]), that if $q = \overline{q}$ and $q \in L^2(B_a)$ is compactly supported, then the resolvent kernel $G(x, y, k)$ of the Schrödinger operator $-\nabla^2 + q(x) - k^2$ is a meromorphic function of k on the whole complex plane k, analytic in Im $k \geq 0$, except, possibly, of a finitely many simple poles at the points ik_j, $k_j > 0$, $1 \leq j \leq n$, where $-k_j^2$ are negative eigenvalues of the self-adjoint operator $-\nabla^2 + q(x)$ in $L^2(\mathbb{R}^3)$. Consequently, the scattering amplitude $A(\beta, \alpha, k)$, corresponding to the above q, is a restriction to the positive semi-axis $k \in [0, \infty)$ of a meromorphic on the whole complex k-plane function.

It was proved by the author ([295, 356], see also Chapter 8), that the *fixed-energy scattering data* $A(\beta, \alpha) := A(\beta, \alpha, k_0)$, $k_0 =$ const > 0, $\forall \beta \in S_1^2$, $\forall \alpha \in S_2^2$, determine real-valued compactly supported $q \in L^2(B_a)$ uniquely. Here, S_j^2, $j = 1, 2$, are arbitrary small open subsets of S^2 (solid angles). No uniqueness results for the potentials which decay at a power rate are known if the scattering data are known at a fixed energy. If the potentials decay faster than exponentially as $|x| \to \infty$, then a uniqueness result for this problem is obtained in [320]. If the potential decays at a power rate but the scattering data are known for all $k > 0$, all $\alpha \in S^2$ and all $\beta \in S^2$, then a uniqueness results was obtained in [439].

In [345] (see also [356, Chapter 5] and [317]), an analytical formula is derived for the reconstruction of the potential q from exact fixed-energy scattering data, and from noisy fixed-energy scattering data, and stability estimates and error estimates for the reconstruction method are obtained. To the author's knowledge, these are the only known until now theoretical error estimates for the recovery of the potential from noisy fixed-energy scattering data in the three-dimensional inverse scattering problem.

In [356] and also in Chapter 8, the relation of the scattering data and the D–N map is used for proving the uniqueness theorems with the boundary data. Knowing these data is equivalent to knowing the D–N map. These data are overdetermined.

The scattering data $A(\beta, \alpha)$ depend on four variables (two unit vectors), while the unknown $q(x)$ depends on three variables. In this sense, the inverse scattering problem, which consists of finding q from the fixed-energy scattering data $A(\beta, \alpha)$, is overdetermined.

In [324], stability results are obtained for the inverse scattering problem for obstacles.

The first uniqueness theorem for the three-dimensional inverse scattering problem with non-overdetermined data was announced by the author in [389], where the uniqueness of the solution to the three-dimensional inverse scattering problem with backscattering data was studied, and a proof of the uniqueness of its solution was outlined. In Section 8.12, the details of this proof were presented for the data $A(-\beta, \beta, k)$ $\forall \beta \in S^2$, $\forall k > 0$. The goal of this section is to prove a uniqueness theorem for the three-dimensional inverse scattering problem with the scattering data $A(\beta, \alpha_0, k)$ $\forall \beta \in S^2$ and $\forall k > 0$. These data are also non-overdetermined. Our work is based on the method developed in Section 8.12, but the presentation is self-contained. The technical details of our proof differ from these in [389], for instance, in the derivation of the important relation (8.13.30).

Theorem 8.13.1. *If Assumption (A) holds, then the data* $A(\beta, \alpha_0, k)$ $\forall \beta \in S^2$, $\forall k > 0$, *and a fixed* $\alpha_0 \in S^2$, *determine* q *uniquely.*

Remark 8.13.1. The conclusion of Theorem 8.13.1 remains valid if the data $A(\beta, \alpha_0, k)$ are known $\forall \beta \in S_1^2$ and $k \in (k_0, k_1)$, where $(k_0, k_1) \subset [0, \infty)$ is an arbitrary small interval, $k_1 > k_0$, and S_1^2 is an arbitrary small open subset of S^2.

The auxiliary results we need are the same as in Section 8.12.

In Section 8.13.2, proof of Theorem 8.13.1 is given.

8.13.2. *Proof of Theorem* 8.13.1

The scattering data in Theorem 8.13.1 determine uniquely these data $\forall \beta \in S^2$ and all $k > 0$.

Let us outline the ideas of the proof of Theorem 8.13.1.

This outline is similar to the one in Section 8.12. Our arguments are self-contained.

Assume that potentials q_j, $j = 1, 2$, generate the same scattering data:

$$A_1(\beta, \alpha_0, k) = A_2(\beta, \alpha_0, k), \qquad \forall \beta \in S^2,\ \forall k > 0,$$

and let

$$p(x) := q_1(x) - q_2(x).$$

Then by Lemma 8.13.2, one gets

$$0 = \int_{B_a} p(x) u_1(x, \alpha_0, k) u_2(x, -\beta, k) dx, \quad \forall \beta \in S^2,\ \forall k > 0. \tag{8.13.3}$$

One can rewrite (8.13.3) as

$$\int_{B_a} e^{i\kappa\zeta\cdot x}[1 + \epsilon(x, k)] p(x) dx = 0, \quad \forall \zeta \in S^2_+,\ \forall k > 0, \tag{8.13.4}$$

where

$$\epsilon(x,k) := \epsilon := \epsilon_1(x,k) + \epsilon_2(x,k) + \epsilon_1(x,k)\epsilon_2(x,k),$$

and we have denoted $|\alpha_0 - \beta| := \tau$, $\zeta := (\alpha_0 - \beta)/\tau$, $\kappa := \tau k$. Without loss of generality, one may assume that α_0 is the unit vector along x_3-axis. Then τ runs through $[0, 2]$ and the unit vector ζ runs through S^2_+, the upper-half of the unit sphere S^2. Since $k \in [0, \infty)$ is arbitrary in (8.13.4), so is $\kappa = \tau k$. Because the left-hand side of (8.13.4) depends on ζ analytically on the variety $\mathcal{M}$, one concludes that relation (8.13.4) holds for any $\zeta \in S^2$ if it holds for $\zeta \in S^2_+$. So, from now on we will use formula (8.13.4) with $\zeta \in S^2$ being arbitrary.

The relations (8.13.3) and (8.13.4) hold for complex k,

$$\tau k = \kappa + i\eta, \quad \eta \geq 0. \tag{8.13.5}$$

As in Section 8.12, one derives from (8.13.4) the relation

$$\tilde{p}((\kappa + i\eta)\zeta) + \frac{1}{(2\pi)^3}(\tilde{\epsilon} * \tilde{p})((\kappa + i\eta)\zeta) = 0, \qquad \forall\zeta \in S^2, \ \forall\kappa \in \mathbb{R}, \tag{8.13.6}$$

where the notation $(f * g)(z)$ means that the convolution $f * g$ is calculated at the argument $z = (\kappa + i\eta)\zeta$.

One has

$$\sup_{\zeta\in S^2} |\tilde{\epsilon} * \tilde{p}| := \sup_{\zeta\in S^2} \left| \int_{\mathbb{R}^3} \tilde{\epsilon}((\kappa + i\eta)\zeta - s)\tilde{p}(s)ds \right| \leq \nu(\kappa, \eta) \sup_{s\in\mathbb{R}^3} |\tilde{p}(s)|, \tag{8.13.7}$$

where

$$\nu(\kappa, \eta) := \sup_{\zeta\in S^2} \int_{\mathbb{R}^3} |\tilde{\epsilon}((\kappa + i\eta)\zeta - s)|ds.$$

We prove that if $\eta = \eta(\kappa) = O(\ln \kappa)$ is suitably chosen, namely as in (8.13.12) below, then the following inequality holds:

$$0 < \nu(\kappa, \eta(\kappa)) < 1, \qquad \kappa \to \infty. \tag{8.13.8}$$

We also prove that

$$\sup_{\zeta\in S^2} |\tilde{p}((\kappa + i\eta(\kappa))\zeta)| \geq \sup_{s\in\mathbb{R}^3} |\tilde{p}(s)|, \quad \kappa \to \infty, \tag{8.13.9}$$

and then it follows from (8.13.6)–(8.13.9) that $\tilde{p}(s) = 0$, so $p(x) = 0$, and Theorem 8.13.1 is proved. Indeed, it follows from (8.13.6) and (8.13.9) that, for sufficiently large κ and a suitable $\eta(k) = O(\ln k)$, one has

$$\sup_{s\in\mathbb{R}^3} |\tilde{p}(s)| \leq \frac{1}{(2\pi)^3}\nu(\kappa, \eta(\kappa)) \sup_{s\in\mathbb{R}^3} |\tilde{p}(s)|.$$

If (8.13.8) holds, then the above equation implies that $\tilde{p} = 0$. This and the injectivity of the Fourier transform imply that $p = 0$.

This completes the outline of the proof of Theorem 8.13.1.

Let us now give a detailed proof of estimates (8.13.8) and (8.13.9) that completes the proof of Theorem 8.13.1. We denote ζ by β in what follows, since both unit vectors run through all of S^2.

We assume that $p(x) \not\equiv 0$, because otherwise there is nothing to prove. Let

$$\max_{s\in\mathbb{R}^3} |\tilde{p}(s)| := \mathcal{P} \neq 0.$$

Lemma 8.13.2. *If Assumption* (A) *holds and* $\mathcal{P} \neq 0$, *then*

$$\limsup_{\eta\to\infty} \max_{\beta\in S^2} |\tilde{p}((\kappa + i\eta)\beta)| = \infty, \tag{8.13.10}$$

where $\kappa > 0$ *is arbitrary but fixed. For any* $\kappa > 0$, *there is an* $\eta = \eta(\kappa)$, *such that*

$$\max_{\beta\in S^2} |\tilde{p}((\kappa + i\eta(\kappa))\beta)| = \mathcal{P}, \tag{8.13.11}$$

where the number $\mathcal{P} := \max_{s\in\mathbb{R}^3} |\tilde{p}(s)|$, *and*

$$\eta(\kappa) = a^{-1}\ln\kappa + O(1) \quad \textit{as} \quad \kappa \to +\infty. \tag{8.13.12}$$

Proof. By formula (8.12.17), one gets

$$\tilde{p}((\kappa + i\eta)\beta) = \int_{B_a} p(x)e^{i(\kappa+i\eta)\beta\cdot x}dx = \int_{-a}^{a} e^{i\kappa\lambda-\eta\lambda}\hat{p}(\beta, \lambda)d\lambda. \tag{8.13.13}$$

The function $\hat{p}(\beta, \lambda)$ is compactly supported, real-valued, and satisfies relation (8.12.18). Therefore,

$$\max_{\beta\in S^2} |\tilde{p}((\kappa + i\eta(\kappa))\beta)| = \max_{\beta\in S^2} |\tilde{p}((\kappa - i\eta(\kappa))\beta)|. \tag{8.13.14}$$

Indeed,

$$\begin{aligned}\max_{\beta\in S^2}|\tilde{p}((\kappa+i\eta(\kappa))\beta)| &= \max_{\beta\in S^2}\left|\int_{-a}^{a} e^{i\kappa\lambda-\eta\lambda}\hat{p}(\beta,\lambda)d\lambda\right|\\ &= \max_{\beta\in S^2}\left|\int_{-a}^{a} e^{-i\kappa\mu+\eta\mu}\hat{p}(\beta,-\mu)d\mu\right|\\ &= \max_{\beta'\in S^2}\left|\int_{-a}^{a} e^{-i\kappa\mu+\eta\mu}\hat{p}(-\beta',-\mu)d\mu\right|\\ &= \max_{\beta'\in S^2}\left|\int_{-a}^{a} e^{-i\kappa\mu+\eta\mu}\hat{p}(\beta',\mu)d\mu\right|\\ &= \max_{\beta\in S^2}|\tilde{p}((\kappa-i\eta)\beta)|. \qquad (8.13.15)\end{aligned}$$

At the last step, we took into account that $\hat{p}(\beta,\lambda)$ is a real-valued function, so

$$\begin{aligned}\max_{\beta\in S^2}\left|\int_{-a}^{a} e^{-i\kappa\mu+\eta\mu}\hat{p}(\beta,\mu)d\mu\right| &= \max_{\beta\in S^2}\left|\int_{-a}^{a} e^{i\kappa\mu+\eta\mu}\hat{p}(\beta,\mu)d\mu\right|\\ &= \max_{\beta\in S^2}|\tilde{p}((\kappa-i\eta)\beta)|. \qquad (8.13.16)\end{aligned}$$

If $p(x)\not\equiv 0$, then (8.13.13) and (8.13.14) imply (8.13.10). Let us give a detailed proof of this statement.

Consider the function h of the complex variable $z:=\kappa+i\eta$:

$$h:=h(z,\beta):=\int_{-a}^{a} e^{iz\lambda}\hat{p}(\beta,\lambda)d\lambda. \qquad (8.13.17)$$

If (8.13.10) is false, then

$$|h(z,\beta)|\le c \quad \forall z=\kappa+i\eta, \quad \eta\ge 0, \quad \forall\beta\in S^2, \qquad (8.13.18)$$

where $\kappa\ge 0$ is an arbitrary fixed number and the constant $c>0$ does not depend on β and η.

Thus, $|h|$ is bounded on the ray $\{\kappa=0,\eta\ge 0\}$, which is part of the boundary of the right angle $\mathcal{A}$, and the other part of its boundary is the ray $\{\kappa\ge 0,\eta=0\}$. Let us check that $|h|$ is bounded on this ray also.

One has

$$|h(\kappa,\beta)| = \left|\int_{-a}^{a} e^{i\kappa\lambda}\hat{p}(\beta,\lambda)d\lambda\right| \leq \int_{-a}^{a} |\hat{p}(\beta,\lambda)|d\lambda \leq c, \qquad (8.13.19)$$

where c stands in this paper for *various* constants. From (8.13.18)–(8.13.19), it follows that on the boundary of the right angle $\mathcal{A}$, namely, on the two rays $\{\kappa \geq 0, \eta = 0\}$ and $\{\kappa = 0, \eta \geq 0\}$ the entire function $h(z,\beta)$ of the complex variable z is bounded, $|h(z,\beta)| \leq c$, and inside $\mathcal{A}$ this function satisfies the estimate

$$|h(z,\beta)| \leq e^{a|\eta|}\int_{-a}^{a} |\hat{p}(\beta,\lambda)|d\lambda \leq ce^{a|\eta|}, \qquad (8.13.20)$$

where c does not depend on β. Therefore, by Lemma 8.12.4, $|h(z,\beta)| \leq c$ in the whole angle $\mathcal{A}$.

By (8.13.14), the same argument is applicable to the remaining three right angles, the union of which is the whole complex z-plane $\mathbb{C}$. Therefore,

$$\sup_{z\in\mathbb{C},\beta\in S^2} |h(z,\beta)| \leq c. \qquad (8.13.21)$$

This implies, by the Liouville theorem, that $h(z,\beta) = c\ \forall z \in \mathbb{C}$.

Since $\hat{p}(\beta,\lambda) \in L^1(-a,a)$, the relation

$$\int_{-a}^{a} e^{iz\lambda}\hat{p}(\beta,\lambda)d\lambda = c \quad \forall z \in \mathbb{C}, \qquad (8.13.22)$$

and the Riemann–Lebesgue lemma imply that $c = 0$, so $\hat{p}(\beta,\lambda) = 0$ $\forall\beta \in S^2$ and $\forall\lambda \in \mathbb{R}$. Therefore, $p(x) = 0$, contrary to our assumption. Consequently, relation (8.13.10) is proved.

Relation (8.13.11) follows from (8.13.10) because for large η the left-hand side of (8.13.11) is larger than $\mathcal{P}$ due to (8.13.10), while for $\eta = 0$ the left-hand side of (8.13.11) is not larger than $\mathcal{P}$ by the definition of the Fourier transform.

Let us derive estimate (8.13.12).

From the assumption $p(x) \in H_0^\ell(B_a)$, it follows that

$$|\tilde{p}((\kappa + i\eta)\beta)| \le c\frac{e^{a|\eta|}}{(1+\kappa^2+\eta^2)^{\ell/2}}. \tag{8.13.23}$$

This inequality is proved in Lemma 8.13.3 below.

The right-hand side of this inequality is of the order $O(1)$ as $\kappa \to \infty$ if $|\eta| = a^{-1}\ln\kappa + O(1)$ as $\kappa \to \infty$. This proves relation (8.13.12) and we specify $O(\ln\kappa)$ as in this relation.

Let us now prove inequality (8.13.23).

Lemma 8.13.3. *If* $p \in H_0^\ell(B_a)$, *then estimate* (8.13.23) *holds.*

Proof. Consider $\partial_j p := \frac{\partial p}{\partial x_j}$. One has

$$\begin{aligned}\left|\int_{B_a} e^{i(\kappa+i\eta)\beta\cdot x}\partial_j p dx\right| &= \left|-i(\kappa+i\eta)\beta_j\int_{B_a} p(x)e^{i(\kappa+i\eta)\beta\cdot x}\partial_j p\, dx\right| \\ &= (\kappa^2+\eta^2)^{1/2}|\tilde{p}((\kappa+i\eta)\beta)|.\end{aligned} \tag{8.13.24}$$

The left-hand side of the above formula admits the following estimate

$$\left|\int_{B_a} e^{i(\kappa+i\eta)\beta\cdot x}\partial_j p\, dx\right| \le ce^{a|\eta|},$$

where the constant $c > 0$ is proportional to $\|\partial_j p\|_{L^2(B_a)}$. Therefore,

$$|\tilde{p}((\kappa+i\eta)\beta)| \le c[1+(\kappa^2+\eta^2)]^{-1/2}e^{a|\eta|}. \tag{8.13.25}$$

Repeating this argument, one gets estimate (8.13.23). Lemma 3.2 is proved. □

Estimate (8.13.25) implies that if relation (8.13.12) holds and $\kappa \to \infty$, then the quantity $\sup_{\beta\in S^2}|\tilde{p}((\kappa+i\eta)\beta)|$ remains bounded as $\kappa \to \infty$.

If η is fixed and $\kappa \to \infty$, then $\sup_{\beta\in S^2}|\tilde{p}((\kappa+i\eta)\beta)| \to 0$ by the Riemann–Lebesgue lemma. This, the continuity of $|\tilde{p}((\kappa+i\eta)\beta)|$ with respect to η, and relation (8.13.10), imply the existence of $\eta = \eta(\kappa)$, such that equality (8.13.11) holds, and, consequently, inequality (8.13.9) holds. This $\eta(\kappa)$ satisfies (8.13.12) because $\mathcal{P}$ is bounded.

Lemma 8.13.3 is proved. □

To complete the proof of Theorem 8.13.1, one has to establish estimate (8.13.8). This estimate will be established if one proves the following relation:

$$\lim_{\kappa\to\infty} \nu(\kappa) := \lim_{\kappa\to\infty} \nu(\kappa, \eta(\kappa)) = 0, \tag{8.13.26}$$

where $\eta(\kappa)$ satisfies (8.13.12) and

$$\nu(\kappa, \eta) = \sup_{\beta\in S^2} \int_{\mathbb{R}^3} |\tilde{\epsilon}((\kappa + i\eta)\beta - s)| ds. \tag{8.13.27}$$

Our argument is valid for ϵ_1, ϵ_2 and $\epsilon_1\epsilon_2$, so we will use the letter ϵ and equation (8.12.13) for $\tilde{\epsilon}$.

Below we denote $2k := \kappa + i\eta$ and we choose $\eta = \eta(\kappa) = a^{-1}\ln \kappa + O(1)$ as $\kappa \to \infty$.

We prove that equation (8.12.12) can be solved by iterations if $\eta \geq 0$ and $|\kappa + i\eta|$ is sufficiently large, because for such $\kappa + i\eta$ the operator T^2 has small norm in $C(B_a)$, the space of functions, continuous in the ball B_a, with the sup-norm. Since equation (8.12.12) can be solved by iterations and the norm of T^2 is small, the main term in the series, representing its solution, as $|\kappa + i\eta| \to \infty$, $\eta \geq 0$, is the free term of equation (8.12.12). The same is true for the Fourier transform of equation (8.12.12), i.e., for equation (8.12.13). Therefore, the main term of the solution $\tilde{\epsilon}$ to equation (8.12.13) as $|\kappa + i\eta| \to \infty$, $\eta \geq 0$, is obtained by using the estimate of the free term of this equation. Thus, it is sufficient to check estimate (8.13.26) for the function $\nu(\kappa, \eta(\kappa))$ using in place of $\tilde{\epsilon}$ the function $\tilde{q}(\xi)(\xi^2 - 2k\beta\cdot\xi)^{-1}$, with $2k$ replaced by $\kappa + i\eta$ and $\eta = a^{-1}\ln \kappa + O(1)$ as $\kappa \to \infty$.

Recall that the operator

$$T\epsilon = \int_{B_a} G(x - y, k)q(y)\epsilon(y, \beta, k)dy,$$

has the property that the norm $\|T^2\|$ in the space $C(B_a)$ tends to zero as $|\kappa + i\eta| \to \infty$, $\eta \geq 0$.

Thus, let us estimate the modulus of the factor $\nu(\kappa, \eta)$ in (8.13.7) with $\eta = \eta(\kappa)$ as in (8.13.12). Using inequality (8.13.23), and denoting $\xi = (\kappa + i\eta)\beta$, where $\beta \in S^2$, one obtains:

$$\begin{aligned} I &:= \sup_{\beta \in S^2} \int_{\mathbb{R}^3} \frac{|\tilde{q}((\kappa + i\eta)\beta - s)|ds}{|[(\kappa + i\eta)\beta - s)^2 - (\kappa + i\eta)\beta \cdot ((\kappa + i\eta)\beta - s)]|} \\ &\leq ce^{a|\eta|} \sup_{\beta \in S^2} \int_{\mathbb{R}^3} \frac{ds}{|s^2 - (\kappa + i\eta)\beta \cdot s|[1 + (\kappa\beta - s)^2 + \eta^2]^{\ell/2}}. \\ &:= ce^{a|\eta|} J. \end{aligned} \tag{8.13.28}$$

Let us prove that

$$J = o\left(\frac{1}{\kappa}\right), \quad \kappa \to \infty.$$

If this estimate is proved and $\eta = a^{-1} \ln \kappa + O(1)$, then $I = o(1)$ as $\kappa \to \infty$, therefore relation (8.13.26) follows, and Theorem 8.13.1 is proved.

Let us write the integral J in the spherical coordinates with x_3-axis directed along vector β. We have

$$|s| = r, \qquad \beta \cdot s = r\cos\theta := rt, \quad -1 \leq t \leq 1.$$

Denote

$$\gamma := \kappa^2 + \eta^2.$$

Then

$$\begin{aligned} J &\leq 2\pi \int_0^\infty drr \int_{-1}^1 \frac{dt}{[(r - \kappa t)^2 + \eta^2 t^2]^{1/2}(1 + \gamma + r^2 - 2r\kappa t)^{\ell/2}} \\ &:= 2\pi \int_0^\infty drr B(r), \end{aligned} \tag{8.13.29}$$

where

$$B := B(r) = B(r, \kappa, \eta)$$
$$:= \int_{-1}^{1} \frac{dt}{[(r-\kappa t)^2 + \eta^2 t^2]^{1/2}(1+\gamma+r^2-2r\kappa t)^{\ell/2}}.$$

Estimate of J we start with the observation

$$\tau := \min_{t\in[-1,1]} [(r-\kappa t)^2 + \eta^2 t^2] = \min\{r^2\eta^2/\gamma,\ (r-\kappa)^2 + \eta^2\}.$$

Let $\tau = r^2\eta^2/\gamma$, which is always the case if r is sufficiently small. In the case when $\tau = (r-\kappa)^2 + \eta^2$, the proof is considerably simpler and is left for the reader. If $\tau = r^2\eta^2/\gamma$, then

$$J \leq 2\pi\gamma^{1/2}\eta^{-1} \int_0^\infty dr \int_{-1}^{1} dt[1+\gamma+r^2-2\kappa rt]^{-\ell/2}.$$

Integrating over t yields

$$J \leq 2\pi\gamma^{1/2}\eta^{-1}[(\ell-2)\kappa]^{-1}\mathcal{J},$$

where

$$\mathcal{J} := \int_0^\infty drr^{-1}[(1+\gamma+r^2-2\kappa r)^{-b} - (1+\gamma+r^2+2\kappa r)^{-b}],$$

and $b := \ell/2 - 1$.

Since $\eta = O(\ln\kappa)$, one has $\frac{\eta}{\kappa} = o(1)$ as $\kappa \to \infty$. Therefore,

$$\gamma^{1/2}\eta^{-1}\kappa^{-1} = O(\eta^{-1}) \qquad \text{as } \kappa \to \infty.$$

Since $\ell > 3$, one has $b > \frac{1}{2}$, and, as we prove below,

$$\mathcal{J} = o\left(\frac{1}{\kappa}\right) \qquad \text{as } \kappa \to \infty. \tag{8.13.30}$$

This relation implies the desired inequality:

$$J \leq o\left(\frac{1}{\kappa}\right) \qquad \text{as } \kappa \to \infty. \tag{8.13.31}$$

Let us derive relation (8.13.30). One has

$$\mathcal{J} = \int_0^1 + \int_1^\infty := J_1 + J_2,$$

$$J_1 \le \int_0^1 dr r^{-1} \frac{(w^2 + 2r\kappa + r^2)^b - (w^2 - 2r\kappa + r^2)^b}{(w^2 + 2r\kappa + r^2)^b (w^2 - 2r\kappa + r^2)^b},$$

where

$$w^2 := 1 + \gamma = 1 + \eta^2 + \kappa^2.$$

Furthermore,

$$(w^2 + 2r\kappa + r^2)^b - (w^2 - 2r\kappa + r^2)^b \le \frac{4br\kappa}{(w^2 - 2r\kappa + r^2)^{1-b}}.$$

Thus,

$$J_1 \le 4b\kappa \int_0^1 dr \frac{1}{(w^2 + 2r\kappa + r^2)^b (w^2 - 2r\kappa + r^2)}.$$

This implies the following estimate

$$J_1 \le O(\kappa / w^{2+2b}) \le O(\kappa^{-(1+2b)}),$$

because $w = \kappa[1 + o(1)]$ as $\kappa \to \infty$. Furthermore,

$$J_2 \le \int_1^\infty dr r^{-1} [(1+\eta^2+(r-\kappa)^2)^{-b} - (1+\eta^2+(r+\kappa)^2)^{-b}] := J_{21} - J_{22}.$$

One has $J_{22} \le J_{21}$.

Let us estimate J_{21}. One obtains

$$J_{21} = \int_1^{\kappa/2} + \int_{\kappa/2}^\infty := j_1 + j_2,$$

and

$$j_1 \le \frac{1}{[W^2 + \frac{\kappa^2}{4}]^b} \ln \kappa = o\left(\frac{1}{\kappa}\right), \qquad W^2 := 1 + \eta^2, \qquad b > \frac{1}{2}.$$

Furthermore

$$j_2 \le \frac{2}{\kappa} \int_{\kappa/2}^\infty \frac{dr}{[W^2 + (r-\kappa)^2]^b} \le \frac{2}{\kappa} \int_{-\infty}^\infty \frac{dy}{[W^2 + y^2]^b} = o\left(\frac{1}{\kappa}\right).$$

Thus, if $b > \frac{1}{2}$, then $J_2 = o(\frac{1}{\kappa})$ and $\mathcal{J} = J_1 + J_2 = o(\frac{1}{\kappa})$. Thus, relation (8.13.30) is proved.

Relation (8.13.30) yields the desired estimate

$$J = o\left(\frac{1}{\kappa}\right).$$

Thus, both estimates (8.13.30) and (8.13.31) are proved.

Estimate (8.13.28) implies

$$I \leq ce^{a|\eta|}o\left(\frac{1}{\sqrt{\kappa^2+\eta^2}}\right), \quad \kappa \to \infty, \quad \eta = a^{-1}\ln\kappa + O(1). \tag{8.13.32}$$

The quantity $\eta = \eta(k) = a^{-1}\ln\kappa + O(1)$ was chosen so that if $\kappa \to \infty$, then the quantity $\frac{e^{|\eta|a}}{\sqrt{\kappa^2+\eta^2}}$ remains bounded as $\kappa \to \infty$. Therefore, esimate (8.13.32) implies

$$\lim_{\kappa\to\infty,\eta=a^{-1}\ln\kappa+O(1)} I = 0. \tag{8.13.33}$$

Consequently, estimate (8.13.26) holds.

Theorem 8.13.1 is proved. □

8.14. A Method for Numerical Solution of Inverse Scattering Problem

The contents of Section 8.14 is based on the paper [423]. The problem we are considering consists of finding a potential $q \in Q$ from the scattering amplitude $A(\beta, k) := A(\beta, \alpha_0, k)$, where $\alpha_0 \in S^2$ is fixed, S^2 is the unit sphere in $\mathbb{R}^3$, and $A(\beta, k)$ is known for all $\beta \in S^2_\beta$, where S^2_β is an open subset of S^2 and $k \in (a, b), 0 \leq a < b$. The set Q consists of real-valued compactly supported continuous potentials. The continuity assumption will be used only for numerical reconstruction: the uniqueness theorem $A(\beta, k) \to q \in Q$ was proved in Section 8.13, under the assumption $q \in L^2(B_R)$, where $B_R := \{x : |x| \leq R\}$, which is a weaker assumption than continuity.

Let us describe a numerical method for recovery of q from $A(\beta, k)$.

The integral equation for the scattering solution is

$$u = e^{ik\alpha_0 \cdot x} - \int g(x,y,k)q(y)u(y,\alpha_0,k)dy, \quad g(x,y,k) = \frac{e^{ik|x-y|}}{4\pi|x-y|}, \tag{8.14.1}$$

where the integral is taken over the support of q. Let

$$h(y,k) = q(y)u(y,\alpha_0,k).$$

The dependence on α_0 is dropped in what follows.

Multiply (8.14.1) by $q(x)$ and get:

$$h(x,k) = q(x)e^{ik\alpha_0 \cdot x} - q(x)\int g(x,y,k)h(y,k)dy. \tag{8.14.2}$$

The scattering amplitude is:

$$-4\pi A(\beta,k) = \int e^{-ik\beta \cdot y}h(y,k)dy. \tag{8.14.3}$$

Let us discretize (8.14.2) and (8.14.3):

$$h_{pm} = q(x_p)e^{ik_m \alpha_0 \cdot x_p} - q(x_p)\sum_{p'=1, p'\neq p}^{P} g(x_p, y_{p'}, k_m)h_{p'm}\Delta \tag{8.14.4}$$

$$-4\pi A_{jm} = \sum_{p=j}^{P} e^{-ik_m \beta_j \cdot y_p} h_{pm}\Delta, \tag{8.14.5}$$

where

$$h_{pm} := h(y_p, k_m), \quad A_{jm} := A(\beta_j, k_m) \tag{8.14.6}$$

and Δ is the element of volume.

Equation (8.14.5) is a linear algebraic system for P^2 unknown h_{pm} for every $j, 1 \leq j \leq P, \beta_i \neq \beta_{j'}$ if $j \neq j'$. If $\beta_j \cdot y_p \neq \beta_{j'} \cdot y_p$ for $j \neq j'$, then $\det\{e^{-ik_m\beta_j \cdot y_p}\} \neq 0 \forall j \in (1,\dots,P)$.

The linear algebraic system (8.14.5) is very ill-posed. Numerical solution to this system requires a regularization [369, 404].

If one solves system (8.14.5), then one can find the potential by the formula:

$$q(x_p) = \frac{h_{pm}}{e^{ik_m\alpha_0\cdot x_p} - \sum_{p'=1,p'\neq p}^{P} g(x_p, y_{p'}, k_m)h_{p'm}}. \tag{8.14.7}$$

The knowledge of $q(x_p)$ defines $q(x)$ if the set of points x_p is sufficiently dense.

Equation (8.14.7) holds for all $m, j, 1 \leq m, j \leq P$, as a compatibility condition. This condition is automatically satisfied for the limiting integral equation (8.14.2). One has:

$$q(x) = \frac{h(x,k)}{e^{ik\alpha_0\cdot x} - \int g(x,y,k)h(y,k)dy}. \tag{8.14.8}$$

The method, proposed in this section, is based on Theorem 8.13.1.

It is remarkable that our inversion procedure is linear although the inverse scattering problem is highly nonlinear. Stable method for solving linear Fredholm integral equation (8.14.3) of the first kind for $h(y, k)$ can be developed using DSM (Dynamical System Method), see [369, 404].

Part III

Modified Rayleigh Conjecture (MRC)

Chapter 9

Modified Rayleigh Conjecture

9.1. Modified Rayleigh Conjecture and Applications

9.1.1. *Introduction*

Modified Rayleigh conjecture (MRC) in scattering theory is proposed by the author and justified. MRC allows one to develop numerical algorithms for solving direct scattering problems related to acoustic wave scattering by soft and hard obstacles of arbitrary shapes. It gives an error estimate for solving the direct scattering problem. It suggests a numerical method for finding the shape of a starshaped obstacle from the scattering data.

Consider a bounded domain $D \subset \mathbb{R}^n$, $n = 3$, with a boundary S. The exterior domain is $D' = \mathbb{R}^3 \backslash D$. Assume that S is smooth and starshaped, that is, its equation can be written as

$$r = f(\alpha), \tag{9.1.1}$$

where $\alpha \in S^2$ is a unit vector and S^2 denotes the unit sphere in $\mathbb{R}^3$. Smoothness of S is used in (9.1.18) below. For solving the direct scattering problem by the method described in the beginning of Section 9.1.2, the boundary S can be Lipschitz. The acoustic wave scattering problem by a soft obstacle D consists in finding the (unique) solution to the problem (9.1.2)–(9.1.3):

$$(\nabla^2 + k^2)u = 0 \text{ in } D', \quad u = 0 \text{ on } S, \tag{9.1.2}$$

$$u = u_0 + v := u_0 + A(\beta, \alpha)\frac{e^{ikr}}{r} + o\left(\frac{1}{r}\right),$$
$$r := |x| \to \infty, \quad \beta := \frac{x}{r}. \tag{9.1.3}$$

Here, $u_0 := e^{ik\alpha\cdot x}$ is the incident field, v is the scattered field, $A(\beta, \alpha)$ is called the scattering amplitude, its k-dependence is not shown, and $k > 0$ is the wavenumber. Denote

$$A_\ell(\alpha) := \int_{S^2} A(\beta, \alpha)\overline{Y_\ell(\beta)}d\beta, \tag{9.1.4}$$

where $Y_\ell(\alpha)$ are the orthonormal spherical harmonics, $Y_\ell = Y_{\ell m}, -\ell \leq m \leq \ell$. Let $h_\ell(r)$ be the spherical Hankel functions, normalized so that $h_\ell(kr) \sim \frac{e^{ikr}}{r}$ as $r \to +\infty$. Let the ball $B_R := \{x : |x| \leq R\}$ contain D.

In the region $r > R$, the solution to (9.1.2)–(9.1.3) is:

$$u(x, \alpha) = e^{ik\alpha\cdot x} + \sum_{\ell=0}^{\infty} A_\ell(\alpha)\psi_\ell, \quad \psi_\ell := Y_\ell(\beta)h_\ell(kr),$$
$$r > R, \quad \beta = \frac{x}{r}, \tag{9.1.5}$$

summation includes summation with respect to m, $-\ell \leq m \leq \ell$, and $A_\ell(\alpha)$ are defined in (9.1.4).

Rayleigh conjecture (RC). *The series* (9.1.5) *converges up to the boundary* S (*originally RC dealt with periodic structures, gratings*).

This conjecture is wrong for many domains, although it holds, for example, for a ball [13]. If $n = 2$ and D is an ellipse, then the series analogous to (9.1.5) converges in the region $r > a$, where $2a$ is the distance between the foci of the ellipse [13]. In the engineering literature, there are numerical algorithms, based on the Rayleigh conjecture. Our aim is to give a formulation of a theorem we call *Modified Rayleigh conjecture* (*MRC*). It can be used in numerical solution of the direct and inverse scattering problems. We discuss the Dirichlet condition but similar argument is applicable to the Neumann boundary condition, corresponding to acoustically hard obstacles.

Fix $\epsilon > 0$, an arbitrary small number.

Lemma 9.1.1. *There exist $L = L(\epsilon)$ and $c_\ell = c_\ell(\epsilon)$ such that*

$$\|u_0 + \sum_{\ell=0}^{L(\epsilon)} c_\ell(\epsilon)\psi_\ell\|_{L^2(S)} \leq \epsilon. \tag{9.1.6}$$

If (9.1.6) and the boundary condition (9.1.2) hold, then

$$\|v_\epsilon - v\|_{L^2(S)} \leq \epsilon, \quad v_\epsilon := \sum_{\ell=0}^{L(\epsilon)} c_\ell(\epsilon)\psi_\ell. \tag{9.1.7}$$

Lemma 9.1.2. *If* (9.1.7) *holds, then*

$$\|v_\epsilon - v\| = O(\epsilon) \quad \epsilon \to 0, \tag{9.1.8}$$

where $\|\cdot\| := \|\cdot\|_{H^m_{\text{loc}}(D')} + \|\cdot\|_{L^2(D';(1+|x|)^{-\gamma})}$, $\gamma > 1$, $m > 0$ *is an arbitrary integer, and* H^m *is the Sobolev space.*

In particular, (9.1.8) implies

$$\|v_\epsilon - v\|_{L^2(S_R)} = O(\epsilon) \quad \epsilon \to 0, \tag{9.1.9}$$

where $S_R := \{x : |x| = R\}$.

Lemma 9.1.3. *One has*:

$$\lim_{\epsilon \to 0} c_\ell(\epsilon) = A_\ell(\alpha) \quad \forall \ell. \tag{9.1.10}$$

The MRC is formulated as a theorem, which follows from the above three lemmas:

Theorem 9.1.4 (MRC). *For an arbitrary small $\epsilon > 0$ there exist $L(\epsilon)$ and $c_\ell(\epsilon)$, $0 \leq \ell \leq L(\epsilon)$, such that* (9.1.6), (9.1.8) *and* (9.1.10) *hold.*

The difference between RC and MRC is:

(9.1.7) does not hold if one replaces v_ϵ by $\sum_{\ell=0}^{L} A_\ell(\alpha)\psi_\ell$, and let $L \to \infty$ (instead of letting $\epsilon \to 0$).

For the Neumann boundary condition, one minimizes $\|\frac{\partial[u_0 + \sum_{\ell=0}^{L} c_\ell\psi_\ell]}{\partial N}\|_{L^2(S)}$ with respect to c_ℓ. Analogs of Lemmas 9.1.1–9.1.3 are valid and their proofs are essentially the same.

In Section 9.1.2, we discuss the usage of MRC in solving the direct scattering problem, in Section 9.1.3 its usage in solving the inverse scattering problem, and in Section 9.1.4 proofs are given.

9.1.2. *Direct scattering problem and MRC*

The direct problem consists in finding the scattered field v given S and u_0. To solve it using MRC, fix a small $\epsilon > 0$ and find $L(\epsilon)$ and $c_\ell(\epsilon)$ such that (9.1.6) holds. This is possible by Lemma 9.1.1 and can be done numerically by minimizing $\|u_0 + \sum_0^L c_\ell \psi_\ell\|_{L^2(S)} := \phi(c_1, \ldots, c_L)$. If the minimum of ϕ is larger than ϵ, then increase L and repeat the minimization. Lemma 9.1.1 guarantees the existence of such L and c_ℓ that the minimum is less than ϵ. Choose the smallest L for which this happens and define $v_\epsilon := \sum_{\ell=0}^L c_\ell \psi_\ell(x)$. Then v_ϵ is the approximate solution to the direct scattering problem with the accuracy $O(\epsilon)$ in the norm $\|\cdot\|$ by Lemma 9.1.2.

In [347], representations of v and v_ϵ are proposed, which greatly simplified minimization of ϕ. Namely, let Ψ_ℓ solve problem

$$(\nabla^2 + k^2)\Psi_\ell = 0 \text{ in } D', \quad \Psi_\ell = f_\ell \text{ on } S, \tag{9.1.11}$$

and Ψ_ℓ satisfies the radiation condition. Here, $\{f_\ell\}_{\ell \geq 0}$ is an arbitrary orthonormal basis of $L^2(S)$. Denote

$$v(x) := \sum_{\ell=0}^{\infty} c_\ell \Psi_\ell(x), \quad u(x) := u_0 + v(x), \quad c_l := (-u_0, f_\ell)_{L^2(S)}. \tag{9.1.12}$$

The series (9.1.12) on S is a Fourier series which converges in $L^2(S)$. It converges pointwise in D' by the argument given in the proof of Lemma 9.1.2. A possible choice of f_ℓ for star-shaped S is $f_\ell = Y_\ell/\sqrt{w}$ where $w := dS/d\alpha$. Here, dS and $d\alpha$ are respectively the elements of the surface areas of the surface S and of the unit sphere S^2.

9.1.3. *Inverse scattering problem and MRC*

Inverse obstacle scattering problems IOSP(a)–IOSP(c) consist of finding S and the boundary condition on S from the knowledge of:

IOSP(a): the scattering data $A(\alpha', \alpha, k_0)$ for all $\alpha', \alpha \in S^2$, $k = k_0 > 0$ being fixed, or,

IOSP(b): $A(\alpha', \alpha_0, k)$, known for all $\alpha' \in S^2$ and all $k > 0$, $\alpha = \alpha_0 \in S^2$ being fixed,

or,

IOSP(c): $A(\beta, \alpha_0, k_0)$ known for all $\beta \in S^2, \alpha = \alpha_0, k = k_0$ being fixed.

Uniqueness of the solution to IOSPa) is proved by Ramm in 1985, see [269] for the Dirichlet, Neumann and Robin boundary conditions, and of IOSPb) by Schiffer (1964), who *assumed a priori* the Dirichlet boundary condition. M. Schiffer did not publish his result. A proof of this result and A. G. Ramm's results appeared in the monographs [269] and [356]. The proofs are given in [269]. A. G. Ramm has also proved that not only S but the boundary condition as well is uniquely defined by the above data in both cases, and gave stability estimates for the solution to IOSP [324]. Later he gave a different method of proof of the uniqueness theorems for these problems which covered the rough boundaries (Lipschitz and much rougher boundaries: the ones with finite perimeter [329], see also [338]). In [339], the uniqueness theorem for the solution of inverse scattering problem is proved for a wide class of transmission problems. It is proved that not only the discontinuity surfaces of the refraction coefficient but also the coefficient itself inside the body and the boundary conditions across these surfaces are uniquely determined by the fixed-frequency scattering data. For any strictly convex, smooth, reflecting obstacle D, analytical formulas for finding S from the high-frequency asymptotics of the scattering amplitude are proposed by A. G. Ramm, who gave error estimates of his inversion formula also [269]. The uniqueness theorems in the above references hold if the scattering data are given not for all $\alpha', \alpha \in S^2$, but only for α' and α in arbitrary small solid angles, i.e., in arbitrary small open subsets of S^2. The inverse scattering problem with the data $\alpha' \in S^2$, $k = k_0$ and $\alpha = \alpha_0$ being fixed, is open. If *a priori* one knows that D is sufficiently small, so that $k_0 > 0$ is not a Dirichlet eigenvalue of the

Laplacian in D, then uniqueness of the solution with the above non-overdetermined data holds [269]. There are many parameter-fitting schemes for solving IOSP [314, 444].

Let us describe a new such scheme, based on MRC, its idea is similar to the one in [272]. Suppose that the scattered field v is observed on a sphere S_R. Calculate

$$c_\ell := (v, Y_\ell)_{L^2(S^2)}/h_\ell(kR).$$

If v is known exactly, then $c_\ell = A_\ell(\alpha)$. If v_δ are noisy data, $\|v - v_\delta\|_{L^2(S_R)} \leq \delta$, then $c_\ell = c_{\ell\delta}$. Choose some L, say $L = 5$, and find $r = r(\alpha')$ as a positive root of the equation

$$u_0 + v_L := e^{ik\alpha\cdot\alpha' r} + \sum_{\ell=0}^{L} c_{\ell\delta}\psi_\ell(kr, \alpha') := p(r, \alpha', \alpha, k) = 0.$$

Here, α and $k > 0$ are fixed, and we are looking for the root $r = r(\alpha')$ which is positive and is stable under changes of k and α. In practice, equation $p(r, \alpha', \alpha, k) = 0$ may have no such root, the root may have small imaginary part. If for the chosen L, such a root (that is, a root which is positive, or has a positive real part and a small imaginary part, and stable with respect to changes of k and α) is not found, then increase L (or decrease L), and repeat the search of the root. Stop the search at a smallest L for which such a root is found. The MRC justifies this method: for a suitable L, the function $p(r, \alpha', \alpha, k)$ approximately equals zero on S, that is, for $r = r(\alpha')$, and this $r(\alpha')$ does not depend on k and α. Moreover, by the uniqueness theorem for IOSPa) and IOSPb) there is only one such $r = r(\alpha')$. Numerically, one expects to find a root of the equation $p(r, \alpha', k) = 0$ which is close to positive semi-axis $r > 0$ and stable with respect to changes of k and α.

If one uses the above scheme for solving the inverse scattering problem for an acoustically hard body (the Neumann boundary condition on S), then one gets not a transcendental equation $p(r, \alpha', \alpha, k) = 0$ for finding the equation of S, $r = r(\alpha')$, but a differential equation for $r = r(\alpha')$, which comes from the equation $\frac{\partial p(r,\alpha',\alpha,k)}{\partial N} = 0$ at $r = r(\alpha')$. One has to write the normal derivative on S in the spherical coordinates and then substitute $r = r(\alpha')$ into the result to get a differential equation for the unknown function

$r = r(\alpha')$. For example, if $n = 2$ (the two-dimensional case), then the role of α' plays the polar angle φ' and the equation for $r = r(\varphi')$ takes the form $\frac{dr}{d\varphi'} = (r^2\frac{dp}{dr}/\frac{dp}{d\varphi'})|_{r=r(\varphi')}$, and $r(\varphi') = r(\varphi' + 2\pi)$.

9.1.4. *Proofs*

Proof of Lemma 9.1.1. This Lemma follows from the results in [269], (p. 162, Lemma 1), see also Lemma 4.2.10 in Chapter 4.

Proof of Lemma 9.1.2. By Green's formula, one has

$$v_\epsilon(x) = \int_S v_\epsilon(s)G_N(x,s)ds, \quad \|v_\epsilon(s) + u_0\|_{L^2(S)} < \epsilon, \tag{9.1.13}$$

where N is the unit normal to S, pointing into D', and G is the Dirichlet Green's function of the Laplacian in D':

$$(\nabla^2 + k^2)G = -\delta(x-y) \quad \text{in} \;\; D', \quad G = 0 \quad \text{on} \;\; S, \tag{9.1.14}$$

$$\lim_{r\to\infty} \int_{|x|=r} \left|\frac{\partial G}{\partial |x|} - ikG\right|^2 ds = 0. \tag{9.1.15}$$

From (9.1.13), one gets (9.1.8) with $H^m_{loc}(D')$-norm immediately by the Cauchy inequality, and with the weighted norm from the estimate

$$|G_N(x,s)| \le \frac{c}{1+|x|}, \quad |x| \ge R. \tag{9.1.16}$$

From local elliptic estimates for $w_\epsilon := v_\epsilon - v$, one gets

$$\|w_\epsilon\|_{L^2(B_R\setminus D)} \le c\epsilon. \tag{9.1.17}$$

Let us recall the elliptic estimate we use. Let $D'_R := B_R\setminus D$, S_R be the boundary of B_R, and choose R such that k^2 is not a Dirichlet eigenvalue of $-\Delta$ in D'_R. The elliptic estimate we have used is [129, p. 189]:

$$\|w_\epsilon\|_{H^m(D'_R)} \le c[\|(\Delta + k^2)w_\epsilon\|_{\mathcal{H}^{m-2}(D'_R)} + \|w_\epsilon\|_{H^{m-0.5}(S_R)} + \|w_\epsilon\|_{H^{m-0.5}(S)}]. \tag{9.1.18}$$

Take $m = 0.5$ in (9.1.18), use the equation $(\Delta + k^2)w_\epsilon = 0$ in D', the estimate $\|w_\epsilon\|_{H^m(S_R)} = O(\epsilon)$, proved above, the estimate

$\|w_\epsilon\|_{H^0(S)} = O(\epsilon)$, and get (9.1.8). For $m = 0.5$, the space in the first term on the right-hand side of formula (9.1.18) differs from the usual Sobolev space $H^{m-2}(D'_R)$ (cf [129, p. 189]), but $(\Delta + k^2)w_\epsilon = 0$, so this term vanishes anyway.

Lemma 9.1.2 is proved. □

Proof of Lemma 9.1.3. Lemma 9.1.2 yields convergence of v_ϵ to v in the norm $\|\cdot\|$. In particular, $\|v_\epsilon - v\|_{L^2(S_R)} \to 0$ as $\epsilon \to 0$. On S_R one has $v = \sum_{\ell=0}^{\infty} A_\ell(\alpha)\psi_\ell$ and $v_\epsilon = \sum_{\ell=0}^{L(\epsilon)} c_\ell \psi_\ell$. Multiply $v_\epsilon(R,\alpha') - v(R,\alpha')$ by $\overline{Y_\ell(\alpha')}$, integrate over S^2 and then let $\epsilon \to 0$. The result is (9.1.10). □

9.2. MRC for Static Problems

9.2.1. *Introduction*

MRC in scattering theory was proposed and justified by the author in Section 9.1. MRC allows one to develop efficient numerical algorithms for solving boundary value problems. It gives an error estimate for solutions. In this section, the MRC is formulated and proved for static problems.

Consider a bounded domain $D \subset \mathbb{R}^n$, $n = 3$ with a boundary S. The exterior domain is $D' = \mathbb{R}^3 \backslash D$. Assume that S is Lipschitz. Let S^2 denotes the unit sphere in $\mathbb{R}^3$. Consider the problem:

$$\nabla^2 v = 0 \text{ in } D', \quad v = f \text{ on } S, \tag{9.2.1}$$

$$v := O\left(\frac{1}{r}\right) \quad r := |x| \to \infty. \tag{9.2.2}$$

Let $\frac{x}{r} := \alpha \in S^2$. Denote by $Y_\ell(\alpha)$ the orthonormal spherical harmonics, $Y_\ell = Y_{\ell m}, -\ell \le m \le \ell$. Let $h_\ell := \frac{Y_\ell(\alpha)}{r^{\ell+1}}$, $\ell \ge 0$, be harmonic functions in D'. Let the ball $B_R := \{x : |x| \le R\}$ contain D.

In the region $r > R$, the solution to (9.2.1)–(9.2.2) is:

$$v(x) = \sum_{\ell=0}^{\infty} c_\ell h_\ell, \qquad r > R, \tag{9.2.3}$$

the summation in (9.2.3) and below includes summation with respect to m, $-\ell \le m \le \ell$, and c_ℓ are some coefficients determined by f.

The series (9.2.3) in general does not converge up to the boundary S. Our aim is to give a formulation of an analog of the MRC from [348], which can be used in numerical solution of the boundary value problems. The author hopes that the MRC method for static problems can be used as a basis for an efficient numerical algorithm for solving boundary value problems for Laplace equations in domains with complicated boundaries. In [354], such an algorithm was developed on the basis of MRC for solving boundary value problems for the Helmholtz equation. Although the boundary integral equation methods and finite elements methods are widely and successfully used for solving these problems, the method based on MRC, proved to be competitive and often superior to the currently used methods.

We discuss the Dirichlet condition but a similar argument is applicable to the Neumann and Robin boundary conditions. Boundary value problems and scattering problems in rough domains were studied in [356].

Let us present the basic results on which the MRC method is based.

Fix $\epsilon > 0$, an arbitrary small number. The following Lemmas and Theorems are similar to the ones in Section 9.1.

Lemma 9.2.1. *There exist $L = L(\epsilon)$ and $c_\ell = c_\ell(\epsilon)$, such that*

$$\left\| \sum_{\ell=0}^{L(\epsilon)} c_\ell(\epsilon) h_\ell - f \right\|_{L^2(S)} \le \epsilon. \tag{9.2.4}$$

If (9.2.4) and the boundary condition (9.2.1) hold, then

$$\|v_\epsilon - v\|_{L^2(S)} \le \epsilon, \quad v_\epsilon := \sum_{\ell=0}^{L(\epsilon)} c_\ell(\epsilon) h_\ell. \tag{9.2.5}$$

Lemma 9.2.2. *If* (9.2.4) *holds, then*

$$\|v_\epsilon - v\| = O(\epsilon) \quad \epsilon \to 0, \tag{9.2.6}$$

where $\|\cdot\| := \|\cdot\|_{H^m_{\text{loc}}(D')} + \|\cdot\|_{L^2(D';(1+|x|)^{-\gamma})}$, $\gamma > 1$, $m > 0$ *is an arbitrary integer, and* H^m *is the Sobolev space.*

In particular, (9.2.6) implies

$$\|v_\epsilon - v\|_{L^2(S_R)} = O(\epsilon) \quad \epsilon \to 0. \tag{9.2.7}$$

Let us formulate an analog of the MRC:

Theorem 9.2.3 (MRC). *For an arbitrary small* $\epsilon > 0$, *there exist* $L(\epsilon)$ *and* $c_\ell(\epsilon), 0 \leq \ell \leq L(\epsilon)$, *such that* (9.2.4) *and* (9.2.6) *hold.*

Theorem 9.2.3 follows from Lemmas 9.2.1 and 9.2.2.

For the Neumann boundary condition, one minimizes $\|\frac{\partial[\sum_{\ell=0}^{L} c_\ell \psi_\ell]}{\partial N} - f\|_{L^2(S)}$ with respect to c_ℓ. Analogs of Lemmas 9.2.1–9.2.2 are valid and their proofs are essentially the same.

If the boundary data $f \in C(S)$, then one can use $C(S)$-norm in (9.2.4)–(9.2.7), and an analog of Theorem 9.2.3 then follows immediately from the maximum principle.

In Section 9.2.2, we discuss the usage of MRC in solving boundary value problems. In Section 9.2.3, proofs are given.

9.2.2. *Solving boundary value problems by MRC*

To solve problem (9.2.1)–(9.2.2) using MRC, fix a small $\epsilon > 0$ and find $L(\epsilon)$ and $c_\ell(\epsilon)$ such that (9.2.4) holds. This is possible by Lemma 9.2.1 and can be done numerically by minimizing $\|\sum_0^L c_\ell h_\ell - f\|_{L^2(S)} := \phi(c_1, \ldots, c_L)$. If the minimum of ϕ is larger than ϵ, then increase L and repeat the minimization. Lemma 9.2.1 guarantees the existence of such L and c_ℓ that the minimum is less than ϵ. Choose the smallest L for which this happens and define $v_\epsilon := \sum_{\ell=0}^{L} c_\ell h_\ell$. Then, by Lemma 9.2.2, v_ϵ is the approximate solution to problem (9.2.1)–(9.2.2) with the accuracy $O(\epsilon)$ in the norm $\|\cdot\|$.

9.2.3. *Proofs*

Proof of Lemma 9.2.1. We start with the claim:

Claim: *The restrictions of harmonic functions h_ℓ on S form a total set in $L^2(S)$.*

Lemma 9.2.1 follows from this claim. Let us prove the claim. Assume the contrary. Then there is a function $g \neq 0$ such that $\int_S g(s)h_\ell(s)ds = 0\ \forall \ell \geq 0$. This implies $V(x) := \int_S g(s)|x-s|^{-1}ds = 0\ \forall x \in D'$. Thus, $V = 0$ on S, and since $\Delta V = 0$ in D, one concludes that $V = 0$ in D. Thus, $g = 0$ by the jump formula for the normal derivatives of the single layer potential V. This contradiction proves the claim. Lemma 9.2.1 is proved. □

Proof of Lemma 9.2.2. By Green's formula, one has

$$w_\epsilon(x) = \int_S w_\epsilon(s)G_N(x,s)ds, \quad \|w_\epsilon\|_{L^2(S)} < \epsilon, \quad w_\epsilon := v_\epsilon - v. \tag{9.2.8}$$

Here, N is the unit normal to S, pointing into D', and G is the Dirichlet Green's function of the Laplacian in D':

$$\nabla^2 G = -\delta(x-y) \text{ in } D', \quad G = 0 \text{ on } S, \tag{9.2.9}$$

$$G = O\left(\frac{1}{r}\right), \quad r \to \infty. \tag{9.2.10}$$

From (9.2.8), one gets (9.2.7) and (9.2.6) with $H^m_{loc}(D')$-norm immediately by the Cauchy inequality. Estimate (9.2.6) in the region $B'_R := \mathbb{R}^3 \setminus B_R$ follows from the estimate

$$|G_N(x,s)| \leq \frac{c}{1+|x|}, \quad |x| \geq R. \tag{9.2.11}$$

In the region, $B_R \backslash D$ estimate, (9.2.6) follows from local elliptic estimates for $w_\epsilon := v_\epsilon - v$, which imply that

$$\|w_\epsilon\|_{L^2(B_R\backslash D)} \leq c\epsilon. \tag{9.2.12}$$

Let us recall the elliptic estimate we have used. Let $D'_R := B_R\backslash D$ and S_R be the boundary of B_R. Recall the elliptic estimate for the solution to the homogeneous Laplace equation in D'_R ([129, p. 189]):

$$\|w_\epsilon\|_{H^{0.5}(D'_R)} \leq c[\|w_\epsilon\|_{L^2(S_R)} + \|w_\epsilon\|_{L^2(S)}]. \tag{9.2.13}$$

The estimates $\|w_\epsilon\|_{L^2(S_R)} = O(\epsilon)$, $\|w_\epsilon\|_{L^2(S)} = O(\epsilon)$, and (9.2.13) yield (9.2.6). Lemma 9.2.2 is proved. □

9.3. MRC for Scattering by Periodic Structures

9.3.1. *Introduction*

This section contains a brief presentation of the scattering theory for periodic structures. Its main result is a theorem (the MRC), which gives a rigorous foundation for a numerical method for solving the direct scattering problem for periodic structures. The presentation is based on the paper [354].

For simplicity, we consider a 2-D setting, but our arguments can be as easily applied to n-dimensional problems, $n \geq 2$. Let $f : \mathbb{R} \to \mathbb{R}$, $f(x + L) = f(x)$ be an L-periodic Lipschitz continuous function, and let D be the domain

$$D = \{(x, y) \ : \ y \geq f(x), \ x \in \mathbb{R}\}.$$

Without loss of generality, we assume that $f \geq 0$. If it is not, one can choose the origin so that this assumption is satisfied, because $M := \sup_{0 \leq x \leq L} |f(x)| < \infty$.

Let $\mathbf{x} = (x, y)$ and $u(\mathbf{x})$ be the total field satisfying

$$(\Delta + k^2)u = 0, \quad \mathbf{x} \in D, \quad k = \text{const} > 0 \tag{9.3.1}$$

$$u = 0 \quad \text{on } S := \partial D, \tag{9.3.2}$$

$$u = u_0 + v, \quad u_0 := e^{ik\alpha\cdot\mathbf{x}}, \tag{9.3.3}$$

where the unit vector $\alpha = (\cos\theta, -\sin\theta)$, $0 < \theta < \pi/2$, and $v(\mathbf{x})$ is the scattered field, whose asymptotic behavior as $y \to \infty$ will be specified below, and

$$u(x + L, y) = \nu u(x, y), \quad u_x(x + L, y) = \nu u_x(x, y) \text{ in } D,$$
$$\nu := e^{ikL\cos\theta}. \tag{9.3.4}$$

Conditions (9.3.4) are the *qp* (**quasiperiodicity**) conditions. To find the proper radiation condition for the scattered field $v(\mathbf{x})$, consider the spectral problem

$$\varphi'' + \ell^2\varphi = 0, \quad 0 < x < L, \tag{9.3.5}$$

$$\varphi(L) = \nu\varphi(0), \quad \varphi'(L) = \nu\varphi'(0) \tag{9.3.6}$$

arising from the separation of variables in (9.3.1)–(9.3.4). This problem has a discrete spectrum, and its eigenfunctions form a basis in $L^2(0,L)$. One has

$$\varphi = Ae^{i\ell x} + Be^{-i\ell x}, \quad A, B = \text{const},$$

$$Ae^{i\ell L} + Be^{-i\ell L} = \nu(A+B), \quad i\ell Ae^{i\ell L} - i\ell Be^{-i\ell L} = i\ell\nu(A-B).$$

Thus,

$$\begin{vmatrix} e^{i\ell L} - \nu & e^{-i\ell L} - \nu \\ i\ell(e^{i\ell L} - \nu) & -i\ell(e^{-i\ell L} - \nu) \end{vmatrix} = 0.$$

So, $i\ell(e^{i\ell L} - \nu)(e^{-i\ell L} - \nu) = 0$. If $\ell = 0$, then $\varphi = A + Bx$, $A + BL = \nu A$, $B = \nu B$. Since $\nu = e^{ikL\cos\theta}$, one has no eigenvalue $\ell = 0$ unless $kL\sin\theta = 2\pi m$, $m > 0$ is an integer. Let us assume that $kL\cos\theta \neq 2\pi m$. Then

$$e^{i\ell L} = e^{ikL\cos\theta} \quad \text{or} \quad e^{-i\ell L} = e^{ikL\cos\theta},$$

that is,

$$\ell_j^+ = k\cos\theta + \frac{2\pi j}{L}, \quad \text{or} \quad \ell_j^- = -k\cos\theta + \frac{2\pi j}{L}, \quad j = 0, \pm 1, \pm 2, \ldots.$$

The corresponding eigenfunctions are $e^{i\ell_j^+ x}$ and $e^{-i\ell_j^- x}$. We will use the system $e^{i\ell_j^+ x}$, which forms an orthogonal basis of $L^2(0,L)$. One has:

$$\int_0^L e^{i\ell_j^+ x} e^{-i\ell_m^+ x}\, dx = \int_0^L e^{\frac{2\pi i}{L}(j-m)}\, dx = 0, \quad j \neq m.$$

The normalized eigenfunctions are

$$\varphi_j(x) = \frac{e^{i\ell_j^+ x}}{\sqrt{L}}, \quad j = 0, \pm 1, \pm 2, \dots.$$

These functions form an orthonormal basis of $L^2(0, L)$. Let us look for $v(\mathbf{x}) = v(x, y)$ of the form

$$v(x, y) = \sum_{j=-\infty}^{\infty} c_j v_j(y) \varphi_j(x), \quad y > M, \quad c_j = \text{const.} \tag{9.3.7}$$

For $y > M$, equation (9.3.1) implies

$$v_j'' + (k^2 - \ell_j^2) v_j = 0. \tag{9.3.8}$$

Let us assume that $\ell_j^2 \neq k^2$ for all j. Then

$$v_j(y) = e^{i\mu_j y}, \tag{9.3.9}$$

where, for finitely many j, the set of which is denoted by J, one has:

$$\mu_j = (k^2 - \ell_j^2)^{1/2} > 0 \quad \text{if } \ell_j^2 < k^2, \ j \in J, \tag{9.3.10}$$

and

$$\mu_j = i(\ell_j^2 - k^2)^{1/2} \quad \text{if } \ell_j^2 > k^2, \ j \notin J. \tag{9.3.11}$$

The **radiation condition** at infinity requires that the scattered field $v(x, y)$ be representable in the form (9.3.7) with $v_j(y)$ defined by (9.3.9)–(9.3.11).

The **periodic scattering problem** consists of finding the solution to (9.3.1)–(9.3.4) satisfying the radiation condition (9.3.7), (9.3.9)–(9.3.11).

The existence and uniqueness for such a scattering problem is established in Section 9.3.2. Our presentation is essentially self-contained. In [6], the scattering by a periodic structure was considered earlier, and was based on a uniqueness theorem from [57]. Our proofs differ from the proofs in [6]. There are many papers on scattering by periodic structures, of which we mention a few [6, 26, 27, 35, 99, 100, 148, 149, 175, 427]. The Rayleigh conjecture is discussed in several of the above papers. It was shown [14, 175], that this conjecture is incorrect, in general. The MRC is a theorem proved in Section 9.1 for scattering by bounded obstacles. See also [360], Chapter 12, [363, 368]. A numerical method for solving obstacle scattering problems, based on the MRC is developed in [349]. The main results of Section 9.3 are: (1) the MRC for periodic structures (Theorem 9.3.5) and a rigorous numerical method for solving scattering problems by periodic structures, based on the MRC (Section 9.3.4). (2) The proof of the limiting absorption principle (LAP) and the rigorous and self-contained development of the plane wave scattering theory by periodic structures is also of interest for broad audience. This theory is based partly on the ideas developed in [269, 332, 338]. The proof of the key Lemma 9.3.2 is based on a version of Ramm's identity (9.3.27). Numerical implementation of the method for solving scattering problems by periodic structures, based on the MRC, used the approach developed in [349, 359]. Applications to inverse problems are discussed in [348, 356].

9.3.2. *Periodic scattering problem*

Existence and uniqueness of solutions of the periodic scattering problem can be proved easily, if one establishes first the existence and uniqueness of the resolvent kernel $G(x, y, \xi, \eta, k)$ of the Dirichlet Laplacian in D:

$$(\Delta + k^2)G(x, y, \xi, \eta, k) = -\delta(x - \xi)\delta(y - \eta),$$

$$G = 0 \quad \text{on } S, \tag{9.3.12}$$

$$G(x + L, y, \xi, \eta, k) = \nu G(x, y, \xi, \eta, k),$$

$$G(x, y, \xi + L, \eta, k) = \overline{\nu} G(x, y, \xi, \eta, k), \tag{9.3.13}$$

$$G_x(x + L, y, \xi, \eta, k) = \nu G_x(x, y, \xi, \eta, k),$$

$$G_x(x, y, \xi + L, \eta, k) = \overline{\nu} G_x(x, y, \xi, \eta, k), \tag{9.3.14}$$

and G satisfies the LAP, see (9.3.16) below. The overbar here and below stands for the complex conjugation.

Indeed, if such a function G exists, then v can be found by Green's formula

$$v(x,y) = -\int_{S_L} u_0(\xi,\eta) G_N(x,y,\xi,\eta,k)\, ds, \tag{9.3.15}$$

where N is the unit-normal vector to S pointing into D.

To prove the existence and uniqueness of $G(x,y,\xi,\eta,k)$, define

$$\ell_0 = -\Delta$$

to be the Laplacian on the set of $C^2(D)$ quasiperiodic functions vanishing on the boundary S, and vanishing near infinity. Let

$$D_L := \{(x,y) \;:\; 0 \le x \le L, \quad (x,y) \in D\}.$$

Then D_L is a section of D, and ℓ_0 is a symmetric operator in $L^2(D_L)$. This operator is nonnegative, and therefore [98] there exists its unique self-adjoint Friedrichs' extension, which will be denoted by ℓ.

Let $\operatorname{Im}(k^2) > 0$. Then there exists a unique resolvent operator $(\ell - k^2)^{-1}$. Thus, its kernel $G(x,y,\xi,\eta,k)$ also exists and it is unique. To establish the existence and uniqueness of the kernel for $k > 0$, we are going to prove the following:

Limiting Absorption Principle (LAP). Let $k > 0$, $\epsilon > 0$ and assume that k^2 is not equal to λ_j^2. Then the limit

$$\lim_{\epsilon \to 0+} G(x,y,\xi,\eta,k+i\epsilon) = G(x,y,\xi,\eta,k), \tag{9.3.16}$$

exists for all $(x,y) \in D$, $x \neq y$. The proof is based on the following two lemmas:

Lemma 9.3.1. *Let* $0 < \epsilon < 1$, *and* $a > 2$. *Then*

$$\int_{D_L} \frac{|G(x,y,\xi,\eta,k+i\epsilon)|^2}{(1+\xi^2+\eta^2)^{a/2}}\, d\xi d\eta \le c, \tag{9.3.17}$$

where $c = \text{const} > 0$ *does not depend on* ϵ, *and* (x,y) *is running on compact sets.*

Proof. It is sufficient to prove that the solution to the problem

$$(\Delta + k^2 + i\epsilon)w_\epsilon = F, \quad \text{in } D_L,\ w_\epsilon \in L^2(D_L),$$

$$w_\epsilon = 0 \quad \text{on } S_L, \tag{9.3.18}$$

$$w_\epsilon(x+L,y) = \nu w_\epsilon(x,y), \quad w_{\epsilon x}(x+L,y) = \nu w_{\epsilon x}(x,y), \tag{9.3.19}$$

satisfies the estimate

$$N_\epsilon^2 := \sup_{0<\epsilon<1} \int_{D_L} \frac{|w_\epsilon(x,y)|^2}{(1+x^2+y^2)^{a/2}}\, dxdy := N^2(w_\epsilon) \le c, \tag{9.3.20}$$

where $F \in C_0^\infty(D_L)$ is arbitrary, and $c = \text{const} > 0$ is independent of $\epsilon > 0$.

If (9.3.20) fails, then $N_{\epsilon_n} \to \infty$, $\epsilon_n \to 0$. Define $\psi_\epsilon := w_\epsilon/N_\epsilon$, where $\epsilon := \epsilon_n$. Then $N(\psi_\epsilon) = 1$, ψ_ϵ solves (9.3.18) (with F replaced by $F_\epsilon := F/N_\epsilon$), and satisfies (9.3.19). From $N(\psi_\epsilon) = 1$, it follows that $\psi_\epsilon \rightharpoonup \psi$ as $\epsilon \to 0$, where $\rightharpoonup$ denotes the weak convergence in $L^2(D_L, 1/(1+x^2+y^2)^{a/2}) := L_a^2$. By elliptic estimates, $\psi_\epsilon \rightharpoonup \psi$ in $H^2_{\text{loc}}(D_L)$, and therefore $\psi_\epsilon \to \psi$ in $L^2_{\text{loc}}(D_L)$. This and (9.3.18)-(9.3.19) imply $\psi_\epsilon \to \psi$ in $H^2_{\text{loc}}(D_L)$. Thus, ψ solves the homogeneous ($F = 0$) problem (9.3.18)–(9.3.19). If we prove that $\psi = 0$, then we get a contradiction, which shows that (9.3.20) holds. The contradiction comes from the relationship $0 = N(\psi) = \lim_{\epsilon\to 0} N(\psi_\epsilon) = 1$. One proves that

$$\lim_{\epsilon\to 0} N(\psi_\epsilon) = N(\psi) \tag{9.3.21}$$

as follows. If

$$(x,y) \in D_R := \{(x,y) \ : \ f(x) \le y \le R,\ 0 \le x \le L\},$$

where $R > M$ is an arbitrary large fixed number, then $\lim_{\epsilon\to 0} N(\psi_\epsilon \eta_R) = N(\psi\eta_R)$, where

$$\eta_R := \begin{cases} 1, & f(x) < y < R, \\ 0, & y > R. \end{cases}$$

In the region $D'_R = \{(x, y) \ : \ y > R, \ 0 \leq x \leq L\}$, one has $|\psi_\epsilon(x, y)| \leq c, \ (x, y) \in D'_R$. Thus,

$$\sup_{0<\epsilon<1} N(\psi_\epsilon(\chi_L - \eta_R)) \leq O\left(\frac{1}{R^\gamma}\right), \ 0 < \gamma < a - 2.$$

The desired result (9.3.21) follows.

To complete the proof, let us show that the problem (9.3.18)–(9.3.19), with $F = 0$, and $\epsilon = 0$, has only the trivial solution w, provided that w is "outgoing" in the sense

$$w_{jy} - i\mu_j w_j = o(1), \ \text{as } y \to \infty, \ w_j := \int_0^L w\overline{\varphi_j}\, dx.$$

One has

$$\lim_{R\to\infty} \int_{S_R} (w\overline{w_y} - w_y\overline{w})\, ds = 0, \tag{9.3.22}$$

where $S_R := \{(x, y) \ : \ y = R, \ 0 \leq x \leq L\}$, $ds = dx$ is the element of the arclength of S_R, and the overbar stands for the complex conjugate.

Let us outline the steps of the further argument.

Step 1: We prove that (9.3.22) implies

$$w \in L^2(D_L), \ |w| + |\nabla w| \leq ce^{-\gamma|y|}, \ \gamma = \text{const} > 0, \tag{9.3.23}$$

if w is outgoing.

Step 2: We prove that if $w \in L^2(D_L)$ solves (2.7)–(2.8), with $F = \epsilon = 0$, then $w = 0$. Then we conclude that (2.9) (and (2.6)) holds, and, therefore, (2.5) holds.

Let us prove (2.12). One has

$$\begin{aligned} 0 &= \int_{D_{LR}} [\bar{w}(\Delta + k^2)w - w(\Delta + k^2)\bar{w}]\, dxdy \\ &= -\int_{S_L} (\bar{w}w_N - w\bar{w}_N)\, ds + \int_{S_R} (\bar{w}w_N - w\bar{w}_N)\, ds \\ &= \int_{S_R} (\bar{w}w_N - w\bar{w}_N)\, ds, \end{aligned} \tag{9.3.24}$$

where the Dirichlet condition (9.3.18) was used, and the integrals over the lines $x = 0$ and $x = L$ are cancelled due to the qp conditions (9.3.19):

$$\int_{x=0} (-\bar{w}w_x + w\bar{w}_x)\, dy + \int_{x=L} (\bar{w}w_x - w\bar{w}_x)\, dy$$
$$= \int_{x=0} (w\bar{w}_x - \bar{w}w_x)\, dy - \int_{x=0} \nu\bar{\nu}(w\bar{w}_x - \bar{w}w_x)\, dy = 0.$$

Here, we have used the relation $\nu\bar{\nu} = 1$. Thus, (9.3.24) implies

$$0 = \int_{S_R} (\bar{w}w_y - w\bar{w}_y)\, dx, \quad \forall R > M. \tag{9.3.25}$$

If w is outgoing, then (9.3.25) implies $w_j(y) = 0$ for $j \in J$, and $|w_j(y)| \le e^{-\gamma|y|}$, $\gamma = \text{const} > 0$, so (9.3.23) holds. □

Lemma 9.3.2. *Assume that $w \in L^2(D_L)$, w solves (9.3.18) with $\epsilon = 0$ and $F = 0$, and w satisfies (9.3.19). Then $w = 0$.*

Proof. If w solves equation (9.3.18) with $\epsilon = 0$ and $F = 0$, then $w = \sum_j w_j(y)\varphi_j(x)$. Since $\{\varphi_j(x)\}$ is an orthonormal basis and $w \in L^2(D_L)$, it follows that $w_j(y) = 0$ for all $j \in J$, and (9.3.23) holds. Let us use a version of Ramm's identity (see [332, p. 92]), which is valid for any solution w of equation (9.3.1) which is outgoing in the sense that

$$w = \sum_j c_j v_j(y)\varphi_j(x), \ c_j = \text{const}, \ j \notin J. \tag{9.3.26}$$

Note that $v_j(y) = \overline{v_j(y)}$ for $j \notin J$. The identity is:

$$0 = (x_2\bar{w}_{,2}w_{,j})_{,j} + \frac{(k^2|w|^2x_2 - |\nabla w|^2x_2)_{,2}}{2} + \frac{|\nabla w|^2 - k^2|w|^2}{2} - |w_{,2}|^2, \tag{9.3.27}$$

where $w_{,j} := \partial w/\partial x_j$, $j = 1, 2$, $x_1 = x$, $x_2 = y$, over the repeated indices one sums up, $|w|^2 := w\bar{w}$. The right-hand side of (9.3.27) equals to

$$\frac{1}{2}[x_2(\bar{w}_{,2j}w_{,j} - w_{,2j}\bar{w}_{,j}) + k^2x_2(w_{,2}\bar{w} - \bar{w}_{,2}w)] = 0,$$

because $w_{,2}\bar{w} = \bar{w}_{,2}w$ for outgoing w.

One has

$$|w| + |\nabla w| \le ce^{-\gamma|y|}, \ \gamma = \text{const} > 0, \ c = \text{const} > 0. \tag{9.3.28}$$

Let $R > \max f(x)$. Integrate (9.3.27) over $D_{LR} := \{(x, y) \ : \ (x, y) \in D_L, \ y \le R\}$ and use Green's formula to get:

$$0 = -\lim_{R\to\infty}\int_{S_L\cup S_R}\left[x_2\bar{w}_{,2}w_{,j}N_j + \frac{(k^2|w|^2x_2 - |\nabla w|^2x_2)N_2}{2}\right]ds$$
$$-\lim_{R\to\infty}\int_{D_{LR}}|w_{,2}|^2\,dx_1dx_2, \tag{9.3.29}$$

where N is the normal pointing into D_{LR}, and we have used the relation

$$\lim_{R\to\infty}\int_{D_{LR}}|\nabla w|^2\,dx_1dx_2 = k^2\lim_{R\to\infty}\int_{D_{LR}}|w|^2\,dx_1dx_2, \tag{9.3.30}$$

which follows from the equation $\Delta w + k^2w = 0$, boundary condition $w = 0$ on S, quasiperiodicity of w, and from (9.3.28). We have also used the relation $\bar{w}_{,2}w_{,j}N_j = x_2|\nabla w|^2N_2$, which follows from the condition $u = 0$ on S. From (9.3.29), one gets:

$$\lim_{R\to\infty}\int_{D_{LR}}|w_{,2}|^2\,dx_1dx_2 = -\frac{1}{2}\int_{S_L}x_2N_2|\nabla w|^2\,ds. \tag{9.3.31}$$

Since $f(x)$ is a graph, one has $N_2x_2 \ge 0$, and it follows from (9.3.31) that $w_{,2} = 0$, so $w = \text{const}$, and $\text{const} = 0$ because $w|_S = 0$. Lemma 9.3.2 is proved. □

Remark. Condition of the type

$$N_2x_2 \ge 0 \text{ on } S_L \tag{9.3.32}$$

was also used in [332].

The proof of Lemma 9.3.2 *is not valid if the Neumann boundary condition is imposed on S*.

9.3.3. *Integral equations method*

In this section, we present another proof of the existence and uniqueness of the resolvent kernel G. We want to construct a scattering theory quite similar to the one for the exterior of a bounded obstacle [269]. The *first step* is to construct an analog to the half-space Dirichlet Green's function. The function $g = g(\mathbf{x}, \xi, k)$ can be constructed analytically ($\mathbf{x} = (x_1, x_2), \xi = (\xi_1, \xi_2)$):

$$g(\mathbf{x}, \xi) = \sum_j \varphi_j(x_1)\overline{\varphi_j(\xi_1)}g_j(x_2, \xi_2, k), \tag{9.3.33}$$

$$g_j := g_j(x_2, \xi_2, k) = \begin{cases} v_j(x_2)\psi_j(\xi_2), & x_2 > \xi_2, \\ v_j(\xi_2)\psi_j(x_2), & x_2 < \xi_2, \end{cases}$$

$$\psi_j = (\mu_j)^{-1}e^{i\mu_j b}\sin[\mu_j(\xi_2 + b)],$$

$$\mu_j = [k^2 - \lambda_j^2]^{1/2}, \quad v_j(x_2) = e^{i\mu_j x_2},$$

where

$$\psi_j'' + (k^2 - \ell_j^2)\psi_j = 0, \ \psi_j(-b) = 0, \ W[v_j, \psi_j] = 1, \ \lambda_j = k\cos(\theta) + \frac{2\pi j}{L},$$

and $W[v, \psi]$ is the Wronskian.

The function g is analytic with respect to k on the complex plain with cuts along the rays $\lambda_j - i\tau$, $0 \le \tau < \infty, j = 0, \pm 1, \pm 2, \ldots$, in particular, in the region $\Im k > 0$, up to the real positive half-axis except for the set $\{\lambda_j\}_{j=0,\pm 1,\pm 2,\ldots}$.

Choose $b > 0$ such that $k^2 > 0$ is not an eigenvalue of the problem:

$$(\Delta + k^2)\psi = 0, \quad \text{in } D_{-b} := \{(x, y) \ : \ -b \le y \le f(x), \ 0 \le x \le L\}. \tag{9.3.34}$$

$$\begin{aligned} &\psi|_{y=-b} = 0, \quad \psi_N = 0 \text{ on } S, \\ &\psi(x + L, y) = \nu\psi(x, y), \quad \psi_x(x + L, y) = \nu\psi_x(x, y). \end{aligned} \tag{9.3.35}$$

One has

$$(\Delta + k^2)g = -\delta(\mathbf{x} - \xi), \ \mathbf{x} = (x_1, x_2), \ \xi = (\xi_1, \xi_2),$$
$$\mathbf{x} \in \{(x, y) \ : \ -b < y < \infty, \quad 0 \leq x \leq L\}, \tag{9.3.36}$$

$$g|_{y=-b} = 0, \tag{9.3.37}$$

and

$$(\Delta + k^2)G = -\delta(\mathbf{x} - \xi), G = 0 \text{ on } S, \tag{9.3.38}$$

G satisfies the *qp* condition and the radiation condition (it is outgoing at infinity).

Multiply (9.3.36) by G, (9.3.38) by g, subtract from the second equation the first one, integrate over D_{LR}, and take $R \to \infty$, to get

$$G = g + \int_{S_L} (Gg_N - G_N g)ds = g - \int_{S_L} g\mu \ ds, \quad \mu := G_N|_{S_L}. \tag{9.3.39}$$

The *qp* condition allows one to cancel the integrals over the lateral boundary ($x = 0$ and $x = L$), and the radiation condition allows one to have

$$\lim_{R \to \infty} \int_{S_R} (Gg_N - G_N g)ds = 0.$$

Differentiate (9.3.39) to get

$$\mu = -A\mu + 2\frac{\partial g}{\partial N} \text{ on } S_L, \ A\mu := 2\int_{S_L} \frac{\partial g(s, \sigma)}{\partial N_s}\mu(\sigma) \ d\sigma. \tag{9.3.40}$$

This is a Fredholm equation for μ in $L^2(S_L)$, if S_L is $C^{1,m}$, $m > 0$. The homogeneous equation (9.3.40) has only the trivial solution: if $\mu + A\mu = 0$, then the function $\psi := \int_{S_L} g\mu \ ds$ satisfies the condition $\psi_N^+|_{S_L} = 0$, where $\psi_N^+(\psi_N^-)$ is the normal derivative of ψ from $D_{-b}(D_L)$, and we use the known formula for the normal derivative of the single-layer potential at the boundary. The ψ satisfies also (9.3.34) and (9.3.35), and, by the choice of b, one has $\psi = 0$ in D_{-b}. Also $\psi = 0$ in D_L, because $(\Delta + k^2)\psi = 0$ in D_L, $\psi|_{S_L} = 0$ (by the continuity of the single-layer potential), ψ satisfies the quasiperiodicity condition (because g satisfies it), and ψ is outgoing (because g is).

Since $\psi = 0$ in D_{-b} and in D_L, one concludes that $\mu = \psi_N^+ - \psi_N^-$, where $\psi_N^+(\psi_N^-)$ is the normal derivative of ψ from $D_{-b}(D_L)$, and we use the jump relation for the normal derivative of the single-layer potential.

Thus, we have proved the existence and uniqueness of μ, and, therefore, of G, and got a representation formula

$$G = g - \int_{S_L} g\mu \, ds. \tag{9.3.41}$$

This representation shows that the rate of decay of G as $y \to \infty$ is essentially the same as that of g.

The G is analytic with respect to k on the complex plain with cuts along the rays $\lambda_j - i\tau$, $0 \le \tau < \infty, j = 0, \pm 1, \pm 2, \ldots$, in particular, in the region $\Im k > 0$, up to the real positive half-axis except for the set $\{\lambda_j\}_{j=0,\pm 1,\pm 2,\ldots}$. This follows from (9.3.40) and (9.3.41), and the general result (see Lemma 1.4.4 in [268]) concerning analyticity of the solution to a Fredholm equation with respect to a parameter.

Suppose a bounded obstacle D_0 is placed inside D_L, $u = 0$ on $S_0 = \partial D_0$, S_0 is a Lipschitz boundary. If qp condition is imposed, then Green's function G_0 in the presence of the obstacle satisfies equations similar to (9.3.41) and (9.3.40):

$$G_0(x, y) = G(x, y) - \int_{S_0} G(x, s)\mu_0(s, y) \, ds, \quad \mu_0 = G_{0N}, \tag{9.3.42}$$

where N is the unit-normal to S_0 pointing into D_L, and

$$\mu_0 = -A_0\mu_0 + 2\frac{\partial G}{\partial N} \text{ on } S_0, \; A_0\mu_0 := 2\int_{S_0} \frac{\partial G(s, \sigma)}{\partial N_s}\mu_0(\sigma) \, d\sigma. \tag{9.3.43}$$

This is a Fredholm equation (with index zero). If k^2 is not an eigenvalue of the Neumann Laplacian in D_0 (=not exceptional), then equation (9.3.43) is uniqueley solvable and, by (9.3.42), G_0 exists and is unique for this $k > 0$. It is not known what are nontrivial sufficient conditions for $k > 0$ to be not exceptional. The exceptional k form a discrete countable set on the positive semi-axis $k > 0$. If the Neumann boundary condition is imposed on S_L, then, even in the absence of the obstacle D_0, it is not known if LAP holds, because the proof of Lemma 2.2 *is not valid for the Neumann boundary condition on* S_L.

9.3.4. *Modified Rayleigh Conjecture (MRC)*

Rayleigh conjectured [427] ("Rayleigh hypothesis") that the series (9.3.7) converges up to the boundary S_L. This conjecture is wrong (see [175]) for some $f(x)$. Since the Rayleigh hypothesis has been widely used for numerical solution of the scattering problem by physicists and engineers, and because these practitioners reported high instability of the numerical solution, and there are no error estimates for the solution, we propose a modification of the Rayleigh conjecture, which is a Theorem. This MRC can be used for a numerical solution of the scattering problem, and it gives an error estimate for this solution. Our arguments are very similar to the ones in Section 9.1.

Rewrite the scattering problem (9.3.1)–(9.3.4) as

$$(\Delta + k^2)v = 0 \text{ in } D, \quad v = -u_0 \text{ on } S_L, \tag{9.3.44}$$

where v satisfies (9.3.4), and v has representation (9.3.7), that is, v is "outgoing", it satisfies the radiation condition. Fix an arbitrary small $\epsilon > 0$, and assume that

$$\left\| u_0 + \sum_{|j| \le j(\epsilon)} c_j(\epsilon) v_j(y) \varphi_j(x) \right\| \le \epsilon, \quad 0 \le x \le L,\ y = f(x), \tag{9.3.45}$$

where $\|\cdot\| = \|\cdot\|_{L^2(S_L)}$.

Lemma 9.3.3. *For any* $\epsilon > 0$*, however small, and for any* $u_0 \in L^2(S_L)$*, there exists* $j(\epsilon)$ *and* $c_j(\epsilon)$ *such that* (9.3.45) *holds.*

Proof. Lemma 9.3.3 follows from the completeness of the system

$$\{\varphi_j(x)v_j(f(x))\}_{j=0,\pm1,\pm2,\ldots}, \quad x \in S_L$$

in $L^2(S_L)$. Let us prove this completeness. Assume that there is an $h \in L^2(S_L)$, $h \not\equiv 0$ such that

$$\int_{S_L} h\overline{\varphi_j(x)}v_j(f(x))ds = 0 \tag{9.3.46}$$

for any j. From (9.3.46), one derives (cf. [269, pp. 162–163])

$$\psi(\mathbf{x}) := \int_{S_L} hg(\mathbf{x},\xi)d\xi = 0, \ \mathbf{x} \in D_{-b}. \tag{9.3.47}$$

Thus, $\psi = 0$ in D_L, and $h = \psi_N^+ - \psi_N^- = 0$. Lemma 9.3.3 is proved. □

Lemma 9.3.4. *If* (9.3.45) *holds, then*

$$\left|\left|\left| v(\mathbf{x}) - \sum_{|j|\le j(\epsilon)} c_j(\epsilon)v_j(y)\varphi_j(x) \right|\right|\right| \le c\epsilon, \ \forall x,y \in D_L,$$
$$0 \le x \le L, \ y \ge f(x),$$

where c = const > 0 *does not depend on* ϵ, x, y, *and* R; $R > M$ *is an arbitrary fixed number, and* $\|w\| = \sup_{\mathbf{x}\in D\setminus D_{LR}} |w(\mathbf{x})| + \|w\|_{H^{1/2}(D_{LR})}$.

Proof. Let $w := v - \sum_{|j|\le j(\epsilon)} c_j(\epsilon)v_j(y)\varphi_j(x)$. Then w solves equation (9.3.1), w satisfies (9.3.4), w is outgoing, and $\|w\|_{L^2(S_L)} \le \epsilon$. One has (cf. (9.3.15))

$$w(\mathbf{x}) = -\int_{S_L} wG_N(\mathbf{x},\xi)\ ds. \tag{9.3.48}$$

Thus, (9.3.45), i.e., $\|w\| := \|w\|_{L^2(S_L)} \leq \epsilon$, implies

$$|w(\mathbf{x})|_{y=R} \leq \|w\|_{L^2(S_L)} \|G_N(\mathbf{x},\xi)\|_{L^2(S_L)} \leq c\epsilon, \quad c = \text{const} > 0, \tag{9.3.49}$$

where c is independent of ϵ, and $R > \max f(x)$ is arbitrary. Now let us use the elliptic inequality

$$\|w\|_{H^m(D_{LR})} \leq c(\|w\|_{H^{m-0.5}(S_L)} + \|w\|_{H^{m-0.5}(S_R)}), \tag{9.3.50}$$

where we have used the equation $\Delta w + k^2 w = 0$, and assumed that k^2 is not a Dirichlet eigenvalue of the Laplacian in D_{LR}, which can be done without loss of generality, because one can vary R. The integer $m \geq 0$ is arbitrary if S_L is sufficiently smooth, and $m \leq 1$ if S_L is Lipschitz. Taking $m = 0.5$ and using (9.3.45) and (9.3.49), one gets

$$\|w\|_{H^{1/2}(D_{LR})} \leq c\epsilon. \tag{9.3.51}$$

Thus, in a neighborhood of S_L, we have proved estimate (9.3.51), and in a complement of this neighborhood in D_L we have proved estimate (9.3.49). Lemma 9.3.4 is proved. □

Remark. In (9.3.50), there are no terms with boundary norms over the lateral boundary (lines $x = 0$ and $x = L$) because of the quasi-periodicity condition.

From Lemma 9.3.4, the basic result, Theorem 9.3.5, follows immediately:

Theorem 9.3.5 (MRC). *Fix $\epsilon > 0$, however small, and choose a positive integer p. Find*

$$\min_{c_j} \|u_0 + \sum_{|j|\leq p} c_j \varphi_j(x) v_j(y)\| := m(p). \tag{9.3.52}$$

Let $\{c_j(p)\}$be the minimizer of (9.3.52). If $m(p) \leq \epsilon$, then

$$v(p) = \sum_{|j|\leq p} c_j(p)\varphi_j(x)v_j(y) \tag{9.3.53}$$

satisfies the inequality

$$|||v - v(p)||| \leq c\epsilon, \tag{9.3.54}$$

where $c = \text{const} > 0$ *does not depend on* ϵ. *If* $m(p) > \epsilon$, *then there exists* $j = j(\epsilon) > p$ *such that* $m(j(\epsilon)) < \epsilon$. *Denote* $c_j(j(\epsilon)) := c_j(\epsilon)$ *and* $v(j(\epsilon)) := v_\epsilon$. *Then*

$$|||v - v_\epsilon||| \leq c\epsilon. \tag{9.3.55}$$

Appendices

Appendix A

Summary of Some Results in Potential Theory and Embedding Theorems

1. Let $D \subset \mathbb{R}^n$ be a compact domain with a closed boundary $\partial D = \Gamma$. We discuss the basic facts from potential theory first under the classical assumptions: Γ is Lyapunov-type, and the densities of the potentials of the single and double layers are Hölder continuous or continuous, and afterwards for not so smooth boundaries and densities: the boundaries of class C^1 and densities of class $L^p(\Gamma), p > 1$.

We write $\Gamma \in C^{1,\alpha}$ if and only if in the local coordinates $x_3 = F(x_1, x_2)$ the function $F \in C^{1,\alpha}$, that is, F has first derivatives which are Hölder-continuous. A function $h(x), x \in \mathbb{R}^n$ is Hölder-continuous $h \in C^{0,\alpha}(D)$ in an open set $D \in \mathbb{R}^n$ if and only if $|h(x) - h(y)| \leq c|x - y|^\alpha, x, y \in D, c = c(D, h, \alpha), 0 < \alpha \leq 1$. We write $C(D)$ for $C^0(D)$, and $C^1(D)$ for $C^{1,0}(D)$. Since the normal to Γ is defined by the formula $N(s) = (e_1 F_{x_1} + e_2 F_{x_2} - e_3)/\sqrt{|\nabla F|^2 + 1}$, where e_l are unit vectors of the local cartesian coordinate system and $s \in \Gamma$, one sees that $N(s)$ is Hölder-continuous if $\Gamma \in C^{1,\alpha}$. A compact closed surface $\Gamma \in C^{1,\alpha}, \alpha > 0$ is called a Lyapunov surface.

Consider an integral operator

$$Kh = \int_D K(s,t)h(t)dt, \tag{A.0.1}$$

where $D \in \mathbb{R}^n$ is a bounded domain. Let $s \in D' \subset \mathbb{R}^m$. We write a.e. for almost every, $|D|$ for meas D, i.e., for volume of $D \subset \mathbb{R}^n$.

Lemma A.0.1 ([96, p. 417]). *Let*

$$\int_D |K(s,t)|^r dt \le c_1^r \quad \text{for a.e. } s \in D',$$

$$\int_{D'} |K(s,t)|^\sigma ds \le c_2^\sigma \quad \text{for a.e. } t \in D,$$

$$q \ge p,\ q \ge \sigma,\ (1-\sigma q^{-1})p' \le r,\ p,q \ge 1,\ p' = p(p-1)^{-1}.$$

Then $K : L^p(D) \to L^q(D')$ *is bounded,* $\|K\| \le c_1^{1-\sigma q^{-1}} c_2^{\sigma q^{-1}} c(|D|)$. *Here* c_1 *and* c_2 *are given positive constants and* $c(|D|)$ *is a positive constant which depends only on* D.

Lemma A.0.2. *Let* $K(s,t) = B(s,t)/|s-t|^\lambda$, *where* $B(s,t)$ *is a continuous function on* $D' \times D \subset \mathbb{R}^m \times \mathbb{R}^n$. *Then* K *is bounded as an operator from* $L^p(D)$ *into* $L^q(D')$ *provided that*

$$q < \frac{mp}{n-(n-\lambda)p}, \quad m > n-(n-\lambda)p. \tag{A.0.2}$$

Lemma A.0.3. *Let* $n = m, D = D'$, *and* $K(s,t)$ *is defined as in Lemma* A.0.2. *Then the operator* $K : L^p(D) \to C(D)$ *is bounded if* $\lambda < n(1-p^{-1})$. *The operator* $K : L^p(D) \to C^\alpha(D)$ *is bounded, where* $\alpha = n - np^{-1} - \lambda$.

2. In this section, the operators S, A, and A' are studied in the Hölder spaces, $\Gamma \in C^{1,\alpha}$.

Lemma A.0.4. *Let* $n = 3, g(x,y)$ *is given by* (1.1.1), $\Gamma \in C^{1,\alpha}$, *and* $S\sigma$, *the single-layer potential, be defined by* (1.1.4). *Then* $S : C(\Gamma) \to C^\alpha(\mathbb{R}^3)$ *is a bounded operator. The operators* $D : C(\Gamma) \to C(D_\pm)$, *defined by* (1.1.1) (*with* $D_+ = D$, $D_- = \Omega$) *are bounded and formula* (1.1.3) *holds. The operators* $A', A : C(\Gamma) \to C^\alpha(\Gamma)$, *defined by* (1.1.4) *and* (1.1.2) *are bounded, the operator* $S : C^\beta(\Gamma) \to C^{1,\beta}(D_\pm)$ *is bounded, and formula* (1.1.3) *holds. Formula* (1.1.3) *holds even if*

$\sigma \in C(\Gamma)$, moreover, the limits are attained uniformly on Γ. Let

$$\mu \in C^{1,\alpha}(\Gamma), \quad B\mu = \frac{\partial A'\mu}{\partial N} = \lim_{\epsilon\to\infty} \frac{\partial}{\partial N_s}(D\mu), \tag{A.0.3}$$

$$x = s + \epsilon N_s, \quad s \in \Gamma, \quad \epsilon \to \pm 0.$$

The Lyapunov–Tauber theorem says that the limits as $\epsilon \to +0$ and, $\epsilon \to -0$ are equal. Notice that according to the definitions (1.1.1)–(1.1.4), the operators A, A', and B depend on the parameter k.

Lemma A.0.5. *The operator $B = B(k) : C^{1,\alpha}(\Gamma) \to C^{\alpha}(\Gamma)$ is bounded. The operator $B(k_1) - B(k_2)$ is compact in $C^{1,\alpha}(\Gamma)$. If k^2 is neither an eigenvalue of the Dirichlet Laplacian nor Neumann Laplacian in D, then $B^{-1}(k) = [(A' + I)/2]^{-1}Q[(A - I)/2]^{-1}$, where Q is defined by (1.1.8).*

Let us explain the last formula. Suppose $B(k)\mu = f$. Then $\partial v/\partial N = f$, where $v = D\mu$ is defined by (1.1.1). If this v can be written in the form $v = S\sigma$, where S is defined by (1.1.1), then $(A - I)\sigma/2 = f$. If k^2 is not an eigenvalue of the Dirichlet Laplacian in D, then the operator $A - I$ is invertible and $\sigma = [(A - I)/2]^{-1}f$. We have $Q\sigma = v_- = (A' + I)\mu/2$. The operator $A' + I$ is invertible if k^2 is not an eigenvalue of the Neumann Laplacian in D. Thus, $\mu = [(A' + I)/2]^{-1}Q[(A - I)/2]^{-1}f = B^{-1}(k)f$. If k^2 is not an eigenvalue of the Dirichlet Laplacian in D, the operator $Q : C^{\alpha}(\Gamma) \to C^{1,\alpha}(\Gamma)$ is invertible. Thus

$$B(k) = \frac{A - I}{2} Q^{-1} \frac{A' + I}{2}. \tag{A.0.4}$$

3. In this section, the operators S, A and A' are studied in the Lebesgue spaces. In Lemmas A.0.6 and A.0.7, below, we assume that

$$\Gamma \in C^1, \quad 1 < p < \infty. \tag{A.0.5}$$

Let

$$A'_\epsilon \mu = 2\int_{\Gamma_\epsilon} \frac{\partial g(s,t)}{\partial N_t}\mu(t)dt, \quad \Gamma_\epsilon = \{t : |t - s| \geq \epsilon > 0, \ t \in \Gamma\} \tag{A.0.6}$$

and

$$A'\mu = \lim_{\epsilon\to\infty} A'_\epsilon\mu \quad (\mathrm{a}), \qquad A'_*\mu = \sup_{\epsilon>\infty} |A'_\epsilon\mu| \quad (\mathrm{b}). \tag{A.0.7}$$

The operator A' is defined by (A.0.7) which coincides on $C(\Gamma), \Gamma \in C^{1,\alpha}$, with the operator A' defined by (1.1.2) as an improper integral.

Lemma A.0.6. *The operator $A'_* : L^p(\Gamma) \to L^p(\Gamma)$ is bounded. For any $\mu \in L^p(\Gamma)$, the limit (1.1.7a) exists in $L^p(\Gamma)$ and pointwise a.e. The operator $A' : L^p(\Gamma) \to L^p(\Gamma)$ is compact. Let $v(x) = D\mu, \mu \in L^p(\Gamma), x \in D, 0 < a < 1$. Then there exists a $\delta = \delta(a, D) > 0$ such that the function*

$$v_*(t) = \sup\{|v(x)|\, x \in D, \quad |x-t| < \delta, \quad (x-t)\cdot N_t < -a|x-t|\} \tag{A.0.8}$$

belongs to $L^p(\Gamma)$,

$$\|v_*\|_{L^p(\Gamma)} \le c\|\mu\|_{L^p(\Gamma)}, \tag{A.0.9}$$

and for almost every $t \in \Gamma$ the pointwise limit exists:

$$\lim v(x) = \frac{A'\mu + \mu}{2} \quad \text{as} \quad x \to t, \quad x \in D, \quad (x-t)\cdot N_t < -a|x-t|. \tag{A.0.10}$$

Let $\mu \in W^{1,p}(\Gamma), 1 < p < \infty$, where $\|\mu\|_{W^{1,p}(\Gamma)} = \|\mu\|_{L^p(\Gamma)} + \|\nabla_t\mu\|_{L^p(\Gamma)}$, and $\nabla_t\mu$ is the surface gradient of μ on Γ. The operator $A' : W^{1,p}(\Gamma) \to W^{1,p}(\Gamma)$ is compact. For any $a, 0 < a < 1$, there exists a $\delta = \delta(a, D)$, such that the function

$$(\nabla v)_*(t) = \sup\{|\nabla v| : x \in D,\, |x-t| < \delta,\, (x-t)\cdot N_t < -a|x-t|\} \tag{A.0.11}$$

belongs to $L^p(\Gamma)$ and

$$\|\nabla v_*(t)\|_{L^p(\Gamma)} \le c\|\mu\|_{L^p(\Gamma)}. \tag{A.0.12}$$

Lemma A.0.7. *Let*

$$A_\epsilon\sigma = 2\int_{|s-t|>\epsilon} \sigma(t)\frac{\partial g(s,t)}{\partial N_s}dt, \quad s \in \Gamma. \tag{A.0.13}$$

Then the limit $\lim_{\epsilon\to 0} A_\epsilon \sigma = A\sigma$ *exists in* $L^p(\Gamma)$ *(and pointwise a.e.), and* $A : L^p(\Gamma) \to L^p(\Gamma)$ *is compact. Let* $w(x)$ *be defined by* (1.1.4), $x \in D_\pm, \sigma \in L^p(\Gamma)$, *and* $0 < a < 1$. *Then there exists a* $\delta = \delta(a, D)$ *such that the functions*

$$(\nabla w)_+(t) = \sup\{|\nabla w| : x \in D_+,\ |x-t| < \delta,\ (x-t)\cdot N_t < -a|x-t|\} \tag{A.0.14}$$

and

$$(\nabla w)_-(t) = \sup\{|\nabla w| : x \in D_-,\ |x-t| < \delta,\ (x-t)\cdot N_t > a|x-t|\} \tag{A.0.15}$$

belong to $L^p(\Gamma)$

$$\|(\nabla w)_\pm\|_{L^p(\Gamma)} < c\|\sigma\|_{L^p(\Gamma)}, \tag{A.0.16}$$

and the limit

$$\lim_{x\to t^\pm} \frac{\partial w(x)}{\partial N_t} = \lim \nabla w \cdot N_t = \frac{A\sigma \pm \sigma}{2} \tag{A.0.17}$$

exists pointwise a.e. Here $x \to t^+ (x \to x^-)$ *means that* $x \to t$ *and* x *belongs to the domain indicated by the inequality in* (A.0.14) ((A.0.15)). *The operator* $I + A$ *is a continuous bijection* $L^p(\Gamma)$ *onto* $L^p(\Gamma)$, *and* $W^{1,p}(\Gamma)$. *The operators* $A : L^p(\Gamma) \to L^p(\Gamma)$ *and* $A : W^{1,p}(\Gamma) \to W^{1,p}(\Gamma)$ *are compact.*

4. In this section, the operators S, A and A' are studied in the Sobolev spaces $H^q(\Gamma), \Gamma \in C^{q+1}$.

Lemma A.0.8. *The operator* $Q : H^q(\Gamma) \to H^{q+1}(\Gamma)$ *is Fredholm-type with index* 0. *If and only if* k^2 *is not an eigenvalue of the Dirichlet Laplacian in* D, *the operator* Q *is an isomorphism of* $H^q(\Gamma)$ *onto* $H^{q+1}(\Gamma)$. *The operators* A *and* A' *are continuous from* $H^q(\Gamma)$ *into* $H^{q+1}(\Gamma)$.

A linear operator $A : B_1 \to B_2$ from a Banach space B_1 into a Banach space B_2 is called Fredholm-type if and only if im A is closed, $n = n(A) = \dim \ker A < \infty$, and $n^* = n^*(A) = \dim$ coker $A < \infty$. Here, im A is the range of A, ker A is the null space of A, i.e., ker $A = \{f : Af = 0\}$, and coker $A = B_2/\text{im } A$, i.e., the quotient space. The index of A, ind A or $i(A)$, is defined by the formula $i(A) = n(A) - n^*(A)$.

5. Consider a compact set $D \subset \mathbb{R}^n, n \geq 2$, and the set $M(D)$ of measures μ whose support lies in D. Let

$$u_\mu(x) = \int g_0(x,y)d\mu, \quad g_0(x,y) = (\omega_n(n-2)|x-y|^{n-2})^{-1}, \tag{A.0.18}$$

where ω_n is the surface area of the unit sphere in $\mathbb{R}^n, n > 2$, $g_0 = -(2\pi)^{-1}\ln|x-y|$ if $n = 2$.

The Wiener capacitance of D is defined by the formula

$$\operatorname{cap} D = C(D) = \sup \mu(D), \quad \mu \in M(D), \quad u_\mu(x) \leq 1, \quad x \in \mathbb{R}^n. \tag{A.0.19}$$

The capacitance has the following properties: $C(D) \geq 0, C(D)$ is invariant under translations and rotations of $D, C(D_1) \leq C(D_2)$ if $D_1 \subseteq D_2, C(D) \leq \sum_{j=1}^\infty C(D_j)$ if $D = \bigcup_{j=1}^\infty D_j, D_j \cap D_i = \emptyset$ if $i \neq j$. If $\mu \in M(D), u_\mu(x) \leq m_2$ if $x \in \mathbb{R}^n$, and $u_\mu(x) \geq m_1$ if $x \in D$, then $m_1 C(D) \leq \mu(D) \leq m_2 C(D)$. One can prove that

$$\operatorname{cap} D = \left(\inf_{\mu \in M_1(D)} \int_D \int_D g_0(x,y)d\mu(x)d\mu(y)\right)^{-1},$$
$$M_1(D) = \left\{\mu : \mu \in M(D), \int_D d\mu = 1\right\} \tag{A.0.20}$$

and

$$\operatorname{cap} D = \inf \int_{\mathbb{R}^n} |\nabla^2 u|^2 dx, \quad u \in C_0^\infty(\mathbb{R}^n), \quad u = 1 \text{ in } D. \tag{A.0.21}$$

The supremum in (A.0.19) is attained at a measure γ which is called the equilibrium measure of the compact D, and $u_\gamma(x) = 1$ a.e. in D. Formula (A.0.21) is easily understood if one recalls that $CU^2 = 2E$, where C is the capacitance of the perfect conductor D, U is the potential of this conductor, and E is the electrostatic energy of the field around D. If $U = 1$ (see (A.0.21)), then $C = 2E$ and the equilibrium is attained if $E = \min$. Since $2E = \int_{\mathbb{R}^n} |\nabla^2 u|^2 dx$ where u is the potential of a charge distribution in D, one obtains (A.0.21).

Formula (A.0.20) is easily understood if one recalls that $Q^2/2C = E$, where Q is the total charge of the conductor D, $Q = \int_D d\mu$. According to (A.0.20), $Q = 1$, and the equilibrium distribution of the charge in D is attained when $E = \min$. Since $2E = \int_D \int_D g_0(x,y)d\mu(x)d\mu(y)$, one obtains (A.0.20).

Let $G \subset \mathbb{R}^n$ be an open set, E be a compact subset of G,

$$M(E,G) = \{u : u \in C_0^\infty(G), \quad 0 \le u \le 1, \quad u = 1 \text{ on } \tilde{E} \supset E\}, \tag{A.0.22}$$

where $\tilde{E}$ is an open set containing E, $\tilde{E}$ depends on u. Let $M(E) = M(E, \mathbb{R}^n)$. If $p > 1$, one defines

$$(p,l)\text{cap}(E,G) = \inf\left\{\int_G \sum_{|\alpha|=l} |D^\alpha u|^p dx, \quad u \in M(E,G)\right\}. \tag{A.0.23}$$

If $l = 1$, then we write $(p,1)\text{cap}(E,G) = (p)\text{cap}(E,G)$,

$$(p)\text{cap}(E,G) = \inf\left\{\int_G |\nabla u|^p dx, \quad u \in M(E,G)\right\}. \tag{A.0.24}$$

Clearly $(2)\text{cap}(E, \mathbb{R}^n) = \text{cap}\, E$ (see (A.0.21)).

6. *Embedding theorems*

Theorem A.0.9. *Let $D \subset \mathbb{R}^n$ be a bounded domain with a smooth boundary $\Gamma, 1 < p < \infty$, and $0 < l < s - np^{-1}$. Then the embedding operator $i : W^{s,p}(D) \to C^l(D)$ is compact. If $l \ge s - np^{-1}, m > n - (s-l)p$, and $1 < q < mp/(n - (s-l)p)$, then the embedding operator $i : W^{s,p}(D) \to W^{l,q}(\Gamma_m)$ is compact. Here Γ_m is any m-dimensional C manifold in D, $\|u\|_{W^{s,p}(D)} \ge c\|u\|_{W^{l,q}(\Gamma_m)}$, where here and below $c > 0$ denotes various constants which do not depend on $u \in W^{s,p}(D)$. The trace operator $r : W^{s,p}(D) \to W^{s-p^{-1},p}(\Gamma)$ is continuous. For any given $u \in W^{s-p^{-1},p}(\Gamma)$, there exists a $v \in W^{s,p}(D)$ such that $v = u$ on Γ, and $\|v\|_{W^{s,p}(D)} \le c\|u\|_{W^{s-p^{-1},p}(\Gamma)}$, so that an extension operator from $W^{s-p^{-1},p}(\Gamma)$ into $W^{s,p}(D)$ is bounded. In particular, if $p = 2$ then $W^{s,2}(D) = H^s(D)$ and the operator $r : H^s(D) \to H^{s-1/2}(\Gamma)$ is bounded.*

7. *Potential theory and boundary value problems in domains with nonsmooth boundaries*

If the boundary condition is homogeneous, then the Dirichlet problem in Ω is formulated as

$$\int_\Omega (\nabla u \nabla h - k^2 uh)dx = \int_\Omega fh\,dx, \quad \forall h \in \mathring{H}^1(\Omega), \quad k>0, \quad u \in \mathring{H}^1(\Omega), \tag{A.0.25}$$

while the Neumann problem is

$$\int_\Omega (\nabla u \nabla h - k^2 uh)dx = \int_\Omega fh\,dx, \quad \forall h \in H^1(\Omega), \quad k>0, \quad u \in H^1(\Omega), \tag{A.0.26}$$

and in both problems u should satisfy the radiation condition (0.0.5). If the boundary Γ is not sufficiently smooth, so that Green's formula may be not applicable, then the usual proof of the uniqueness theorem, based on Green's formula, should be modified slightly. Namely, if $f=0$ then (A.0.25) with $h = \bar{u}\alpha((R-|x|)/\delta), \delta > 0$, $D \subset B_R$, yields

$$\int_\Omega |\nabla u|^2 \alpha dx + \int_\Omega \nabla u \cdot \bar{u} \nabla \alpha dx - k^2 \int_\Omega |u|^2 \alpha dx = 0, \tag{A.0.27}$$

where

$$\alpha(x) = \alpha(|x|) \in C^\infty, 0 \le \alpha \le 1, \alpha(r) = 0, r \le 0,$$
$$\alpha(r) = 1, r \ge 1, \alpha' \ge 0.$$

Also

$$\int_{\Omega_R} (|\nabla u|^2 - k^2|u|^2)\alpha dx + \int_\Omega \overline{\nabla u} \cdot u \nabla \alpha dx = 0. \tag{A.0.28}$$

Thus

$$\int_\Omega (\nabla u \bar{u} - \overline{\nabla u} u)\nabla \alpha dx = 0. \tag{A.0.29}$$

Let $\delta \to 0$. Then (A.0.29) becomes

$$\int_{S_R} \left(\frac{\partial u}{\partial r} \bar{u} - u \frac{\partial \bar{u}}{\partial r} \right) ds = 0. \tag{A.0.30}$$

The radiation condition and (A.0.30) yield

$$\lim_{R \to \infty} \int_{S_R} |u|^2 ds = 0. \tag{A.0.31}$$

Equation (A.0.31) and Lemma 1.2.1 imply that $u = 0$ in Ω. This argument is applicable as well for the Neumann boundary value problem. It does not require any smoothness from the boundary. Existence of the solution to the Dirichlet boundary value problem does not require any smoothness of Γ, but there are some requirements on Γ in the case of the Neumann boundary condition. This can be explained as follows. In Section 1.2.2, we showed how to reduce a boundary value problem in Ω to a boundary value problem for a compact domain. Therefore, we can discuss the role of the smoothness of Γ assuming that the domain B, in which the boundary value problem is being studied, is compact. In this case, the existence of the solution to the problem

$$-\nabla^2 u = f \text{ in } B, \quad u = 0 \text{ on } \partial B, \quad f \in L^2(B), \tag{A.0.32}$$

in $\mathring{H}^1(B)$ is equivalent to the validity of the inequality

$$\|h\|_{L^2(B)} \leq c \|\nabla h\|_{L^2(B)}, \quad \forall h \in \mathring{H}^1(B), \tag{A.0.33}$$

where $c > 0$ does not depend on h. Indeed, the problem (A.0.32) in the generalized formulation is equivalent to the validity of the following integral identity:

$$\int_B \nabla u \nabla h dx = \int_B f h dx, \quad \forall h \in \mathring{H}^1(B). \tag{A.0.34}$$

The right side of (A.0.34) is a linear bounded functional $F(h)$ in $\mathring{H}^1$ for any $f \in L^2$ if and only if (A.0.33) holds. The left side of (A.0.34) is a linear bounded functional in $\mathring{H}^1(B)$. Therefore, the problem (A.0.32) is uniquely solvable in $\mathring{H}^1(B)$ for any $f \in L^2(B)$ if and only if (A.0.33) holds. But (A.0.33) holds for any bounded open set $B \subset \mathbb{R}^3$ and in this case the embedding $i : \mathring{H}^1(B) \to L^2(B)$ is compact (see [444, p. 331]). The spectrum of the Dirichlet Laplacian in $L^2(B)$ is therefore discrete for any bounded open set B.

A similar argument shows that the Neumann boundary value problem in B is solvable in $H^1(B)$ if and only if the following inequality holds

$$\inf_{a\in\mathbb{R}^2}\{\|h-a\|_{L^2(B)}\} \le c\|\nabla h\|_{L^2(B)}, \quad \forall h \in H^1(B), \qquad \text{(A.0.35)}$$

where $c > 0$ does not depend on h. This is Poincare's inequality. A necessary and sufficient condition on B for (A.0.35) to hold is known. A necessary and sufficient condition on B for the compactness of the embedding $i : H^1(B) \to L^2(B)$ is known.

Let us formulate these conditions. First we should introduce some definitions. Let F be a closed set, G be an open set, $F \subset G \subset B$. The set $G \setminus F = M$ is called the conductor. Let $V(M) = \{f : f \in C^\infty(B), f = 1 \text{ on } F, f = 0 \text{ on } B \setminus G\}$. Define the conductivity of M by the formula

$$\operatorname{cond} M = \inf\left\{\int_M |\nabla f|^2 dx, \ f \in V(M)\right\}. \qquad \text{(A.0.36)}$$

Let $\{M\}$ be the set of the conductors for which $2\operatorname{mes} G \le \operatorname{mes} B$, where $\operatorname{mes} B$ is the Lebesgue measure of B. Then (A.0.35) holds if and only if

$$\sup_{M\in\{M\}} \frac{\operatorname{mes} F}{\operatorname{cond} M} < \infty. \qquad \text{(A.0.37)}$$

The embedding $i : H^1(B) \to L^2(B)$ is compact if and only if

$$\lim_{c\to 0}\left\{\sup_{\operatorname{mes} G\le c} \frac{\operatorname{mes} F}{\operatorname{cond} M}\right\} = 0. \qquad \text{(A.0.38)}$$

Condition (A.0.38) is necessary and sufficient for the spectrum of the Neumann Laplacian in $L^2(B)$ to be discrete. These results are formulated in [444], where one can find numerous references to the original papers. An often used sufficient condition for compactness of the embedding $i : H^1(B) \to L^2(B)$ for a bounded domain B is the so-called "finite tiling property" which is defined in [480].

In [104] and [31], the integral equation of potential theory are studied for the surface which are not smooth. In this case, the operators A' and A defined by (1.1.2) and (1.1.4) are not necessarily compact. If one defines the essential norm of a linear operator $T : B \to B$ on a Banach space B by the formula

$$|T|_{\text{ess}} = \inf_{Q \in K} \|T - Q\| \tag{A.0.39}$$

where K is the set of all compact linear operators on B, then the equation

$$(I + T)h = f \tag{A.0.40}$$

in B is of Fredholm's type if $|T|_{\text{ess}} < 1$. Therefore, if the boundary Γ is not smooth but $|A|_{\text{ess}} < 1, |A'|_{\text{ess}} < 1$, then the usual potential theory is still valid. The essential norm of the operators A and A' is calculated in [30, 104].

Appendix B

Summary of Some Results in Operator Theory

1. Let A be a linear operator on a Hilbert space H. This operator is called *densely defined* if its domain of definition $D(A)$ is dense in H. The adjoint operator A^* is defined by $A^*h = h^*$, where $D(A^*)$ consists of all h such that $(Af, h) = (f, h^*), f \in D(A)$. If $(Af, h) = (f, Ah), \forall f, h \in D(A)$ then A is called *symmetric* on $D(A)$. If $A = A^*$, then A is called self-adjoint. A linear operator is called *closed* if $h_n \to h$ and $Ah_n \to f$ imply that $h \in D(A)$ and $Ah = f$. This means that $D(A)$ is complete in the *graph norm* $\|h\|_{\Gamma(A)} = \|h\| + \|Ah\|$. If $R_\lambda(A) = (A - \lambda I)^{-1} = (A - \lambda)^{-1}$ is a bounded operator defined on all of H, then λ is called a *regular point* of A and one writes $\lambda \in \rho(A)$. Otherwise λ belongs to the *spectrum* of A, and one writes $\lambda \in \sigma(A)$. The operator $R_\lambda(A)$ is called the resolvent of A. The set of regular points is open, $\sigma(A) = C \setminus \rho(A)$, where C is the set of all complex numbers. If $\lambda \in \sigma(A)$ and $Af = \lambda f, f \in H, f \neq 0$, then λ is an eigenvalue of A, f is the eigenvector, $N_\lambda = N(A - \lambda I) = \{f : Af = \lambda f\}$ is the eigenspace of A corresponding to λ.

The *geometric multiplicity* of λ is $\dim N(A - \lambda I)$. The set $M_\lambda = \bigcup_{s=1}^{\infty} N\{(A - \lambda I)^s\}$ is the *rootspace* of A corresponding to λ, $\dim M_\lambda$ is the *algebraic multiplicity* of λ. If $A = A^*$, then $M_\lambda = N_\lambda$. An eigenvalue λ is called normal if (i) $\dim M_\lambda < \infty$ and (ii) $H = M_\lambda \oplus M'_\lambda$, where $\oplus$ is the sign of the direct sum and $A - \lambda I$ is invertible on its invariant subspace M'_λ. The dimension of A is the number $r(A) = \dim \overline{R(A)}$, where the bar denotes closure and $R(A)$ is the range of A; $r(A) = r(A^*) = r\{(A^*A)^{1/2}\} = r\{(AA^*)^{1/2}\}$. If A

is a linear bounded operator with $r(A) < \infty$, then A is called *finite-dimensional*. In this case, $Af = \sum_{j=1}^{r(A)} (f, h_j) f_j$, where $\{f_j\}$ is a basis of $R(A)$ and $\{h_j\}$ is a biorthogonal to the $\{f_j\}$ system.

2. If $A = A^*$, then there is a uniquely determined family of orthogonal projections E_λ associated with A, the resolution of the identity for A, such that $F(A) = \int_{-\infty}^{\infty} F(\lambda) dE_\lambda$, the integral converges strongly on $D(F(A)) = \{h : \int_{-\infty}^{\infty} |F(\lambda)|^2 d(E_\lambda h, h) < \infty\}, E(\lambda) \le E(\lambda')$ if $\lambda \le \lambda', E(-\infty) = 0, E_\lambda E_\mu = E_{\min(\lambda,\mu)}, E(+\infty) = I, E_\lambda F(A) = F(A) E_\lambda, E_\mu \underset{s}{\to} E_\lambda$ as $\mu \to \lambda + 0$, where $\underset{s}{\to}$ means strong convergence. One has

$$((E_{\lambda_2} - E_{\lambda_1}) h, f) = \lim_{\delta \to +0} \lim_{\epsilon \to +0} \frac{1}{2\pi i} \int_{\lambda_1 + \delta}^{\lambda_2 + \delta} (R_{\lambda + i\epsilon} - R_{\lambda - i\epsilon} h, f) d\lambda \tag{B.0.1}$$

for any $h, f \in H$. If the weak derivative $dE_\lambda / d\lambda$ exists and is an integral operator, then its kernel is

$$\frac{dE_\lambda(x, y)}{d\lambda} = \frac{1}{\pi} \operatorname{Im} R_{\lambda + i0}(x, y) \tag{B.0.2}$$

where $\operatorname{Im} R_{\lambda+i0}(x, y) = \lim_{\epsilon \to +0} \operatorname{Im} R_{\lambda + i\epsilon}(x, y)$, and $R_{\lambda+i\epsilon}(x, y)$ is the kernel of the operator $R_{\lambda + i\epsilon} = (A - \lambda - i\epsilon)^{-1}$. Let $A = A^*$. We write $A \ge m$ if $(Af, f) \ge m(f, f), f \in D(A)$, and call A *semi-bounded from below*. One defines the *multiplicity* $N(\lambda_1, \lambda_2)$ *of the spectrum* of $A = A^*$ in the interval $(\lambda_1, \lambda_2]$ as $\dim(E_{\lambda_2} - E_{\lambda_1}) H$.

Lemma B.0.1. *Let A be semi-bounded from below and $N(\lambda) = \sum_{\lambda_n \le \lambda} 1$, where λ_n are the eigenvalues of A counted according to their multiplicities, $\lambda_1 \le \lambda_2 \le \cdots$. Then*

$$N(\lambda) = \sup_M \dim M, \quad M \subseteq \{f : f \in D(A),\ (Af, f) \le \lambda(f, f)\}, \tag{B.0.3}$$

where M are linear manifolds. More generally, if $N(\lambda_1, \lambda_2)$ is the multiplicity of the spectrum of A in the interval $(\lambda_1, \lambda_2]$ then

$$\begin{aligned} &N(\lambda_1, \lambda_2) = \sup_M \dim M, \\ &M \subseteq \{f : f \in D(A), \lambda_1(f, f) < (Af, f) \le \lambda_2(f, f)\}. \end{aligned} \tag{B.0.4}$$

Sketch of the Proof. We prove (B.0.4) since (B.0.3) is a particular case of (B.0.4). $N(\lambda_1, \lambda_2) = \infty$ if and only if there exists a linear manifold $M \subset D(A)$ of arbitrary dimension d, such that $\lambda_1(f, f) < (Af, f) \le \lambda_2(f, f), f \in M$. For example, M can be constructed as a linear span of d linearly independent vectors from $(E_{\lambda_2} - E_{\lambda_1})H = E_{\lambda_1\lambda_2}H$. If $N(\lambda_1, \lambda_2) = d < \infty$, then there exists d linearly independent vectors f_j in $E_{\lambda_1\lambda_2}H$. One has $(Af_j, f_j) = (AE_{\lambda_1\lambda_2}f_j, E_{\lambda_1\lambda_2}f_j) = \int_{\lambda_1}^{\lambda_2} \lambda d(E_\lambda f_j, f_j)$. Thus, $\lambda_1\|f_j\|^2 < (Af, f_j) \le \lambda_2\|f_j\|^2$. Conversely, if there is an M_0, such that $\lambda_1\|f\|^2 < (Af, f) \le \lambda_2\|f\|^2, f \in M_0, \dim M_0 = d = \sup \dim M$, where M is defined in (B.0.4), then $\dim E_{\lambda_1\lambda_2}H = d$. Indeed, $\dim E_{\lambda_1\lambda_2}H \ge d$ because $M_0 \subset E_{\lambda_1\lambda_2}H$, (if $M_0 \not\subset E_{\lambda_1\lambda_2}H$ then there is an $f \in M_0, f \perp E_{\lambda_1\lambda_2}H, \|f\| = 1$, for which $(Af, f) \notin (\lambda_1, \lambda_2])$; and $\dim E_{\lambda_1\lambda_2}H \le d$, because otherwise one could find at least $d+1$ linearly independent vectors f_j such that $\lambda_1\|f_j\|^2 < (Af_j, f_j) \le \lambda_2\|f_j\|^2, 1 \le j \le d+1$, which is a contradiction. □

A point λ is called a point of the *continuous spectrum* of a self-adjoint operator $A, \lambda \in \sigma_c(A)$ if and only if $N(\lambda - \epsilon, \lambda + \epsilon) = \infty$ for any $\epsilon > 0$. If $0 < N(\lambda - \epsilon, \lambda + \epsilon) < \infty$ for some $\epsilon > 0$, then λ is a point of the *discrete spectrum* of A, it is an isolated eigenvalue of finite multiplicity. If A is not necessarily self-adjoint, then define continuous spectrum of A, $\sigma_c(A)$, by the formula: $\sigma_c(A) = \{\lambda : \exists f_n \in D(A), \|Af_n - \lambda f_n\| \to 0 \text{ as } n \to \infty, \{f_n\} \text{ is a bounded noncompact sequence}\}$. Some authors call $\sigma_c(A)$ *essential spectrum* of A.

Lemma B.0.2. *$N(\lambda - \epsilon, \lambda + \epsilon) < \infty, \epsilon > 0$, if and only if there exists a subspace F, $\operatorname{codim} F < \infty$, such that $\|(A - \lambda)f\| \ge \epsilon\|f\|$, $f \in F \cap D(A)$.*

Lemma B.0.3. *If $A = A^*$, then $\lambda \in \sigma_c(A)$ if and only if there exists a noncompact sequence $f_n, \|f_n\| = 1$, such that $\|(A - \lambda)f_n\| \to 0$, $n \to \infty$.*

The set of $h \in D(A)$ for which the function $(E_\lambda h, h)$ is absolutely continuous in λ is an invariant subspace of A, $H_{ac}(A)$. The restriction A_{ac} of A onto $H_{ac}(A)$ is the *absolutely continuous part of A*,

$\sigma(A_{ac}) = \sigma_{ac}(A)$, its spectrum is the absolutely continuous spectrum of $A, \sigma_{ac}(A) \subset \sigma_c(A)$. One says that $\tilde{A}$ is a *closed extension* of A if $D(\tilde{A}) \supset D(A)$ and $\tilde{A} = A$ on $D(A)$. If $\dim D(A) < \infty \,(\mathrm{mod}\, D(A))$, then the extension is called finite-dimensional.

Here, $\dim M \,(\mathrm{mod}\, L) = \dim M/L$, where $M \supset L$ are linear manifolds, and M/L is the quotient set. One has $\sigma_c(\tilde{A}) \supset \sigma_c(A)$ and the following lemma holds:

Lemma B.0.4. *If $\tilde{A}$ is a finite-dimensional extension of A, then $\sigma_c(\tilde{A}) = \sigma_c(A)$. Let M_1 and M_2 be orthogonal linear manifolds invariant with respect to A. If A_j is the restriction of A onto M_j, then one defines $(A_1 \oplus A_2)(f_1 + f_2) = A_1 f_1 \oplus A_2 f_2$, where $f_j \in M_j, D(A_1 \oplus A_2) = M_1 \oplus M_2$. Therefore, A is an extension of $A_1 \oplus A_2, \sigma_c(A) \supset \{\sigma_c(A_1) \cup \sigma_c(A_2)\}$.*

3. Let A be a linear symmetric closed operator on H, λ be a complex number, $\mathrm{Im}\,\lambda \neq 0, N_\lambda^* = N(A^* - \lambda)$, and the bar denotes complex conjugate. One has $D(A^*) = D(A) \dot{+} N_\lambda^* \dot{+} N_{\bar{\lambda}}^*$, where $\dot{+}$ is the symbol of the direct sum. This means that any vector from $D(A^*)$ can be represented as a sum of the vectors from $D(A), N_\lambda^*$, and $N_{\bar{\lambda}}^*$, and this representation is unique. The numbers $\dim N_\lambda^* = n(\lambda)$ and $\dim N_{\bar{\lambda}}^* = n(\bar{\lambda})$ are called the *defect numbers* of A, the pair $(n(\lambda), n(\bar{\lambda}))$ is the *defect index* of A. These numbers do not depend on $\lambda, \mathrm{Im}\,\lambda \neq 0$. A closed symmetric operator A is self-adjoint if and only if $n(\lambda) = n(\bar{\lambda}) = 0$. A closed symmetric operator has self-adjoint extension if and only if $n(\lambda) = n(\bar{\lambda})$. If A is symmetric and its closure $\bar{A}$ is self-adjoint, then A is called essentially self-adjoint. If A is symmetric and bounded from below, then $\bar{A}$ has equal defect numbers and it has self-adjoint extensions. If $A > m$, then $A - m > 0$. Therefore, for a study of symmetric extensions of A one can assume that A is positive definite, i.e., $A \geq m > 0$. In this case, the closure of $D(A)$ in the norm $\|f\|_A = (Af, f)^{1/2}$ is a Hilbert space $H_A \subset H$. One can prove that on $D(A^*) \cap H_A$ the operator A^* is self-adjoint and is, therefore, a self-adjoint extension $\tilde{A}_m$ of A. This extension is called *Friedrichs's extension*.

It is the minimal self-adjoint extension of A in the sense that $\tilde{A}_m \geq m$ and $(\tilde{A}_m^{-1}f, f) \leq (\tilde{A}^{-1}f, f)$ for any self-adjoint positive definite extension of A, so that $B = \tilde{A}^{-1} - \tilde{A}_m^{-1} \geq 0$. The operator $B = 0$ on $R(A)$, so one can consider B as an operator on the orthogonal complement to $\overline{R(A)}$, i.e., on $N(A^*)$. If $\tilde{A}$ is an arbitrary self-adjoint positive definite extension of A, then $D(\tilde{A}) = D(A) \oplus (\tilde{A}_m^{-1} + B)N(A^*)$, where $B \geq 0$ is a self-adjoint bounded operator on $N(A^*)$, $D(\tilde{A}^{1/2}) = D(A_m^{1/2}) \oplus R(B^{1/2})$, $D(\tilde{A}_m^{1/2}) = H_A$.

4. If A is a linear closed operator and K is a linear compact operator, then $\sigma_c(A) = \sigma_c(A + K)$.

If $A \geq m > 0$, K is symmetric, $D(K) \subset D(A)$, and $\tilde{A}_m^{-1}K$ is compact in H_A, then K is called *relatively compact* in H_A with respect to A. If $K \geq 0$, then K is relatively compact in H_A with respect to A if and only if the embedding $i : H_A \to H_K$ is compact. If $A \geq m > 0$ and K is symmetric, relatively compact in H_A with respect to A, then $\sigma_c(A) = \sigma_c(A + K)$.

If A is self-adjoint and $D(A) \subset D(T)$, where T is a linear operator, then T is called relatively compact with respect to A if and only if $T(A + i)^{-1}$ is compact. If A is a closed linear densely defined operator in H and $\lambda \notin \sigma(A)$, then $(A - \lambda)^{-1}$ is compact if and only if $(I + A^*A)^{-1}$ is compact.

5. If λ is an isolated eigenvalue of finite algebraic multiplicity r of a linear operator A and Γ is a simple closed curve around λ, then the operator

$$P = -(2\pi i)^{-1} \int_\Gamma (A - zI)^{-1} dz \tag{B.0.5}$$

is a projection, $P^2 = P$. If there are no other points of the spectrum of A inside Γ, then $\dim PH = r$.

If L is a self-adjoint operator with a discrete spectrum, $0 < \lambda_1 \leq \lambda_2 < \ldots, \lambda_n \to +\infty$, then $\|P_n f - f\| \to 0, n \to \infty$, for any $f \in H, P_n$ is the orthoprojection on the linear span of the eigenvectors of L corresponding to the eigenvalues $\lambda_1 \ldots \lambda_n$.

6. Lemma B.0.5. *If $A > mI$ is a self-adjoint, bounded from below, operator in a Hilbert space H, and $H_A \subset H$ is the Hilbert space which is the completion of $D(A)$ in the norm generated by the inner product $((A - mI + I)f, f)$, then the spectrum of A is discrete if and only if the embedding $i\colon H_A \to H$ is compact. The spectrum of A is called discrete if it consists of isolated eigenvalues of finite multiplicity with the only possible accumulation point at infinity.*

Proof. Without loss of generality, assume that $A \geq I$, so that A^{-1} is bounded and defined on all of H. Then A has a discrete spectrum if and only if A^{-1} is compact. Since A^{-1} is self-adjoint, it is compact if and only if $A^{-1/2}$ is. If $A > 0$, then $(Au, u) = \|A^{1/2}u\|^2$. Therefore, compactness of the embedding $i : H_A \to H$ is equivalent to compactness of $A^{-1/2}$, i.e., to the discreteness of the spectrum of A. □

Appendix C

Eigenfunction Expansions and Asymptotic of the s-Values for Some Nonself-adjoint Operators

1. In this section, we introduce the notions of the *Riesz basis* and *Riesz basis with brackets*, and give conditions for a root system of an operator to form a Riesz basis with brackets.

A system $\{f_j\}$ of the elements of a Hilbert space H is called a Riesz basis of H, and we write $\{f_j\} \in R$ or $\{f_j\} \in R(H)$ if and only if $f_j = Th_j$, where T is a linear isomorphism of H onto H, and $\{h_j\}$ is an orthonormal basis of H. Let P_F denote an orthoprojection onto a subspace F.

A system $\{f_j\}$ is called a Riesz basis of H with brackets, $\{f_j\} \in R_b$, if and only if there exists a sequence of integers $n_1 \leq n_2 \leq \cdots \to \infty$, and a linear isomorphism T of H onto H such that $TF_j = H_j, j = 1, 2, \ldots$. Here, F_j is a linear span of the elements $f_{n_j}, f_{n_j+1}, \ldots, f_{n_{j+1}-1}$, $H_j = \text{span}\{h_{n_j}, \ldots, h_{n_{j+1}-1}\}$ and $\{h_j\}$ is an orthonormal basis of H. The sequence $n_1, n_2, \ldots$ defines the bracketing rule. If $\{f_j\} \in R_b$ and $\sum_{j=1}^{\infty} \|P_{F_j} - P_{H_j}\| < \infty$, then we write $\{f_j\} \in B_b$, and call $\{f_j\}$ a *Bari basis* with brackets.

Theorem C.0.1. *A system $\{f_j\} \in R(H)$ if and only if the system $\{f_j\}$ is complete in H and one of the following conditions holds:*

(1) *There exist $a_1 > 0$ and $a_2 \geq a_1$, such that for any n and any c_j one has*

$$a_1 \sum_{j=1}^{n} |c_j|^2 \leq \| \sum_{j=1}^{n} c_j f_j \|^2 \leq a_2 \sum_{j=1}^{n} |c_j|^2. \tag{C.0.1}$$

(2) *The Gram matrix $(f_i, f_j), 1 \leq i, j < \infty$, generates an isomorphism of l^2 onto l^2.*

(3) *For any $f \in H$ one has*

$$\sum_{j=1}^{\infty} |(f, f_j)|^2 < \infty \quad \text{and} \quad \sum_{j=1}^{\infty} |(f, \tilde{f}_j)|^2 < \infty, \tag{C.0.2}$$

where $\{\tilde{f}_j\}$ is the biorthogonal system, i.e., $(f_j, \tilde{f}_j) = \delta_{ij}$.

(4) *The system $\{f_j\}$ is an unconditional basis of H and*

$$0 < m \leq \|f_j\| < M < \infty. \tag{C.0.3}$$

A system $\{f_j\}$ is a basis (*Schauder's basis*) of M if and only if any $f \in H$ can be represented by a convergent series $f = \sum_{j=1}^{n} c_j f_j$, and this representation is unique. A basis is *unconditional* if and only if it remains a basis after an arbitrary transposition of its terms, i.e., $\{f_{n_j}\}$ is a basis for any transposition $(n_1, n_2, \dots)$ of the sequence $(1, 2, 3, \dots)$.

We write $L \in R_b(H)(L \in B_b(H))$ if the root vectors of the operator L form a Riesz (Bari) basis with brackets.

2. Let $L \geq m > 0$ be a self-adjoint operator on H with a discrete spectrum $0 < \lambda_1 \leq \lambda_2 \leq \cdots$,

$$\lambda_n = cn^p(1 + O(n^{-p_1})), \quad n \to \infty,$$

$$p_1 > 0, \, p > 0, \, c > 0, \tag{C.0.4}$$

$$M = L + T, \quad D(T) \supset D(L), \tag{C.0.5}$$

$$|Tf| \leq c|L^a f|, \quad a < 1, \quad D(T) \supset D(L^a). \tag{C.0.6}$$

By c here and below, one denotes various positive constants, $L^a = \int_m^{\infty} \lambda^a dE_\lambda$, where E_λ is the resolution of the identity for L.

Theorem C.0.2. *Let* (C.0.4)–(C.0.6) *hold. If* $p(1-a) = 1$, *then* $M \in R_b(H)$. *If* $p(1-a) > 1$, *then* $M \in B_b(H)$. *The spectrum of* M *is discrete and* $\sigma(M) \in \bigcup_{j=1}^{\infty} C_j, C_j = \{\lambda : |\lambda - \lambda_j| \le c_0 \lambda_j^a\}, c_0 > 0$.

It is often convenient to reduce the problem of basisness for the operator M to the same problem for a compact operator. We say that an operator M has a basisness property if $M \in R_b(H)$. One writes $M = (I + TL^{-1})L$ and

$$M^{-1} = L^{-1}(I + TL^{-1})^{-1} = A(I + Q), \quad A = L^{-1}. \tag{C.0.7}$$

Here, we assume that M is invertible, and use the formula

$$(I + S)^{-1} = I + Q, \quad Q = -(I + S)^{-1}S, \quad S = TL^{-1} \tag{C.0.8}$$

which holds if $I + S$ is invertible.

We assume that

$$\|TL^{-a}\| \le c, \quad a < 1. \tag{C.0.9}$$

In this case, the operator $S = TL^{-a}L^{-1+a}$ is compact since TL^{-a} is bounded and L^{-1+a} is compact. The operator $L^{-b}, b > 0$, is compact because of (C.0.4). Therefore, the basisness property for M holds if and only if it holds for the operator $B = A(I + Q)$. Here,

$$A = A^* > 0 \text{ is compact}, \ \lambda_n(A) \sim c^{-1} n^{-p}(1 + O(n^{-p_1})),$$
$$p > 0, \quad p_1 > 0 \tag{C.0.10}$$

$$Q \text{ is compact}, \ I + Q \text{ is invertible}, \ \|Qf\| \le c\|A^{1-a}f\|,$$
$$f \in D(L^{1-a}) \tag{C.0.11}$$

where $a < 1$ is the same as in (C.0.6).

Theorem C.0.2′ can be reformulated as follows.

Theorem C.0.2′. *Let $B = A(I + Q)$ and conditions* (C.0.10) *and* (C.0.11) *hold. If $p(1 - a) = 1$, then $B \in R_b(H)$, and if $p(1 - a) > 1$, then $B \in B_b(H)$.*

Completeness of the root system of an operator $B = A(I + Q)$ can be checked as follows. Let A be a compact linear operator in $H, s_n(A) = \lambda_n\{(A^*A)^{1/2}\}$ be its s-values. We say that $A \in \sigma_p$ if $\sum_{n=1}^{\infty} |s_n(A)|^p < \infty(*)$.

Let $p(A) = \inf p$, where the infimum is taken over all p for which $(*)$ holds. If $p(A) = 1$, then A is called trace-class operator, if $p(A) = 2$ then A is called Hilbert–Schmidt class operator. We write $A \in \sigma_\infty$ if A is compact.

Theorem C.0.3. *Let $B = A(I + Q), A = A^*, p(A) < \infty, Q \in \sigma_\infty$ and $N(B) = \{0\}$. Then the root systems of B and B^* are complete in H.*

Theorem C.0.4. *Let $M = L+T, L = L^*, L^{-1} \in \sigma_\infty, p(L^{-1}TL^{-1}) < \infty$. Then the root system of M is complete in H.*

It is interesting to know when the root vector expansion for the operator L in Theorem C.0.2′ converges or diverges simultaneously with that for the operator M, in other words, when the root vector expansions for L and M are equiconvergent. The equiconvergence is uniform on bounded sets if for any $R > 0$ the estimate (C.0.45) (see below) holds.

Theorem C.0.5. *The root vector expansions for L and M are equiconvergent uniformly on bounded sets if the assumptions* (C.0.4)–(C.0.6) *hold, $p(1 - a) > 2$, and the proper bracketing is used.*

Proofs of Theorems C.0.1, C.0.3, and C.0.4 can be found in [71]. The results in Theorems C.0.2 and C.0.2′ were obtained in [95, 140] (see also [1]). We prove Theorem C.0.2 and recover part of the result of Theorem C.0.2 in Section 3 below. The proof of Theorem C.0.2 follows closely the arguments in [226, p. 306]. Lemma C.0.13 (see below) was used in Kacnelson [95].

3. In this section, we prove some theorems about asymptotic of the s-values of perturbed compact operators.

We will need the following lemma. This lemma and the result below belong to the author, see [365], Chapter 8.

Lemma C.0.6. *A linear operator Q on H is compact if and only if $\sup_{h \perp L_n} \|Qh\|/\|h\| \to 0$, as $n \to \infty$, where L_n is an n-dimensional subspace of $H, L_n \subset L_{n+1}$ and the sequence L_n is limit dense, i.e., $\rho(f, L_n) \to 0, n \to \infty$ for any $f \in H$, where $\rho(f, L_n)$ is the distance from f to L_n.*

Theorem C.0.7 ([227, 241]). *Let $A \in \sigma_\infty, Q \in \sigma_\infty, N(I+Q) = \{0\}$, and $B = A(I+Q)$. Then*

$$\lim_{n \to \infty} s_n(B) s_n^{-1}(A) = 1. \tag{C.0.12}$$

Proof. One has

$$\begin{aligned} s_{n+1}(b) &= \min_{L_n} \max_{f \perp L_n} \frac{\|Bf\|}{\|f\|} \\ &\leq \max_{(I+Q)f \perp L_n(A)} \left\{ \frac{\|A(I+Q)f\|}{\|(I+Q)f\|} \frac{\|(I+Q)f\|}{\|f\|} \right\} \\ &\leq s_{n+1}(A) \left(1 + \max_{(I+Q)f \perp L_n(A)} \frac{\|Qf\|}{\|f\|} \right) \\ &\leq s_{n+1}(A)(1+\epsilon_n), \quad \epsilon_n \to 0. \end{aligned} \tag{C.0.13}$$

Here, L_n is an n-dimensional subspace, $L_n(A)$ is the linear span of the first n eigenvectors (counted according to their multiplicities) of the operator A^*A, so that

$$s_{n+1}(A) = \max_{h \perp L_n(A)} \frac{\|Ah\|}{\|h\|}, \tag{C.0.14}$$

and Lemma C.0.6 was used at the last step in (C.0.13). The invertibility of $I + Q$ guarantees that the sequence $(I + Q^*)L_n(A)$ is limit dense in H.

Since $A = B(I+S)$, where $I+S = (I+Q)^{-1}$, and S is compact if Q is, one concludes as above that

$$s_{n+1}(A) \le s_{n+1}(B)(1+\epsilon'_n), \quad \epsilon'_n \to 0, \quad n \to \infty. \tag{C.0.15}$$

Inequalities (C.0.13) and (C.0.15) yield (C.0.12). □

One can estimate the remainder under some additional assumptions, see [241].

Theorem C.0.8. *Let* $B = A(I+Q), A \in \sigma_\infty, Q \in \sigma_\infty, N(I+Q) = \{0\}$,

$$s_n(A) = cn^{-p}(1+O(n^{-p_1})), \quad p>0,\ p_1>0, \tag{C.0.16}$$

and

$$\|Qf\| \le c\|Af\|^a\|f\|^{1-a}, \quad a>0. \tag{C.0.17}$$

Then

$$s_n(B) = s_n(A)(1+O(n^{-q})), \quad q = \min\left(p_1, \frac{pa}{1+pa}\right). \tag{C.0.18}$$

The estimate of the remainder in (C.0.18) *is sharp in the sense that it is attained for some* Q.

Proof. One has

$$\begin{aligned} s_{n+m+1}(B) \le & \max_{\substack{(I+Q)f\perp L_n(A)\\ f\perp L_m(A)}} \frac{\|A(I+Q)f\|}{\|(I+Q)f\|} \\ & \cdot \left(1 + \max_{\substack{(I+Q)f\perp L_n(A)\\ f\perp L_m(A)}} \frac{\|Qf\|}{\|f\|}\right) \\ & \le s_{n+1}(A)\left(1 + c\max_{f\perp L_m(A)} \frac{\|Af\|^a}{\|f\|^a}\right) \\ & \le s_{n+1}(A)(1+cs^a_{m+1}(A)), \quad n,m\to\infty. \end{aligned} \tag{C.0.19}$$

The symmetry argument used in the proof of Theorem C.0.5 should be modified. We have $A = B(I+S), S = -(I+Q)^{-1}Q$

$$\begin{aligned} s_{n+m+1}(A) &\le \max_{\substack{(I+S)f \perp L_n(B) \\ f \perp L_m(A)}} \frac{\|B(I+S)f\|}{\|(I+S)f\|} \cdot \left(1 + \max_{f \perp L_m(A)} \frac{\|Sf\|}{\|f\|}\right) \\ &\le s_{n+1}(B)\left(1 + c_1 \max_{f \perp L_m(A)} \frac{\|Af\|^a}{\|f\|^a}\right) \\ &\le s_{n+1}(B)(1 + c_1 s^a_{m+1}(A)), \quad n, m \to \infty. \end{aligned} \tag{C.0.20}$$

Therefore, (C.0.19) and (C.0.20) imply

$$\begin{aligned} s_{n+2m+1}(A) &\le s_{n+m+1}(B)(1 + cs^a_{m+1}(A)) \\ &\le s_{n+1}(A)(1 + c_1 s^a_{m+1}(A)). \end{aligned} \tag{C.0.21}$$

Thus,

$$\begin{aligned} \frac{s_{n+2m+1}(A)}{s_{n+m+1}(A)}(1 + O(s^a_{m+1}(A))) &\le \frac{s_{n+m+1}(B)}{s_{n+m+1}(A)} \\ &\le \frac{s_{n+1}(A)}{s_{n+m+1}(A)}(1 + O(s^a_m(A))). \end{aligned} \tag{C.0.22}$$

Let $m = n^{1-x}, x > 0$. It follows from (C.0.16) that

$$\begin{aligned} \frac{s_{n+m}(A)}{s_n(A)} &= \left(\frac{n+m}{n}\right)^{-p} (1 + O(n+m)^{-p_1} + O(n^{-p_1})) \\ &= 1 + O(n^{-p_1}) + O\left(\frac{m}{n}\right) = 1 + O(n^{-p_1}) + O(n^{-x}). \end{aligned} \tag{C.0.23}$$

It follows from (C.0.22) and (C.0.23) that

$$\frac{s_{n+m+1}(B)}{s_{n+m+1}(A)} = 1 + O(n^{-p_1}) + O(n^{-x}) + O(n^{-(1-x)pa}). \tag{C.0.24}$$

Since $\min(x, (1-x)pa) = pa(1+pa)^{-1}$, one has

$$q = \min(p_1, x, (1-x)pa) = \min(p_1, pa(1+pa)^{-1}). \tag{C.0.25}$$

This and (C.0.24) imply (C.0.18). The estimate of the remainder is sharp: it is attained if $p_1 > pa(1+pa)^{-1}$, $A > 0$ and $Q = A^a$. □

4. Here, we prove Theorem C.0.5.

Lemma C.0.9. *If* (C.0.4) *holds, then*

$$N(\lambda) = \sum_{\lambda_n < \lambda} 1 = (\lambda c^{-1})^{1/p}(1 + O(\lambda^{-p_1/p})). \tag{C.0.26}$$

Proof. The function λ_N and $N(\lambda)$ are reciprocal, thus (C.0.4) implies that

$$(\lambda c^{-1})^{1/p} = n(1 + O(n^{-p_1})), \quad n = (\lambda c^{-1})^{1/p}(1 + O(\lambda^{-p_1/p})). \quad \square$$

Lemma C.0.10. *If* (C.0.4) *holds, then for sufficiently large* n *and* $m, 0 < q_1 \leq nm^{-1} \leq q_2 < 1$, *there exist eigenvalues* $\lambda^{(1)}$ *and* $\lambda^{(2)}$ *such that the interval* $(\lambda^{(1)}, \lambda^{(2)})$ *is free from the eigenvalues,* $\lambda_n \leq \lambda^{(1)} < \lambda^{(2)} \leq \lambda_m$, *and* $\lambda^{(2)} - \lambda^{(1)} \geq cm^{p-1}$.

Proof. There are $m - n$ eigenvalues (counting multiplicities) on the segment $(\lambda_n, \lambda_m]$. So, there exists at least two eigenvalues $\lambda_n \leq \lambda^{(1)} < \lambda^{(2)} \leq \lambda_m$ such that

$$\lambda^{(2)} - \lambda^{(1)} \geq \frac{\lambda_m - \lambda_n}{m-n} \geq c\frac{m^p - n^p}{m-n} - O\left(\frac{m^{p-p_1}}{m-n}\right) \geq c_1 m^{p-1}. \quad \square$$

Lemma C.0.11. *Let* $m = m(n), 1 - b_2(n) \leq m^{-1}(n)n \leq 1 - b_1(n), b_1(n)b_2^{-1}(n) \geq c_2, n^{p_1}b_1(n) \to \infty$, *and* $b_2(n) \to 0$, *as* $n \to \infty$. *Then the conclusion of Lemma* C.0.9 *holds.*

Proof. It is similar to that of Lemma C.0.9. One has

$$\begin{aligned}\frac{m^p - n^p - O(m^{p-p_1})}{m-n} &= m^{p-1}\frac{1 - (m^{-1}(n)n)^p - O(m^{-p_1})}{1 - m^{-1}(n)n} \\ &\geq m^{p-1}\frac{1 - (1 - b_1(n))^p - O(m^{-p_1})}{b_2(n)}\end{aligned}$$

$$\geq m^{p-1}\frac{b_1(n)\{1-O(m^{-p_1}b_1^{-1}(n))\}}{b_2(n)}$$

$$\geq c_2 m^{p-1}(1-o(1)) \geq c_1 m^{p-1}. \qquad \square$$

Remark C.0.1. If $b_1(n) = b_2(n) = n^{-p_2}, p_2 < p_1$ then the conditions of Lemma C.0.10 are satisfied.

Lemma C.0.12. *If* (C.0.4) *holds,* $R^0_\lambda = (L-\lambda I)^{-1}$ *and* $\lambda^{(m)} = \frac{1}{2}(\lambda^{(2)} - \lambda^{(1)})$, *where* $\lambda^{(1)}$ *and* $\lambda^{(2)}$ *are defined in Lemma* C.0.9, *then*

$$\|R^0_{\lambda^{(m)}} L^a\| \leq cm^{1-p(1-a)}. \tag{C.0.27}$$

Proof. One has

$$\|R^0_{\lambda^{(m)}} L^a\| = \sup_{\lambda_j} \frac{|\lambda_j|^a}{|\lambda^{(m)} - \lambda_j|} \leq c\frac{\lambda^a_m}{m^{p-1}} \leq cm^{1-p(1-a)},$$

where Lemma C.0.9 was used. $\square$

It follows from (C.0.27) that

$$\|R^0_{\lambda^{(m)}} L^a\| \to 0, \quad m \to \infty, \quad \text{if} \quad p(1-a) > 1. \tag{C.0.28}$$

Furthermore,

$$\|R^0_{\lambda^{(m)}}\| \leq \sup_j \frac{1}{|\lambda^{(m)} - \lambda_j|} \leq \frac{c}{m^{p-1}}. \tag{C.0.29}$$

For large m, one has

$$\begin{aligned} R_{\lambda^{(m)}} &= (M - \lambda^{(m)} I)^{-1} \\ &= R^0_{\lambda^{(m)}}(I + TR^0_{\lambda^{(m)}})^{-1}, \quad \|TR^0_{\lambda^{(m)}}\| \leq \|R^0_{\lambda^{(m)}} L^a\| \cdot \|TL^{-a}\|. \end{aligned} \tag{C.0.30}$$

It follows from (C.0.30), (C.0.27), and (C.0.6) that

$$\|R_{\lambda^{(m)}}\| \leq c\|R^0_{\lambda^{(m)}}\| \leq cm^{1-p}. \tag{C.0.31}$$

Furthermore, (C.0.27) and (C.0.31) imply that

$$\|R_{\lambda^{(m)}} - R^0_{\lambda^{(m)}}\| = \|R_{\lambda^{(m)}} TR^0_{\lambda^{(m)}}\|$$

$$\leq c\|R_{\lambda(m)}\| \cdot \|R^0_{\lambda(m)} L^a \leq cm^{2-p(2-a)} \tag{C.0.32}$$

$$\begin{aligned} \|R_{\lambda(m)}T\| &\leq \|(R_{\lambda(m)} - R^0_{\lambda(m)})T\| + \|R^0_{\lambda(m)}T\| \\ &\leq cm^{1-p(1-a)} \quad \text{if } 1 < p(1-a) \end{aligned} \tag{C.0.33}$$

and

$$\|(R_{\lambda(m)} - R^0_{\lambda(m)})L\| = \|R_{\lambda(m)} T R^0_{\lambda(m)} L\| \leq cm^{2-p(1-a)}. \tag{C.0.34}$$

Proof of Theorem C.0.5. Let $m(n) = n/(1-n^{-p_2}), p_2 < p_1$. Then the conditions of Lemma C.0.10 are satisfied. Let the sequence $m(n)$ define the bracketing rule, and $\lambda^{(n)} = (\lambda^{(1)} + \lambda^{(2)})/2$, where $\lambda^{(1)}$ and $\lambda^{(2)}$ are defined in Lemma C.0.9 with $\lambda_n = \lambda_{m(n)}$ and $\lambda_m = \lambda_{m(n+1)}$. Let $C_n = \{\lambda : |\lambda| = \lambda^{(n)}\}$. Integrating the identity

$$\frac{1}{2\pi i}\lambda^{-1} f = (-R_\lambda f + \lambda^{-1} R_\lambda L f + \lambda^{-1} R_\lambda f)\frac{1}{2\pi i} \tag{C.0.35}$$

over C_n yields

$$f = f_n + a_n(f) + b_n(f) \tag{C.0.36}$$

where

$$\begin{aligned} f_n = -\frac{1}{2\pi i}\int_{C_n} R_\lambda f d\lambda, \quad a_n(f) &= \frac{1}{2\pi i}\int_{C_n} \lambda^{-1} R_\lambda f d\lambda, \\ b_n(f) &= \frac{1}{2\pi i}\int_{C_n} \lambda^{-1} R_\lambda T f d\lambda. \end{aligned} \tag{C.0.37}$$

In a similar way,

$$\begin{aligned} f = f_n^0 + a_n^0(f), \quad f_n^0 &= -\frac{1}{2\pi i}\int_{C_n} R^0_\lambda f d\lambda, \\ a_n(f) &= \frac{1}{2\pi i}\int_{C_n} \lambda^{-1} R^0_\lambda L f d\lambda. \end{aligned} \tag{C.0.38}$$

According to formula (A2.5), $f_n^0 = P_n^0 f$, where P_n^0 is the orthoprojection in H onto the linear span of eigenvectors of L corresponding to the eigenvalues lying in the circle $|\lambda| < \lambda^{(n)}$. The spectral theorem for self-adjoint operators yields $\|P_n^0 f - f\| \to 0$,

$$\begin{aligned} &\|a_n^0(f)\| \to 0, \quad n \to \infty, \quad \text{for any } f \in H, \\ &\|P_n^0\| \le 1, \quad \|a_n^0\| \le \|I - P_n^0\| \le 2. \end{aligned} \tag{C.0.39}$$

We have $f_n = P_n f$, where P_n is a projection on the linear span of the root vectors of M corresponding to the eigenvalues of M lying in the circle $|\lambda| < \lambda^{(n)}$. Thus, (C.0.36) and (C.0.38) yield

$$\|P_n f - P_n^0 f\| \le \|(a_n - a_n^0) f\| + \|b_n(f)\|. \tag{C.0.40}$$

To show that

$$\|P_n f - P_n^0 f\| \to 0, \quad n \to \infty, \quad f \in H, \tag{C.0.41}$$

it is sufficient to show that each term in the right-hand side of (C.0.40) goes to zero as $n \to \infty$.

It follows from (C.0.4), (C.0.33), and (C.0.37) that

$$\begin{aligned} \|b_n(f)\| &\le \|R_{\lambda^{(n)}} T f\| \\ &\le c n^{1-p(1-a)} \|f\| \quad \text{if} \quad p(1-a) > 1. \end{aligned} \tag{C.0.42}$$

Furthermore,

$$\begin{aligned} \|a_n(f) - a_n^0(f)\| &= \left\| \frac{1}{2\pi i} \int_{C_n} \lambda^{-1} (R_\lambda - R_\lambda^0) L f d\lambda \right\| \\ &\le \sup_{|\lambda| = \lambda^{(n)}} \|(R_\lambda - R_\lambda^0) L f \lambda\| \le c n^{2-p(1-a)} \|f\|, \end{aligned} \tag{C.0.43}$$

where we use (C.0.34). It follows from (C.0.42), (C.0.43), and (C.0.40) that

$$\|P_n f - P_n^0 f\| \le \epsilon(n) \|f\|, \quad \epsilon(n) = c(n^{1-p(1-a)} + n^{2-p(1-a)}). \tag{C.0.44}$$

If $p(1-a) > 2$, then $\epsilon(n) \to 0$. Therefore, the partial sums $P_n f$ and $P_n^0 f$ of the root vector expansion for M and eigenvector expansion for L are not only equiconvergent at any $f \in H$ but this equiconvergence is uniform on bounded sets $\|f\| <$ const in the sense that

$$\sup_{\|f\| \le R} \|P_n f - P_n^0 f\| \to 0 \quad \text{as} \quad n \to \infty, \quad \text{if} \quad p(1-a) > 2. \tag{C.0.45}$$

□

Remark C.0.2. Assume that (C.0.4) holds with $p_1 > 1$. Then

$$\lambda_{n+1} - \lambda_n \ge c_1 n^{p-1} \tag{C.0.46}$$

and one can take $m(n) = n$. In this case, the Riesz basis with brackets becomes a Riesz basis ([226], p. 306).

Lemma C.0.13. *The last statement of Theorem* C.0.2 *holds.*

Proof. It follows from (C.0.30) that $R_{\lambda(m)}$ is compact (since $R^0_{\lambda(m)}$ is compact and $(I + R^0_{\lambda(m)} T)^{-1}$ is bounded). Therefore, the spectrum of M is discrete. Furthermore, it follows from (C.0.30) that if $\lambda \notin \sigma(L)$ and $\|TR^0_\lambda\| \le q < 1$ then $\lambda \notin \sigma(M)$. One has $\|TR^0_\lambda\| \le \|TL^{-a}\| \, \|L^a R^0_\lambda\| \le c \max \|\lambda_j^a|/|\lambda_j - \lambda| \le q < 1$ provided that $\max_{\lambda_j} |\lambda_j^a|/|\lambda_j - \lambda| \le qc^{-1}$. Therefore, if $|\lambda_j - \lambda| \ge c_0|\lambda_j|^a, c_0 = cq^{-1}$, then $\lambda \notin \sigma(M)$. This is equivalent to the last statement of Theorem C.0.2. □

Appendix D

Stable Numerical Differentiation

If one looks for an optimal (by criterion of minimal variance) linear estimate of $s'(t)$ from the observation of $u(t) = s(t) + n(t)$, where $n(t)$ is noise and $s(t)$ is useful signal, then one can derive an integral equation for the weight function of optimal estimate. This integral equation is often difficult to solve and, even if one can solve it, it is difficult to construct the corresponding filter. In Appendix D, an optimal estimate of s' on a subset of all linear estimates is sought and it is shown that this quasioptimal estimate is easy to calculate, the corresponding filter is easy to construct, and the error of this estimate differs little from the error of optimal estimates. It is also shown that among all estimates (linear and nonlinear) of s' for $|n| \leq \delta$ and $|s''| \leq M$, the best estimate is given by $\Delta_h u = (2h)^{-1}[u(t+h) - u(t-h)]$ with $h = (2\delta/M)^{1/2}$. About stable numerical differentiation see [197], where for the first time, a stable differentiation formula was given, see also [356], pp. 71–84, [365], pp. 169–181, [369], pp. 197–234, [404], pp. 213–231, [235, 341, 351].

D.1. Deterministic Case

Let $u = s + n, |n| \leq \delta, |s''| \leq M$. Let us assume that $u(t)$ and $s(t)$ are continuous functions defined on $\mathbb{R}^1$ and $\|\cdot\|$ denotes the norm in $C(\mathbb{R}^1)$. Let

$$\Delta_h u = (2h)^{-1}[u(t+h) - u(t-h)], \tag{D.1.1}$$

$$h(\delta) = (2\delta/M)^{1/2}, \quad \epsilon(\delta) = (2M\delta)^{1/2}. \tag{D.1.2}$$

Let A denote the set of all operators $T : C(\mathbb{R}^1) \to C(\mathbb{R}^1)$.

Theorem D.1.1. *Under the above assumptions, the following estimates hold*

$$\|\Delta_{h(\delta)}u - s'\| \leq \epsilon(\delta), \tag{D.1.3}$$

$$\inf_{T\in A} \sup_{\substack{|s''|\leq M \\ |n|\leq\delta}} \|Tu - s'\| = \epsilon(\delta), \tag{D.1.4}$$

and infimum in (D.1.4) *is attained on* $T = \Delta_{h(\delta)}$. *Here and below* $\epsilon(\delta)$ *and* $h(\delta)$ *are defined in* (D.1.2).

Remark D.1.1. If $u(t)$ and $s(t)$ are defined on a bounded domain, e.g., on $I = [0, 1]$, and if

$$\tilde{\Delta}_h u = \begin{cases} h^{-1}[u(x+h) - u(x)], & 0 \leq x \leq \frac{h}{2}, \\ h^{-1}\left[u\left(x+\frac{h}{2}\right) - u\left(x-\frac{h}{2}\right)\right], & \frac{h}{2} \leq x \leq 1-\frac{h}{2}, \\ h^{-1}[u(x) - u(x-h)], & 1-\frac{h}{2} \leq x \leq 1, \end{cases} \tag{D.1.5}$$

then

$$|\tilde{\Delta}_{h_1(\delta)}u - s'| \leq 2(M\delta)^{1/2} = \sqrt{2}\epsilon(\delta), \tag{D.1.6}$$

where

$$h_1(\delta) = 2(\delta M^{-1})^{1/2} = h(\delta)/\sqrt{2}. \tag{D.1.7}$$

Remark D.1.2. The function $n(t)$ can be considered as a random function uniformly distributed on the segment $[-\delta, \delta]$.

Remark D.1.3. Theorem D.1.1 and Remark D.1.1 combine the results of [197] and [365].

Proof. We have

$$|\Delta_h u - s'| \leq |\Delta_h(u - s)| + |\Delta_h s - s'| \leq \frac{\delta}{h} + \frac{Mh}{2}. \tag{D.1.8}$$

The second term in the right-hand side of (D.1.8) is obtained by expansion of $\Delta_h s$ in Taylor series in powers of h, while the first term follows from the obvious estimates $\|\Delta_h\| \le h^{-1}$, where $\|\cdot\|$ is the norm of an operator in $C(\mathbb{R}^1)$. Minimizing in h, the right-hand side in (D.1.8) we obtain (D.1.3) with $h_{\min} = h(\delta)$. Let us prove (D.1.4). We have

$$\sup_{\substack{|s''|\le M\\ |n|\le\delta}} \|Tu - s'\| = \sup_{\substack{s_j, j=1,2\\ |s_j''|\le M\\ |n|\le\delta}} \|Tu - s_j'\| \ge \sup_{\substack{s_j, j=1,2\\ |s_j''|\le M\\ |n|\le\delta}} |Tu(0) - s_j'(0)|$$

$$\ge \inf_a \sup_{\substack{s_j, j=1,2\\ |s_j''|\le M\\ |n|\le\delta}} |a - s_j'(0)| \ge \frac{1}{2}|s_2'(0) - s_1'(0)| \ge \epsilon(\delta). \tag{D.1.9}$$

The last inequality is valid if we make the following choice of s_j:

$$\begin{aligned} s_1(t) &= -\frac{M}{2}t(t - 2h(\delta)), \\ s_2(t) &= -s_1(t) \quad \text{for } 0 \le t \le 2h(\delta), \end{aligned} \tag{D.1.10}$$

and define $s_j(t), j = 1, 2$ outside the segment $[0, 2h(\delta)]$ so that the condition, $|s_j''| \le M, |s_j| \le \delta$ will be satisfied. It is easy to check that for $0 \le t \le 2h(\delta), |s_j''| \le M, |s_j| \le \delta, j = 1, 2$. Thus, both functions s_1 and s_2 are of the form $u - n$ with $|n| \le \delta$ and $u = 0$. Taking the infimum over $T \in A$, we conclude that (D.1.4) holds with $\ge$ and from (D.1.3) it follows that for $T = \Delta_{h(\delta)}$ we have equality. This completes the proof of Theorem D.1.1. The proof of Remark D.1.1 is similar (see [197]). □

D.2. Stochastic Case

Assume that $u(t) = s(t) + n(t), -\infty < t < \infty$, s is a useful signal and n is noise. For simplicity, assume that s and n are uncorrelated and n has a zero mean value. Let D denote variance. Assume that the random functions s and n are stationary processes, $R_s(t)$ is the covariance function of s and $n = \delta v$, where $D[v] = 1, \delta > 0, \delta^2 = D[n], R_v(t)$ is the covariance function of v. The problem is to find a linear estimate of s' from the observed signal u. The optimal estimate is the solution of the variational problem

$$D[Lu - s'] = \min . \tag{D.2.1}$$

The solution of this problem was obtained in [479] for the case of infinite time of observation. Even for simple covariance functions, this solution is often difficult to calculate (one has to solve some integral equation) and difficult to construct a filter which implements the estimate. That is why it is natural to look for a quasioptimal estimate which satisfies the following requirements: (1) it is easy to calculate and to implement; (2) the error of the quasioptimal estimate is approximately the same as the error of optimal estimate. The idea to look for a quasioptimal estimate in the problem of differentiation of random functions was suggested, probably for the first time, in [197].

Let

$$D[s^{(m)}] \leq M_m, \tag{D.2.2}$$

$$\Delta_h^{(Q)} s = h^{-1} \sum_{k=-Q}^{Q} A_k^{(Q)} s\left(t + \frac{kh}{Q}\right). \tag{D.2.3}$$

If $m = 2q$ or $2q+1$, let us take $Q = q$ in (D.2.3). Expanding (D.2.3) in powers of h and requiring that the order of $\Delta_h^{(Q)} s - s'$ be minimal, we obtain the following system for the coefficients $A_k^{(Q)}$

$$\begin{gathered} \sum_{k=-Q}^{Q} \left(\frac{k}{Q}\right)^j A_k^{(Q)} = \delta_{1j}, \quad 0 \leq j \leq 2Q, \\ \delta_{1j} = \begin{cases} 0, & j \neq 1, \\ 1, & j = 1. \end{cases} \end{gathered} \tag{D.2.4}$$

The system (D.2.4) is uniquely solvable since its matrix is nondegenerate (Vandermonde determinant). We have (see [235])

$Q = 1$: $A_0^{(1)} = 0,\ A_{\pm 1}^{(1)} = \pm \frac{1}{2}$;

$Q = 2$: $A_0^{(2)} = 0,\ A_{\pm 1}^{(2)} = \pm \frac{4}{3},\ A_{\pm 2}^{(2)} = \mp \frac{1}{6}$;

$$Q = 3\colon A_0^{(3)} = 0,\ A_{\pm 1}^{(3)} = \pm\frac{9}{4},\ A_{\pm 2}^{(3)} = \mp\frac{9}{20},\ A_{\pm 3}^{(3)} = \pm\frac{1}{20};$$

$$Q = 4\colon A_0^{(4)} = 0,\ A_{\pm 1}^{(4)} = \pm\frac{16}{5},\ A_{\pm 2}^{(4)} = \mp\frac{4}{5},\ A_{\pm 3}^{(4)} = \pm\frac{16}{105},\ A_{\pm 4}^{(4)} = \mp\frac{1}{70}. \tag{D.2.4$'$}$$

Lemma D.2.1. *Let $m = 2q + 1$. We have*

$$D[\Delta_h^{(Q)} s - s'] \leq C_m M_m h^{2m-2} = \gamma_m h^{2m-2} \tag{D.2.5}$$

if the coefficients $A_k^{(Q)}$ in (D.2.3) *satisfy* (D.2.4) *(with $0 \leq j \leq 2q$) and*

$$C_m = \frac{(2Q+1)}{(m!)^2 Q^{2m}} \sum_{k=-Q}^{Q} |A_k^{(Q)}|^2 k^{2m}, \quad m = 2q+1, \quad Q = q. \tag{D.2.6}$$

In order to prove Lemma D.2.1, we need the following lemma.

Lemma D.2.2. *Let g_j be some random variables and a_j be numbers, $D[g_j] \leq M$. Then*

$$D\left[\sum_{k=1}^{n} c_k g_k\right] \leq nM \sum_{k=1}^{n} |c_k|^2. \tag{D.2.7}$$

Proof. Since $D[cg] = |c|^2 D[g]$, we can prove (D.2.7) for $c_k = 1$. If $b_k \geq 0$, then

$$\left(\sum_{k=1}^{n} b_k\right)^2 \leq n \sum_{k=1}^{n} b_k^2. \tag{D.2.8}$$

This elementary inequality follows immediately from the Cauchy inequality. Now, if $f(x_1, \ldots, x_n)$ is the probability density of the joint distribution of random variables $g_k, 1 \leq k \leq n$, and if we assume (without any loss of generality) that $\bar{g}_j = 0$, and set

$dx = dx_1, \ldots, dx_n, \int = \int_{\mathbb{R}^n}, x = (x_1, \ldots, x_n)$, then

$$D\left[\sum_{k=1}^{n} g_k\right] = \int \left|\sum_{k=1}^{n} x_k\right|^2 f(x)dx$$
$$\le n \int \sum_{k=1}^{n} |x_k|^2 f(x)dx = n \sum_{k=1}^{n} D[g_k].$$

Lemma D.2.2 is proved. □

Proof of Lemma D.2.1. We have

$$\Delta_h^{(Q)} s - s' = h^{-1} \sum_{k=-Q}^{Q} A_k^{(Q)} \frac{h^m k^m}{m! Q^m} s^{(m)}(t_k),$$

where t_k are some points in the remainders of the Taylor formula. Now we apply Lemma D.2.2 with $g_k = s^{(m)}(t_k), D[g_k] \le M_m$. □

Lemma D.2.3. *We have*

$$D[\Delta_h^{(Q)} u = s']$$
$$\le \gamma_m h^{2m-2} + h^{-2}\delta^2 \sum_{k,j=-Q}^{Q} A_k^{(Q)} A_j^{(Q)} R\left(\frac{k-j}{Q} h\right) = \phi(h), \tag{D.2.9}$$

where $m = 2Q + 1$, *and*

$$R(t - \tau) = \overline{v^*(t)v(\tau)} \tag{D.2.10}$$

is the covariance function of $v(t), n(t) = \delta v(t), D[v] = 1.$

Theorem D.2.4. *The quasioptimal estimate of* s' *has the form*

$$\hat{s}' = \Delta_h^{(Q)} u, \tag{D.2.11}$$

where the hat is the symbol of estimate (so that the right-hand side of (D.2.11) is just a definition of $\hat{s}'$*), the operator* $\Delta_h^{(Q)}$ *is defined in (D.2.3), the coefficients* $A_k^{(Q)}$ *are uniquely defined from the system*

(D.2.4) *and $h = h_m > 0$ is defined as a minimizer of the right-hand side of* (D.2.9):

$$\phi(h) = \phi_{\min} \quad \textit{for } h = h_m. \tag{D.2.12}$$

In order to calculate h_m explicitly, we need some assumptions about $R(t)$ or its spectral density (Fourier transform).

Example D.2.1. Let $m = 2$. From (D.2.4′)–(D.2.6), we obtain

$$\Delta_h^{(1)} u = \frac{u(t+h) - u(t-h)}{2h}, \quad C_2 = \frac{3}{8}.$$

$D[\Delta_h^{(1)} s - s'] \leq \frac{3}{8} M_2 h^2$, where $D[s''] \leq M_2$. Let $R(t) = \exp(-a|t|), a > 0$. Then from (D.2.9) we obtain

$$\phi(h) = \frac{3}{8} M_2 h^2 + \frac{\delta^2 h^{-2}}{2}(1 - \exp(-ah)).$$

Assuming $h \ll 1$, we have

$$\phi(h) = bh^2 + dh^{-1}, \quad b = \frac{3}{8} M_2, \quad d = \frac{\delta^2 a}{2}.$$

Thus,

$$h_m = (d/2b)^{1/3}, \tag{D.2.13}$$

$$\begin{aligned} \phi_{\min} &= (2^{-2/3} + 2^{1/3}) d^{2/3} b^{1/3} \\ &= (2^{-2/3} + 2^{1/3}) \left(\frac{3}{8}\right)^{1/3} \left(\frac{1}{2}\right)^{2/3} M_2^{1/3} \delta^{4/3} a^{2/3}. \end{aligned} \tag{D.2.14}$$

Remark D.2.1. Calculations similar to those given in Example D.2.1 can be carried through in most practical cases. Note that the behavior of the spectral density $\tilde{R}(\lambda)$ at infinity $\lambda \to \infty$ defines the behavior of $R(t)$ near $t = 0$. Since only the behavior of R near $t = 0$ is essential for $h \ll 1$ for calculation of h_m and $\phi_{\min}$, we can make the following conclusion:

The above recipe for calculating $h_{\min}$ is stable towards perturbations of the spectral density $\tilde{R}(\lambda)$ preserving its asymptotics at infinity.

Remark D.2.2. If we assume that $D[s^{(m)}] \leq M_m, m > 2$ and find h_m and $\phi_{\min}$, then we can see that for large m the error $\phi_{\min}$ of the quasioptimal estimate (D.2.3) is of the same order in δ as $\delta \to 0$ as the error of optimal Wiener estimate [479]. It is clear the estimate (D.2.3) is easy to implement as a filter: only arithmetic operations and delay lines are used. It is easy also to find the quasioptimal estimate, i.e., to find h_m. We only need to know the asymptotics of the spectral density of noise and the estimation constant M_m.

D.3. Search of Extremum of a Random Function

In applications (e.g., in navigation theory), the following problem is of interest. A function $u = s + n$ is given, where $s(t)$ is a univalent smooth deterministic unknown function defined on the segment $I = [0, 1]$ (univalent means that $s(t)$ has exactly one maximum on $[0, 1]$), and $n(t)$ is noise, $|n| \leq \delta$. The problem is to find the point t^*, where $s(t)$ is maximal. We assume that $|s''| \leq M$ and M is known. To solve this problem, let us find $h = h_1(\delta)$ (see (D.1.7)), divide the segment by points $t_k = kh$, and calculate $\hat{u}_k = 2h^{-1}[u(t_k + h) - u(t_k - h)]$. If $\hat{u}_k \hat{u}_{k+1} < 0$ and $|\hat{u}_j| > 2(M\delta)^{1/2}, \forall j$, then we conclude with the probability 1 that $t_k \leq t^* \leq t_k + h$. Indeed from (D.1.6), it follows that $\hat{u}_k = 2(M\delta)^{1/2} < s'(t_k) < \hat{u}_k + 2(M\delta)^{1/2}$. If $|\hat{u}_j| > 2(M\delta)^{1/2}$, then $s'(t_k)$ has the same sign as $\hat{u}_k$. Therefore, from $\hat{u}_k \hat{u}_{k+1} < 0$, it follows that s' changes sign on the segment $[t_k, t_{k+1}]$. Since $s(t)$ is univalent, the point t^* lies in the segment.

Appendix E

Limit of the Spectra of the Interior Neumann Problems when a Solid Domain Shrinks to a Plane One

Let $(\nabla^2 + \lambda)u = 0$ in $\mathcal{D} \subset \mathbb{R}^d, \partial u/\partial N = 0$ on $\partial\mathcal{D}$. *How do the eigenvalues* λ_j *behave when* $\mathcal{D}$ *shrinks to a domain* $\Delta \subset \mathbb{R}^{d-1}$? The answer depends not only on Δ but on the way $\mathcal{D}$ shrinks to Δ. The limit of λ_j is found. Examples are given. The results in this Appendix are taken from [261].

E.1. Introduction

Consider the problem

$$\nabla^2 u + \lambda u = 0 \text{ in } \mathcal{D} \subset \mathbb{R}^2, \quad \frac{\partial u}{\partial N} = 0 \text{ on } \Gamma. \tag{E.1.1}$$

Here, $\mathcal{D} = \mathcal{D}(\epsilon)$ is a bounded domain with a piecewise smooth boundary Γ, N is the unit outer normal to Γ, and ∇^2 is the Laplacian. Let us assume that the equation of Γ is of the form

$$y = \begin{cases} \epsilon f(x), \ a \leq x \leq b, & y > 0, \\ \epsilon g(x), & y < 0, \end{cases} \tag{E.1.2}$$

where $\epsilon > 0$ is a parameter, f and g are piecewise smooth functions. If $f(a) > 0, g(a) < 0$ then the segment $x = a, g(a) \leq y \leq f(a)$ belongs to Γ, and the same holds for $x = b$. Let $\lambda_j(\epsilon)$ be the eigenvalues of (E.1.1)

$$0 = \lambda_0 \leq \lambda_1(\epsilon) \leq \lambda_2(\epsilon) \leq \dots.$$

Problem. *How do $\lambda_j(\epsilon)$ behave when $\epsilon \to 0$?*

The limit of $\mathcal{D}(\epsilon)$ is the segment $\Delta : a \leq x \leq b, \mathcal{D}(\epsilon) \subset \mathbb{R}^2$, $\Delta \subset \mathbb{R}^1$. The same problem can be formulated for $\mathcal{D}(\epsilon) \subset \mathbb{R}^d$, $\Delta \subset \mathbb{R}^{d-1}, d > 2$. One can also consider a general second order self-adjoint elliptic operator $\mathcal{L}$ instead of Δ. From the results of this appendix, it follows that

$$\lim_{\epsilon \to \infty} \lambda_j(\epsilon) = \lambda_j, \tag{E.1.3}$$

where λ_j are the eigenvalues of a differential operator on $L^2(\Delta) = L^2([a, b])$ with coefficients which depend on f and g, and with boundary conditions which are $u'(a) = u'(b) = 0$ if $f(a) - g(a) \neq 0, f(b) - g(b) \neq 0$ and $|u(a)| < \infty$ if, for example, $f(a) - g(a) = 0$. Similar results hold for the case $\mathcal{D}(\epsilon) \subset \mathbb{R}^d, d > 2, \Delta \subset \mathbb{R}^{d-1}$.

Examples:

1. Let $\mathcal{D}(\epsilon)$ be a rectangle: $g = 0, f = 1$. Then $\lambda_{n_1 n_2} = \pi^2(n_1^2/(b - a)^2 + n_2^2/\epsilon^2)$. Therefore, $\lambda_{n_1 n_2}(\epsilon) \xrightarrow[\epsilon \to 0]{} \infty$ if $n_2 \neq 0$, $\lambda_{n_1 0} = \lambda_{n_1}(\epsilon)$ $= (\pi^2/(b-a)^2)n_1^2$ does not depend on ϵ, $\lim_{\epsilon \to 0} \lambda_n(\epsilon) = \pi^2 n^2/(b^2 - a^2) = \lambda_n$, where the set $\{\lambda_n\}$ is the spectrum of the limiting problem in Δ:

$$u'' + \lambda u = 0, \quad a \leq x \leq b, \quad u'(a) = u'(b) = 0.$$

2. Let $f = \epsilon\sqrt{1 - x^2}, g = -f, \Gamma$ is the ellipse $x^2 + y^2/\epsilon^2 = 1$. By separation of variables, one can find $\lambda_j(\epsilon)$ and check that

$$\lim_{\epsilon \to 0} \lambda_j(\epsilon) \neq \pi^2 n^2/(b - a)^2.$$

3. Let $\mathcal{D}(\epsilon)$ be the sector: $g = 0, 0 \le x \le 1, a = 0, b = 1$,

$$f = \begin{cases} x \tan \epsilon, & 0 \le x \le \cos \epsilon, \\ \sqrt{1 - x^2}, & \cos \epsilon \le x \le 1. \end{cases}$$

One can calculate easily the spectrum $\lambda_j(\epsilon)$ and find that $\lim_{\epsilon \to 0} \lambda_j(\epsilon) = x_{0j}^2$ are the roots of the equation $J_0'(x) = 0, x_{00} = 0 < x_{01} < \cdots$.

From these examples, one can see that the limits of $\lambda_j(\epsilon)$, $\lambda_1(\epsilon) \le \lambda_2(\epsilon) \le \ldots$ as $\epsilon \to 0$ depend on f and g. In Section E.2, it will be explained how to calculate these limits. Notice that for the Dirichlet boundary condition and for the third boundary condition $\partial u / \partial N + hu|_\Gamma = 0, h > 0$ the limits $\lim_{\epsilon \to 0} \lambda_j(\epsilon) = \infty$.

E.2. Calculation of the Limiting Spectra

1. The starting point is the variational definition of $\lambda_j(\epsilon)$ as the consecutive minima of the ratio

$$J(\epsilon) = \frac{\int_{\mathcal{D}(\epsilon)} |\nabla u|^2 dx\, dy}{\int_{\mathcal{D}(\epsilon)} |u|^2 dx\, dy} = \frac{\int_a^b dx \int_{\epsilon g(x)}^{\epsilon f(x)} (|u_x^2| + |u_y^2|) dy}{\int_a^b dx \int_{\epsilon g(x)}^{\epsilon f(x)} |u|^2 dy}. \tag{E.2.1}$$

One has $\int_{\epsilon g}^{\epsilon f} |u|^2 dy = \epsilon (f - g)|u^2(x, \zeta)|, \epsilon g < \zeta < \epsilon f, \zeta \to 0$ when $\epsilon \to 0$. Since $\partial u / \partial N \to \partial u / \partial y$ as $\epsilon \to 0$ (because $N_x = -\epsilon f' / \sqrt{1 + \epsilon^2 |f'|^2} \to 0$ as $\epsilon \to 0$) one passed to the limit $\epsilon \to 0$ in (E.2.1) and obtains

$$J = \frac{\int_a^b dx (f(x) - g(x)) |v^2|}{\int_a^b dx (f(x) - g(x)) |v|^2} = \frac{\int_a^b h |v'|^2 dx}{\int_a^b h |v|^2 dx}, \tag{E.2.2}$$

where

$$f(x) - g(x) = h(x), \quad v = u(x, 0), \quad v' = \frac{dv}{dx}. \tag{E.2.3}$$

The spectrum corresponding to the variational problem (E.2.2) is the spectrum of the prolem

$$(h(x)v')' + \lambda hv = 0, \quad a \le x \le b, \tag{E.2.4}$$

$$h(a)v'(a) = 0, \quad h(b)v'(b) = 0. \tag{E.2.5}$$

The boundary conditions (E.2.5) are the natural boundary conditions for the variational problem (E.2.2). If $h(a) \ne 0, h(b) \ne 0$ then (E.2.5) is equivalent to the Neumann boundary conditions for the limit problem: $v'(a) = v'(b) = 0$. If, e.g., $h(a) = 0$, then equation (E.2.4), which is of the form

$$v'' + \lambda v + \frac{h'(x)}{h(x)} v = 0, \quad a \le x \le b, \tag{E.2.6}$$

has a unbounded coefficient at $x = a$ and the boundary condition at $x = a$ is $|v(a)| < \infty$. One can assume that the parameter ϵ enters nonlinearly in the equation of Γ, e.g., $y = f(x, \epsilon), y > 0, y = g(x, \epsilon), y < 0$. In this case, the role of $h(x)$ is played by $\frac{\partial f(x,\epsilon)}{\partial \epsilon}\big|_{\epsilon=0} - \frac{\partial g(x,\epsilon)}{\partial \epsilon}\big|_{\epsilon=0}$.

2. If $\mathcal{D}(\epsilon) \subset \mathbb{R}^d, \Gamma$ is given by the equations

$$x_d = \epsilon f(x_1, \ldots, x_{d-1}), \quad x_d > 0, \quad x_d = \epsilon g(x_1, \ldots, x_{d-1}), \quad x_d < 0,$$

and $\mathcal{D}(0) = \Delta \subset \mathbb{R}^{d-1}$, then the argument in Section E.2.1 shows that $\lim_{\epsilon \to 0} \lambda_j(\epsilon) = \lambda_j$, where $\lambda_j, 0 = \lambda_0 < \lambda_1 \le \lambda_2 \le \cdots$ are the eigenvalues of the problem

$$\operatorname{div}(h(p)\nabla v) + \lambda h(p) v = 0 \text{ in } \Delta,\ p = (x_1, \ldots, x_{d-1}), \tag{E.2.7}$$

$$h\frac{\partial v}{\partial \nu} = 0 \quad \text{on } \partial\Delta,\ h = f(p) - g(p). \tag{E.2.8}$$

Here, $\partial\Delta$ is the boundary of Δ, ν is the unit outer normal to $\partial\Delta, \nabla v = \sum_{j=1}^{d=1} (\partial v / \partial x_j) e_j, e_j$ are the unit vectors of the orthonormal cartesian basis in $\mathbb{R}^{d-1}$, $\operatorname{div} a = \sum_{j=1}^{d-1} \partial a_j / \partial x_j$. If $h \ne 0$ on $\partial\Delta$, then (E.2.8) is the Neumann boundary condition on $\partial\Delta$. If $h = 0$ on $\partial\Delta$, then the problem (E.2.7) is degenerate near $\partial\Delta$.

Examples:

4. In Example 2, one has $h = 2\sqrt{1-x^2}$. The problem (E.2.4)–(E.2.5) is of the form

$$(\sqrt{1-x^2}v')' + \lambda\sqrt{1-x^2}v = 0, \quad -1 \le x \le 1, \tag{E.2.9}$$
$$|v(\pm 1)| < \infty. \tag{E.2.10}$$

One can write (E.2.9) as

$$v'' + \lambda v - \frac{x}{1-x^2}v = 0, \quad -1 \le x \le 1. \tag{E.2.11}$$

5. In Example 3, one has $h = \partial(x,0)/\partial\epsilon = x, h'/h = 1/x$, and equation (E.2.6) is of the form

$$v'' + \lambda v + \frac{v}{x} = 0. \tag{E.2.12}$$

The boundary condition at $x = 0$ is $|v(0)| < \infty$. The boundary condition at $x = 1$ is $v'(1) = 0$, because $h(1) \ne 0$. (Note also that for any $\epsilon > 0$ the normal to the line $y = \sqrt{1-x^2}$ at $x = 1$ is directed along the x-axis). The solution of (E.2.12) bounded at $x = 0$ is $J_0(\sqrt{\lambda}x)$, where $J_0(x)$ is the Bessel function. The eigenvalues λ_j are to be found from the equation $J_0'(\sqrt{\lambda}) = 0$, which follows from the boundary condition $v'(1) = 0$. Thus, $\lambda_j = x_{0j}^2$ where x_{0j} are zeros of the equation $J_0'(x) = 0$.
6. Let $\mathcal{D}(\epsilon)$ be the ellipsoid $x^2+y^2+z^2/\epsilon^2 = 1, \mathcal{D}(0) = \Delta$ is the circle $x^2 + y^2 = 1, h = 2\sqrt{1-x^2-y^2}$. Then (E.2.7) takes the form

$$\nabla^2 v + \lambda v - \frac{x\frac{\partial v}{\partial x} + y\frac{\partial v}{\partial y}}{1-x^2-y^2} = 0 \text{ in } \Delta.$$

Since $h = 0$ on $\partial\Delta$, one requires that v be bounded near $\partial\Delta$.

E.3. Additional Remarks

1. If $h(p)$ vanishes at the boundary $\partial\Delta$, one has a self-adjoint but not uniformly elliptic operator $\mathrm{div}(h(p)\nabla v) = h\Delta v + (\nabla h, \nabla v)$. Let us prove that the spectrum of the ratio $(\int_\Delta h|\nabla v|^2 dp)/(\int_\Delta h|v|^2 dp)$ is still discrete regardless of the rate at which $h(p)$ goes to zero when p approaches $\partial\Delta$. It is sufficient to prove that the embedding operator $i : H_1 \to H_0$ is compact, where $H_1 = H_1(\Delta)$(respectively $H_0(\Delta)$) is the Hilbert space of functions defined on Δ with the inner product $(u,v)_1 = \int_\Delta (h\nabla u\overline{\nabla v} + hu\bar{v})dp$ (respectively $(u,v)_0 = \int_\Delta hu\bar{v}dp$). Let $\Delta' \subset \Delta$ be any strictly inner subdomain of Δ, $\inf_{p\to\Delta'} h(p) > 0$. It is well known that the embedding: $H_1(\Delta) \to H_0(\Delta')$ is compact. One proves that $i : H_1(\Delta) \to H_0(\Delta)$ is compact if the estimate:

$$\int_\delta h|v|^2 dp < \epsilon \int_\Delta h\{|v'|^2 + |v|^2\}dp,$$

holds for any $\epsilon > 0$ where δ is a sufficiently thin strip near $\partial\Delta$, e.g., $\delta = \{p : \rho(p,\partial\Delta)\} < \eta, \eta > 0$ is a sufficiently small number, and $\rho(p,\partial\Delta)$ is the distance between p and $\partial\Delta$. But the above estimate is obvious since $\int_\Delta h|v|^2 dp < \infty$. Note that the degeneracy of h at $\partial\Delta$ could lead to a continuous spectrum of the ratio $\int_\Delta h|\nabla v|^2 dp \cdot (\int_\Delta |v|^2 dp)^{-1}$ if h goes to zero fast enough as $p \to \partial\Delta$.

2. Let us show that

$$\lambda_j(\epsilon) \to \lambda_j, \quad \epsilon \to 0, \tag{E.3.1}$$

where $\lambda_j(\epsilon)$ and λ_j are the consecutive minima of the ratio (E.2.1) and (E.2.2), respectively. From the argument after formula (E.2.1), one derives that

$$J(\epsilon) = J + o(1) \quad \text{as } \epsilon \to 0, \tag{E.3.2}$$

where J is defined in (E.2.2). Therefore, the equality

$$\lambda_j(\epsilon) \to \lambda_j = o(1) \quad \text{as} \quad \epsilon \to 0 \tag{E.3.3}$$

holds because $o(1)$ is uniform in u running through an appropriate set dense in the unit ball in the Sobolev space H^1. Indeed, in this case

$$\min J - \max o(1) \le \min J(\epsilon) \le \min J + \max o(1),$$

and $o(1)$ is uniform in $u \in M$, where M is a unit ball in H^q, H^q is the Sobolev space $W_2^q(\mathcal{D}_1), \mathcal{D}_1 = \mathcal{D}(1)$, and q is so chosen that the embedding $i: H^q(\mathcal{D}_1) \to C^2(\mathcal{D}_1)$ is bounded. For example, this is the case if $q > 2+d/2$, where d is the dimension of the space. Since H^q is dense in H^1, $\min_{u\in H^1} J = \inf_{u\in H^q} J$. If the boundary $\partial\mathcal{D}$ is smooth enough, one knows that the $\inf_{u\in H^q} J$ is attained and therefore inf can be substituted by min. To see that $o(1)$ is uniform in $u \in M$, one can estimate $o(1)$ as follows:

$$\begin{aligned} o(1) &= \int_a^b dx|u^2(x,\zeta) - u^2(x,0)| \\ &\le c\max|u| \cdot \max|\nabla u| \cdot \epsilon \max|h(x)| \le c_1\epsilon \end{aligned}$$

since $\max|u| + \max|\nabla u| \le \text{const}$ if $|u|_{H^q} \le 1$. Similarly, one estimates the terms with the derivatives $\nabla u(x,\zeta) - \nabla u(x,0)$.

3. It is clear from our argument that the eigenvalues of the problem

$$\nabla^2 u + \lambda u = 0 \text{ in } \mathcal{D}(\epsilon), \quad \frac{\partial u}{\partial N} + hu = 0 \text{ on } \Gamma(\epsilon) \tag{E.3.4}$$

go to infinity as $\epsilon \to 0$ provided for instance that h is a continuous function on $\Gamma, h \neq 0$. Indeed, the ratio corresponding to problem (E.3.4) is of the form

$$\frac{\int_{\mathcal{D}(\epsilon)} |\nabla u|^2 dx\, dy + \int_{\Gamma(\epsilon)} h|u|^2 ds}{\int_{\mathcal{D}(\epsilon)} |u|^2 dx\, dy}. \tag{E.3.5}$$

As $\epsilon \to 0$, the denominator in (E.3.5) is of order ϵ while the term $\int_{\Gamma(\epsilon)} h|u|^2 ds$ is of order $O(1)$. Thus, the ratio (E.3.5) is unbounded as $\epsilon \to 0$.

Remark. The results of this appendix allow one to prove easily that the eigenvalues of the Neumann Laplacian do not increase monotonically when the domain decreases. This is in contrast with the Dirichlet Laplacian $-\Delta_D$ for which $D \supset D'$ implies $\lambda_n(D) \le \lambda_n(D')$, where $\lambda_n(D)$ are the eigenvalues of $-\Delta_D$. For the Neumann Laplacian, the above implication does not hold. Indeed, if $\mu_n(D)$ are the eigenvalues of the Neumann Laplacian, in D, and if

$$D \supset D' \text{ implies that } \mu_n(D) \leq \mu_n(D'), \qquad (*)$$

then the limit of the spectrum of the Neumann Laplacian for the sector in Example 3, Section E.1, should be the set $\{n^2\pi^2\}$ as in Example 1 in this section. *Since in fact this limit is different, we conclude that* (*) *is false.*

Let us show that (*) implies that the limit of the spectrum in Example 3 is the set $\{n^2\pi^2\}$. Take a rectangular R inscribed into the sector S with height η and base $1-\delta$, where $\eta > 0$ and $\delta > 0$ are small numbers. If (*) holds, then $\mu_n(R) \geq \mu_n(S)$. Pass to the limit $\eta \to 0$ and $\epsilon \to 0$, where $\epsilon > 0$ is the same as in Example 3, Section 1. The result is $\pi^2 n^2/(1-\delta)^2 \geq x_{0n}^2$, where x_{0n} are the roots of $J_0'(x) = 0$. The same argument with the circumscribed rectangular R with base 1 and height η shows that $x_{0n}^2 \leq \pi^2 n^2$. Since $\delta > 0$ is arbitrarily small, we conclude that $x_{0n}^2 = \pi^2 n^2$, which is absurd. This contradiction shows that (*) is false. The author did not see any published proof of the fact that (*) is false.

Appendix F

Recovering of the Surface from the Knowledge of Its Principal Curvatures

F.1. Introduction

The results of the Appendix are taken from [150]. Let Γ be a closed compact oriented surface in $\mathbb{R}^3$ and $\psi : \Gamma \to S^2$ be such a map that the vector $\psi(s)$ when placed at the point s is the positive unit normal to Γ. Here, $\psi(s)$ is the vector with the initial point at the origin (the center of the unit ball with the boundary S^2) and the end point at the $\psi(s)$. Let us assume that ψ is a diffeomorphism of Γ onto some part $\tilde{S}^2$ of S^2 and let $\Phi : \tilde{S}^2 \to \Gamma$ be the inverse mapping ψ^{-1}. Let $(e_1(s), e_2(s), e_3(s))$ be a moving orthonormal frame on $\Gamma, e_3(s) = \psi(s), e_1(s)$ and $e_2(s)$ be the eigenvectors of $-de_3(s)$ with eigenvalues $K_1(s)$ and $K_2(s)$, respectively, where $K_j(s), j = 1, 2$ are the principal curvatures of Γ at the point s. Let us define differential forms ω_1, ω_2 by the equations $\omega_j(e_j) = \delta_{ij}, 1 \le i, j \le 2$. Let $de_i = \omega_{ij} e_j$. Here and below one should sum over the repeated indices. It is known Hsiung [82] that $d\omega_{ij} = \omega_{im} \wedge \omega_{mj}, d\omega_1 = -\omega_2 \wedge \omega_{12}, d\omega_2 = \omega_1 \wedge \omega_{12}$, where $\wedge$ denotes the wedge product of the differential forms, $\omega_{13} = K_1\omega_1, \omega_{23} = K_2\omega_2$, and the fundamental forms on Γ are $\mathrm{I} = \omega_1^2 + \omega_2^2, \mathrm{II} = K_1\omega_1^2 + K_2\omega_2^2, \mathrm{III} = K_1^2\omega_1^2 + K_2^2\omega_2^2$. Let us pull back to S^2 the basic differential forms using the mapping $\Phi = \psi^{-1}$. The pulled back forms $\omega_i^*, \omega_{ij}^*$ satisfy the equations

$$\begin{aligned} &d\omega_{ij}^* = \omega_{im}^* \wedge \omega_{mj}^*,\ d\omega_1^* = -\omega_2^* \wedge \omega_{12}^*,\ d\omega_2^* = \omega_1^* \wedge \omega_{12}^*, \\ &\omega_{ij}^* = -\omega_{ji}^*,\ \omega_{13}^* = K_1\omega_1^*,\ \omega_{23}^* = K_2\omega_2^*. \end{aligned} \tag{F.1.1}$$

Here, $K_i = K_i(\theta, \phi) = K_i(\Phi(\theta, \phi))$. These equations imply that

$$dK_1 \wedge \omega_1^* = \sigma_1(K_1 - K_2)\omega_1^* \wedge \omega_2^*, \tag{F.1.2a}$$

$$dK_2 \wedge \omega_2^* = \sigma_2(K_1 - K_2)\omega_1^* \wedge \omega_2^*, \tag{F.1.2b}$$

where σ_1 and σ_2 are defined by the equation

$$\omega_{12}^* = \sigma_1\omega_1^* + \sigma_2\omega_2^*. \tag{F.1.3}$$

Assume that $K_1(\theta, \phi)$ and $K_2(\theta, \phi)$ are given. If one finds the forms ω_i^*, then determine vectors e_i^* such that

$$\omega_i^*(e_j^*) = \delta_{ij}, \quad 1 \le i, j \le 2, \tag{F.1.4}$$

and finally uses the equation

$$d\Phi = -\frac{\omega_1^* e_1^*}{K_1} - \frac{\omega_2^* e_2^*}{K_2} \tag{F.1.5}$$

then, assuming that K_1 and K_2 are the principal curvatures of a surface, one concludes that $d\Phi$ is a closed differential. Integrating this differential, one recovers the surface. In what follows the equations for the forms $\omega_i^*, 1 \le i \le 2$, are derived and analyzed, and a constructive method for finding the surface is given under the assumption that K_1 and K_2 are the principal curvatures of a surface. Let

$$\omega_1^* = a_1 d\phi + a_2 d\theta, \quad \omega_2^* = b_1 d\phi + b_2 d\theta. \tag{F.1.6}$$

The e_j^* are the eigenvectors of the mapping $d\Phi$ with the eigenvalues $-1/K_j$. One can check, using the orthonormality of the vectors e_j^*/K_j with respect to III* metric, (or the fact that $\text{III}^* = \sin^2\theta d(\phi)^2 + (d\theta)^2$

(see Section 2)) that

$$b_1 = \mp a_2 \frac{K_1}{K_2} \sin\theta, \quad b_2 = \pm a_1 \frac{K_1}{K_2} \frac{1}{\sin\theta} \tag{F.1.7}$$

$$\frac{K_1^2 a_1^2}{\sin^2\theta} + K_1^2 a_2^2 = 1. \tag{F.1.8}$$

One has

$$\begin{aligned} \omega_{13}^* &= K_1 \omega_1^*, \quad \omega_{23}^* = K_2 \omega_2^*, \\ \omega_{12}^* &= \sigma_1 \omega_1^* + \sigma_2 \omega_2^* = \alpha d\phi + \beta d\theta, \end{aligned} \tag{F.1.9}$$

where α and β are to be found. Furthermore,

$$\begin{aligned} d\omega_{12}^* &= \omega_{13}^* \wedge \omega_{32}^* = -K\omega_1^* \wedge \omega_2^* = -K(a_1 b_2 - a_2 b_1) d\phi \wedge d\theta \\ &= -K \frac{\sin\theta}{K} d\phi \wedge d\theta = -\sin\theta \, d\phi \wedge d\theta, \end{aligned} \tag{F.1.10}$$

where $K = K_1 K_2$. If one uses the definition of ω_i^* in terms of a_i and b_i, then one finds

$$a_1\sigma_1 - \frac{K_1}{K_2} \sin\theta \, a_2 \sigma_2 = \alpha,$$

$$a_2\sigma_2 + \frac{K_1}{K_2 \sin\theta} a_1 \sigma_2 = \beta.$$

Solving this system for σ_1 and $-\sigma_2$, one obtains:

$$\begin{pmatrix} \sigma_1 \\ -\sigma_2 \end{pmatrix} = T \begin{pmatrix} a_1 \\ a_2 \end{pmatrix}, \quad T = \begin{pmatrix} \frac{K_1^2 \alpha}{\sin^2\theta} & K_1^2 \beta \\ -\frac{K\beta}{\sin\theta} & \frac{K\alpha}{\sin\theta} \end{pmatrix}. \tag{F.1.11}$$

Here, the equation

$$\det \begin{vmatrix} a_1 & -\frac{K_1}{K_2} \sin\theta \, a_2 \\ a_2 & \frac{K_1}{K_2 \sin\theta} a_1 \end{vmatrix} = \frac{\sin\theta}{K}$$

was used. Using the equation (F.0.2), one obtains

$$Q\begin{pmatrix}a_1\\a_2\end{pmatrix}=\frac{K_2-K_1}{K}\sin\theta\begin{pmatrix}\sigma_1\\-\sigma_2\end{pmatrix}, \tag{F.1.12}$$

where

$$Q=\begin{pmatrix}-\dfrac{\partial K_1}{\partial\theta} & \dfrac{\partial K_1}{\partial\phi}\\ \dfrac{K_1}{K_2\sin\theta}\dfrac{\partial K_2}{\partial\phi} & \dfrac{K_1\sin\theta}{K_2}\dfrac{\partial K_2}{\partial\theta}\end{pmatrix}. \tag{F.1.13}$$

Therefore,

$$\left(Q-\frac{K_2-K_1}{K}(\sin\theta)T\right)\begin{pmatrix}a_1\\a_2\end{pmatrix}=0. \tag{F.1.14}$$

This equation can be written as

$$\begin{pmatrix}A & B\\C & D\end{pmatrix}\begin{pmatrix}a_1\\a_2\end{pmatrix}=0,\qquad \begin{aligned}A&=-\frac{\partial K_1}{\partial\theta}-\frac{K_2-K_1}{K}\frac{K_1^2\alpha}{\sin\theta}\\ B&=\frac{\partial K_1}{\partial\phi}-\frac{K_2-K_1}{K}\sin\theta\,K_1^2\beta\end{aligned} \tag{F.1.15}$$

where A and D are linear in α, and B and C are linear in β. Since a_1 and a_2 are not zeros simultaneously, one has

$$AD-BC=0. \tag{F.1.16}$$

If $K_2-K_1\neq 0$, this is a quadratic equation in α,β. Equation (F.1.15) can be written as

$$Aa_1+Ba_2=0,\quad Ca_1+Da_2=0. \tag{F.1.17}$$

Therefore,

$$A^2a_1^2=B^2a_2^2,\quad K_1^2\sin^2\theta\,A^2a_1^2=B^2\sin^2\theta\left(1-\frac{K_1^2a_1^2}{\sin^2\theta}\right). \tag{F.1.18}$$

Assume that $B\neq 0$. Then

$$a_1=\pm\frac{\sin\theta}{K_1}\frac{1}{[1+(A^2/B^2)\sin^2\theta]^{1/2}},\quad a_2=-\frac{A}{B}a_1. \tag{F.1.19}$$

Therefore, a_1,a_2 (and b_1,b_2) are determined by α,β.

From (F.1.10), it follows that

$$\omega_{12}^* = -\cos\theta\, d\phi + dg, \tag{F.1.20}$$

where $g(\theta,\phi)$ is a scalar function. Therefore,

$$\alpha = \frac{\partial g}{\partial \phi} - \cos\theta, \quad \beta = \frac{\partial g}{\partial \theta}, \tag{F.1.21}$$

so that α and β are determined by g. If one solves the quadratic equation (F.1.16) for α and write $\alpha = H(\theta,\phi,\beta)$, then (F.1.19) can be written as

$$a_j = F_j\left(\theta,\phi,\frac{\partial g}{\partial \theta}\right), \quad j = 1,2. \tag{F.1.22}$$

The equation $d\omega_1^* = -\omega_2^* \wedge \omega_{12}^*$ can be written as

$$\frac{\partial a_2}{\partial \phi} - \frac{\partial a_1}{\partial \theta} = \frac{K_1}{K_2}\sin\theta\left(a_2\beta + a_1\alpha\frac{1}{\sin^2\theta}\right) = G\left(\theta,\phi,\frac{\partial g}{\partial \theta}\right). \tag{F.1.23}$$

From (F.1.22) and (F.1.23), one obtains

$$\frac{\partial F_2}{\partial \theta} + \frac{\partial F_2}{\partial \theta}\frac{\partial}{\partial \theta}\left(\frac{\partial g}{\partial \phi}\right) - \frac{\partial F_1}{\partial \theta} - \frac{\partial F_1}{\partial \beta}\frac{\partial^2 g}{\partial \theta^2} = G\left(\theta,\phi,\frac{\partial g}{\partial \theta}\right). \tag{F.1.24}$$

From (F.1.21), it follows that

$$\frac{\partial}{\partial \theta}\frac{\partial g}{\partial \theta} = -\sin\theta + \frac{\partial H}{\partial \theta}\left(\theta,\phi,\frac{\partial g}{\partial \theta}\right) + \frac{\partial H}{\partial \beta}\cdot\frac{\partial^2 g}{\partial \theta^2}. \tag{F.1.25}$$

From (F.1.24) and (F.1.25), one obtains

$$\left(\frac{\partial F_2}{\partial \beta}\frac{\partial H}{\partial \beta} - \frac{\partial F_1}{\partial \beta}\right)\frac{\partial^2 g}{\partial \theta^2} + \frac{\partial F_2}{\partial \phi} - \frac{\partial F_1}{\partial \theta} + \left(\frac{\partial H}{\partial \theta} - \sin\theta\right) \cdot \frac{\partial F_2}{\partial \beta} = G. \tag{F.1.26}$$

This equation can be written as

$$\frac{\partial^2 g}{\partial \theta^2} = G_1\left(\theta, \phi, \frac{\partial g}{\partial \theta}\right). \tag{F.1.27}$$

Similarly, the equation $d\omega_2^* = \omega_1^* \wedge \omega_{12}^*$ leads to

$$\frac{\partial^2 g}{\partial \theta^2} = G_2\left(\theta, \phi, \frac{\partial g}{\partial \theta}\right), \tag{F.1.28}$$

where G_2 is explicitly determined. From (F.1.27) and (F.1.28), one obtains the equation for $\partial g/\partial \theta$

$$G_1\left(\theta, \phi, \frac{\partial g}{\partial \theta}\right) = G_2\left(\theta, \phi, \frac{\partial g}{\partial \theta}\right). \tag{F.1.29}$$

Assuming that this equation is solvable one finds $\partial g/\partial \theta = \beta$, then $\alpha = H(\theta, \phi, \beta)$, then A and B from (F.1.15), then a_1 and a_2 from (F.1.19), then b_1 and b_2 from (F.1.7), then ω_1^* and ω_2^* from (F.1.6), then e_1^* and e_2^* from (F.1.4), and then $d\Phi$ from (F.1.5). Integrating (F.1.5), one finds the mapping Φ which determines the parametric equation of the surface Γ (parameters are θ, ϕ). *The above construction is valid under the assumption that K_1, K_2 are the principal curvatures of a surface.*

It was assumed above that $K_1 \neq K_2$. If $K_1 = K_2$, then it is easily seen that $K_1 = K_2 = \text{const}$. Indeed (F.1.12) shows that in this case

$$Q\begin{pmatrix} a_1 \\ a_2 \end{pmatrix} = 0.$$

Since a_1 and a_2 both are not zero one concludes that

$$0 = \det Q = -(\partial K_1/\partial \theta)^2 \sin\theta - (\partial K_1/\partial \phi)^2 (1/\sin\theta).$$

Thus, $\partial K_1/\partial \theta = \partial K_1/\partial \phi = 0, K_1 = \text{const}$.

If $K_1 = K_2 = \text{const}$, then the surface is spherical, its radius is K_1^{-1}.

In conclusion of this section, let us formulate some open problems:

(1) Characterization problem: *What are the necessary and suficient conditions on $K_1(\theta,\phi), K_2(\theta,\phi)$ in order that these functions be principal curvatures of a closed surface of class C^2?*

The following problem is closely connected with characterization problem:

(2) If K_1 and K_2 are the principal curvatures of a surface, then $d\Phi$ (see (F.1.5)) is a closed differential. Is the converse true? What is the relationship between the condition that $d\Phi$ is closed and the solvability conditions for equations (F.1.16) and (F.1.29)?

F.2. Derivation of Formulas (F.1.7) and (F.1.8)

The identity $\text{III}^* = \sin^2\theta(d\phi)^2 + (d\theta)^2 = K_1^2(a_1 d\phi + a_2 d\theta)^2 + K_2^2(b_1 d\phi + b_2 d\theta)^2$ is the starting point. One obtains the system

$$K_1^2 a_1^2 + K_2^2 b_1^2 = \sin^2\theta,$$

$$K_1^2 a_1 a_2 + K_2^2 b_1 b_2 = 0,$$

$$K_1^2 a_2^2 + K_2^2 b_2^2 = 1.$$

Let $K_1/K_2 = \lambda, K_2 \neq 0, b_i/\lambda = c_i$. Then the system takes the form

$$a_1^2 + c_1^2 = \sin^2\theta, \quad a_1 a_2 + c_1 c_2 = 0, \quad a_2^2 + c_2^2 = 1.$$

One has

$$\frac{a_1}{c_2} = -\frac{c_1}{c_2}, \quad c_1^2\left(1 + \frac{c_2^2}{a_2^2}\right) = \sin^2\theta, \quad \frac{c_1^2}{a_2^2} = \sin^2\theta, \tag{F.2.1}$$

$$c_1 = \mp a_2 \sin\theta, \quad c_2 = \pm\frac{a_1}{\sin\theta}. \tag{F.2.2}$$

Thus,

$$b_1 = \mp\frac{K_1 \sin\theta}{K_2} a_2, \quad b_2 = \pm\frac{K_1}{K_2 \sin\theta} a_1, \quad \frac{K_1^2 a_1^2}{\sin^2\theta} + K_1^2 a_2^2 = 1.$$

The choice of the sign in (F.2.2) is $(-+)$. One could choose $(+-)$ without any changes in the argument.

Remark. The unit sphere is the only compact closed surface such that $\mathrm{III}^* = \mathrm{I}^*$, that is the pull-backs of the third and first fundamental forms are equal on this surface. Indeed, $\mathrm{III}^* = \mathrm{I}^*$ if and only if $K_1^2 \omega_1^{*2} + K_2^2 \omega_2^{*2} = \omega_1^{*2} + \omega_2^{*2}$. The last equation implies that $K_1^2 = 1, K_2^2 = 1$. This means that $K_1 = 1, K_2 = 1$, i.e., the surface is the unit sphere. The case $K_1 K_2 = -1$ does not correspond to any compact closed smooth surface as is clear from the geometrical reasons.

Appendix G

Calculating Resonances (Natural Frequencies) and Extracting them from Transient Fields

Mathematical formulation and analysis of numerical methods for calculating the natural frequencies (resonances) are given. Stability of these methods towards roundoff errors and small perturbations of the obstacles is established. Some formulas for the variations of the natural frequencies due to small perturbations of the surface of the obstacle are given. A simple new method for extraction of resonances from transient fields is given.

G.1. Introduction

Let $\mathcal{D}$ be a finite obstacle with a smooth surface Γ, and let Ω be the exterior domain. The obstacle (scatterer) is three-dimensional. The smoothness of Γ is of the type that ensures the applicability of Green's formulas. Roughly speaking, cusps-type singular points of the surface are not admissible, but edges (as in a cube) or conical points are admissible. We will discuss for simplicity scalar wave scattering, but the results and arguments are valid for electromagnetic wave scattering. The Green's function for a reflecting obstacle satisfies the equations

$$(-\nabla^2 - k^2)G(x, y, k) = \delta(x - y) \quad \text{in } \Omega,$$

$$k > 0, \quad x = (x_1, x_2, x_3), \tag{G.1.1}$$

$$G = 0, \quad \text{if } x \in \Gamma \tag{G.1.2}$$

$$r\left(\frac{\partial G}{\partial r} - ikG\right) \to 0 \text{ as } r := |x| \to \infty. \tag{G.1.3}$$

Here, y is the position vector of the source and ∇^2 is the Laplacian. The function G is uniquely determined by the conditions (G.1.1)–(G.1.3) and can be continued analytically on the whole complex plane of k as a meromorphic function of k. Its poles lie in the half-plane $\operatorname{Im} k < 0$ and are called resonances, natural frequencies, or complex poles. The meromorphic nature of G as a function of k and the (closely connected with it) behavior of solutions of the time-dependent wave equation as $t \to +\infty$ were studied in the series of papers starting with [184]. In [240], there is a bibliography of the subject. In [49], there are reviews of the subject for engineers. The connection of the complex poles asymptotic with the behavior of solutions to the time-dependent wave equation is the foundation of the singularity expansion method (SEM). If

$$\begin{aligned} &\nabla^2 u = u_{tt} \text{ in } \Omega,\ t > 0,\ u = 0 \text{ on } \Gamma, \\ &u(x, 0) = 0,\ u_t(x, 0) = f(x), \end{aligned} \tag{G.1.4}$$

then the function v, defined as

$$v(x, k) = \int_0^\infty \exp(ikt) u(x, t) dt, \tag{G.1.5}$$

satisfies the equations

$$\begin{aligned} (\nabla^2 + k^2) v = -f, \quad v = 0 \text{ on } \Gamma, \\ r\left(\frac{\partial v}{\partial r} - ikv\right) \to 0, \quad r \to \infty, \end{aligned} \tag{G.1.6}$$

$$v = \int_\Omega G(x, y, k) f\, dy, \tag{G.1.7}$$

$$u = (2\pi)^{-1} \int_{-\infty}^\infty \exp(-ikt) v(x, k) dk. \tag{G.1.8}$$

Assume that f is a smooth function which vanishes outside of a bounded domain (compactly supported). In the engineering literature, the complex variable $s = -ik$ is often used. In the physical and mathematical literature, k is usually the complex variable. The half-plane $\operatorname{Im} k < 0$ (which we use in this book) corresponds to the half-plane $\operatorname{Re} s < 0$ on the s plane. If one knows that

$$v \text{ is meromorphic (and analytic if } \operatorname{Im} k \geq 0), \tag{G.1.9}$$

$$|v| \leq C(b)(1+|k|)^{-a},\ a > \frac{1}{2},\ b = \operatorname{Im} k,\ |\operatorname{Re} k| \to \infty, \tag{G.1.10}$$

$$|\operatorname{Im} k_j| < |\operatorname{Im} k_{j+1}| \to \infty \text{ as } j \to \infty, \tag{G.1.11}$$

then one can move the contour of integration in (G.1.8) in the k plane down and obtain the SEM expansion

$$u(x,t) = \sum_{j=1}^{N} c_j(x,t)e^{-ik_jt} + O(e^{-|\operatorname{Im} k_{N+1}|t}) \text{ as } t \to \infty. \tag{G.1.12}$$

Here, k_j are the complex poles of $v(x,k), c_j(x,t) = i\operatorname{Res}_{k=k_j}\{e^{-ikt} v(x,k)\}$ and N is the number of poles in the strip $0 > \operatorname{Im} k \geq \operatorname{Im} k_N$. Usually, it is assumed by engineers that the poles k_j are simple, in which case $c_j(x,t) = c_j(x)$. If $m_j + 1$ is the multiplicity of the pole k_j, then $c_j(x,t) = O(t^{m_j})$, and one can write (G.1.12) as

$$u(x,t) = \sum_{j=1}^{N}\sum_{m=0}^{m_j} c_{jm}(x)t^m e^{-ik_jt} + O(e^{-|\operatorname{Im} k_{N+1}|t}) \text{ as } t \to \infty. \tag{G.1.13}$$

Expansions (G.1.12) and (G.1.13) were called in [240] as asymptotic SEM expansions. These expansions are proved under the assumption that $\mathcal{D}$ is strictly convex. The expansion

$$u = \sum_{j=1}^{\infty} c_j(x,t)e^{-ik_jt}, \tag{G.1.14}$$

which one can see in the literature, is not proved and is not valid in general. Formula (G.1.13) is not valid, in general, for nonconvex obstacles. For example, in [83], it is proved that if the obstacle consists of two strictly convex bodies, then $G(x, y, k)$ has countably many complex poles approximately on the line $\operatorname{Im} k = c_0$. These poles asymptotically are equidistant and the distance between the poles depends on the distance between the bodies and the curvatures and principal directions of the surfaces Γ_1 and Γ_2 of the bodies $\mathcal{D}_1$ and $\mathcal{D}_2$ at the two closest points $a_1 \in \Gamma_1$ and $a_2 \in \Gamma_2$. (That is, $|a_1 - a_2| = \min_{s_1 \in \Gamma_1, s_2 \in \Gamma_2} |s_1 - s_2|$.) This result shows that SEM expansion (G.1.13) is not valid for two convex bodies. It also suggests that (G.1.13) is not valid for a single body with nonconvex boundary which can hold a trapping mode (i.e., a standing wave in the geometrical optics approximation). This, however, is not proved yet.

In principle, one can tell the difference between convex obstacles and nonconvex obstacles, capable of holding a trapping mode, by the behavior of complex poles k_j for large j: for convex obstacles (G.1.11) holds and $|\operatorname{Im} k_j| \to +\infty$, while in the other case there exist infinitely many poles on a line $\operatorname{Im} k_j = c_0 < 0$.

The significance of the complex poles is manifold. We mention only two areas important in applications. First, one can tabulate the complex poles and use them for target identification. Practically, it is expected that different scatterers produce different sets of complex poles. Although this is not proved, there are some supporting arguments ([240, pp. 585–586]). Second, system theory uses representations of impulse responses as sums of exponentials. The problem is to find these exponentials from transient fields.

It is a long-standing open problem to prove that there exist infinitely many complex not purely imaginary poles of G for any reflecting obstacles. It was proved that there exist infinitely many purely imaginary poles, Theorem G.5.1.

The objectives of this appendix include:

(1) formulation of the mathematical methods for numerically calculating the complex poles,
(2) analysis of convergence and stability of these methods, and
(3) formulation of a simple technique for extracting resonances (natural frequencies) from transient fields.

The techniques used in the literature were based mostly on Prony's method. Some other methods were also used. Here we present a very simple numerical technique which seems to be new and does not require solving nonlinear or even linear equations. The question most difficult in this problem, that of noisy data, is discussed. This chapter is organized as follows: Section G.2 deals with questions (1) and (2). Section G.3 deals with question (3).

Appendix I contains some formulas for perturbations of complex poles under perturbations of the surface of the scatterer. Numeration of formulas is separate in each of the sections. The material in Section G.3 appeared in [262, 263, 269].

G.2. Methods of Calculating Natural Frequencies

A. *Basic equations*

From the Green's formula, one obtains

$$G(x, y, k) = g(x, y, k) - \int_\Gamma g(x, s')h(s', y)ds', \qquad \text{(G.2.1)}$$

$$g = e^{ik|x-y|}/(4\pi|x - y|), \qquad \text{(G.2.2)}$$

$$h = \frac{\partial G}{\partial N_s}, \qquad \text{(G.2.3)}$$

where N_s is the outer normal to Γ at the point s, and the dependence on k is suppressed in some of the function for brevity. Let $x = s \in \Gamma$ in (G.2.1). Then

$$\int_\Gamma gh\, ds' = g. \qquad \text{(G.2.4)}$$

If k_j is a pole of G, then it is a pole of h, so that $h = \psi/(k - k_j)^m$. Multiply (G.2.4) by $(k - k_j)^m$ and let $k = k_j$ to obtain

$$Q\psi = \int_\Gamma g(s, s', k_j)\psi\, ds' = 0. \qquad \text{(G.2.5)}$$

Therefore, the complex poles are the points k_j at which equation (G.2.5) has a nontrivial solution.

Let us differentiate (G.2.1) in the direction N_s from inside of $\mathcal{D}$ and then take $x \to s \in \Gamma$ to obtain

$$[I + A(k)]h = h + Ah = 2\frac{\partial g}{\partial N_s}, \tag{G.2.6}$$

$$Ah = \int_\Gamma 2\frac{\partial g}{\partial N_s} h\, ds'. \tag{G.2.7}$$

This gives the second way to characterize the complex poles: They are the points at which the equation

$$B\psi = [I + A(k)]\psi = 0 \tag{G.2.8}$$

has a nontrivial solution.

B. *Projection methods for calculating the poles*

First, consider equation (G.2.8). Take a complete, in $L^2(\Gamma) = H$, set of linearly independent functions $\{\phi_j\}$. The linear span of the first n functions is a linear subspace $H_n, H_n \subset H_{n+1}$. Since the system $\{\phi_j\}$ is complete in H, one concludes that the system of subspaces H_n is limit dense in H, that is, $\operatorname{dist}(\psi, H_n) \to 0$ as $n \to \infty$ for any $\psi \in H$. This property is crucial for the analysis below. Here, dist is the distance between the element ψ and the subspace H_n. Let $\psi_n = \sum_{j=1}^n c_j^{(n)} \phi_j$. Consider the projection method for solving (G.2.8):

$$\begin{aligned} &(B\psi_n, \phi_m) = 0, \quad 1 \le m \le n, \\ &\sum_{j=1}^n (B\phi_j, \phi_m) c_j = 0, \quad 1 \le m \le n. \end{aligned} \tag{G.2.9}$$

A necessary and sufficient condition for (G.2.9) to have a nontrivial solution is

$$\begin{aligned} &\det b_n(k) = 0, \\ &b_n(k) = [b_{jm}(k)]_{j,m=1,\dots,n}, \quad b_{jm} = (B\phi_j, \phi_m). \end{aligned} \tag{G.2.10}$$

The parentheses denote the inner product in $L^2(\Gamma), (u, v) = \int_\Gamma u\bar{v}ds$. The elements $b_{jm}(k)$ are entire functions of k since the operator A in (G.2.7) is an entire analytic operator function of k. Therefore, (1) it is not obvious that equation (G.2.10) has zeros [e.g., $\exp(k)$ does not have zeros] and (2) if (G.2.10) has zeros $k_j^{(n)}$, $j = 1, 2, \ldots,$ then one should prove convergence of this method, that is, one should prove that

$$\lim_{n\to\infty} k_j^{(n)} = k_j, \tag{G.2.11}$$

where k_j are the complex poles of G, and that all of the complex poles can be obtained in this way. This will be done later.

Consider equation (G.2.5). In the same way as in the previous section, one can derive the equation $\det Q_n(k) = 0$,

$$Q_n(k) = [Q_{jm}(k)]_{j,m=1,\ldots,n}, \quad Q_{jm} = (Q\phi_j, \phi_m). \tag{G.2.12}$$

This equation is of the same structure as (G.2.10), and the same questions (1) and (2) should be investigated for equation (G.2.12). The difference between operators Q and B is that Q is compact while B is of the Fredholm type, so that equation (G.2.5) is of the first kind, while (G.2.8) is of the second kind. The elements of Q_{jm} are easier to compute than b_{jm}.

C. *Variational methods for calculating the poles*

Consider the problem

$$\|Qf\| = \min, \quad \|f\| = 1, \tag{G.2.13}$$

where $\|f\|$ is the $L^2(\Gamma)$ norm, $\|f\|_p$ is the Sobolev space $W^{2,p}(\Gamma) = H^p$ norm, $\|f\| = \|f\|_0$, $\|f\|_p^2 = \int_\Gamma \{|u|^2 + |Du|^2 + \cdots + |D^p u|^2\} ds$, and D denotes the first-order derivative on Γ. For $p < 0$, the space $W^{2,p}$ is defined as a dual to $W^{2,|p|}$.

Take $f_n = \sum_{j=1}^n c_j^{(n)} \phi_j$, substitute in (G.2.13), and obtain the problem

$$\sum_{j=1}^n q_{mj}^{(n)} c_j^{(n)} = \lambda c_m^{(n)},\ 1 \le m \le n,\ q_{mj}^{(n)} = (Q\phi_j, Q\phi_m), \tag{G.2.14}$$

where λ is an eigenvalue of the matrix $q_{mj}^{(n)}$. This matrix is an entire function of k. Its minimal eigenvalue $\lambda_1^{(n)}(k)$ is the minimum of the function $|Qf|$ under the constraint (G.2.13). The points $k_j^{(n)}$, which are zeros of $\lambda_1^n(k)$,

$$\lambda_1^{(n)}(k_j^{(n)}) = 0, \tag{G.2.15}$$

converge to the complex poles of $G(x, y, k)$ and all of the complex poles can be obtained as limits of $k_j^{(n)}$ as $n \to \infty$,

$$k_j = \lim_{n\to\infty} k_j^{(n)}. \tag{G.2.16}$$

A similar idea was used in [232] and [242]. Convergence of the methods given in Section G.2.B and a study of their stability are given in the next subsection.

D. *Convergence and stability of the methods for calculating the poles*

The basic ideas and methods of the analysis and proofs are taken from [226]. The basic results consist of a proof of convergence and stability of the methods given in Section G.2.B towards roundoff errors and perturbations of the data.

(1) We start with the method given in the first part of Section G.2.B. Let us assume that there exists a countable discrete set $\mathcal{P}$ of points k_j at which equation (G.2.8) has a nontrivial solution. In Section G.2.A, we proved that any complex pole of G belongs to $\mathcal{P}$. Let us show that any point $k_0 \in \mathcal{P}$ is a complex pole of G. Let ψ be a nontrivial solution to (G.2.8). Define the simple layer potential $v = \int_\Gamma g\psi ds'$. From the known formula (see Section 1.1.1) $\partial v/\partial N_i = (A\psi + \psi)/2$ [in which $\partial/\partial N_i$ denotes the limit value of the normal derivative on Γ from the interior and A is given in (G.2.7)] and from equation (G.2.8), it follows that $\partial v/\partial N_i = 0$. We know that $(\nabla^2 + k_0^2)v = 0$ in $\mathcal{D}$. Since k_0^2 is a complex number and the spectrum of the interior Neumann Laplacian consists of positive numbers only, we conclude that $v = 0$ in $\mathcal{D}$. Therefore, $v = 0$ on Γ. If G does not have a pole at $k = k_0$, then the problem $(\nabla^2 + k_0^2)v = 0$ in $\Omega, v = 0$ on $\Gamma, v(x, k_0)$ is the limit value of a function $v(x, k)$ analytic in k in a neighborhood of k_0 and belonging to $L^2(\Omega)$ when $\operatorname{Im} k > 0$, has only the trivial solution. Thus, $v = 0$ in Ω if k_0 is not a pole of G.

Therefore, $\psi = \partial v/\partial N_i - \partial v/\partial N_e = 0$, where $\partial v/\partial N_e$ is the limit value of the normal derivative on Γ from the exterior domain. This contradicts the assumption that $\psi \neq 0$. Therefore, k_0 is a pole of G.

Let us prove that for sufficiently large n, (1) Equation (G.2.10) has solutions, (2) Equation (G.2.11) holds, (3) all the complex poles can be obtained as limits (G.2.11), (4) complex poles k_j are stable towards small perturbations of the data; the notion of small perturbation will be specified.

Equation (G.2.9) can be written as an operator equation $P_n B P_n \psi = 0$, or

$$(I + P_n A)\psi_n = 0, \tag{G.2.17}$$

where P_n is the orthoprojection onto $H_n, \psi_n = P_n\psi \in H_n$, and I is the identity. Since $A(k)$ is compact in H for any k and $P_n \to I$ as $n \to \infty$, where the arrow denotes strong convergence, one has $\|B - P_n B\| \to 0$ as $n \to \infty$. Therefore, the operator $I + P_n A = I + A(k) - P^{(n)}A(k)$ is invertible for sufficiently large n in a neighborhood of any point k_0 at which $I + A(k_0)$ is invertible. Here $P^{(n)} = I - P_n, P^{(n)} \to 0$ as $n \to \infty$. This argument shows that if k_0 is not a complex pole, then there are no roots $k_j^{(n)}$ of equation (G.2.10) in a neighborhood of k_0. It remains to be proved that if k_0 is a pole of G, then for sufficiently large n there exists a root $k_j^{(n)}$ of equation (G.2.10) which lies in the circle $C_\delta : |k - k_0| \leq \delta$, where $\delta > 0$ is an arbitrary small number. Suppose that for some $\delta > 0$ and all n there are no roots $k_k^{(n)}$ of equation (G.2.10) in the circle C_δ. Then the operator $I + P_n A(k)$ is invertible in C_δ, the operator $(I + P_n A(k))^{-1}$ is analytic in k in C_δ, and therefore $\|(I + P_n A(k))^{-1}\| \leq c$, where c is a constant which does not depend on n. On the other hand, $(I + A(k))^{-1} = (I + P_n A(k) + P^{(n)}A(k))^{-1} = (I + P_n A(k))^{-1}(I + P^{(n)}A(k)(I + P_n A(k))^{-1})^{-1}$. Since $\|P^{(n)}A(k)\| \to 0$ and $\|(I + P_n A)^{-1}\| \leq c$, one concludes that $(I + A(k))^{-1}$ is a bounded operator in C_δ. This is a contradiction since k_0 is a pole of the operator $(I + A(k))^{-1}$. The contradiction proves that for any $\delta > 0$ and sufficiently large n, there is a root $k_j^{(n)}$ of equation (G.2.10) in the circle C_δ.

(2) The above argument settles also the question about stability of the poles towards small perturbations of the data and roundoff errors. Indeed, small perturbations of the data and roundoff errors are equivalent to small perturbations of the matrix $b_{jm}(k)$.

Let us assume that a small perturbation of the matrix $b_{jm}(k)$ is caused by a small perturbation of the operator $B = I + A$. Let us denote $\tilde{B} = I + \tilde{A} = I + A + T$ as the perturbed operator. In this formulation, the perturbed matrix $\tilde{b}_{jm}^{(n)}$ is the matrix of the operator $P_n \tilde{B} P_n$. The perturbation T can describe both the perturbation of Γ and the roundoff errors in the computing matrix $b_{jm}^{(n)}$. Our aim is to prove that in any finite domain on the k plane the poles $\tilde{k}_j^{(n)}$ of the perturbed operator $(I + P_n \tilde{B}(k) P_n)^{-1}$ differ from the poles $k_j^{(n)}$ of the unperturbed operator $(I + P_n B(k) P_n)^{-1}$ a little: $|k_j^{(n)} - \tilde{k}_j^{(n)}| \leq \epsilon(n, \|T\|), \epsilon \to 0$ if $\|T\| \to 0$ and $n \to \infty$. Since we have already established the convergence property (G.2.11), it is sufficient to prove that

$$|\tilde{k}_j - k_j| \to 0, \text{ if } \|T\| \to 0, T = \tilde{B} - B. \tag{G.2.18}$$

Let k_j be a pole of $(I + A(k))^{-1}$ and there are no other poles of this operator in the circle C_δ. One has

$$(I + A(k) + T)^{-1} = (I + A(k))^{-1}(I + T(I + A(k))^{-1})^{-1}.$$

Suppose k_j is a pole of multiplicity m. Then

$$\|(I + A(k))^{-1}\| \leq c/|k - k_j|^m, \, k \in C_\delta, \, c = \text{const}.$$

Thus,

$$\|T(I + A(k))^{-1}\| \leq c\|T\| \, |k - k_j|^{-m}.$$

If $c\|T\|\delta^m < 1$, then the perturbed pole $\tilde{k}_j$, corresponding to the unperturbed pole k_j, lies inside the circle C_δ, that is, $|\tilde{k}_j - k_j| < \delta$. In other words, $|\tilde{k}_j - k_j| = O(\|T\|^{1/m})$ where m is the multiplicity of the pole k_j and $O(\|T\|^{1/m})$ means $\leq \text{const}\|T\|^{1/m}$.

The smallness of the perturbation of the surface is described in terms of the smallness of the norm $\|T\|$. One can give a relationship between the equation of the perturbed surface and the norm of T. This is cumbersome and is done in Appendix I.

Let us study the method based on equation (G.2.12). The results will be the same:

(1) Equation (G.2.12) has roots $k_j^{(n)}$ for sufficiently large n,
(2) Equation (G.2.11) holds,
(3) all the complex poles can be obtained as limits (G.2.11), and
(4) small perturbations of the data lead to small perturbations of the complex poles uniformly on any bounded domain on the complex k plane.

Analysis of equations (G.2.5) and (G.2.12) is more complicated than that of equations (G.2.8) and (G.2.10) because (G.2.5) is an equation of the first kind. The basic tool in our analysis is the factorization formula

$$Q(k) = Q_0(I + V), \quad Q_0 = Q(0), \quad V = Q_0^{-1}(Q(k) - Q_0). \quad \text{(G.2.19)}$$

Here, $Q_0 f = \int_\Gamma f\, ds'/(4\pi|s - s'|)$ is a self-adjoint positive definite operator on $H = L^2(\Gamma)$. This operator is an isomorphism between $H = H^0$ and H^1, while the operator V is compact in any space H^p. Therefore, the bilinear form $(Q_0 u, v)$ defines an inner product equivalent to the inner product $(u, v)_{-1/2}$ in $H^{-1/2}$. The matrix

$$Q_{jm} = (Q_0(I + V)\phi_j, \phi_m) = ((I + V)\phi_j, \phi_m)_{-1/2}.$$

Our previous arguments in Section G.2.D are fully applicable to this matrix because:

(1) V is compact in $H^{-1/2}$ and depends analytically on k, and
(2) If a system $\{\phi_j\}$ is complete in H^0, then it is complete in H^p for any $p < 0$.

Compactness of V was already mentioned. To explain the second statement, assume that $f_p \in H^p, p < 0$. It is known that $H^p \subset H^q$ if $p > q, H^p$ is dense in H^q (that is, for any $\epsilon > 0$ and any $f \in H^q$ there exists an $f_\epsilon \in H^p$ such that $\|f_\epsilon - f\|_q < \epsilon$, where $\|\cdot\|_q$ is the norm in H^q), and $\|f\|_p \leq \|f\|_q$ if $p < q$. Let $f \in H^p, p < 0$, and $\epsilon > 0$ is fixed. Find $f_\epsilon \in H^0$ such that $|f - f_\epsilon|_p < \epsilon/2$. Use completeness of the system $\{\phi_j\}$ in H^0 to find $h_\epsilon = \sum_{j=1}^{n(\epsilon)} c_j(\epsilon)\phi_j$ such that $\|h_\epsilon - f_\epsilon\|_0 < \epsilon/2$. Then

$$\begin{aligned}\|f - h_\epsilon\|_p &\leq \|f - f_\epsilon\|_p + \|f_\epsilon - h_\epsilon\|_p \\ &\leq \epsilon/2 + \|f_\epsilon - h_\epsilon\|_0 < \epsilon, \quad \text{if} \quad p < 0.\end{aligned}$$

Therefore, the system $\{\phi_j\}$ is complete in $H^{-1/2}$ and the matrix Q_{jm} is a matrix of the operator $I + V(k)$ in $H^{-1/2}$, where $V(k)$ is compact in $H^{-1/2}$ and analytic in k. The rest of the argument is the same as in Sections G.2.D(1) and G.2.D(2) and the conclusions are formulated in the beginning of this section.

(3) In this section, we study the variational method given in Section G.2.C. If $\lambda_1^{(n)}(k_j^{(n)}) = 0$, then equation (G.2.14) corresponds to the projection method for the equation $Q^*Qf = 0$. The factorization in Section G.2.D is sufficient for the arguments of that section to hold for the operator Q^*Q (the reason is that $Q^*Q = (I + V^*)Q_0^2(I + V) = Q_0^2(I + V_1)$ where V_1 is compact). Here, we used the self-adjointness of Q_0. Compactness of V_1 follows from simple properties of pseudodifferential operators: $\operatorname{ord} Q_0 = -1, \operatorname{ord} V = -3, V_1 = Q_0^{-2}V^*Q_0^2 + V + Q_0^{-2}V^*Q_0^2V$, and $\operatorname{ord} V_1 \leq \operatorname{ord} V < 0$. Here, $\operatorname{ord} Q$ is the order of the pseudodifferential operator Q. One can find properties of pseudodifferential operators, e.g., in [113]. Thus, one concludes that the results (G.2.1)–(G.2.4) in Section G.2.D(3) hold for the variational method described in Section G.2.C.

G.3. Extracting Natural Frequencies from Transient Fields

A. *Preliminaries*

Consider the problem

$$\begin{aligned}u_{tt} = \nabla^2 u,\ t \geq 0,\ x \in \Omega \subset \mathbb{R}^3,\ u = 0 \text{ on } \Gamma, \\ u(x, 0) = 0,\ u_t(x, 0) = f(x),\end{aligned} \tag{G.3.1}$$

where $\mathcal{D} = \mathbb{R}^3 \setminus \Omega$ is a bounded connected domain with a smooth strictly convex boundary, $f \in C_0^\infty$. In Chapter 3, the basic results on the asymptotic behavior of u as $t \to +\infty$ are described. In particular, the following asymptotic SEM (singularity expansion method) formula holds:

$$u = \sum_{j=1}^{N} \sum_{m=0}^{m_j} c_{jm}(x) t^m e^{-ik_j t} + O(e^{-|\operatorname{Im} k_{N+1}|t}),\ t \to +\infty, \quad \text{(G.3.2)}$$

where $c_{jm}(x)$ do not depend on t, $k_j = a_j - ib_i, b_i > 0$, are complex poles of the resolvent kernel G of the Dirichlet Laplacian in Ω

$$(\nabla^2 + k^2)G(x, y, k) = -\delta(x - y) \text{ in } \Omega,\ G = 0 \text{ on } \Gamma,$$
$$|x| \left(\frac{\partial G}{\partial |x|} - ikG \right) \to 0 \text{ as } |x| \to \infty,$$

and $m_j + 1$ is the multiplicity of the pole $k_j, \operatorname{Im} k_j < 0, |\operatorname{Im} k_j| \to \infty$ as $j \to \infty$. The poles k_j are called resonances or natural frequencies. The signal (G.3.2) is the transient field that can be observed experimentally. Knowledge of the resonances k_j may serve for target identification: the scatterers of various shapes produce various sets of resonances, and this is one of the reasons to be interested in resonances.

The other reason is that in systems theory one often models impulse responses as a sum of the type given in formula (G.3.2). The important problem of system identification can be formulated as follows: From the observation of the transient field (G.3.2), find the numbers k_j and m_j. There is an extensive literature on the subject. Many researchers contributed to the field (Prony, Bruns, Lagrange, Kühnen, and quite a few modern researchers). A large bibliography can be found in [448]. Only the case $m_j = 0$ (simple poles) was treated in the literature.

The purpose of the following subsection is

(1) to give a simple numerical procedure for computing the number m_j and $k_j, 1 \leq j \leq N$ for any fixed N from the exact transient data,

(2) to discuss this problem for noisy data, and
(3) to briefly review the classical methods (Prony, Bruns, Lagrange).

B. *A simple method for extracting resonances from the transient field*

Assume first that the scatterer is a strictly convex reflecting body so that (G.3.2) holds. By $u(n)$, let us denote the sequence $u(x, nh)$, where $h > 0$ is a fixed number. It follows from (G.3.2) that

$$\begin{aligned} &u(n) = c_{1m_1} h^{m_1} n^{m_1} e^{-ia_1 nh} e^{-b_1 nh}(1 + O(1/n)) \\ &\text{as } n \to \infty, \text{ provided that } b_1 < b_2 < \cdots . \end{aligned} \tag{G.3.3}$$

From (G.3.3), one obtains

$$\frac{u(n+1)}{u(n)} = e^{-ia_1 h} e^{-b_1 h}(1 + O(1/n)) \text{ as } n \to \infty. \tag{G.3.4}$$

Thus,

$$b_1 = \frac{1}{h} \ln \left| \frac{u(n)}{u(n+1)} \right| + O\left(\frac{1}{n}\right) \quad \text{as } n \to \infty, \tag{G.3.5}$$

$$b_1 + ia_1 = \frac{1}{h} \ln \frac{u(n)}{u(n+1)} + O\left(\frac{1}{n}\right) \quad \text{as } n \to \infty. \tag{G.3.6}$$

Suppose that k_j, m_j, and $c_{jm}, 1 \le m \le m_j, 1 \le j \le N$, are computed. In what follows, a method for computing c_{jm} is given. Let u_N denote the sum in formula (G.3.2), $u - u_N = \omega_N$. Then, as above,

$$b_{N+1} = \frac{1}{h} \ln \left| \frac{\omega_N(n)}{\omega_N(n+1)} \right| = O\left(\frac{1}{n}\right) \quad \text{as } n \to \infty, \tag{G.3.7}$$

$$\begin{aligned} b_{N+1} + ia_{N+1} &= \frac{1}{h} \ln \frac{\omega_N(n)}{\omega_N(n+1)} \\ &= O\left(\frac{1}{n}\right) \quad \text{as } n \to \infty. \end{aligned} \tag{G.3.8}$$

If a_1 and b_1 are found, then m_1 can be found by the formula

$$m_1 = \frac{\ln\{u(n) e^{ia_1 nh + b_1 nh}\}}{\ln n} + O\left(\frac{1}{\ln n}\right) \quad \text{as } n \to \infty. \tag{G.3.9}$$

Similarly,

$$m_{j+1} = \frac{\ln\{\omega_j(n)e^{ia_{j+1}nh+b_{j+1}nh}\}}{\ln n} + O\left(\frac{1}{\ln n}\right) \quad \text{(G.3.10)}$$

as $n \to \infty$, $j = 1, 2, \ldots$.

If a_1, b_1, and m_1 are found, then

$$c_{1m_1} = \frac{u(n)e^{ia_1nh+b_1nh}}{(nh)^{m_1}} + O\left(\frac{1}{nh}\right) \quad \text{as } n \to \infty. \quad \text{(G.3.10}'\text{)}$$

In the literature, the case of simple resonances ($m_j = 0$) only was discussed. In this case, $O(1/n)$ in (G.3.11) can be substituted by $O(e^{-(b_2-b_1)nh})$ as $n \to \infty$.

In a similar way, all the coefficients c_{1m} can be computed. Practically, one takes n large, neglects the remainder in formulas of the type (G.3.10′), and uses the main term in the right-hand side of (G.3.10′) as the formula for c_{1m}, etc.

If k_1, a_1, b_1, and $c_{1m}, 1 \le m \le m_1$, are found, then one works with $\omega_1 = u - u_1$ and so on. This is a method for computing the coefficients c_{jm} from the transient field.

An alternative method is the least squares method. If m_1, a_1, and b_1 are found, then $c_{1m}, 1 \le m \le m_1$, can be found from the requirement

$$\sum_{n=1}^{\infty}\left|u(n) - \sum_{m=0}^{m_1} c_{1m}(nh)^m e^{(-ia_1-b_1)nh}\right|^2 = \min.$$

This leads to a uniquely solvable linear system for the coefficients $c_{1m}, 1 \le m \le m_1$. If k_1, a_1, b_1, m_1, and $c_{1m}, 1 \le m \le m_1$ are found, then one works with $u - u_1 = \omega_1$ and considers ω_1 as the transient field. The method previously described is quite simple and does not require solving linear systems.

Formulas (G.3.5)–(G.3.11) give a simple method for extracting resonances and their multiplicities from the exact transient field. The much more complicated case of noisy data is discussed below.

In systems theory u, being an impulse response of a system, often does not depend on x.

We will now discuss the case of *noisy data*. Assume that $y(n) = u(n) + \epsilon(n)$ is measured instead of $u(n)$. Here $\epsilon(n)$ is noise. Let us assume that $\epsilon(n)$ is uniformly distributed on the interval $[-\epsilon, \epsilon]$ and $\epsilon > 0$ is a given number. In practice, the level of noise is not known exactly since the noise comes not only from the errors in measurements but also from the unknown background noise in the environment of the scatterer. But without some assumptions about the noise nothing can be derived. One has

$$\begin{aligned} y(n) &= c_{1m_1} h^{m_1} n^{m_1} e^{-ia_1nh-b_1nh}(1+O(1/n)) + \epsilon(n) \\ &= c_{1m_1} h^{m_1} e^{-ia_1nh-b_1nh}\{1+O(1/n) \\ &\quad +\epsilon(n) c_{1m_1}^{-1}(hn)^{-m_1} e^{ia_1nh+b_1nh}\}. \end{aligned} \tag{G.3.11}$$

From (G.3.11), it follows that, regardless of the method used, the extraction of the complex poles k_j from noisy data is highly unstable and depends on the magnitude of $\alpha_n = O(1/n) + \epsilon(hn)^{-m_1} e^{b_1nh} c_{1m_1}^{-1}$. If there exists n such that $\alpha_n \ll 1$ (say $\alpha_n < 0.1$), then the pole $k_1 = a_1 - ib_1$ can be computed by formulas (G.3.5) and (G.3.6) in which $y(n)(\alpha_n)$ should take the place of $u(n)(O(1/n))$. Similar considerations hold for other poles. Since $b_1 > 0$, the factor e^{b_jnh} is growing as $n \to \infty$. Therefore, ϵ should be small in order that α_n be small and k_j could be computed. In this case, it is not advisable to take n too large because for large n the second term in α_n becomes large. Since the bound on $O(1/n)$ is not available, it is not worthwhile to compute the optimal n, but practically n should be taken as a value for which $(1/h)\ln|y(n)/y(n+1)|$ is stationary when one computes b_1, and for which $(1/h)\ln|z_j(n)/z_j(n+1)|$ is stationary when one computes b_{j+1}.

If the constant c_{1m_1} is small, then the second term in α_n is large unless ϵ is sufficiently small. Therefore, it is difficult to compute resonance with small Laurent coefficients (coupling coefficients) in front of the singular terms $(k-k_j)^{-m}$. All these arguments are very simple but they show clearly the nature of the difficulties for which noise is responsible and the limitations of any method of resonances extraction from noisy data.

We assumed that the scatterer was convex. This assumption implies the basic result: the validity of equation (G.3.2). In [83], it is proved that for the scatterer consisting of two strictly convex reflecting bodies (G.3.2) does not hold: these exist countably many poles k_j on some line $\operatorname{Im} k = \text{const}$. Therefore, one cannot order the poles by the rule $|\operatorname{Im} k_j| \leq |\operatorname{Im} k_{j+1}|$ in the case of two disjoint convex reflecting bodies. If the scatterer is just one strictly convex reflecting body, then it is known that $|\operatorname{Im} k_j| \to +\infty$ and (G.3.2) holds.

C. *A brief review of the existing methods for the resonances extraction*

The most popular method is Prony's method. One assumes that $u = u(t) = \sum_{j=1}^{N} c_j e^{s_j t}, s_j = -ik_j, c_j = \text{const}$. One observes $u(t)$ experimentally and wants to compute s_j and c_j. If the data is exact (there is no noise), then Prony's method consists in the following. Let $f_n = u(nh)$ where $h > 0$ is a fixed number, $e^{s_j h} = z_j$. Then $f_n = \sum_{j=1}^{N} c_j z_j^n, n \geq 0$. An obvious linear algebra argument shows that $\det A_{pq}^{(m)} = 0, m \geq 0, 0 \leq p, q \leq N$, where $A_{p0}^{(m)} = f_{p+m}, \quad A_{pq}^{(m)} = z_q^p, \quad 0 \leq p \leq N, \quad 1 \leq q \leq N$. Therefore,

$$0 = \sum_{p=0}^{N} f_{p+m} A_p, \quad m \geq 0, \tag{G.3.*}$$

where A_p are the cofactors corresponding to the elements f_{p+m} of the matrix $A_{pq}^{(m)}$. Notice that A_p do not depend on m. Write $N+1$ equations (G.3.*) taking $m = 0, 1, \ldots, N$, and find a nontrivial solution $(A_0, A_1, \ldots, A_N)$ to the $N+1$ simultaneous equations (G.3.*). Consider the equation

$$\sum_{p=0}^{N} A_p z^p = 0. \tag{G.3.**}$$

From the structure of the matrix $A_{pq}^{(0)}$, it is clear that equation (G.3.**) has solutions $z_j = e^{s_j h}$. Thus, $s_j = h^{-1} \ln z_j$. If one does not know the number N (and this is usually the case in practice), then there is a problem of choosing the right N. In [448], there is a method (due to Kühnen) for choosing N. If the data are noisy, then one faces the difficulties explained in Section G.3.B and reported in the literature. If the data are noisy, then the matrix $f_{p+m}, 0 \leq p, m \leq N$ is nonsingular and system (G.3.*) with $0 \leq m \leq N$ has only the trivial solution $A_p = 0$, which cannot be used since equation (G.3.**) in this case gives no information. Therefore, in practice, one takes $A_p, 0 \leq p \leq N$, the components of an eigenvector corresponding to the minimal eigenvalue of the matrix F^*F. Here, F^* is the adjoint matrix, and F is the matrix of the noisy data, $F_{p+m} = f_{p+m} + \epsilon_{p+m}$, where ϵ_{p+m} is noise. If there are several eigenvectors corresponding to the minimal eigenvalue, one has no rule to pick up any particular eigenvector. But this situation is not generic in the sense that a small perturbation of the matrix will split up the multiple eigenvalue into a number of simple ones. However, the simple eigenvalues will be close to each other and it will be difficult to find the minimal eigenvalue numerically. Recall that an eigenvalue is called simple if there is only one linearly independent eigenvector corresponding to this eigenvalue. One can find an extensive bibliography and a discussion of Prony's method in [448]. The Bruns' method described in [448] is essentially the Prony method for real resonances.

Let us outline another method for extracting the resonances. Let $f(t) = \sum_{j=1}^{N} c_j e^{s_j t}$ and N is assumed known. Then $f^{(m)}(t) = \sum_{j=1}^{N} c_j s_j^m e^{s_j t}$. Taking $t = 0$ and $m = 0, \ldots, N, \ldots$, one obtains

$$\begin{vmatrix} f(0) & 1 & \ldots & 1 \\ f'(0) & s_1 & & s_N \\ \vdots & \vdots & & \vdots \\ f^{(N)}(0) & s_1^N & & s_N^N \end{vmatrix} = 0, \quad \begin{vmatrix} f'(0) & 1 & 1 \\ \vdots & \vdots & \vdots \\ f^{(N+1)} & s_1^N & S_N^N \end{vmatrix} = 0, \ldots. \tag{G.3.12}$$

Therefore,

$$\sum_{p=0}^{N} f^{(m+p)}(0) A_p = 0, \quad m = 0, 1, 2, \ldots. \tag{G.3.13}$$

Here, A_p is the cofactor corresponding to the element $f^{(p)}(0)$ in the first matrix in (G.3.12). The argument above is very similar to that given in the previous method and $f^{(m+p)}(0)$ plays the role of f_{m+p} in (G.3.*) in that section. Taking $m = 0, 1, \ldots, N$ in (G.3.13), one obtains a homogeneous system of linear equations. If $A_0, \ldots, A_N$ is a solution to this system, then the N roots of equation (G.3.**) for this method are equal to $s_j, 1 \leq j \leq N$. This gives a method of extracting the resonances s_j from the exact transient data. It is interesting that only the data near $t = 0$ are used in this procedure, while the procedure in the previous method needs the data at large t. On the other hand, the procedure based on equation (G.3.13) is very sensitive to the noise in the data because one needs to differentiate the data.

We assumed above that the number N of resonances was known. If N is not known, then one can find N as the smallest number for which $\det f_{p+m} = 0, \quad 0 \leq p, m \leq N$.

The simple algorithm in Section G.3.B requires that $b_1 < b_2 < b_3 < \ldots$. In practice, the poles $a_j - ib_j$ occur in pairs $\pm a_j - ib_j$ and the measured transient field is a real-valued function. Assuming that the poles are simple, i.e., $m_j = 0$, one has

$$u(t) = \sum_{j=1}^{N} c_j \exp(-b_j t) \cos(a_j t + \phi_j) + O(e^{-b_{N+1} t})$$

as $t \to +\infty$. Therefore, for large t, one obtains $u(t) = c_1 \exp(-b_1 t) \cos(a_1 t + \phi_1) + O(e^{-b_2 t}), t \to +\infty$. If the values $u_n = u(nh)$ are measured, then the values c_1, b_1, a_1, and $\phi_1, 0 \leq \phi_1 < 2\pi$, can be numerically obtained from the requirement

$$F(c_1, b_1, a_1, \phi_1) = \frac{1}{m} \sum_{n=n_0}^{n_0+m} |c_1 \exp(-b_1 hn) \times \cos(a_1 hn + \phi_1) - u_n|^2 = \min . \tag{G.3.14}$$

Here, n_0 is a large number such that $\exp(-b_1 hn_0) \gg \exp(-b_2 hn_0)$, $m > 4$, is a fixed number, and the function $F(c_1, b_1, a_1, \phi_1)$ is to be minimized numerically. If this minimization problem is solved, one can consider $\omega_1 = u(t) - u_1(t)$, where $u_1(t) = c_1 \exp(-b_1 t) \cos(a_1 t + \phi_1)$, and apply the same procedure for finding c_2, b_2, a_2, ϕ_2. Each step requires minimization of a function of four variables only. The basic new idea in our method is to use the asymptotic behavior of the transient field as $t \to +\infty$. One should have in mind that the basic asymptotic SEM expansion (G.3.2) is proved only if the scatterer is convex (or, more generally, star shaped). It does not hold, for example, when the scatterer consists of two complex obstacles. In this case, there exists infinitely many poles k_j, such that $|k_j - ic_0 - \pi d^{-1} j| \leq c(1+|j|)^{-1/2}$ for all large $j, j = \pm j_0, \pm(j_0+1), \ldots$. Here, d is the distance between the two obstacles and c_0 depends on d, on the principal curvatures, and principal directions of the surfaces Γ_1 and Γ_2 of the two obstacles at the points $s_1 \in \Gamma_1$ and $s_2 \in \Gamma_2$, such that $|s_1 - s_2| = d$. This result is based on [83].

G.4. Bibliographical Remarks

First we mention some of the old papers. Of these, only Prony's paper is often mentioned by modern authors. Bruns used practically the same idea as Prony. His work is discussed in [448]. There are several authors, astronomers mostly, who were interested in detection of hidden periodicities. Although only the case $m_j = 0, b_j = 0$ was discussed in these papers, the basic questions (extracting the resonances from the transient field, determining the number N of resonances, etc.) were actually identical with the questions discussed in a review [448] of the state of art in this field.

There are many reasons for being interested in the extracting of resonances. We mention only two major theories: SEM and system identification.

G.5. Appendices

Appendix I: Perturbation of resonances

1. *Abstract scheme*

First, let us present an abstract scheme. Assume that a compact operator function $A(k)$ on a Hilbert space H is analytic in k in a domain on the complex plane k, and $A(k_0)$ has an eigenvalue -1. Then $(I + A(k))^{-1}$ has a pole at $k = k_0$. Suppose that $A(k, \epsilon)$ is a compact operator function such that $A(k, 0) = A(k)$, which is analytic in k and $\epsilon, |k - k_0| < \delta_0, |\epsilon| < \delta_1$. Assume that k_0 is an isolated pole of $I = A(k)$. This is the case if $I + A(k)$ is invertible for at least one k in the disk $|k - k_0| < \delta_0$. Then, in a neighborhood of k_0, there exist a finite number m_0 of points $k_j(\epsilon) = k_0 + \sum_{m=1}^{\infty} a_{mh}\epsilon^{m/p}, 1 \leq j \leq j_0$, such that the operator $(I + A(k, \epsilon))^{-1}$ has poles $k(\epsilon)$. Here, j_0 is the multiplicity of the pole k_0, and the meaning of the integer p will be explained in the proof which is based on the idea in [240], p. 582.

Let $\phi_1, \ldots, \phi_n$ be an orthonormal basis of $N(I + A(k))$, where $N(A)$ is the null space of an operator A. Let $\psi_1, \ldots, \psi_n$ be an orthonormal basis of $N(I + A^*(k_0))$, where the star denotes the adjoint operator. Let $Th = \sum_{j=1}^{n}(h, \phi_j)\psi_j$. The operator $I + A(k_0) + T$ is invertible in H. Indeed, $(I + A(k_0) + T)h = 0$ implies that $(Th, \psi_j) = 0, 1 \leq j \leq n$. This leads to $(h, \phi_j) = 0, 1 \leq j \leq n$, i.e., $Th = 0$ and $(I + A(k_0))h = 0$. Thus, $h \in N(I + A(k_0))$ and $h \perp N(I + A(k_0))$. Therefore, $h = 0$, and by Fredholm's alternative, $(I + A(k_0) + T)^{-1} = \Gamma$ is bounded. Consider $(I + A(k, \epsilon))^{-1} = (I + A(k_0) + T + A(k, \epsilon) - A(k_0) - T)^{-1} = (I + a(k, \epsilon))^{-1}$. Then $\Gamma(k, \epsilon)$, where $\Gamma(k, \epsilon) = (I + A(k_0) + T + A(k, \epsilon) - A(k_0))^{-1}$, is analytic in k and ϵ in a neighborhood Δ of $(k_0, 0)$, and $a(k, \epsilon) = -\Gamma(k, \epsilon)T$ is a finite-dimensional operator analytic in k and ϵ in $\delta, ah = -\sum_{j=1}^{n}(h, \phi_j)\Gamma(k, \epsilon)\psi_j$. Since $\Gamma(k, \epsilon)$ is an isomorphism of H onto H for $k, \epsilon \in \Delta$, the elements $\psi_j(k, \epsilon) = -\Gamma(k, \epsilon)\psi_j$ are linearly independent and analytic in $k, \epsilon \in \Delta$. Therefore, the operator $(I + a(k, \epsilon))^{-1}$ can be constructed explicitly. If $(I + a)h = f$, then $h + \sum_{j=1}^{n} h_j\psi_j(k, \epsilon) = f, h_j = (h, \phi_j)$. Multiply by ϕ_m to obtain $h_m + \sum_{j=1}^{n} c_{mj}h_j = f_m$, where $f_m = (f, \phi_m), c_{mj} = (\psi_j(k, \epsilon), \phi_m)$, c_{mj} are analytic in $k, \epsilon \in \Delta$. Thus, $h_m = d_m(k, \epsilon)/d(k, \epsilon)$, where d_m and $d = \det(\delta_{mj} + c_{mj})$ are analytic in Δ. One has $(I + a(k, \epsilon))^{-1}f = f - \frac{1}{d}\sum_{j=1}^{n} d_j(k, \epsilon)\psi_j(k, \epsilon)$. From this formula, it is clear that the poles of $(I + a(k, \epsilon))^{-1}$ can occur only at the zeros of $d(k, \epsilon)$. Thus, the equation for the perturbed poles is

$$\begin{aligned} d(k,\epsilon) &= \det[\delta_{jm} - (\Gamma(k,\epsilon)\psi_j, \phi_m] = 0, \\ \Gamma(k,\epsilon) &= (I + T + A(k,\epsilon))^{-1}. \end{aligned} \tag{G.5.1}$$

For $\epsilon = 0$, the function $d(k) = d(k, 0)$ has, by assumption, a zero of multiplicity j_0. By the Weierstrass' preparation theorem (see [240], p. 583), one has $d(k,\epsilon) = [k^{j_0} + \sum_{j=1}^{j_0-1} c_j(\epsilon)k^j]g(k,\epsilon)$, where $g(0,0) \neq 0, c_j(0) = 0$, and c_j and $g(k,\epsilon)$ are holomorphic functions. Therefore, equation (G.5.1) is equivalent to

$$k^{j_0} + \sum_{j=1}^{j_0-1} c_j(\epsilon)k^j = 0. \tag{G.5.2}$$

This equation has j_0 roots. These roots can be divided into several groups so that the p roots $(k_1(\epsilon), \ldots, k_p(\epsilon))$ in the νth group can be expanded in a Puiseux series $k_j(\epsilon) = k_0 + \sum_{m=1}^{\infty} a_{mj}\epsilon^{m/p\nu}$, $\sum_\nu p_\nu = j_0$. The number of groups and the integers p_ν can be computed by the algorithm known as Newton's diagram method (see [465]). Let us summarize our arguments: a method for computing the poles $k_j(\epsilon)$ of the perturbed operator $(I + A(k,\epsilon))^{-1}$ is given. The method is valid under the following assumptions: $A(k,\epsilon)$ is a compact analytic (in k and ϵ) linear operator function on a Hilbert space, the operator $(I + A(k,0))^{-1}$ has an isolated pole at $k = k_0$. To use the method computationally, one needs to compute (or to know) the bases in the subspaces $N(I + A(k_0))$ and $N(I + A^*(k_0))$. This is a linear algebra problem.

2. *Reduction of a concrete perturbation problem to the abstract one*

Suppose that the surface Γ of the obstacle is perturbed. Let $x_j = x_j(u,v), 1 \leq j \leq 3$ be a parametric equation of Γ, and $z_j = x_j(u,v) + \epsilon y_j(u,v)$ be the equation of the perturbed surface Γ_ϵ, where ϵ is a small parameter. Assume that the functions x_j and $y_j, 1 \leq j \leq 3, u,v \in S = \{u,v : 0 \leq u,v \leq 1\}$ are smooth. Consider equation (G.2.8). Assume that k_0 is a pole of the operator $(I + A(k))^{-1}$. Suppose that the bases of the subspaces $N(I + A(k_0))$ and $N(I + A^*(k_0))$ are computed. Consider the problem corresponding to the perturbed surface Γ_ϵ. The operator $A(k,\epsilon)$, associated with this problem, is of the form

$$\int_{\Gamma_\epsilon} 2\frac{\partial g}{\partial N_s} h\, ds = \int_S A(u,v,u',v',k,\epsilon) h\, du'\, dv'.$$

Here, $A(u,v,u',v',k,\epsilon)$ is the kernel of the integral operator in the variables u,v,u',v'. If $\epsilon = 0$, then $\Gamma_\epsilon = \Gamma$ and we assume that the sets $\{\phi_j\}$ and respectively $\{\psi_j\}, 1 \le j \le n$ of all linearly independent solutions of the equation $\phi + \int_S A(u,v,u',v',k_0,0)\phi du'\, dv' = 0$ and respectively $(\psi + \int_S \overline{A(u',v',u,v,k_0,0)}\psi du'\, dv' = 0)$ are known. Then the abstract scheme is applicable.

Since small perturbations of the kernel cause small perturbations of the poles, one can approximate the kernel A by a degenerate kernel and consider the corresponding matrix problem.

As an example, consider a simple case when the matrix is 2×2. Let

$$\begin{pmatrix} 1 & e^{\pi k} \\ e^{\pi k} & 1 \end{pmatrix} \begin{pmatrix} h_1 \\ h_2 \end{pmatrix} = \begin{pmatrix} f_1 \\ f_2 \end{pmatrix}.$$

Then the inverse matrix is

$$\begin{vmatrix} 1 & -e^{\pi k} \\ -e^{\pi k} & 1 \end{vmatrix} \frac{1}{1 - e^{2\pi k}}.$$

It has simple poles $k_m = im, m = 0, \pm 1, \ldots$. Consider, for example, pole $k_0 = 0$. The set $\{\phi_j\}$ corresponding to this pole is the set of linearly independent solutions to the equation

$$\begin{cases} c_1 + c_2 = 0, \\ c_1 + c_2 = 0. \end{cases}$$

Thus, there exists one linearly independent solution

$$\phi_1 = \begin{pmatrix} -1 \\ 1 \end{pmatrix}.$$

In our example, the matrix

$$\begin{pmatrix} 1 & e^{\pi k} \\ e^{\pi k} & 1 \end{pmatrix}$$

is self-adjoint for $k = 0$. Thus,

$$\psi_1 = \phi_1 = \begin{pmatrix} -1 \\ 1 \end{pmatrix}.$$

Consider the perturbed matrix

$$\begin{pmatrix} 1 + \epsilon a_{11} & e^{\pi k} + \epsilon a_{12} \\ e^{\pi k} + \epsilon a_{21} & 1 + \epsilon a_{22} \end{pmatrix}.$$

The poles of this matrix are the roots of the equation $(1 + \epsilon a_{11})(1 + \epsilon a_{22}) - (e^{\pi k} + \epsilon a_{12})(e^{\pi k} + \epsilon a_{21}) = 0$. We are interested in the root $k(\epsilon)$ such that $k(\epsilon) \to 0$ as $\epsilon \to 0$. One has $1 - e^{2\pi k} + \epsilon[a_{11} + a_{22} - e^{\pi k}(a_{12} + a_{21})] + \epsilon^2(a_{11}a_{22} - a_{12}a_{21}) = 0$. Let $e^{2\pi k} = z, a_{11} + a_{22} = a, a_{12} + a_{21} = b, a_{11}a_{22} - a_{12}a_{21} = c$. Then

$$\begin{aligned} z^2 + z\epsilon b - 1 - \epsilon a - \epsilon^2 c &= 0, \\ z(\epsilon) &= -\epsilon b/2 + \sqrt{(\epsilon^2 b^2/4) + 1 + \epsilon a + \epsilon^2 c} \\ &\approx 1 + (\epsilon/2)(a - b). \end{aligned}$$

The plus sign in front of the radical is chosen because $z = 1$ if $k = 0$. Thus, in this example, the perturbation of the pole $k_0 = 0$ can be computed for small ϵ as $k(\epsilon) = \pi^{-1} \ln z(\epsilon) = (a - b)\epsilon/2\pi$. Depending on the values of a and b the perturbed pole can move in any direction. If $a \neq b$, then $k(\epsilon) = O(\epsilon)$. If $a = b$, then $k(\epsilon) = O(\epsilon^2)$. The bifurcation theory and the Newton diagram method solve the following problem. Given an equation $F(k, \epsilon) = 0$, find its solutions $k(\epsilon)$ such that $k(\epsilon) \to 0$ as $\epsilon \to 0$. It is assumed that $F(0, 0) = 0$. If $F_k(0, 0) \neq 0$, then the solution is well known and is given by the standard implicit function theorem. If $F_k(0, 0) = 0$, then the solution is more complicated. Methods and algorithms for solving this problem can be found in the book [465].

Appendix II: Extraction of resonances from the transient field

The case of simple poles symmetrical with respect to the imaginary axis is discussed here.

Let the poles be simple and occur in pairs $\pm a_j - ib_j, 0 < b_1 < b_2 < \dots$. Then the transient field is $u = \sum_{j=1}^{N} u_j + O(\exp(-b_{N+1}t))$, as $t \to +\infty, u_j = c_j \exp(-b_j t)\cos(a_j t + \phi_j), c_j > 0, a_j > 0, b_j > 0, 0 \leq \phi_j < 2\pi$. We sketch a method for finding c_j, b_j, a_j, and ϕ_j, which is simpler than the one given by formula (G.3.14).

Step 1: One has $u(t) = u_1 + O(\exp(-b_2 t))$, so $u(t) \approx u_1(t)$ as $t \gg 1$. Thus,

$$\frac{1}{mh} \ln \frac{u[(n+m)h]}{u(nh)} \approx -b_1 + \frac{1}{mh} \ln \frac{\cos(a_1 nh + a_1 mh + \phi_1)}{\cos(a_1 nh + \phi_1)} \cdot h > 0. \quad \text{(G.5.3)}$$

Therefore for $m \gg 1$ one can find b_1 from (G.5.3). Some caution is needed: if $\cos(a_1 nh + a_1 mh + \phi_1) \approx 0$ then the ln term is large. But this can happen rarely. One can compute the left side of (G.5.3) for a number of consecutive m and neglect large values of this quantity.

Step 2: If b_1 is found, then a_1 can be found by the formula $a_1 \approx (-\ddot{v}/v)^{1/2}, t \gg 1$, where $v(t) = u(t)\exp(b_1 t)$, and $\ddot{v} = d^2v/dt^2$.

Step 3: If a_1 and b_1 are found, then $c_1 \approx (\dot{v}^2/a_1^2 + v^2)^{1/2}, \quad t \gg 1$.

Step 4: If a_1, b_1, and c_1 are found, then ϕ_1 can be found from the equations

$$c_1^{-1} v(2\pi n/a_1) = \cos\phi_1, \; -c_1^{-1} v(2\pi n/a_1 + \pi/2a_1) = \sin\phi_1, \quad n \gg 1.$$

Since $0 \leq \phi_1 < 2\pi$, these equations determine ϕ_1 uniquely. If b_1, a_1, c_1, and ϕ_1 are found, one can use $u - u_1$ for finding u_2, etc. The basic idea is the same as in Section G.3. The difficulties are similar to the ones discussed after formula (G.3.11).

Appendix III: Existence of infinitely many purely imaginary resonances

If $-ib, b > 0$, is a resonance, i.e., a complex pole of Green's function of the Dirichlet Laplacian in Ω, then the equation

$$q(b)f = \int_\Gamma \frac{\exp(b|s-t|)}{|s-t|} f(t)dt = 0 \quad \text{(G.5.4)}$$

has a nontrivial solution and vice versa. In order to prove that there are infinitely many purely imaginary resonances, it is sufficient to prove that equation (G.5.4) has nontrivial solutions for infinitely many values of $b, b = b_n, b_n \to +\infty$. The operator $q(b)$ in (G.5.4) is compact in $H = L^2(\Gamma)$, self-adjoint for real-valued b, and analytic in b. Thus, its eigenvalues $\lambda_n(b)$ are analytic in b about the real axis. If $b \leq 0$, then $q(b)$ is positive definite in H, and $\lambda_n(b) > 0$. If we show that there are $N(b) \to +\infty$, as $b \to +\infty$, negative eigenvalues of $q(b)$, then there are $N(b)$ points b_n at which $\lambda_n(b_n) = 0$, i.e., $q(b_n)f = 0, f \neq 0$.

The number of negative eigenvalues of $q(b)$ is

$$N(b) = \sup_M \dim M, \quad M = \{f : q(b)f, f) < 0\} \tag{G.5.5}$$

(see Lemma B.0.1 in Appendix B). Let M be a linear span of the functions $h_j(x) \in H, h_j(x) = 1$ if $x \in \Delta_j, h_j(x) = -1$ if $x \in \Delta_{-j}, h_j(x) = 0$ otherwise. Here, Δ_j is a small neighborhood of the points $s_{\pm j} \in \Gamma, j = 1, 2, \ldots, N, \Delta_j \cap \Delta_i = \emptyset$ if $j \neq i$. The points $s_{\pm j}$ are chosen so that

$$\max_{i \neq -j} |s_i - s_j| < \min_{1 \leq j \leq N} |s_j - s_{-j}|. \tag{G.5.6}$$

Let a be the area of Δ_j and $a^{1/2}$ be the linear dimension of Δ_j. One has (the bar denotes complex conjugate)

$$\left(q(b) \left(\sum_{j=1}^N c_j h_j \right), \sum_{m=1}^N c_m h_m \right) = \sum_{j,m=1}^N q_{jm}(b) \bar{c}_m c_j, \quad q_{jm} = \bar{q}_{mj}, \tag{G.5.7}$$

where

$$\begin{aligned} q_{jm}(b) &= \int_\Gamma \int_\Gamma \frac{\exp(b|s-t|)}{|s-t|} h_j(t) h_m(s) dt\, ds \\ &= \left(\frac{\exp(b|s_j - s_m|)}{|s_j - s_m|} - \frac{\exp(b|s_j - s_{-m}|)}{|s_j - s_{-m}|} \right. \\ &\quad \left. - \frac{\exp(b|s_{-j} - s_m|)}{|s_{-j} - s_m|} + \frac{\exp(b|s_{-j} - s_{-m}|)}{|s_{-j} - s_{-m}|} \right) \end{aligned}$$

$$\times a^2(1+o(1)), \quad a \to 0, \quad j \neq m;$$

$$q_{jj}(b) = -2\frac{\exp(b|s_j - s_{-j}|)}{|s_j - s_{-j}|}a^2(1+o(1)) + O(a^{3/2}\exp(ba^{1/2})), \quad a \to 0. \tag{G.5.8}$$

It follows from (G.5.6)–(G.5.8) that

$$q_{jj}(b) < 0, \text{ and } |q_{jj}| > \sum_{i \neq j, i=1}^{N} |q_{jj}(b)|, \text{ as } b \to +\infty. \tag{G.5.9}$$

Inequalities (G.5.9) imply that the matrix $q_{jm}(b), 1 \leq j, m \leq N$, is negative definite if b is sufficiently large positive. This proves the following theorem.

Theorem G.5.1. *There exists infinitely many purely imaginary poles $k_n = -ib_n, 0 < b_n \to +\infty$, of $G(x,y,k)$, the resolvent kernel of the Dirichlet Laplacian in Ω.*

One can prove similarly [255] that the resolvent kernel of the Schrödinger operator in $L^2(\mathbb{R}^3)$ with a real-valued compactly supported potential $V(x) \in L^2(\mathbb{R}^3), V(x) \geq 0$ has infinitely many poles $k_n = -ib_n, 0 < b_n \to +\infty$.

Appendix IV: Bound on the location of the resonances

Let $V(x) \in L^2(\mathbb{R}^3), V(x) = 0$ if $|x| \geq R$ where $R > 0$ is an arbitrary large fixed number. Then the resolvent kernel $G(x,y,k)$ of the Schrödinger operator $-\nabla^2 + V(x)$, i.e., the kernel of $(-\nabla^2 + V(x) - k^2)^{-1}$, is meromorphic in k on the whole complex k-plane. If $V = \bar{V}$, then the poles of $G(x,y,k)$ are symmetric with respect to the imaginary axis, and if, in addition, $V \geq -c$, where $c = \text{const}$, then there exist infinitely many poles $k_n = -ib_n, b_n > 0, b_n \to +\infty$. In the half-plane $\operatorname{Im} k > 0$, there may be only a finite number of poles $k_n = -iE_n^{1/2}, E_n > 0$ is the energy of the nth bound state. The point $k = 0$ may or may not be a pole of G.

The aim of this appendix is to prove the following theorem.

Theorem G.5.2. *If* $\nabla V \in L^1(\mathbb{R}^3), V = 0$ *for* $|x| \geq R$, *then* $G(x, y, k)$ *has no poles in the region*

$$\Delta(a, b, d) = \{-a \ln |k| - b < \operatorname{Im} k < 0,\ |\operatorname{Re} k| > 2\} \\ \cup \{-d < \operatorname{Im} k < 0,\ |\operatorname{Re} k| \leq 2\},$$

where a, b, d *are positive constants.*

Proof. There are no poles in the region where the equation

$$u = -\int_{|y| \leq R} \frac{\exp(ik|x - y|)}{4\pi|x - y|} V(y) u \, dy = T(k)u$$

has only the trivial solution. Thus, there are no poles in the region where

$$\|T^2(k)\| \leq \max_{|y| \leq R, |x| \leq R} \left| \int_{|z| \leq R} \frac{\exp\{ik(|x - z| + |z - y|)\}}{16\pi^2 |x - z| \, |z - y|} \right. \\ \left. \cdot V(z) dz \right| \int_{|y| \leq R} |V(y)| dy < 1. \tag{G.5.10}$$

Here, the operator norm is taken in $C(B_R), B_R := \{x : |x| \leq R\}$.

If we prove that

$$A(x, y, k) = \left| \int_{|z| \leq R} \frac{\exp\{ik(|x - z| + |z - y|)\}}{|x - z| \, |z - y|} V(z) dz \right| \to 0, \tag{G.5.11}$$

where

$$|k| \to \infty, \quad k \in \Delta(a, b, d),$$

then (G.5.10) holds in a region $\Delta(a, b, d)$ with a sufficiently small d, b, and a.

To prove (G.5.11), one uses the coordinates s, t, ψ defined by the formulas

$$z_1 = lst + \frac{x_1 + y_1}{2}, \quad z_2 = l\sqrt{(s^2-1)(1-t^2)}\cos\psi + \frac{x_2 + y_2}{2},$$
$$z_3 = l\sqrt{(s^2-1)(1-t^2)}\sin\psi + \frac{x_3 + y_3}{2},$$

where $l = |x-y|/2$ and $|x-z|+|z-y| = 2ls$, $|x-z|-|z-y| = 2lt$, and $J = l^3(s^2-t^2)$ is the Jacobian of the transformation $(z_1, z_2, z_3) \to (s, t, \psi)$. In the new coordinates, one has

$$A(x, y, k) = l\int_0^{2\pi} d\psi \int_{-1}^{1} dt \int_1^{\infty} \exp(2ikls)q(s, t, \psi)ds,$$

where $q(s, t, \psi)$ is $V(z)$ in the new coordinates. Define

$$p(s) = \int_0^{2\pi} d\psi \int_{-1}^{1} dt\, q(s, t, \psi).$$

Then

$$A(x, y, k) = l\int_0^{\infty} \exp(2ikls)p(s)ds. \qquad \text{(G.5.12)}$$

The function $p(s)$ is compactly supported and differentiable since $v(z)$ is. Thus,

$$A(x, y, k) = -\frac{\exp(2ikl)}{2ik}p(1) - \frac{1}{2ik}\int_1^{\infty} \exp(2ikls)p'(s)ds. \quad \text{(G.5.13)}$$

Let $k = k_1 - ik_2, k_2 > 0$. Then

$$|A(x, y, k)| \leq c\frac{\exp(2k_2 l) + \exp(clk_2)}{(k_1^2 + k_2^2)^{1/2}}, \qquad \text{(G.5.14)}$$

where $c > 0$ denotes various constants which do not depend on k, x, and y. It follows from (G.5.14) that, uniformly in x and y running through any fixed compact set, the estimate holds for $|k_1| \to \infty$:

$$|A(x, y, k)| \to 0 \quad \text{if } k_2 < c_1 \ln |k_1|, \quad |k_1| \to \infty,$$

and

$$c_1 < \max\left(\frac{1}{2l}, \frac{1}{cl}\right). \tag{G.5.15}$$

From (G.5.15) and (G.5.10), the conclusion of the theorem follows for sufficiently small a, b, and d.

This argument is taken from [194]. □

Bibliographical Notes

Chapter 1. Wave scattering by obstacles is very important in applications and from the mathematical point of view. Therefore, the literature of the subject is enormous. Integral equation methods were used by Kupradze [110] in 1934 and in his later works, Vekua [467] in 1943, by Werner [474], Leis [126], Müller [159], Ramm [269], and many others. The properties of the potentials of single and double layers are well known for domains with Lyapunov boundaries. Many of these properties were established by Lyapunov (see [79]). For nonsmooth boundaries, potential theory in $\mathbb{R}^2$ was developed by J. Radon [179] in 1920, and in $\mathbb{R}^n$ by Burago, Maz'ya and Sapozhnikova in 1962–1966 (see [30]), by Kral [104], and by Fabes *et al.* [61] (see also [107]). Existence and uniqueness of the solutions to exterior boundary value problems were established by Kupradze [110] in 1934, and Vekua [467] in 1943. Kupradze used the classical approach, but modified the integral equations for the exceptional values of k for which the adjoint homogeneous integral equation has nontrivial solutions. He obtains the solution of the exterior boundary value problems at the exceptional values of k as a sum of the single- and double-layer potentials (see also [478]). His analysis is complicated. Brakhage and Werner [29], Leis [126], and Panich [174] used representation (1.1.15) for the solution of the exterior Dirichlet problem. One can look for a solution of the exterior Neumann problem in this form but the resulting equation is complicated because the normal derivative of the double-layer potential is an unbounded operator. The radiation condition was introduced in [447], in 1912. Until 1943, this condition was used in the form $r(v_r - ikv) = o(1), v = O(r^{-1}), r \to \infty$. In 1943, Vekua [467] proved that the second equation, $v = O(r^{-1})$, is redundant, and that both equations follow from the radiation condition of the form $\int_{S_r} |v_r - ikv|^2 ds \to 0, r \to \infty$.

Werner [474] used representation (1.1.20) and proved existence of the solution to the exterior boundary value problems. For computational purposes, the two-dimensional boundary integral equations are used mostly. A review of the boundary integral equations is presented in [103]. Numerical methods based on the boundary integral equations were suggested by Greenspan and Werner [76], Burton and Miller [31], Kussmaul [115], Terai [457], and others.

The short proof of the existence of solutions to the exterior boundary value problems in Lemma 1.1.3 is given by the author. The other short proof, in Section 1.1.3, is taken from [187]. The existence proof in Section 1.1.6 was discussed by Ursell [461]. The existence proof in Section 1.1.7 was suggested in [106]. Lemma 1.2.1 was proved in [68, 110, 467]. The second proof is given in [269]. The result of Lemma 1.2.4 is given in [131]. Our argument is taken from [125]. The result in Section 1.2.6 is taken from [269]. The argument in Section 1.2.8 is similar to the one in [121]. The argument in Section 1.2.9 is standard. Lemma 1.2.11 is taken from [193]. In Section 1.2.11, we followed [269]. The limiting absorption principle (LAP) was first established by the argument used in the proof of Theorem 1.2.3 in [56]. Similar arguments were used afterwards by many authors (see [435]) in the study of the general second order elliptic operators. This argument has been known for a long time in the Fredholm–Riesz theory. The operator $(-\nabla^2 - k^2 - i0)^{-1}$, where $-\nabla^2 = -\Delta_D$, is unbounded in $L^2(\Omega)$, but is bounded from $L^2(\Omega, p(x))$ into $L^2(\Omega, p^{-1}(x))$ with a suitable weight $p(x)$, e.g., $p(x) = (1 + |x|)^a, a > 1$.

The solution to the problem $(-\nabla^2 - k^2)v = f$ in Ω, v satisfies conditions (0.0.2) and (0.0.5), where $f \in L^2_0(\Omega)$, can be obtained as follows:

$v = \lim_{n\to\infty} v_n$, where v_n solves the problem $[-\nabla^2 - k^2 + q_n(x)]v_n = 0$ in $\mathbb{R}^3$, v_n satisfies (0.0.5),

$$q_n(x) = \begin{cases} 0 & \text{in } \Omega, \\ n & \text{in } D,\ n > 0. \end{cases}$$

The limit

$$\lim_{n\to\infty} v_n = \begin{cases} 0 & \text{in } D, \\ v & \text{in } \Omega \end{cases}$$

is attained in the norm $L^2(\mathbb{R}^3, p^{-1}(x))$. This result was proved first in [189] and later in [121]. In [269], a generalization to the case of interface problems is given. Wilcox [480] used the LAP for a proof of the existence of solutions to the scattering problem with the homogeneous Neumann boundary condition. This approach allows one to assume less about the smoothness of the boundary than the integral equation approach. In particular, in [480], a certain condition on Γ called the local compactness property of Γ, is used. This condition is sufficient for the proof of the existence result in L^2_{loc} (see Appendix A, Section 7). Necessary and sufficient conditions on the boundary for the existence of solutions to elliptic equations of the second (and higher) order for various boundary conditions, including the Neumann condition, in L^p and $W^{p,1}$ classes were given in [142, 143] (see also [338, 342, 361, 444], and Appendix A). The arguments in [480] depend crucially on the assumption that the boundary condition is homogeneous. If it is not, then some additional smoothness are required from Γ so that the embedding theorems, in particular, the trace theorems, be valid. The presentation of the eigenfunction expansion follows that in [182, 183, 187]. The first proof along these lines was given in [178]. A different proof can be found in [85] (see also [442, 443, 480]). The properties of the scattering amplitude given in Theorem 1.4.1 are well known [269]. The properties (1.4.5) and (1.4.6) are called "optical theorem" (see [165] for historical remarks). The proof of the meromorphic character of the scattering amplitude in k-variable was first given in [184]. Lemma 1.4.2 is taken from [269]. A weaker lemma was used in the proof of the meromorphicity in [199]. The meromorphic nature of the scattering amplitude for the Schrödinger operator in $\mathbb{R}^3$ was proved independently in [450] by means of a lemma similar to that in [199, p. 462]. (see also [52, 71, 428, 432, 452, p. 107]). Lemma 1.4.5 is the classical Schmidt lemma [465]. It was used in [269] for the proof of the meromorphic dependence of the scattering amplitude on k (see also [240]). Another method of proving the meromorphicity of the scattering amplitude is given in [120]. The argument in Section 1.4.3 is taken from [259]. The material in Section 1.5 is standard (see [114]). Theorem 1.6.1 is proved in [4]. We follow the argument in [4]. Inequality (1.6.30), a maximum principle for solutions to Helmholtz's equation, is proved in [269]. The results and proofs in Section 1.7 are taken from [269]. Boundary value problems for pseudodifferential elliptic equations are studied in [59].

There is an extensive literature on the high-frequency asymptotics of solutions to Helmholtz's equation, on justification of the geometrical optics [4, 5, 12, 32, 66, 67, 78, 81, 135, 136, 157, 158, 176, 204, 269, 462]. We do not discuss wave scattering by moving bodies (see [42] and references therein).

Chapter 2. The inverse diffraction problem is of great practical and theoretical interest and therefore the literature of the subject is extensive. By the diffraction problem, we mean the scattering problem for a bounded obstacle. The basic questions to consider are: (1) uniqueness of the solution to the inverse problem and (2) actual reconstruction of the obstacle from the data. In the uniqueness problem, the basic result is due to M. Schiffer, who did not publish it. His argument is used in the proof of Lemma 2.1.1. The scattering data in this lemma is three-dimensional and the object to recover, the surface in $\mathbb{R}^3$, is two-dimensional. In practice, the data is often given at a fixed frequency. Theorem 2.1.2, which handles this case, is due to the author. The results of Section 2.1.3 are new. These results belong to the author [420]. No results of this nature were known. In [39], a uniqueness theorem is formulated which says that an acoustically soft connected obstacle, which is situated in a known ball B_R of known radius R, is uniquely determined by the scattering data $f(n_j, \nu, k_0)$ where $k_0 > 0$ is fixed, $\nu \in S^2$ and n_j are distinct unit vectors, $j = 1, 2, \ldots, m$, and $m = m(R)$ is the number of eigenvalues of the Dirichlet Laplacian in D which do not exceed k_0^2. The argument in this book is not satisfactory since the authors assume that the other obstacle with the same scattering data lies also in B_R, which is actually an additional assumption. The initial assumption that D lies in the known ball is also not satisfactory. There is an error in the proof of Theorem 2.2 in [39], and in formulas (3.1) and (3.6). It is an open problem if the above data determine the obstacle uniquely or not. Engineers assume implicitly that the above two-dimensional data determine the obstacle uniquely and tried to suggest some numerical procedures for actual reconstruction of the obstacle (see [87]). Some reviews on the inverse scattering by obstacles are presented in [38, 446]. Practically the high frequency data are used for inversion. The first inversion of the high frequency back-scattering data was based on the Kirchhoff approximation. This procedure allows one to compute the Fourier transform of the characteristic function of D. The three-dimensional Fourier inversion gives the characteristic function of D (see [25, 127]). Another idea, practically more useful, was given in [269] (see also [204, 226]). In this book, the support function of the obstacle is calculated from the scattering data $f(n, \nu, k_0)$, where k_0 is fixed and large enough so that the Kirchhoff approximation is applicable, and $n, \nu \in S^2$ run through such subsets of S^2 that the vector $l = (n - \nu)|n - \nu|^{-1}$ runs through all of S^2. This solution allows one to estimate the error of the reconstruction procedure even in the case of noisy data. The results and the arguments in Section 2.2.3 are taken from [269]. Lemma 2.2.6 is from [66]. In all of the above papers, it is assumed that the obstacle is convex. In [135, 136], some of the formulas from [204] were rediscovered and proved. In [145], the Kirchhoff approximation is justified as an asymptotic high-frequency approximation for the solutions of diffraction problems. It was known long ago [67] that the scattering data $f(n, -n, k_0)$, k_0 large fixed, determines the Gaussian curvature of the convex obstacle. The classical Minkowski result says that the smooth convex compact surface is uniquely determined by its Gaussian curvature given as a function of a unit normal. For existence results, see [167], where sufficient conditions for a given function of a unit normal to be the Gaussian curvature of a convex smooth compact surface are formulated. The solution to Minkowski's problem is not constructive: the problem of finding a convex compact smooth surface from its Gaussian curvature is reduced to a nonlinear elliptic boundary value problem. Lemma 2.2.3 can be found, e.g., in [23]. The stable differentiation procedure is 2.2.3 which is taken from [197] (see also [356] and Appendix D). There are very few results on the high-frequency asymptotics of the solutions to the exterior problems for nonconvex obstacles (see [7, 176]). The inverse scattering problem for a class of nonconvex obstacles was solved in [7]. We discuss the inverse scattering problem for potentials in Part II of this book. In [356, pp. 247–252], a critical analysis of the published inversion methods for inverse scattering problem for obstacle is given.

Chapter 3. Behavior of the solutions to the time-dependent problems for large t was discussed in many papers, often in the connection with the limiting amplitude principle and exponential decay of solutions to the wave equation in exterior domains [54, 123, 158]. The results and arguments in Chapter 3 are taken from [269]. The presentation in Section 3.2.5 is based on [269]. It is influenced by the work of [54]. There are some minor inaccuracies in this paper, which resulted in different constants in the asymptotics of $I(t, a)$ (see formula (1.31) in [54] and (3.2.82)). The argument and results in Section 3.2.6 are essentially from [269]. In [10], there is an example of the solution to the exterior Robin problem for the wave equation which does not decay as $t \to \infty$. The arguments in Section 3.4.1 are closely connected with these in Section 3.2.4. Similar arguments can be found in [439]. The singularity expansion method (SEM) associates the analytic properties of the resolvent kernel of a Laplacian on the complex k-plane with the behavior of the solutions to the corresponding time-dependent problem as $t \to \infty$. The resolvent kernel of the Dirichlet Laplacian in $\mathbb{R}^3$ is meromorphic on the k-plane, and in $\mathbb{R}^2$ it is meromorphic on the k-plane with the cut along the negative imaginary semi-axis (or on the Riemannian surface of $\ln k$). This leads to different behavior of the solutions to the exterior time-dependent problems in $\mathbb{R}^2$ and $\mathbb{R}^3$. Indeed, in $\mathbb{R}^3$ for star-shaped (or, more generally, nontrapping) obstacles the solution to problem (0.0.11) decays exponentially as $t \to \infty$. In $\mathbb{R}^2$, the solution decays not faster than $O(t^{-1}), t \to \infty$, in general [269]. In Ramm [188], the diffusion of waves in $\mathbb{R}^2$ was explained by the presence of the logarithmic branch point in the analytic continuation of the resolvent kernel in k-plane as well as the order $O(t^{-1})$ of the decay.

It is not easy to prove exponential decay of the solution to (0.0.11). In [123, 144, 155, 158], the exponential decay was established for the Dirichlet boundary condition for $L = -\nabla^2$. In [144], it is stated that the same holds for the Neumann boundary condition. It is easier to prove such results when $L = -\nabla^2 + q(x), q(x) \in C_0(\mathbb{R}^3)$ (or $L_0^2(\mathbb{R}^3)$) and the wave equation is considered in all of $\mathbb{R}^3$. In [194], it was proved that the poles of the analytic continuation of the resolvent kernel of the Schrödinger operator L lie under a logarithmic curve on the k-plane, $k_2 \leq -a \ln |k| - b$ where a and b are some positive constants, $k = k_1 + ik_2$, and there are no poles in some strip $-b_0 < k_2 < 0, b_0 > 0$. This bound on the location of the poles is sharp: for some spherically symmetric potentials the bound is attained. In [203], the exponential decay was proved for the Dirichlet Schrödinger operator in the exterior of a star-shaped obstacle. The argument in this book depends on the result in [123].

The eigenmode expansion method (EEM) consists of solving equation (3.5.4) using the root function expansion. For the self-adjoint operator B, this method goes back to Picard (1881) (see [74]). If B is non self-adjoint, then the problem is very difficult in general: (1) B might be quasinilpotent (e.g., Volterra operator) and then it has no eigenvalues, (2) the closure of the linear span of the eigenvectors (or root vectors) of B may not coincide with the whole space H, (3) if it does, the root system of B, i.e., the set of all linearly independent root vectors of B, may not form a basis of H. There is a considerable interest in applications to EEM (see [17, 208, 212, 216, 230, 238, 240, 470]). At this time, there is an extensive bibliography of the engineering work on EEM and SEM, hundreds of references can be found in the special issue of the journal "Electromagnetics" 1, N4, (1981). Engineers (see the above references) believed that the complex poles of the analytic continuation of the resolvent kernel of $L = -\nabla^2$ are all simple, and that EEM is applicable to equation (3.5.2) in the form (3.5.6). Mathematical analysis of the problem was initiated in [216, 226, 231, 238]. This work was stimulated by the questions asked by the author's colleague Kacenelenbaum, professor of electrical engineering in Moscow, who initiated the modern applications of the EEM method in electrical engineering in 1968–1969. In the USA, Baum initiated, applications of the SEM and EEM primarily for the target identification problems. Dolph [48] was one of the early papers which promoted interest in applications of the theory of non self-adjoint operators to electrical engineering and physics (see also [49, 50]). Completeness of the root system of the operator (3.5.2) was proved first in [216]. Theorems 3.5.1 and 3.5.3 and their proofs are taken from this paper. Theorem 3.5.3 and its proof are based on the works by Marcus [140] and Kacnelson [95].

Chapter 4. The T-matrix approach is one of the most popular numerical methods for solving the scattering problems. Varadan and Varadan [466] is a good source of references. It is a collection of papers presented at the international symposium on the T-matrix approach held in 1979. In the USA, the T-matrix approach is often associated with the papers [471, 472]. However, Barantsev [13] used this method earlier (see also [14]). Although the T-matrix approach was used at least from 1962, there was no mathematical analysis of its convergence, validity and dependence of the rate of convergence on the geometry of the obstacle. The method was used practically in the acoustic, electromagnetic, and elastic wave scattering problems without theoretical justification. Many speakers at the 1979 symposium called for such a justification of the T-matrix approach and its analysis. Theoretical analysis of the T-matrix approach was given first in [244, 245]. It was refined and illustrated by numerical results in [108]. The results and arguments in Chapter 4 are taken from these papers. Some abstract theorems on convergence of projection methods for solving linear equations in Banach spaces are taken from [70, 96, 105, 146] in a modified form. The first part of the example in Remark 4.2.4 is taken from [89] while the second is new. One can find the notion of the Riesz basis and some related facts from linear functional analysis in [71, 482]. We did not discuss the finite difference and finite elements methods for solving the exterior boundary value problems (see [72] and references). We mention also [11, 128], and parts 3 and 4 of [466].

One of the interesting problems in the T-matrix approach is that of the validity of the Rayleigh hypothesis. A review of this problem is [148] where many references can be found (see also [14, 21, 348, 349, 363]).

Chapter 5. Wave scattering by small bodies is of great interest in many applications. There is enormous literature of the subject. The applications include: light scattering in muddy water, in cosmic dust, in colloidal solutions, radiowave scattering by rain and hail, radiation from small holes, particle distribution in smoke fog, etc. The pioneering work was done by Rayleigh in 1871 and he had been working on this problem until his death [425]. Rayleigh understood that the field scattered by a small body of an arbitrary shape is basically a dipole radiation. J. J. Thomson (in 1898) realized that if the small body is a perfect conductor then the magnetic dipole radiation is of the order of magnitude of the electric dipole radiation. If $ka \sim 0.1$, where k is the wave number and a is the diameter of the body, then the multiple radiations are negligible compared with the dipole radiations. Therefore, physics of the phenomenon has been well understood for a long time (nearly a century). The problem is to compute the dipole moments in terms of the geometry of the body. The body is assumed homogeneous. The other problem is to study wave scattering by many small bodies. This problem is subdivided into two: scattering by r bodies with (1) $r \sim 10$, and (2) $r \sim 10^{23}$. The second case is the case of wave scattering in a medium consisting of many small bodies. The mathematical problem of finding the expansion of the scattering amplitude about $k = 0$ is of interest. The results and the arguments in Section 5.1-3 are taken from [226, 239, 360] and from the author's papers cited in the above monographs. Only a small part of the author's results on wave scattering by small bodies of arbitrary shapes is presented here. Many related problems and results from Ramm [410] and Ramm [360] are not discussed (two-sided estimates of capacitances and polarizability tensors; stable solution of integral equations at an eigenvalue: such equations arise in electro- and magnetostatics, elastostatics, hydrodynamics of incompressible ideal fluid; numerical examples; estimates of the capacitance of a perfect conductor placed in an inhomogeneous medium, the inverse radiation problem for small bodies, radiation from small holes, skin-effect for thin wires, etc). Section 5.4 is the paper [271]. In [134], the low frequency behavior of the solutions to the exterior Dirichlet and Neumann problems was studied for the obstacle whose boundary consists of one connected component. In [477], the case of multiconnected boundaries is treated by a different method. Werner's method is based on some suitable Fredholm's integral equations (see also [62, 161]; the last paper is the only one in which the Robin boundary condition was considered). Kress (see [37], p. 23) proposed in 1979 a method to study the dependence of the solution to the equation $[I - A(k)]u(k) = f(k)$ on k under the assumptions: $||A(k) - A(0)|| \to 0, ||f(k) - f(0)|| \to 0, k \to 0, A(k)$ is a compact operator function, $N(I - A(k)) = \{0\}$ if $k \neq 0, \dim N(I - A(0)) \geq 1$. Kress obtained from his abstract approach some results for the Helmholtz equation in $\mathbb{R}^3$ which were obtained earlier by different methods. In [477], the low frequency behavior of the solutions to Maxwell's equations is studied. The method in Section 5.4 is new and the results obtained include previously known results on this topic.

In [139], some necessary and sufficient conditions are given for the solutions of the Dirichlet and Neumann boundary value problems in the exterior of r obstacles to have a limit when $r \to \infty$ and the obstacles become smaller in a suitable way. There are many possible applications of the results of this type. A related area of research is the homogenization theory for partial differential equations (see [19, 139, 439], and references therein). In [8, 9], the scattering theory is discussed for the Schrödinger operators with point interactions. The inverse radiation problem is solved in [239, p. 110].

We mention papers [94, 101, 102, 441] which deal with low-frequency scattering.

Chapter 6. In this chapter, we are interested in finding coefficients of some elliptic equations from the knowledge of the solutions to these equations on certain manifolds. Such type of problems are of prime interest in explorational geophysics. Various aspects of these problems are discussed in [118, 356]. These inverse problems are often ill-posed. Various aspects of the theory of ill-posed problems are discussed in [64, 88, 453, 458], and in a number of other books and papers (there are more than 300 references in the cited books). A new approach to solving stably ill-posed problems (Dynamical Systems Method (DSM)) is developed in [369, 404] and in a large series of papers by the author, and the author and his student N. S. Hoang, cited in the above books. Newton [166] is a review of some inverse problems of geophysics which can be reduced to the inverse scattering problem for the one-dimensional Schrödinger equation. Weglein [473] is a review of the inverse scattering theory methods in geophysical exploration. Chapter 6 is based on papers [264–267] and book [269].

The presentation uses the works Ramm [249, 264]. The properties of the Clebsch–Gordan coefficients are well known (see [117] or [132]). Expansions of the type (6.1.108) were used in [169]. Some unclear point in [169] is mentioned in [266]. Section 6.2 uses the paper by Ramm, Weglein [421]. Section 6.3 is the work done by Ramm [250]. Formulas and error estimates for approximation by entire functions whose Fourier transform has support in a given ball are taken from [226], where references to his earlier papers on this topic and applications to apodization theory (synthesis theory in optics) are given. Section 6.4 is the paper by Ramm [270]. In

[36, 151, 173], one can find many references on the induction logging problem.

Chapter 7. Physical intuition dictates the statement of the scattering problem by a compact obstacle. If the boundary of the obstacle is infinite, then it is not clear what to take as the "plane wave" u_0 in D. The spectrum of the Dirichlet Laplacian L is not necessarily continuous as in the case of the Laplacian in the exterior of a compact obstacle. Therefore, first one should study the spectral properties of L in $L^2(D)$. The material in 7.2 is well known (see [69]) except Lemma 7.2.3 which is new. Lemma 7.2.4 is taken from [431] and [430]. Lemmas 7.2.5 was used in [187, 269]. In Section 7.3, the results from [431] are given (see also [93]). Our presentation differs from the presentation in [431].

The arguments and results in Section 7.4 are taken from [181, 187]. The arguments follow closely the arguments in [97] but the basic new idea in [181] was to study the operator Sturm–Liouville equation (7.4.3) not in the fixed Hilbert space $L^2(S^2)$ as in [97], but in a family of the Hilbert spaces depending on the parameter r. This idea was generalized by P. A. Mishnaevskii in remarkable papers [152, 153]. He developed a very powerful theory and obtained the strongest results on the spectral analysis of the general second-order elliptic differential operators, not necessarily self-adjoint, in some domains with infinite boundaries. His results contain the results in [41, 53, 54, 86, 90, 455, 469], and other papers.

The results in Section 7.5 are new. The results in Section 7.6 are taken from [187]. We do not discuss wave scattering by periodic structures and random surfaces (see [6, 21, 46, 47, 481], and references therein). Existence of the wave operators for the scattering problems in some domains with infinite boundaries was discussed in [40, 190, 439, 456].

Scattering by Potentials

The theory in Part II, Scattering by potentials, in its major part, dealing with inverse scattering problems, belongs to the author. It is based on the results in [345, 356, 395, 396, 401]. The basic novelty in this theory consists in introducing and using Property

C (completeness of the set of products of linear partial differential equations). The notion of Property C was introduced by the author [304] and applied to many inverse problems, see [337, 356], and the author's papers cited there. There are two topics discussed in Part II:

(1) inverse scattering for problems with fixed-energy data,
and
(2) inverse scattering for problems with *non-overdetermined data.*

The main novelty in the first topic is the author's formulas for inversion of the exact fixed-energy scattering data, algorithm for inversion of noisy fixed-energy scattering data and estimates for the stability of this algorithm. No such results were known. The results were published in a series of author's papers and summarized in [345, 356]. The second topic was not studied in the literature, although it is of basic interest: it describes the minimal scattering data sufficient for a unique recovery of the potential or an obstacle from scattering data. The author obtained basic uniqueness theorems for inverse scattering problems with non-overdetermined data. These results were used for a brief description of an inversion algorithm for such inverse problems.

In [356] for the inverse scattering problem with fixed-energy data, the inversion of the Dirichlet-to-Neumann map (D–N map) was discussed. In [356, pp. 297–298], a comparison of the author's inversion method and the inversion of the D–N map are compared and it is pointed out that the author's method has several principal advantages.

Modified Rayleigh Conjecture (MRC)

The theory in Part III, Modified Rayleigh Conjecture (MRC), belongs to the author, see [348, 363]. This theory deals with the stable numerical method for solving direct scattering obstacle problem. Rayleigh [425] conjectured that certain series, used for solving the obstacle scattering problem, converges up to the boundary of the obstacle. This conjecture is wrong in general, but holds for some obstacles, for example, for a ball. The author proved a theorem, that he called Modified Rayleigh Conjecture (MRC), which allowed him to develop stable numerical procedures for solving direct scattering problems for obstacles. Professor S. Gutman and the author [359] demonstrated practical efficiency of the MRC algorithms. The author also developed MRC method for solving static problems [363]. In [354], MRC method is used for solving problems with periodic boundary conditions.

Appendices.
The material in Appendix A is taken from [61, 79, 96, 365, pp. 265–297]. More material on embedding theorems can be found in [460]. Lemma (A.0.13) is proved, e.g., in Chapter 1 and in [269]. The argument on the top of p. 326 was used in [480]. The results in Appendix B are standard. Their proofs can be found in [69]. The results of Section 3 in Appendix C are from [365], Chapter 8, and the author's papers cited in [269].

Appendix D is the paper [258], see also [369, 404]. Appendix E is the paper Ramm [261]. The results of Appendix F are from [150, 269]. Appendix G is the paper [263] and can be read as a separate paper, see also [262].

Bibliography

[1] Agranovich, M. S. (1976). Summability of series in root vectors of non-self-adjoint elliptic operators, *Funct. Anal. Appl.* 10, pp. 165–174.

[2] Akhiezer, N. (1956). *Theory of Approximation*, Ungar, New York.

[3] Akhiezer, N. (1965). *The Classical Moment Problem and Some Related Questions in Analysis*, Oliver and Boyd, Edinburgh.

[4] Alber, H. (1982). Konvergenz der geometrishen Opticapproximation und Konstruktion des ersten Gliedes einer Parametrix fuer die Wellengleichung bei Tangentialstrahlen hoeherer Ordnung, *J. Reine Angew. Math.* 345, pp. 39–62.

[5] Alber, H. (1981). Justification of geometrical optics for non-convex obstacles, *J. Math. Anal. Appl.* 80, pp. 372–386.

[6] Alber, H. (1979). A quasi-periodic boundary value problem for the Laplacian and the continuation of its resolvent, *Proc. Roy. Soc. Edinburgh* 82a, pp. 251–272.

[7] Alber, H. and Ramm, A. G. (1986). Scattering amplitude and algorithm for solving the inverse scattering problem for a class of non-convex obstacles, *J. Math. Anal. Appl.* 117, pp. 570–597.

[8] Albeverio, S. *et al.* (1983). Low-energy parameters in nonrelativistic scattering theory, *Ann. Phys.* 148, pp. 308–326.

[9] Albeverio, S. *et al.* (1984). Schrödinger operators with point interactions and short range expansions, *Physica A* 124, pp. 11–28.

[10] Asakura, F. (1978). On the Green's function for $\Delta + \lambda$ with the boundary condition of the third kind in the exterior of a bounded obstacle, *J. Math. Kyoto Univ.* 18–3, pp. 615–625.

[11] Aziz, A., Dorr, M. and Kellog, B. (1982). A new approximation method for the Helmholtz equation in an exterior domain, *SIAM J. Numer. Anal.* 19, pp. 899–908.

[12] Babich, V. and Buldyrev, V. (1972). *Asymptotic Methods in the Short-Wave Difraction Problems*, Nauka, Moscow (in Russian).

[13] Barantsev, R. (1963). The method of separation of variables in the problem of scattering from finite bodies of arbitrary shape, *Sov. Phys. Doklady* 7(11), pp. 974–975.

[14] Barantsev, R. *et al.* (1971). Concerning the Rayleigh hypothesis in the problem of scattering from finite bodies of arbitrary shapes, *Vestnik Leningradskogo Univ., Math. Mech. Astron.* 7, pp. 56–62; *Ibid.* 7, pp. 71–77, (1968).

[15] Bass, F. and Fuks, I. (1979). *Wave Scattering from Statistically Rough Surfaces*, Pergamon Press, New York.

[16] Bateman, H. and Erdelyi, A. (1954). *Tables of Integral Transforms*, McGraw-Hill, New York.

[17] Baum, C. (1976). The singularity expansion method, in *Transient Electromagnetic Fields*, L. B. Felsen (ed.), Springer-Verlag, New York, pp. 129–179.

[18] Bellman, R., Kalaba, R. and Lockett J. (1966). *Numerical Inversion of Laplace Transform*, Elsevier, New York.

[19] Bensoussan, A., Lions, J.-L. and Papanicolaou, G. (1978). *Asymptotic Analysis for Periodic Structures*, American Mathematical Society, Providence, RI.

[20] Berezanskii, Ju. M. (1968). *Expansions in Eigenfunctions of Selfadjoint Operators*, American Mathematical Society, Providence, RI.

[21] Berg, P. van der and Fokkema, J. (1979). The Rayleigh hypothesis in the theory of diffraction by a cylindrical obstacle, *IEEE Trans. Antennas Propag.* AP-27, pp. 577–583.

[22] Beylkin, G. (1983). The fundamental identity for iterated spherical means and the inversion formula for diffraction tomography and inverse scattering, *J. Math. Phys.* 24, pp. 1399–1400.

[23] Blaschke, W. (1924). *Vorlesungen Ueber Differentiale Geometrie*, Springer-Verlag, Berlin.

[24] Bloom, C. O. and Kazachinoff, N. D. (1976). *Short Wave Radiation Problems in Inhomogeneous Media: Asymptotic Solutions*, Springer-Verlag, New York.

[25] Bojarski, N. (1982). A survey of the physical optics inverse scattering identity, *IEEE Trans. Antennas Propag.* AP-30, pp. 980–988.

[26] Bonnet-Bendhia, A. (1994). Guided waves by electromagnetic gratings and non-uniqueness examples for the diffraction problem, *Math. Math. Appl. Sci.* 17, pp. 305–338.

[27] Bonnet-Bendhia, A. and Ramdani, K. (2002). Diffraction by an acoustic grating perturbed by a bounded obstacle, *Adv. Comp. Math.* 16, pp. 113–138.

[28] Borovikov, V. and Kinber, B. (1978). *Geometric Theory of Diffraction*, Svjaz, Moscow (in Russian).

[29] Brakhage, H. and Werner, P. (1965). Ueber das Dirichletsche Aussenraum problem fuer die Helmholtzsche Schwingungsgleichung, *Arch. Math.* 16, pp. 325–329.

[30] Burago, Yu. D. and Maz'ya, V. G. (1969). *Potential Theory and Function Theory for Irregular Regions*, Consultants Bureau, New York.

[31] Burton, A. and Miller, G. (1971). The application of integral equation methods to the numerical solution of some exterior boundary-value problems, *Proc. Roy. Soc. London Ser. A.* 323, pp. 201–220.

[32] Buslaev, V. (1975). On the asymptotic behavior of the spectral characteristics of exterior problems for the Schrödinger operator, *Math. USSR* 9, pp. 201–220.

[33] Chadan, K. and Sabatier, P. (1977). *Inverse Problems in Quantum Scattering*, Springer-Verlag, New York.

[34] Choquet-Bruhat, Y., DeWitt-Morette, C. and Dillard-Bleik, U. (1982). *Analysis, Manifolds, and Physics*, North-Holland, Amsterdam.

[35] Christiansen, S. and Kleinman, R. E. (1996). On a misconception involving point collocation and the Rayleigh hypothesis, *IEEE Trans. Antennas Propag.* 44(10), pp. 1309–1316.

[36] Coen, S. and Yu, M. (1981). The inverse problem of the direct current conductivity profile of a layered earth, *Geophysics* 46, pp. 1702–1713.

[37] Colton, D. and Kress, R. (1983). *Integral Equation Methods in Scattering Theory*, Wiley, New York.

[38] Colton, D. (1984). The inverse scattering problem for time-harmonic acoustic waves, *SIAM Rev.* 26, pp. 323–350.

[39] Colton, D. and Sleeman, B. (1983). Uniqueness theorems for the inverse problem of acoustic scattering, *IMA J. Appl. Math.* 31, pp. 253–259.

[40] Combes, J. and Weder, R. (1981). Criterion for existence and completeness of wave operators and applications to scattering by unbounded obstacles, *Comm. Partial Differential Equations* 6, pp. 1129–1263.

[41] Constantin, P. (1981). Scattering for Schrodinger operators in a class of domain with noncompact boundaries, *J. Func. Anal.* 44, pp. 87–119.

[42] Cooper, J. and Strauss, W. (1979). Representation of the scattering operator for moving obstacles, *Indiana Univ. Math. J.* 28, pp. 643–671.

[43] Courant, R. and Hilbert, D. (1962). *Methods of Mathematical Physics*, Vol. 2, Interscience, New York.

[44] Davies, B. and Martin, B. (1979). Numerical inversion of the Laplace transform: A survey and comparison of methods, *J. Comput. Phys.* 33, pp. 1–32.

[45] Devaney, A. (1982). Inversion formula for inverse scattering within the Born approximation, *Opt. Lett.* 7, pp. 111–112.

[46] De Santo, J. (1972). Scattering from a periodic corrugated structure: Thin comb, *J. Math. Phys.* 13, pp. 336–341.

[47] De Santo, J. (1981). Scattering from a perfectly reflecting periodic surface: An exact theory, *Radio Sci.* 16, pp. 1315–1326.

[48] Dolph, C. (1961). Recent developments in some nonselfadjoint problems of mathematical physics, *Bull. Amer. Math. Soc.* 67, pp. 1–69.

[49] Dolph, C. and Scott, R. (1978). Recent developments in the use of complex singularities in electromagnetic theory and elastic wave propagation, in

Electromagnetic Scattering, P. Uslenghi (ed.), Academic Press, New York, pp. 503–570.

[50] Dolph, C. and Cho, K. (1980). On the relationship between the singularity expansion method and the mathematical theory of scattering, *IEEE Trans. Antennas Propag.* AP-28, pp. 888–896.

[51] Dolph, C., McLeod, J., and Thoe, D. (1966). The analytic continuation of the resolvent kernel and scattering operator associated with Schrödinger operator, *J. Math. Anal. Appl.* 16, pp. 311–332.

[52] Dunford, N. and Schwartz, J. T. (1958). *Linear Operators, I*, Interscience, New York.

[53] Eastham, M. and Kalf, H. (1982). *Schrodinger-Type Operators with Continuous Spectra*, Pitman, Boston.

[54] Eidus, D. (1969). The limiting amplitude principle, *Russ. Math. Survey* 24, pp. 51–94.

[55] Eidus, D. (1982). Completeness properties of scattering problem solutions, *Comm. Partial Differential Equations* 7, pp. 55–75.

[56] Eidus, D. (1965). The principle of limiting absorption, *Amer. Math. Soc. Transl.* 47, pp. 157–191.

[57] Eidus, D. (1963). Some boundary-value problems in infinite regions, *Izv. Akad. Nauk SSSR Ser. Mat.* 27, 1055–1080.

[58] Erdelyi, A. *et. al.* (1954). *Tables of Integral Transforms*, McGraw-Hill, New York.

[59] Eskin, G. (1981). *Boundary Value Problems for Elliptic Pseudodifferential Equations*, American Mathematical Society, Providence, RI.

[60] Ewing, W., Jardetsky, W. and Press, F. (1957). *Elastic Waves in Layered Media*, McGraw-Hill, New York.

[61] Fabes, E., Jofeit, M. and Riviere, N. (1978). Potential techniques for boundary value problems on C^1 domains, *Acta Math.* 141, pp. 165–186.

[62] Favorin, V. (1979). Behavior as $k \to 0$ of the Green's function of the exterior Robin problem for the two-dimensional Helmholtz's equation, *Differ. Uravn.*, 15, pp. 709–716 (in Russian); *Differ. English Translation Equ.* 15. pp. 497–502.

[63] Fedorjuk, M. (1977). *The Saddle-Point Method*, Nauka, Moscow.

[64] Fedotov, A. (1982). *Linear Ill-Posed Problems with Random Errors in Data*, Nauka, Novosibirsk.

[65] Filippov, A. (1969). Justification of the high frequency asymptotic in the three dimensional diffraction problem, *Siberian Math. J.* 10, pp. 1406–1421.

[66] Firey, W. (1967). The determination on convex bodies from their mean radius curvature function, *Mathematika* 14, pp. 1–13.

[67] Fock, V. (1965). *Electromagnetic Diffraction and Propagation Problems*, Pergamon Press, New York.

[68] Freudenthal, H. (1938). Ueber ein Beugungsproblem aus der electromagnetischen Lichttheorie, *Comp. Math.* 6, pp. 221–227.

[69] Glazman, I. (1965). *Direct Methods of Qualitative Spectral Analysis of Singular Differential Equations*, Israel Program for Scientific Translations, Davey, New York.

[70] Gohberg, I. and Feldman, I. (1974). *Convolution Equations and Projection Methods for Their Solutions*, Translations of Mathematical Monographs, Vol. 41, American Mathematical Society, Providence, RI.

[71] Gohberg, I. and Krein, M. (1969). *Introduction to the Theory of Linear Nonselfadjoint Operators*, Translations of Mathematical Monographs, Vol. 18, American Mathematical Society, Providence, RI.

[72] Goldstein, C. (1981). The finite element method with nonuniform mesh sizes for unbounded domains, *Math. Comput.* 36, pp. 387–404; *Numer. Math.* 38, pp. 61–82.

[73] Goldstein, C. (1969). Eigenfunction expansions associated with the Laplacian for certain domains with infinite boundaries, *Trans. Amer. Math. Soc.* 143, pp. 283–301.

[74] Goursat, E. (1927). *Course d'analyse Mathematique,* Vol. 3, Dunod, Paris.

[75] Gradshtein, I. and Ryzhik, I. (1980). *Tables of Integrals Series and Products*, Academic Press, New York.

[76] Greenspan, D. and Werner, P. (1966). A numerical method for the exterior Dirichlet problem for the reduced wave equation, *Arch. Ration. Mech. Anal.* 23, pp. 288–316.

[77] Gubernatis, J. (1979). Long wave approximations for the scattering of elastic waves from flaws with applications to ellipsoidal voids and inclusions, *J. Appl. Phys.* 50, pp. 4046–4058.

[78] Guillemin, V. and Sternberg, S. (1977). *Geometric Asymptotics*, American Mathematical Society, Providence, RI.

[79] Günter, N. (1967). *Potential Theory and its Applications to Basic Problems of Mathematical Physics*, Ungar, New York.

[80] Hardy, G. (1956). *Divergent Series*, Oxford University Press, New York.

[81] Hoenl, G., Maue, A. and Westpfahl, K. (1981). Theorie der Beugung, in *Handbuch der Physik*, Springer-Verlag, Berlin, pp. 281–573.

[82] Hsiung, C. (1981). *A First Course in Differential Geometry*, Wiley, New York.

[83] Ikawa, M. (1983). On the poles of the scattering matrix for two strictly convex obstacles, *J. Math. Kyoto Univ.* 23, pp. 127–194.

[84] Ikawa, M. (1983). On the distribution of the poles of the scattering matrix for two strickly convex obstacles, *Hokkaido Math. J.* 12, pp. 343–359.

[85] Ikebe, T. (1960). Eigenfunction expansions associated with the Schrodinger operator and their applications to scattering theory, *Arch. Ration. Mech. Anal.* 5, pp. 1–35.

[86] Ikebe, T. and Uchiyama, J. (1971). On the asymptotic behavior of eigenfunctions of second order elliptic operators, *J. Math. Kyoto Univ.* 11, pp. 425–448.

[87] Imbriale, W. and Mittra, R. (1970). The two-dimensional inverse scattering problem, *IEEE Trans. Antennas Propag.*, AP-18, pp. 633–642.

[88] Ivanov, V. K. *et al.* (1978). *Theory of Linear Ill-Posed Problems and Applications*, Nauka, Moscow.

[89] Ize, J., England, R. and Sabina, F. (1981). Theoretical and numerical study of diffraction of waves by inhomogeneous obstacles, *Tech. Rep.*, Univ. Nac. Aut., Mexico, p. 291.

[90] Jaeger, W. (1970). Ein gewoehnlicher differentialer Operator zweiter Ordnung fuer Funktionen mit Werten in Hilbertraum, *Math. Z.* 113, pp. 68–78.

[91] John, F. (1955). *Plane Waves and Spherical Means Applied to Partial Differential Equations*, Interscience, New York.

[92] John, F. (1982). *Partial Differential Equations*, Springer-Verlag, New York.

[93] Jones, D. (1953). The eigenvalues of the Laplacian when the boundary conditions are given on semi-infinite domains, *Proc. Cambridge Philos. Soc.* 40, pp. 668–683.

[94] Jones, D. (1979). Low frequency electromagnetic radiation, *J. Inst. Math. Appl.* 23, pp. 421–447.

[95] Kacnelson, V. (1967). Conditions under which systems of eigenvectors of some classes of operators form a basis, *Funct. Anal. Appl.* 23, pp. 421–447.

[96] Kantorovich, L. and Akilov, G. (1980). *Functional Analysis*, Pergamon Press, New York.

[97] Kato, T. (1959). Growth properties of solutions of the reduced wave equation with a variable coefficient, *Comm. Pure Appl. Math.*, 12, pp. 403–425.

[98] Kato, T. (1995). *Perturbation Theory for Linear Operators*, Springer-Verlag, Berlin.

[99] Kazandjian, L. (1992). Rayleigh–Fourier and extinction theorem methods applied to scattering and transmission at a rough solid–solid interface, *J. Acoust. Soc. Am.* 92, pp. 1679–1691.

[100] Kazandjian, L. (1995). Comments on "Reflection from a corrugated surface revisited", [*J. Acoust. Soc. Am.* 96, 1116–1129 (1994)]", *J. Acoust. Soc. Am.* 98, pp. 1813–1814.

[101] Keller, J., Kleinman, R. and Senior, T. (1972). Dipole moments in Rayleigh scattering, *J. Inst. Math. Appl.* 9, pp. 14–22.

[102] Kleinman, R. (1967). Low frequency solution of electromagnetic scattering problems, in *Electromagnetic Wave Theory, II*, Brown, J. (ed.), Pergamon Press, New York, pp. 891–905.

[103] Kleinman, R. and Roach, G. (1974). Boundary integral equations for the three-dimensional Helmholtz equation, *SIAM Rev.* 16, pp. 214–236.

[104] Kral, J. (1980). *Integral Operators in Potential Theory*, Springer-Verlag, New York.

[105] Krasnoselskii, M. *et al.* (1972). *Approximate Solutions of Operator Equations* Wolters-Noordhoof, Groningen.

[106] Krein, S. (1982). *Linear Equations in Banach Spaces*, Birkhäuser, Boston.

[107] Kresin, G. and Maz'ya, V. (1979). On the essential norm of an operator of double layer potential type in the space C_m, *Sov. Math. Dokl.* 20, pp. 459–462.

[108] Kristensson, G., Ramm, A. G. and Stroem, S. (1983). Convergence of the T-matrix approach in scattering theory, II, *J. Math. Phys.* 24, pp. 2619–2631.

[109] Krylov, V. and Skoblya, N. (1969). *Handbook of Numerical Inversion of Laplace Transform*, Israel Program for Scientific Translations, Jerusalem.

[110] Kupradze, V. (1956). *Randwertaufgaben der Schwingungstheorie und Integralgleichungen*, VEB, Berlin.

[111] Kupradze, V. (1965). *Potential Methods in the Theory of Elasticity*, Israel Program for Scientific Translations, Jerusalem.

[112] Kupradze, V. (1934). On Sommerfeld's radiation principle, *Dokl. Acad. Nauk SSSR* 2, pp. 1–7.

[113] Kumano-go, H. (1981). *Pseudodifferential Operators*, MIT, Cambridge, MA.

[114] Kuroda, S. (1978). *An Introduction to Scattering Theory*, Lecture Notes, No. 51. Aarhus University.

[115] Kussmaul, R. (1969). Ein numerisches Verfahren fuer Loesung des Neumannischen Assenproblems fuer die Helmholtzsche Schwingungsgleichung, *Computing* 4, pp. 246–273.

[116] Ladyzhenskaya, O. (1957). On the limiting amplitude principle, *Uspechi Math. Nauk* 12, pp. 161–164.

[117] Landau, L. and Lifschitz, E. (1959). *Quantum Mechanics. Nonrelativistic Theory*, Pergamon Press, New York.

[118] Lavrentiev, M. *et al.* (1970). *Multidimensional Inverse Problems for Differential Equations*, Springer-Verlag, New York.

[119] Lavrentiev, M. *et al.* (1986). *Ill-posed Problems of Mathematical Physics and Analysis*, Translations of Mathematical Monographs, Vol. 64, American Mathematical Society, Providence, RI.

[120] Lax, P. and Phillips, R. (1967). *Scattering Theory*, Academic Press, New York.

[121] Lax, P. and Phillips, R. (1969). Decaying modes for the wave equation in the exterior of an obstacle, *Comm. Pure Appl. Math.* 22, pp. 737–787.

[122] Lax, P. and Phillips, R. (1971). A logarithmic bound on the location of the poles of the scattering matrix, *Arch. Ration. Mech. Anal.* 40, pp. 268–280.

[123] Lax, P., Morawetz, C. and Phillips, R. (1963). Exponential decay of solutions of the wave equation in the exterior of a star-shaped obstacle, *Comm. Pure Appl. Math.* 16, pp. 477–486.

[124] Leis, R. (1986). *Initial Boundary Value Problems in Mathematical Physics*, Teubner, Stuttgart.

[125] Leis, R. (1967). Zur Monotonie der Eigenwerte selbstadjungierten elliptischer Differentialoperatoren, *Math. Z.* 96, pp. 26–32.

[126] Leis, R. (1965). Zur Dirichletschen Randwertaufgabe des Assenraums der Schwingungsgleichung, *Math. Z.* 90, pp. 205–211.

[127] Lewis, R. (1969). Physical optics inverse diffraction, *IEEE Trans. Antennas Propag.* AP-17, pp. 308–314, correction AP-18, p. 194 (1970).

[128] Limic, N. (1981). The exterior Neumann problem for the Helmholtz equation, *Glaznik Math.* 16, pp. 51–64, 245–260.
[129] Lions, J. and Magenes, E. (1972). *Non-homogeneous Boundary Value Problems and Applications*, Springer-Verlag, New York.
[130] Ljance, V. (1967). Analog of the inverse scattering problem for a nonselfadjoint operator, *Math. Sb.* 72, pp. 537–557.
[131] Lubic, Ju. (1957). On fundamental solutions of linear elliptic equations, *Math. Sb.* 39, pp. 23–36.
[132] Lyubarskii, G. (1957). *Group Theory and its Applications to Physics*, Fizmatgiz, Moscow.
[133] Lyubarskii, G. (1960). *The Application of Group Theory in Physics*, Pergamon Press, Oxford.
[134] MacCamy, R. (1965). Low frequency acoustic oscillations, *Quart. Appl. Math.* 23, pp. 247–256.
[135] Majda, A. (1976). High-frequency asymptotics of the scattering matrix and the inverse problem of acoustical scattering, *Comm. Pure Appl. Math.* 29, pp. 261–291.
[136] Majda, A. and Taylor, M. (1977). Inverse scattering problems for transparent obstacles and hyperbolic systems, *Comm. Partial Differential Equations* 2, pp. 395–438.
[137] Marchenko, V. (1968). Stability of inverse scattering problem, *Math. Sb.* 77, pp. 139–162.
[138] Marchenko, V. (1969). Stability of inverse scattering problem, *Math. Sb.* 78, pp. 475–484.
[139] Marchenko, V. and Khruslov, E. (1974). *Boundary Value Problems in Domains with Granular Boundary*, Naukova Dumka, Kiev (in Russian).
[140] Marcus, A. (1962). Expansion in root vectors in weakly perturbed selfadjoint operators, *Sov. Math. Doklady* 142, pp. 538–541.
[141] Martin, P. and Ramm, A. G. (1985). Inverse scattering for geophysical problems III. On the velocity-inversion problems of acoustics, *Proc. Roy. Soc. London* A399, pp. 153–166.
[142] Maz'ya, V. (1968). On Neumann's problem in irregular domains, *Siberian Math. J.*, 9, pp. 1322–1350.
[143] Maz'ya, V. and Shaposhnikova (1980). On conditions on the boundary in L_p theory of elliptic problems, *Sov. Math. Dokl.*, 251, pp. 1055–1059.
[144] Melrose, R. (1979). Singularity and energy decay in acoustical scattering, *Duke Math. J.* 46, pp. 43–59.
[145] Melrose, R. (1980). Forward scattering by a convex obstacle, *Comm. Pure Appl. Math.* 23, pp. 461–499.
[146] Mikhlin, S. (1971). *Numerical Performance of Variational Methods*, Walters-Noordhoff, Groningen.
[147] Mikhlin, S. and Prössdorf, S. (1986). *Singular Integral Operators*, Springer-Verlag, Berlin.

[148] Millar, R. (1973). The Rayleigh hypothesis and a related least-squares solution to scattering problems for periodic surfaces and othe scatterers, *Radio Sci.* 8, pp. 785–796.

[149] Millar, R. (1971). On the Rayleigh assumption in scattering by a periodic surface, *Proc. Cambridge Philos. Soc.* 69, 217–225.

[150] Miller, F. (1983). The construction of a surface from its principal curvatures, preprint, Kansas State University.

[151] Minerbo, G. (1983). Radial inversion of TE measurements in a borehole, preprint.

[152] Mishnaevskii, P. (1976). On the spectral theory for the Sturm–Liouville equation with operator coefficient, *Math. USSR-Izv.* 10, pp. 145–180.

[153] Mishnaevskii, P. (1983). A study of elliptic second order operators in domains with infinite boundaries by means of the operator Sturm–Liouville equation, I, II, *Dif. Uravneniya*, 19, pp. 445–457 and 645–653.

[154] Molchanov, A. (1953). Conditions for discreteness of spectra for differential operators of second order, *Proc. Moscow Math. Soc.* 2, pp. 169–200.

[155] Morawetz, C. (1966). Exponential decay of solutions of the wave equation, *Comm. Pure Appl. Math.* 19, pp. 439–444.

[156] Morawetz, C. (1962). The limiting amplitude principle, *Comm. Pure Appl. Math.* 15, pp. 349–361.

[157] Morawetz, C. and Ludwig, D. (1968). An inequality for the reduced wave operator and the justification of geometrical optics, *Comm. Pure Appl. Math.* 21, pp. 187–203.

[158] Morawetz, C., Ralston, J. and Strauss, W. (1977). Decay of solutions of the wave equation outside non-trapping obstacles, *Comm. Pure Appl. Math.* 30, pp. 447–508.

[159] Müller, C. (1969). *Foundations of the Mathematical Theory of Electromagnetic Waves*, Springer-Verlag, Berlin.

[160] Müller, C. (1955). Radiation patterns and radiation fields, *J. Rat. Mech. Anal.* 4, pp. 235–246.

[161] Muravei, L. (1978). Analytic continuation with respect to a parameter of the Green's functions of exterior boundary value problems for the two-dimensional Helmholtz equation, III, *Math. USSR Sbornik* 34, pp. 55–98.

[162] Mushelishvili, N. (1972). *Singular Integral Equations*, Noordhoff, Groningen.

[163] Newton, R. (1966). *Scattering of Waves and Particles*, McGraw-Hill, New York.

[164] Newton, R. (1982). On a generalized Hilbert problem, *J. Math. Phys.* 23, pp. 2257–2265.

[165] Newton, R. (1976). Optical theorem and beyond, *Amer. J. Phys.* 44, pp. 639–642.

[166] Newton, R. (1981). Inversion of reflection of data for layered media: a review of exact methods, *Geophys. J. R. Ast. Soc.* 65, pp. 191–215.

[167] Nirenberg, L. (1953). The Weyl and Minkovski problems in differential geometry in the large, *Comm. Pure Appl. Math.* 6, pp. 337–394.

[168] Noble, B. (1962). Integral equation perturbation methods in low frequency diffraction, pp. 323–360, in *Electromagnetic Waves*, Langer, R. (ed), University of Wisconsin Press, Madison.

[169] Norton, S. and Linzer, M. (1981). Ultrasonic reflectivity imaging in three dimensions: exact inverse scattering solutions for plane, cylindrical, and spherical apertures, *IEEE Trans. Biomed. Eng.*, BME-28, pp. 202–220.

[170] Nussenzveig, H. (1959). The poles of the S-matrix of a rectangular potential well or barrier, *Nucl. Phys.* 11, pp. 499–521.

[171] Olmsted, I. (1958). Completeness and Parseval equation, *Amer. Math. Monthly* 65, pp. 343–345.

[172] Olsson, P. (1984). Elastostatics as a limit of elastodynamics — a matrix formulation, *Appl. Sci. Res.* 41, pp. 125–151.

[173] Oristaglio, M. (1983). Born inversion of the induction log, preprint.

[174] Panich, O. (1965). On the question of the solvability of the exterior boundary value problems for the wave equation and Maxwell's equations, *Russ. Math. Survey* 20, pp. 221–226.

[175] Petit, R. (ed.) (1980). *Electromagnetic Theory of Gratings*, Topics in Current Physics, Vol. 22, Springer-Verlag, Berlin.

[176] Petkov, V. (1980). High frequency asymptotics of the scattering amplitude for non-convex bodies, *Comm. Partial Differential Equations*, 6, pp. 293–329.

[177] Petkov, V. (1982). Distribution of poles of the S-matrix for trapping obstacles, *C. R. Acad. Sci., Paris, I. Math.* 295, pp. 443–445.

[178] Povzner, A. (1953). On the expansion of arbitrary functions in terms of eigenfunctions of the operator $-\Delta+c(x)$, *Math. Sb.* 32, pp. 109–156; *Amer. Math. Transl. Ser.* (2), 60, pp. 1–49 (1967); *Soviet Math. Doklady* 104, pp. 360–363 (1955).

[179] Radon, J. (1920). Ueber die randwertaufgaben beim logarithmischen potential, *Sitzungsber. Akad. Wiss. Math. Naturw. Kl.* 128, 5.B.IIa.

[180] Ralston, J. (1981). Propagation of singularities and the scattering matrix, in *Singularities in Boundary Value Problems*, Garnir, H. (ed.), D. Reidel Publ., Co., Dordrecht.

[181] Ramm, A. G. (1963). Spectral properties of the Schrödinger operator in some domains with infinite boundaries, *Doklady Acad. Sci. USSR* 152, pp. 282–285.

[182] Ramm, A. G. (1963). Investigation of the scattering problem in some domains with infinite boundaries I, II, *Vestnik Leningrad Gos. Univ., Seria Math., Mech. i Astonom.* 7, pp. 45–66; 19, pp. 67–76.

[183] Ramm, A. G. (1964, 1966). Absence of the discrete positive spectrum of the Dirichlet Laplacian in some infinite domains, *Vestnik Leningrad Gos. Univ., Seria Math., Mech. i Astonom.*, 13, pp. 153–156; N 1, p. 176.

[184] Ramm, A. G. (1964). On the analytic continuation of the solution of the Schrödinger equation in the spectral parameter and the behavior of the solution to the nonstationary problem as $t \to \infty$, *Uspechi Mat. Nauk* 19, pp. 192–194.

[185] Ramm, A. G. (1964). Statement of the diffraction problem in domains with infinite boundaries, *Proc. 3rd All-union Wave Diffraction Symposium*, Nauka, Moscow, pp. 28–31.

[186] Ramm, A. G. (1965). Analytic continuation of the resolvent kernel of the Schrödinger operator in the spectral parameter and limiting amplitude principle in some infinite domains, *Doklady Acad. Sci. Azerb. SSR* 21, pp. 3–7.

[187] Ramm, A. G. (1965). Spectral properties of the Schrödinger operator in some infinite domains, *Matem. Sb.* 66, pp. 321–343.

[188] Ramm, A. G. (1965). On wave diffusion, *Math. Izv. VUZ'ov* 2, pp. 136–138.

[189] Ramm, A. G. (1965). On a method of solving the Dirichlet problem in some infinite domains, *Math. Izv. VUZ'ov* 5, pp. 124–127.

[190] Ramm, A. G. (1965). On the conditions under which integral operators are nuclear and existence of the S-matrix in the problem of scalar scattering on a potential and surface, *Ukrainian Math. J.* 17, pp. 92–98.

[191] Ramm, A. G. (1965). Necessary and sufficient conditions for the validity of the limiting amplitude principle, *Doklady Acad. Sci. USSR* 163, pp. 584–586.

[192] Ramm, A. G. (1965). Reconstruction of the domain shape from the scattering amplitude, *Radiotech. i Electron.* 11, pp. 2068–2070.

[193] Ramm, A. G. (1966). Behavior of the solution to a nonstationary problem as $t \to \infty$, *Math. Izv. VUZ'ov* 1, pp. 124–138.

[194] Ramm, A. G. (1966). Domain free from the resonances in the three-dimensional scattering problem, *Doklady Acad. Sci. USSR* 166, pp. 1319–1322.

[195] Ramm, A. G. (1968). On the limiting amplitude principle, *Differ. Equ.* 4, pp. 714–720.

[196] Ramm, A. G. (1968). Optimal solution of the antenna synthesis problem, *Doklady Acad. Sci. USSR* 180, pp. 1071–1074.

[197] Ramm, A. G. (1968). On numerical differentiation, *Math. Izv. VUZ'ov* 11, pp. 131–135.

[198] Ramm, A. G. (1968). Asymptotic distribution of the Schrödinger operator eigenvalues when the potential tends to infinity and the boundary is infinite, *Doklady Acad. Sci. USSR* 183, pp. 780–783.

[199] Ramm, A. G. (1968). Some theorems on analytic continuation of the Schrödinger operator resolvent kernel in the spectral parameter, *Izv. Acad. Nauk Armyan. SSR, Math.* 3, pp. 443–464.

[200] Ramm, A. G. (1969). Green's function study for differentialequation of the second order in domains with infinite boundaries, *Differ. Equ.* 5, pp. 1509–1516.

[201] Ramm, A. G. (1969). Calculation of the scattering amplitude for the wave scattering from small bodies of an arbitrary shape, *Radiofisika* 12, pp. 1185–1197.

[202] Ramm, A. G. (1970). Eigenfunction expansion for nonselfadjoint Schrödinger operator, *Doklady Acad. Sci. USSR* 191, pp. 50–53.

[203] Ramm, A. G. (1970). Exponential decay of solution to hyperbolic equation, *Differ. Equ.* 6, pp. 2099–2100.

[204] Ramm, A. G. (1970). Reconstruction of the shape of a reflecting body from the scattering amplitude, *Radiofisika* 13, pp. 727–732.

[205] Ramm, A. G. (1970). Approximate formulas for polarizability tensors and capacitances of bodies of arbitrary shapes and applications, *Doklady Acad. Sci. USSR* 195, pp. 1303–1306.

[206] Ramm, A. G. (1971). Calculation of the initial field from the scattering amplitude, *Radiotech. i Electron.* 16, pp. 554–556.

[207] Ramm, A. G. (1971). Necessary and sufficient conditions for the validity of the limiting amplitude principle, *Differ. Equ.* 7, pp. 366–367.

[208] Ramm, A. G. (1971). Eigenfunction expansions for exterior boundary problems, *Differ. Equ.* 7, pp. 737–742.

[209] Ramm, A. G. (1971). Approximate formulas for polarizability tensor and capacitances for bodies of an arbitrary shape. *Radiofisika* 14, pp. 613–620.

[210] Ramm, A. G. (1971). Electromagnetic wave scattering by small bodies of an arbitrary shape, in *Proc. 5th All-Union Symp. Wave Diffraction*, Trudy. Math. Inst. Steklova, Leningrad, pp. 176–186.

[211] Ramm, A. G. (1972). Simplified optimal differentiators, *Radiotech. i Electron.* 17, pp. 1325–1328.

[212] Ramm, A. G. (1972). On exterior diffraction problems, *Radiotech. i Electron.* 7, pp. 1362–1365.

[213] Ramm, A. G. (1972). Calculation of the quasistationary states in quantum mechanics, *Doklady Acad. Sci. USSR* 204, pp. 1071–1074.

[214] Ramm, A. G. (1973). Iterative process to solve the third boundary problem, *Diff. Uravneniya* 9, pp. 2075–2079.

[215] Ramm, A. G. (1973). Optimal harmonic synthesis of generalized Fourier series and integrals with randomly perturbed coefficients, *Radiotechnika* 28, pp. 44–49.

[216] Ramm, A. G. (1973). Eigenfunction expansion corresponding the discrete spectrum, *Radiotech. i Electron.* 18, pp. 496–501.

[217] Ramm, A. G. (1974). Light scattering matrix for small particles of an arbitrary shape, *Opt. Spectroscopy*, 37(1), pp. 125–129.

[218] Ramm, A. G. (1974). Scalar scattering by the set of small bodies of an arbitrary shape, *Radiofisika* 17, pp. 1062–1068.

[219] Ramm, A. G. (1975). Approximate solution of some integral equations of the first kind, *Differ. Equ.* 11, pp. 582–586.

[220] Ramm, A. G. (1975). Estimates of some functionals in quasistatic electrodynamics, *Ukrain. Phys. J.* 5, pp. 534–543.

[221] Ramm, A. G. (1977). On simultaneous approximation of a function and its derivative by interpolation polynomials, *Bull. London Math. Soc.* 9, pp. 283–288.

[222] Ramm, A. G. (1978). Necessary and sufficient conditions for the validity of the limiting amplitude principle, *Math. Inv. VUZ'ov* 5, pp. 96–102.

[223] Ramm, A. G. (1978). A uniqueness theorem for the Dirichlet problem, *Siberian Math. J.* 6, pp. 1421–1423.

[224] Ramm, A. G. (1978). Approximation by entire functions, *Math. Inv. VUZ'ov* 10, pp. 72–76.

[225] Ramm, A. G. (1980). Electromagnetic wave scattering by small bodies of arbitrary shapes, in *Acoustic, Electromagnetic and Elastic Scattering-Focus on T-matrix Approach*, V. Varadan (ed.), Pergamon Press, New York, pp. 537–546.

[226] Ramm, A. G. (1980). *Theory and Applications of Some New Classes of Integral Equations*, Springer-Verlag, New York.

[227] Ramm, A. G. (1980). Perturbation preserving asymptotics of spectrum of linear operators, *J. Math. Anal. Appl.* 76, pp. 10–17.

[228] Ramm, A. G. and Mishnaevskii, P. (1980). Uniqueness theorem of use in general relativity, *J. Math. Anal. Appl.* 75, pp. 58–65.

[229] Ramm, A. G. (1980). Two sided estimates of the scattering amplitude at low energies, *J. Math. Phys.* 21, pp. 308–310.

[230] Ramm, A. G. (1980). Nonselfadjoint operators in diffraction and scattering, *Math. Methods Appl. Sci.* 2, 327–346.

[231] Ramm, A. G. (1980). Theoretical and practical aspects of singularity and eigenmode expansion methods, *IEEE Trans. Antennas Propag.* AP-28(6), 897–901.

[232] Ramm, A. G. (1980). A variational principle for resonances, *J. Math. Phys.* 21, pp. 2052–2053.

[233] Ramm, A. G. (1981). Existence uniqueness and stability of solutions to some nonlinear problems, *Appl. Anal.* 11, pp. 223–232.

[234] Ramm, A. G. (1981). Exterior boundary value problems as limits of interface problems, *J. Math. Anal. Appl.* 84, pp. 256–263.

[235] Ramm, A. G. (1981). Stable solutions of some ill-posed problems, *Math. Methods Appl. Sci.* 3, pp. 336–363.

[236] Ramm, A. G. (1981). On some properties of solutions of Helmholtz equation, *J. Math. Phys.* 22, pp. 275–276.

[237] Ramm, A. G. (1981). Spectral properties of some nonselfadjoint operators, *Bull. Amer. Math. Soc.* 5(3), pp. 313–315.

[238] Ramm, A. G. (1981). On the singularity and eigenmode expansion methods, *Electromagnetics*, 1(4), pp. 385–394.

[239] Ramm, A. G. (1982). *Iterative Methods for Calculating Static Fields and Wave Scattering by Small Bodies*, Springer-Verlag, New York.

[240] Ramm, A. G. (1982). Mathematical foundations of the singularity and eigenmode expansion methods, *J. Math. Anal. Appl.* 86, pp. 562–591.

[241] Ramm, A. G. (1982). Perturbations preserving asymptotic of spectrum with a remainder, *Proc. Amer. Math. Soc.* 85(2), pp. 209–212.

[242] Ramm, A. G. (1982). Variational principles for resonances II, *J. Math. Phys.* 23(6), pp. 1112–1114.

[243] Ramm, A. G. (1982). Perturbation of resonances, *J. Math. Anal. Appl.* 88, pp. 1–7.

[244] Ramm, A. G. (1982). Convergence of the T-matrix approach to scattering theory, *J. Math. Phys.* 23(6), pp. 1123–1125.

[245] Ramm, A. G. (1983). Convergence of the T-matrix approach in the potential scattering, *J. Math. Phys.* 23, pp. 2408–2409.

[246] Ramm, A. G. (1983). Eigenfunction expansions for some nonselfadjoint operators and the transport equation, *J. Math. Anal. Appl.* 92, pp. 564–580.

[247] Ramm, A. G. (1983). Convergence of the T-matrix approach in scattering theory II, *J. Math. Phys.* 24(11), pp. 2619–2631.

[248] Ramm, A. G. (1983). Convergence of the T-matrix scheme, in *Proc. IEEE Int. Symp. Antennas and Propagation*, May 1983, Houston.

[249] Ramm, A. G. (1983). Inverse scattering for geophysical problems, *Phys. Letters*, *A*99, pp. 258–260.

[250] Ramm, A. G. (1983). An inversion formula in scattering theory, *Phys. Lett.* A99, pp. 201–204.

[251] Ramm, A. G. (1984). On a property of the set of radiation patterns, *J. Math. Anal. Appl.* 98, pp. 92–98.

[252] Ramm, A. G. (1984). A uniqueness theorem in scattering theory, *Phys. Rev. Lett.* 52(1), p. 13.

[253] Ramm, A. G. (1984). Scattering by a penetrable body, *J. Math. Phys.* 25(3), pp. 469–471.

[254] Ramm, A. G. (1984). Representations of solutions to Helmholtz's equation, *J. Math. Phys.* 25(4), pp. 807–809.

[255] Ramm, A. G. (1984). Existence of infinitely many purely imaginary resonances in the problem of potential scattering, *Phys. Lett. A* 101(4), pp. 187–188.

[256] Ramm, A. G. (1984). A uniqueness theorem in scattering theory, *Phys. Rev. A* 102(5–6), pp. 218–219.

[257] Ramm, A. G. (1984). Description of the degree of nonuniqueness in inverse source problem, *J. Math. Phys.* 25(6), pp. 1791–1793.

[258] Ramm, A. G. (1984). Estimates of the derivatives of random functions, *J. Math. Anal. Appl.* 102, pp. 244–250.

[259] Ramm, A. G. (1984). Remarks about inverse diffraction problem, *J. Math. Phys.* 25(11), pp. 2672–2674.

[260] Ramm, A. G. (1984). On inverse diffraction problem, *J. Math. Anal. Appl.* 103, pp. 139–147.

[261] Ramm, A. G. (1985). Limit of the spectra of the interior Neumann problems when a solid domain shrinks to a plane one, *J. Math. Anal. Appl.* 108, pp. 107–112.

[262] Ramm, A. G. (1985). Extraction of resonances from transient fields, *IEEE Trans. Antennas Propag.* AP-33, pp. 223–226.

[263] Ramm, A. G. (1985). Calculating resonances (natural frequencies) and extracting them from transient fields, *J. Math. Phys.* 26(5), pp. 1012–1020.

[264] Ramm, A. G. (1985). Inverse scattering for geophysical problems, *Inverse Problems* 1(2), pp. 133–172.

[265] Ramm, A. G. and Miller, T. (1985). Estimates of the derivatives of random functions II, *J. Math. Anal. Appl.* 110, pp. 429–435.

[266] Ramm, A. G. (1985). Inversion of the back scattering data and a problem of integral geometry, *Phys. Lett. A* 113, pp. 172–176.

[267] Ramm, A. G. (1986). Inverse scattering in an absorptive medium, *Inverse Prob.* 2, pp. L5–L7.

[268] Ramm, A. G. (1986). Singularities of the inverses of Fredholm operators, *Proc. Roy. Soc. Edinburgh Sect. A* 102, pp. 117–121.

[269] Ramm, A. G. (1986). *Scattering by Obstacles*, D. Reidel, Dordrecht.

[270] Ramm, A. G. (1986). Inverse scattering for geophysical problems IV. Inversion of the induction logging measurements, *Geophys. Prospect.* 34(3), pp. 293–301.

[271] Ramm, A. G. (1986). Behavior of solutions to exterior boundary value problems at low frequencies, *J. Math. Anal. Appl.* 117, pp. 561–569.

[272] Ramm, A. G. (1986). A geometrical inverse problem. *Inverse Probl.* 2, pp. L19–L21.

[273] Ramm, A. G. (1986). Nonuniqueness of the solution to an inverse geophysical problem, *Inverse Prob.* 2, pp. L23–L25.

[274] Ramm, A. G. (1986). On completeness of the products of harmonic functions, *Proc. Amer. Math. Soc.* 99, pp. 253–256.

[275] Ramm, A. G. (1986). Inversion of the Laplace transform from the real axis, *Inverse Prob.* 2, pp. L55–L59.

[276] Ramm, A. G. (1986). Inverse scattering for geophysical problems when the background is variable, *J. Math. Phys.* 27, pp. 2687–2689.

[277] Ramm, A. G. (1987). An inverse problem for the Helmholtz equation in a semi-infinite medium, *Inverse Prob.* 3, pp. 19–22.

[278] Ramm, A. G. (1987). A method for solving inverse diffraction problems, *Inverse Prob.* 3, pp. 23–25.

[279] Ramm, A. G. (1987). Characterization of the low-frequency scattering data in the inverse problems of geophysics, *Inverse Prob.* 3, pp. 33–35.

[280] Ramm, A. G. (1987). Sufficient conditions for zero not to be an eigenvalue of the Schrödinger operator, *J. Math. Phys.* 28, pp. 1341–1343.

[281] Ramm, A. G. (1987). Optimal estimation from limited noisy data, *J. Math. Anal. Appl.* 125, pp. 258–266.

[282] Ramm, A. G. (1987). Signal estimation from incomplete data, *J. Math. Anal. Appl.* 125, pp. 267–271.

[283] Ramm, A. G. (1987). Recovery of the potential from I-function, *Math. Reports Acad. Sci., Canada* 9, pp. 177–182.

[284] Ramm, A. G. (1987). Necessary and sufficient conditions for a function to be the scattering amplitude corresponding to a reflecting obstacle, *Inverse Prob.* 3, pp. 53–57.

[285] Ramm, A. G. (1987). Necessary and sufficient conditions on the scattering data for the potential to be in L^2, *Inverse Prob.* 3, pp. 71–76.

[286] Ramm, A. G. (1987). Completeness of the products of solutions to PDE and uniqueness theorems in inverse scattering, *Inverse Prob.* 3, pp.77–82.

[287] Ramm, A. G. (1988). A uniqueness theorem for two-parameter inversion, *Inverse Prob.* 4, pp. 7–10.

[288] Ramm, A. G. (1988). A uniqueness theorem for a boundary inverse problem, *Inverse Prob.* 4, pp. 1–5.

[289] Ramm, A. G. (1988). Inverse scattering on half-line, *J. Math. Anal. Appl.* 133(2), pp. 543–572.

[290] Ramm, A. G. (1989). Multidimensional inverse problems and completeness of the products of solutions to PDE, *J. Math. Anal. Appl.* 134(1), pp. 211–253; 139, p. 302 (1989).

[291] Ramm, A. G. (1988). A uniqueness theorem for an inverse problem, *Appl. Math. Lett.* 1, pp. 185–187.

[292] Ramm, A. G. and Li, L. (1988). Estimates for Green's functions, *Proc. Amer. Math. Soc.* 103(3), pp. 875–881.

[293] Ramm, A. G. (1988). A criterion for completeness of the set of scattering amplitudes, *Phys. Lett. A* 129, pp. 191–194.

[294] Ramm, A. G. (1988). Conditions for zero not to be an eigenvalue of the Schrödinger operator, *J. Math. Phys.* 29, pp. 1431–1432.

[295] Ramm, A. G. (1988). Recovery of the potential from fixed energy scattering data, *Inverse Prob.* 4, pp. 877–886; 5, p. 255 (1989).

[296] Ramm, A. G. (1988). A simple proof of uniqueness theorem in impedance tomography, *Appl. Math. Lett.* 1(3), pp. 287–290.

[297] Ramm, A. G. (1988). Uniqueness theorems for multidimensional inverse problems with unbounded coefficients, *J. Math. Anal. Appl.* 136, pp. 568–574.

[298] Ramm, A. G. (1989). Multidimensional inverse scattering problems and completeness of the products of solutions to homogeneous PDE, *Zeitschr. Angew. Math. Mech.* 69(4), pp. 13–22.

[299] Ramm, A. G. (1989). Necessary and sufficient condition for a scattering amplitude to correspond to a spherically symmetric scatterer, *Appl. Math. Lett.* 2, pp. 263–265.

[300] Ramm, A. G. (1989). An inverse problem for Maxwell's equations, *Phys. Lett. A* 138, pp. 459–462.

[301] Ramm, A. G. (1990). Stability of the numerical method for solving the 3D inverse scattering problem with fixed energy data, *Inverse Prob.* 6, pp. 7–12.

[302] Ramm, A. G. (1990). Algorithmically verifiable characterization of the class of scattering amplitudes for small potentials, *Appl. Math. Lett.* 3, pp. 61–65.

[303] Ramm, A. G. (1990). Is the Born approximation good for solving the inverse problem when the potential is small? *J. Math. Anal. Appl.* 147, pp. 480–485.

[304] Ramm, A. G. (1990). Completeness of the products of solutions of PDE and inverse problems, *Inverse Prob.* 6, pp. 643–664.

[305] Ramm, A. G. (1990). Calculating singular integrals as an ill-posed problem, *Numer. Math.* 57, pp. 139–145.

[306] Ramm, A. G. (1990). Uniqueness result for inverse problem of geophysics I, *Inverse Prob.* 6, pp. 635–642.

[307] Ramm, A. G. (1990). Uniqueness theorems for geophysical problems with incomplete surface data, *Appl. Math. Lett.* 3(4), pp. 41–44.

[308] Ramm, A. G. and Sjostrand, J. (1991). An inverse problem for the wave equation, *Math. Z.* 206, pp. 119–130.

[309] Ramm, A. G. (1991). Symmetry properties for scattering amplitudes and applications to inverse problems, *J. Math. Anal. Appl.* 156, pp. 333–340.

[310] Ramm, A. G. (1991). Necessary and sufficient condition for a PDE to have property C, *J. Math. Anal. Appl.* 156, pp. 505–509.

[311] Ramm, A. G. (1991). Stability of the numerical method for solving 3D inverse scattering problem with fixed energy data, *J. Reine Angew. Math.* 414, pp. 1–21.

[312] Ramm, A. G. (1991). Can a constant be a scattering amplitude? *Phys. Lett. A* 154, pp. 35–37.

[313] Ramm, A. G. (1991). Solution of some integral equations arising in integral geometry, *Appl. Math. Lett.* 4, pp. 177–181.

[314] Ramm, A. G. (1992). *Multidimensional Inverse Scattering Problems*, Longman/Wiley, New York.

[315] Ramm, A. G. (1992). Stability of the solution to inverse scattering problem with exact data, *Appl. Math. Lett.* 5(1), pp. 91–94.

[316] Ramm, A. G. (1992). Inverse scattering problem with fixed-energy data, *Appl. Math. Lett.* 5(4), pp. 63–67.

[317] Ramm, A. G. (1992). Stability estimates in inverse scattering, *Acta Appl. Math.* 28(1), pp. 1–42.

[318] Ramm, A. G. (1992). Stability of the inversion of 3D fixed-frequency data, *J. Math. Anal. Appl.* 169(2), pp. 329–349.

[319] Ramm, A. G. (1992). Stability of the solution to 3D fixed-energy inverse scattering problem, *J. Math. Anal. Appl.* 170(1), pp. 1–15.

[320] Ramm, A. G. (1993). Fixed-energy inverse scattering for non-compactly supported potentials, *Math. Comp. Modelling* 18(1), (1993), pp. 57–64. (with P. Stefanov).

[321] Ramm, A. G. (1993). New method for proving uniqueness theorems for obstacle inverse scattering problems, *Appl. Math. Lett.* 6(6), pp. 19–22.

[322] Ramm, A. G. (1993). Scattering amplitude as a function of the obstacle, *Appl. Math. Lett.* 6(5), pp. 85–87.

[323] Ramm, A. G. (1994). Inversion of fixed-frequency surface data for layered medium, *J. Inverse Ill-Posed Prob.* 2(3), pp. 263–268.

[324] Ramm, A. G. (1994). Stability of the solution to inverse obstacle scattering problem, *J. Inverse Ill-Posed Prob.* 2(2), pp. 269–275.

[325] Ramm, A. G. (1994). Stability estimates for obstacle scattering, *J. Math. Anal. Appl.* 188(3), pp. 743–751.

[326] Ramm, A. G. (1995). Examples of nonuniqueness for an inverse problems of geophysics, *Appl. Math. Lett.* 8(4), pp. 87–90.

[327] Ramm, A. G. (1995). Continuous dependence of the scattering amplitude on the surface of an obstacle, *Math. Meth. Appl. Sci.* 18, pp. 121–126.

[328] Ramm, A. G. (1995). A formula for inversion of boundary data, *J. of Inverse Ill-Posed Prob.* 3(5), pp. 411–415.

[329] Ramm, A. G. (1995). Uniqueness theorems for inverse obstacle scattering problems in Lipschitz domains, *Appl. Anal.* 59, pp. 377–383.

[330] Ramm, A. G. and Katsevich A. I. (1996). *The Radon Transform and Local Tomography*, CRC Press, Boca Raton.

[331] Ramm, A. G. (1996). Finding potential from the fixed-energy scattering data via D-N map, *J. Inverse Ill-Posed Prob.* 4(2), pp. 145–152.

[332] Ramm, A. G. and Makrakis, G. (1998). Scattering by obstacles in acoustic waveguides, in *Spectral and Scattering Theory*, A. G. Ramm (ed.), Plenum Publishers, New York, pp. 89–110.

[333] Ramm, A. G. (1998). Fundamental solutions to elliptic equations with discontinuous senior coefficients and an inequality for these solutions, *Math. In equal. Appl.* 1(1), pp. 99–104.

[334] Ramm, A. G. (1999). Property C for ODE and applications to inverse scattering, *Zeit. Angew. Anal.* 18(2), pp. 331–348.

[335] Ramm, A. G. (1999). Inverse problem for an inhomogeneous Schrödinger equation, *J. Math. Phys.* 40(8), pp. 3876–3880.

[336] Ramm, A. G. (2000). Finding small inhomogeneities from scattering data, *J. Inverse Ill-Posed Prob.* 8(2), pp. 205–210.

[337] Ramm, A. G. (2000). Property C for ODE and applications to inverse problems, in *Operator Theory and Its Applications*, Fields Institute Communications, Vol. 25, American Mathematical Society, Providence, RI, pp. 15–75.

[338] Ramm, A. G. and Sammartino, M. (2000). Existence and uniqueness of the scattering solutions in the exterior of rough domains, in *Operator Theory and Its Applications*, A. G. Ramm, P. N. Shivakumar and A. V. Strauss (eds.), Fields Institute Communications, Vol. 25, American Mathematical Society, Providence, RI, pp. 457–472.

[339] Ramm, A. G., Pang, P. and Yan, G. (2000). A uniqueness result for the inverse transmission problem, *Int. J. Appl. Math.* 2(5), pp. 625–634.

[340] Ramm, A. G. (2001). A non-over-determined inverse problem of finding the potential from the spectral function, *Int. J. Diff. Eqs.* 3(1), pp. 15–29.

[341] Ramm, A. G. and Smirnova, A. (2001). On stable numerical differentiation, *Math. Comput.* 70, pp. 1131–1153.

[342] Ramm, A. G. and Gol'dshtein, V. (2001). Embedding operators for rough domains, *Math. Inequal. Appl.* 4(1), pp. 127–141.

[343] Ramm, A. G. and Gutman, S. (2001). Piecewise-constant positive potentials with practically the same fixed-energy phase shifts, *Appl. Anal.* 78(1–2), pp. 207–217.

[344] Ramm, A. G. (2001). An inverse problem for the heat equation, *J. Math. Anal. Appl.* 264(2), pp. 691–697.

[345] Ramm, A. G. (2002). Stability of solutions to inverse scattering problems with fixed-energy data, *Milan J. Math.* 70, pp. 97–161.

[346] Ramm, A. G. (2002). Regularization of ill-posed problems with unbounded operators, *J. Math. Anal. Appl.* 271, pp. 547–550.

[347] Ramm, A. G. (2002). Numerically efficient version of T-matrix method, *Appl. Anal.* 80(3), pp. 385–393.

[348] Ramm, A. G. (2002). Modified Rayleigh conjecture and applications, *J. Phys. A* 35, pp. 357–361.

[349] Ramm, A. G. and Gutman, S. (2002). Numerical implementation of the MRC method, *J. Phys. A.* 35, pp. 8065–8074.

[350] Ramm, A. G. (2002). Injectivity of the spherical means operator, *Compt. Rend. Acad. Sci., Paris, Ser. I*, 335(12), pp. 1033–1038.

[351] Ramm, A. G. and Smirnova, A. (2003). Stable numerical differentiation: When is it possible? *J. Korean SIAM* 7(1), pp. 47–61.

[352] Ramm, A. G. (2003). On a new notion of regularizer, *J. Phys. A* 36, pp. 2191–2195.

[353] Ramm, A. G. (2003). Equations for the self-consistent field in random medium, *Phys. Lett. A* 312(3–4), pp. 256–261.

[354] Ramm, A. G. and Gutman, S. (2004). Modified Rayleigh Conjecture for scattering by periodic structures, *Int. J. Appl. Math. Sci.* 1(1), pp. 55–66.

[355] Ramm, A. G. and Gutman, S. (2005). Analysis of a linear sampling method for identification of obstacles, *Acta Appl. Math. Sinica* 21(3), pp. 399–404.

[356] Ramm, A. G. (2005). *Inverse Problems*, Springer, New York.

[357] Ramm, A. G. (2005). Inequalities for the derivatives and stable differentiation of piecewise-smooth discontinuous functions, *Math. Inequal. Appl.* 8(1), pp. 169–172.

[358] Ramm, A. G. (2005). Discrepancy principle for the dynamical systems method, *Commun. Nonlinear Sci. Numer. Simul.* 10(1), pp. 95–101.

[359] Ramm, A. G. and Gutman, S. (2005). Modified Rayleigh Conjecture method for multidimensional obstacle scattering problems, *Numer. Funct. Anal. Optim.* 26(2), pp. 69–80.

[360] Ramm, A. G. (2005). *Wave Scattering by Small Bodies of Arbitrary Shapes*, World Scientific Publishers, Singapore.

[361] Ramm, A. G. (2005). Embedding operators and boundary-value problems for rough domains, *Int. J. Appl. Math. Mech.* 1, 51–72. (with V. Gol'dshtein).

[362] Ramm, A. G. (2005). A new discrepancy principle, *J. Math. Anal. Appl.* 310, pp. 342–345.

[363] Ramm, A. G. (2005). Modified Rayleigh conjecture for static problems, *Appl. Math. Lett.* 18(12), pp. 1396–1399.

[364] Ramm, A. G. (2005). Dynamical systems method (DSM) and nonlinear problems, in *Spectral Theory and Nonlinear Analysis*, J. Lopez-Gomez (ed.), World Scientific Publishers, Singapore, pp. 201–228.

[365] Ramm, A. G. (2005). *Random Fields Estimation*, World Scientific Publishers, Singapore, 2005.
[366] Ramm, A. G. (2005). Uniqueness of the solution to inverse obstacle scattering problem, *Phys. Lett. A* 347(4–6), pp. 157–159.
[367] Ramm, A. G. (2006). Completeness of the set of scattering amplitudes, *Phys. Lett. A* 360(1), pp. 22–25.
[368] Ramm, A. G. and Gutman, S. (2006). Modified Rayleigh Conjecture with optimally placed sources, *J. Appl. Funct. Anal.* 1(2), 223–236.
[369] Ramm, A. G. (2007). *Dynamical Systems Method for Solving Operator Equations*, Elsevier, Amsterdam.
[370] Ramm, A. G. (2007). Electromagnetic wave scattering by many small particles, *Phys. Lett. A* 360(6), pp. 735–741.
[371] Ramm, A. G. (2007). Materials with the desired refraction coefficients can be made by embedding small particles, *Phys. Lett. A* 370(5–6), pp. 522–527.
[372] Ramm, A. G. (2007). Scattering by many small bodies and applications to condensed matter physics, *Eur. Phys. Lett.* 80, p. 44001.
[373] Ramm, A. G. (2007). Many-body wave scattering by small bodies and applications, *J. Math. Phys.* 48(10), p. 103511.
[374] Ramm, A. G. (2007). Wave scattering by small particles in a medium, *Phys. Lett. A* 367, pp. 156–161.
[375] Ramm, A. G. (2007). Wave scattering by small impedance particles in a medium, *Phys. Lett. A* 368(1–2), pp. 164–172.
[376] Ramm, A. G. (2007). A symmetry problem, *Ann. Polon. Math.* 92, pp. 49–54.
[377] Ramm, A. G. (2007). Distribution of particles which produces a desired radiation pattern, *Commun. Nonlinear Sci. Numer. Simul.* 12(7), pp. 1115–1119.
[378] Ramm, A. G. (2007). Distribution of particles which produces a "smart" material, *J. Stat. Phys.* 127(5), pp. 915–934.
[379] Ramm, A. G. (2007). Invisible obstacles, *Ann. Polon. Math.* 90(2), pp. 145–148.
[380] Ramm, A. G. (2008). Inverse scattering problem with data at fixed energy and fixed incident direction, *Nonlinear Anal.* 69(4), pp. 1478–1484.
[381] Ramm, A. G. (2008). Creating wave-focusing materials, *LAJSS* (*Latin-Am. J. Solids Struct.*) 5, pp. 119–127.
[382] Ramm, A. G. and Gutman, S. (2008). Modified Rayleigh Conjecture method and its applications, *Nonlinear Anal.* 68, pp. 3884–3908.
[383] Ramm, A. G. (2008). Electromagnetic wave scattering by many conducting small particles, *J. Phys. A* 41, p. 212001.
[384] Ramm, A. G. (2008). Fixed-energy inverse scattering, *Nonlinear Anal.* 69(3), pp. 971–978.
[385] Ramm, A. G. (2008). A recipe for making materials with negative refraction in acoustics, *Phys. Lett. A* 372(13), pp. 2319–2321.

[386] Ramm, A. G. (2008). Does negative refraction make a perfect lens? *Phys. Lett. A* 372, pp. 6518–6520.

[387] Ramm, A. G. (2008). Wave scattering by many small particles embedded in a medium, *Phys. Lett. A* 372(17), pp. 3064–3070.

[388] Ramm, A. G. (2009). Preparing materials with a desired refraction coefficient, *Nonlinear Anal.* 70(12), pp. 186–190.

[389] Ramm, A. G. (2009). Inverse scattering with non-overdetermined data, *Phys. Lett. A* 373, pp. 2988–2991.

[390] Ramm, A. G. (2009). Creating desired potentials by embedding small inhomogeneities, *J. Math. Phys.* 50(12), 123525.

[391] Ramm, A. G. and Indratno, S. (2009). Inversion of the Laplace transform from the real axis using an adaptive iterative method, *Int. J. Math. Math. Sci.* 898195, 38 pp. doi:10.1155/2009/898195.

[392] Ramm, A. G. (2010). Asymptotic stability of solutions to abstract differential equations, *J. Abst. Diff. Eq. Appl.* 1(1), pp. 27–34.

[393] Ramm, A. G. (2010). Electromagnetic wave scattering by many small particles and creating materials with a desired permeability, *Prog. Electromag. Res.* 14, pp. 193–206.

[394] Ramm, A. G. (2010). On a new notion of the solution to an ill-posed problem, *J. Comp. Appl. Math.* 234, pp. 3326–3331.

[395] Ramm, A. G. (2010). Uniqueness theorem for inverse scattering problem with non-overdetermined data, *J. Phys. A* 43, 112001.

[396] Ramm, A. G. (2010). Uniqueness of the solution to inverse scattering problem with backscattering data, *Eurasian Math. J.* 1(3), pp. 97–111 (Open Access Journal).

[397] Ramm, A. G. (2010). A method for creating materials with a desired refraction coefficient, *Inter. J. Mod. Phys. B* 24(27), pp. 5261–5268.

[398] Ramm, A. G. (2010). Materials with a desired refraction coefficient can be created by embedding small particles into a given material, *Int. J. Struct. Changes Solids.* 2(2), pp. 17–23.

[399] Ramm, A. G. (2011). Wave scattering by many small bodies and creating materials with a desired refraction coefficient, *Afrika Matematika* 22(1), pp. 33–55.

[400] Ramm, A. G. (2011). Electromagnetic wave scattering by a small impedance particle of arbitrary shape, *Opt. Commun.* 284, pp. 3872–3877.

[401] Ramm, A. G. (2011). Uniqueness of the solution to inverse scattering problem with scattering data at a fixed direction of the incident wave, *J. Math. Phys.* 52, p. 123506.

[402] Ramm, A. G. (2011). Stability of solutions to some evolution problems, *Chaotic Model. Simul.* 1, pp. 17–27.

[403] Ramm, A. G. (2011). Scattering of electromagnetic waves by many thin cylinders, *Results Phys.* 1(1), pp. 13–16.

[404] Ramm, A. G. and Noang, N. (2012). *Dynamical Systems Method and Applications. Theoretical Developments and Numerical Examples*, Wiley, Hoboken.

[405] Ramm, A. G. (2012). Stability of solutions to abstract evolution equations with delay, *J. Math. Anal. Appl.* 396, pp. 523–527.

[406] Ramm, A. G. (2012). Electromagnetic wave scattering by many small perfectly conducting particles of an arbitrary shape, *Opt. Commun.* 285(18), pp. 3679–3683.

[407] Ramm, A. G. (2013). Heat transfer in a medium in which many small particles are embedded, *Math. Model. Nat. Phenom.* 8(1), pp. 193–199.

[408] Ramm, A. G. (2013). Stability result for abstract evolution problems, *Math. Methods Appl. Sci.* 36(4), pp. 422–426.

[409] Ramm, A. G. (2013). Stability of the solutions to evolution problems, *Mathematics* 1, pp. 46–64.

[410] Ramm, A. G. (2013). *Scattering of Acoustic and Electromagnetic Waves by Small Bodies of Arbitrary Shapes: Applications to Creating New Engineered Materials*, Momentum Press, New York.

[411] Ramm, A. G. (2014). Inverse scattering with under-determined scattering data, *Math. Meth. Nat. Phenomena* 9(5), pp. 244–253.

[412] Ramm, A. G. (2014). Recovery of the potential from I-function, *Reports Math. Phys.* 74(2), pp. 135–143.

[413] Ramm, A. G. (2014). Electromagnetic wave scattering by small perfectly conducting particles and applications, *J. Math. Phys.* 083505, 55 pp.

[414] Ramm, A. G. and Tran, N. (2015). A fast algorithm for solving scalar wave scattering problem by billions of particles, *J. Algorithms Optim.* 3(1), pp. 1–13 (Open Access: http://www.academicpub.org/jao/Issue.aspx?Abstr=false).

[415] Ramm, A. G. and Schechter, M. (2015). Existence of the solution to electromagnetic wave scattering problem for an impedance body of an arbitrary shape, *Appl. Math. Lett.* 41, pp. 52–55.

[416] Ramm, A. G. (2015). Scattering of EM waves by many small perfectly conducting or impedance bodies, *J. Math. Phys.* 9, 091901, 56 pp.

[417] Ramm, A. G. (2015). EM wave scattering by many small impedance particles and applications to materials science, *Open Optics J.* 9, pp. 14–17 (Open Access: http://benthamopen.com/TOOPTSJ/VOLUME/9/).

[418] Ramm, A. G. (2016). Large-time behavior of solutions to evolution equations, in *Handbook of Applications of Chaos Theory*, C. Skiadas (ed.), Chapman & Hall/CRC, pp. 183–200.

[419] Ramm, A. G. (2016). Heat transfer in a complex medium, in *The Foundations of Chaos Revisited: From Poincare to Recent Advances*, Springer, pp. 119–136.

[420] Ramm, A. G. (2016). Uniqueness of the solution to inverse obstacle scattering with non-over-determined data, *Appl. Math. Lett.* 58, pp. 81–86.

[421] Ramm, A. G. and Weglein, A. (1984). Inverse scattering for geophysical problems, II, *J. Math. Phys.* 25, pp. 3231–3234.

[422] Ramm, A. G. (2017). Completeness of the set $\{e^{ik\beta\cdot s}\}|_{\forall\beta\in S^2}$, *Global J. Math. Anal.* 5(2), pp. 43–44. (Open Access: https://www.sciencepubco.com/index.php/GJMA/article/view/7975).

[423] Ramm, A. G. (2017). A numerical method for solving 3D inverse scattering problem with non-over-determined data, *J. Pure Appl. Math.* 1(1), pp. 1–3.

[424] Ramm, A. G. (2017). Perturbation of zero surfaces, *Global J. Math. Anal.* 5(1), pp. 27–28 (Open Access: http://www.sciencepubco.com/index.php/GJMA/article/view/7474).

[425] Rayleigh, J. (1922). *Scientific Papers*, Cambridge University Press, Cambridge.

[426] Rayleigh, J. (1945). *The Theory of Sound*, Dover, New York.

[427] Rayleigh, J. (1907). On the dynamical theory of gratings, *Proc. Roy. Soc. A* 79, pp. 399–416.

[428] Reed, M. and Simon, B. (1979). *Scattering Theory*, Academic Press, New York.

[429] Reed, M. and Simon, B. (1978). *Analysis of Operators*, Academic Press, New York.

[430] Rellich, F. (1943). Ueber das asymptotische Verhalten der Loesungen von $\Delta u + \lambda u = 0$ in unendlichen Gebieten, *Jaresber. Deutsch. Math. Verein.* 53, pp. 57–65.

[431] Rellich F. (1948). Das Eigenwertproblem von $\Delta u + \lambda u = 0$, in, *Studies and Essays Presented to R. Courant*, Halbroehren (ed.), New York.

[432] Ribaric, M. and Vidav, I. (1969). Analytic propertiesof the inverse $A^{-1}(z)$ of an analytic operator valued function $(A(z))$, *Arch. Ration. Mech. Anal.* 32, pp. 298–310.

[433] Rose, J. (1982). Inverse scattering at long wavelength, in *Review of Progress in Quantitative Nondestructive Evaluation*, Thompson (ed.), Vol. 1, Plenum Publisher New York, pp. 131–135.

[434] Saenz, A. (1981, 1982). Scattering by periodic surfaces, *J. Math. Phys.* 22, pp. 2872–2884; 23, pp. 1115–1122.

[435] Saito, Y. (1971). The principle of limiting absorption for second-order differential operators with operator valued coefficients, *Publ. Res. Inst. Math. Sci.* 7, pp. 581–619.

[436] Saito, Y. (1982). Some properties of the scattering amplitude and the inverse scattering problem, *Osaka J. Math.* 19, pp. 527–547.

[437] Saito, Y. (1983). Extended limiting absorption method and analyticity of the S-matrix, *J. Reine Angew. Math.* 343, pp. 1–22.

[438] Saito, Y. (1987). An inverse problem in potential theory and inverse scattering problem, *Math. Kyoto. Univ.* 22(2), pp. 307–321.

[439] Sanchez-Palencia, E. (1980). *Nonhomogenous Media and Vibration Theory* Lecture Notes in Physics, Vol. 127, Springer-Verlag, New York.

[440] Scotti, T. and Wirgin, A. (1996). Shape reconstruction of an impenetrable body via the Rayleigh hypothesis, *Inverse Prob.* 12, pp. 1027–1055.

[441] Senior, T. (1976). Low frequency scattering by a dielectric body, *Radio Sci.* 11, pp. 477–482.

[442] Shenk, N. (1966). Eigenfunction expansions and scattering theory for the wave equation in an exterior region, *Arch. Ration Mech. Anal.* 22, pp. 120–150.

[443] Shenk, N. and Thoe, D. (1971). Eigenfunction expansions and scattering theory for perturbations of $-\Delta$, *Rocky Mountain J. Math.* 1, pp. 89–125.

[444] Schulze, B. and Wildenhain, G. (1977). *Methoden der Potentialtheorie fuer Elliptische Differentialgleichungen beliebiger Ordnung*, Akad. Verlag, Berlin.

[445] Simon, M. and Ursell, F. (1984). Uniqueness in linearized two-dimensional water-wave problems, *J. Fluid Mech.* 148, pp. 137–154.

[446] Sleeman, B. (1982). The inverse problem of acoustic scattering, *IMA J. Appl. Math.* 29, pp. 113–142.

[447] Sommerfeld, A. (1912). Greensche Funktion der Schwingungsgleichung, *Jahresber. Deutschen Math. Vereins.* 21, pp. 309–353.

[448] Serebrennikov and Pervozvanski, A. (1965). *Detection of Hidden Periodicities*, Nauka, Moscow.

[449] Stechkin, S. and Subbotin, Ju. (1976). *Splines in Numerical Analysis*, Nauka, Moscow (in Russian).

[450] Steinberg, S. (1968). Meromorphic families of compact operators, *Arch. Ration. Mech. Anal.* 31, pp. 372–379.

[451] Stevenson, A. (1953). Solution of electromagnetic problems as power series in the ratio (dimension of scatterer)/(wave-length), *J. Appl. Phys.* 24, pp. 1134–1152.

[452] Tamarkin, J. (1927). On Friedholm's integral equations whose kernels are analytic in a parameter, *Math. Ann.* 28, pp. 127–152.

[453] Tanana, V. (1981). *Methods of Solving Operator Equations*, Nauka, Moscow.

[454] Treves, F. (1980). *Introduction to Pseudodifferential and Fourier Integral Operators*, Plenum Press, New York.

[455] Tayoshi, T. (1972). The asymptotic behavior of the solutions of Helmhotz's equation in a domain with unbounded boundary, *Publ. RIMS, Kyoto Univ.* 8, pp. 375–391.

[456] Tayoshi, T. (1980). Wave operators in the scattering problem related to non-finite obstacles, *Publ. RIMS, Kyoto Univ.* 16, pp. 627–634.

[457] Terai, T. (1980). On calculation of the sound speed around three dimensional objects by integral equation method, *J. Sound Vibr.* 69, pp. 71–100.

[458] Tikhonov, A. and Arsenin, V. (1977). *Solutions of Ill-posed Problems*, Winston, Washington.

[459] Titchmarsh, E. (1948). *Theory of Fourier Integrals*, Oxford University Press, Oxford.

[460] Triebel, H. (1978). *Interpolation Theory, Function Spaces, Differential Operators*, VEB Verlag, Berlin.

[461] Ursell, F. (1973, 1978). On the exterior problems of acoustics, *Proc. Cambridge Philos. Sci.* 74, pp. 117–125; 84, pp. 545–548.

[462] Ursell, F. (1966). On the rigorous foundation of short-wave asymptotics, *Proc. Cambridge Philos. Sci.* 62, pp. 227–244.

[463] Vainberg, B. (1966). Principles of radiation, limiting amplitude, and limiting absorption in the general theory of partial differential equations, *Russian Math. Surveys* 21, pp. 115–194.

[464] Vainberg, B. (1975). Short wave asymptotics of solutions to the stationary problems and asymptotics $t \to \infty$ of solutions to time-dependent problems, *Russian Math. Surveys.* 30, pp. 3–55.

[465] Vainberg, M. and Trenogin, V. (1974). *Bifurcation of Solutions of Nonlinear Equations*, Noordhoff, Leiden.

[466] Varadan, V. and Varadan, V. (ed.), (1980). *Acoustic, Electromagnetic, and Elastic Wave Scattering-Focus on the T-matrix Approach*, Pergamon Press, New York.

[467] Vekua, I. (1967). *New Methods of Solving Elliptic Equations*, North-Holland, Amsterdam; Appendix: Metaharmonic functions, *Trudy Tbilis. Math. Inst.* 12, (1943), 105–174.

[468] Vekua, I. (1953). On completeness of a system of metaharmonic functions, *Doklady Acad. Nauk USSR* 90, pp. 715–718; 495–498.

[469] Vinnik, A. and Eidus, D. (1974). On radiation conditions for domains with infinite boundaries, *Soviet Math. Doklady* 15, pp. 12–15.

[470] Voitovich, N., Kacenelenbaum, B. and Sivov, A. (1977). *Generalized Method of Eigen Oscillations in Diffraction Theory*, Nauka, Moscow.

[471] Waterman, P. (1965). Matrix formulation of electromagnetic scattering, *Proc. IEEE* 53, pp. 805–812.

[472] Waterman, P. (1969). New formulation of Acoustic scattering, *J. Acoust. Soc. Amer.* 45, pp. 1417–1429.

[473] Weglein, A. (1985). The inverse scattering concept and its scientific application, in *Developments in Geophysical Exploration Methods*, Vol. 6, Elsevier, London.

[474] Werner, P. (1960). Zur mathematischen Theorie akustischer Wellenfelder, *Arch. Ration. Mech. Anal.* 6, pp. 231–260.

[475] Werner, P. (1966). On the behavior of stationary electromagnetic wave fields for small frequencies, *J. Math. Anal. Appl.* 15, pp. 447–496.

[476] Werner, P. (1975). Ueber das Verhalten electromagnetischer Felder fuer kleine Frequenzen in mehrfach zusammenhaengenden Gebieten, *J. Reine Angew. Math.* 1(278/279), pp. 365–397; (1976), 2(280), pp. 98–121.

[477] Werner, P. (1986). Low frequency asymptotics for the reduced wave equation in two-dimensional exterior spaces, *Math. Methods in the Appl. Sci.* 8, pp. 134–156.

[478] Weyl, H. (1952). Kapazitaet von Strahlungsfeldern, *Math. Z.* 55, pp. 187–198.

[479] Wiener, N. (1949). *Extrapolation, Interpolation, and Smoothing of Stationary Time Series*, Wiley, New York.

[480] Wilcox, C. (1975). *Scattering Theory for the d'Alembert Equation in Exterior Domains*, Lecture Notes in Mathematics, Vol. 492, Springer-Verlag, New York.

[481] Wilcox, C. (1984). *Scattering Theory for Diffraction Gratings*, Springer-Verlag, New York.

[482] Young, R. (1980). *An Introduction to Nonharmonic Fourier Series*, Academic Press, New York.

List of Symbols

$D, \mathcal{D}$	a domain
$\Gamma = \partial D = S$	its boundary
Ω, D'	the exterior domain, $D \cup \Omega = \mathbb{R}^3$
$\mathbb{R}^n$	the Euclidean n-dimensional space, $\mathbb{R}^n_+ = \{x : x \in \mathbb{R}^n, x_n \geq 0\}$, $\mathbb{R}^1_+ = \mathbb{R}_+$, $\mathbb{R}^n_- = \mathbb{R}^n \setminus \mathbb{R}^n_+$, S^{n-1}-the unit sphere in $\mathbb{R}^n$
S^2	the unit sphere in $\mathbb{R}^3$, $\tilde{S}^2$ — a solid angle, i.e., an open set in $\mathbb{S}^2$
$n(\nu)$ or $\alpha(\beta)$	unit vectors in the direction of propagation of the scattered (incident) wave
B_r	$\{x : \lvert x\rvert \leq r\}, D_r = D \cap B_r, D'_r = D \cap \Omega_r$
Ω_r	$\{x : \lvert x\rvert \geq r\}, \Omega'_r = \{x : \lvert x\rvert \leq r, x \in \Omega\}$, $\Gamma_r = \Gamma \cap B_r, \Gamma'_r = \Gamma \cap \Omega_r$
S_r	$\{x : \lvert x\rvert = r\}, P_r = \{x : x \in D, x_r = r\}$, $S_r(D) = S_r \cap D$
$\tilde{\Gamma}_R$	$\{x : x \in \Gamma, x_3 \geq R\}$
$D\mu, S\sigma, Q\sigma, T\rho$	Section 1.1
g, g_ϵ, g_M	Section 1.1
A, A_ϵ	Section 1.1
A'	Section 1.1
$\tilde{A}$	Section 1.1
ψ_{nm}	Section 1.1
$j_n, h_n^{(j)}$	Section 1.1

$\partial/\partial N_{\pm}$	Section 1.1, N is the exterior normal, N points into Ω
$\ker A$	$N(A) = \{f : Af = 0\}$
$\operatorname{im} A$	$R(A) = \{f : f = Af\}$
$C^m(D)(C_0^m(D))$	the space of m times continuously differentiable (compactly supported) functions in D
$H^s = W^{2,s}(D)$	the Sobolev spaces
$\tilde{H}^s$	$H^s(\Gamma)$
$\mathring{H}^s(D)$	the closure of $C_0^\infty(D)$ in the norm $H^s(D)$, $\mathring{H}^{2s}(D) = H^{2s}(D) \cap \circ H^s(D)$
$H_0^s(D)$	the set of $H^s(D)$ functions with compact support in D
$L^p(D)$	the Lebesgue spaces
$\|\|u\|\|_{L^p(D)}$	$\left(\int_D \|u\|^p dx\right)^{1/p}$
$W^{m,p}(D)$	the Sobolev spaces
Im (Re) or $\Im$ ($\Re$)	imaginary (real) part
$\gamma(r)$	$r^{-1}\exp(ikr)$
l^p	the space of sequence $f = \{f_n\}$ with the norm
$\|\|f\|\|_{l^p}$	$\left(\sum_{n=1}^{\infty} \|f_n\|^p\right)^{1/p}$
$\forall$	for all
$\exists$	there exists
$\equiv$	identically equal, equal by definition
$\Rightarrow$	implies
$\to$	converges
$\rightharpoonup$	weakly converges
lim	limit in mean (in L^2)
δ_{ij}	$\begin{cases} 0, & i \neq j \\ 1, & i = j \end{cases}$
$\alpha_{ij}(\gamma), \beta_{ij}$	polarizability tensors
SEM, EEM	singularity expansion method, eigenmode expansion method
cap	capacity, Appendix A

σ_∞	the set of all linear compact operators
σ_p	Appendix C
W_R	the set of entire functions of exponential type $\leq R$
a.e.	almost every
$u(x, \alpha, k), u(x, n, k)$	scattering solution
Q_a	class of potentials
$A(\beta, \alpha, k), f(-n, \nu, k)$	scattering amplitude
$\alpha, \beta \in S^2$	p. 335
$g(x, y, k) = \frac{e^{ik\|x-y\|}}{4\pi\|x-y\|}$	p. 338
$G(x, y, k)$	Green's function, p. 338
$\delta(x - y)$	delta function, p. 338
$\tilde{q}$	Fourier transform, p. 349
$\mathcal{M}$	p. 354
ρ	p. 355
$Y_\ell(\beta)$	spherical harmonics, p. 356
$j_\ell(r), h_\ell(r)$	spherical Bessel functions, p. 357
$N(\delta)$	p. 365
$A_\delta(\beta, \alpha)$	noisy scattering amplitude, k is fixed p. 365
ψ	special solution, p. 371
Λ	Dirichlet-to-Neumann map, p. 371
$\hat{f}$	the Radon transform, p. 379
$A(-\beta, \beta, k)$	back-scattering amplitude, non-overdetermined data, p. 377
MRC (Modified Rayleigh Conjecture)	p. 409
LAP (Limiting absorption principle)	p. 420
Ramm's identity (9.3.27)	p. 423

Index

L

M

N

P

R

S

T

W

Printed in the United States
By Bookmasters